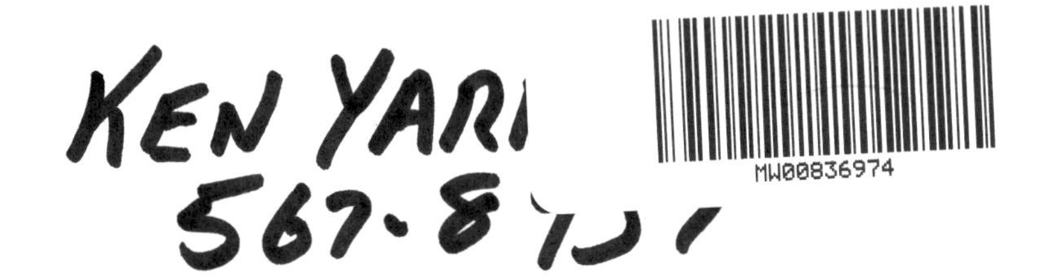

Transportation Depth Reference Manual

for the Civil PE Exam

Norman R. Voigt, PE, PLS

Professional Publications, Inc. • Belmont, California

Benefit by Registering This Book with PPI

- Get book updates and corrections.
- Hear the latest exam news.
- Obtain exclusive exam tips and strategies.
- Receive special discounts.

Register your book at **www.ppi2pass.com/register**.

Report Errors and View Corrections for This Book

PPI is grateful to every reader who notifies us of a possible error. Your feedback allows us to improve the quality and accuracy of our products. You can report errata and view corrections at **www.ppi2pass.com/errata**.

TRANSPORTATION DEPTH REFERENCE MANUAL FOR THE CIVIL PE EXAM

Current printing of this edition: 1

Printing History

edition number	printing number	update
1	1	New book.

Printed in the United States of America.

PPI
1250 Fifth Avenue, Belmont, CA 94002
(650) 593-9119
www.ppi2pass.com

ISBN: 978-1-59126-094-3

Library of Congress Control Number: 2011931021

Table of Contents

Appendices Table of Contents

Preface

While the *Civil Engineering Reference Manual* is absolutely essential for the morning PE exam, the afternoon depth exams require additional, discipline-specific knowledge. The *Transportation Depth Reference Manual* comprehensively covers the civil PE transportation depth exam specifications as presented by the National Council of Examiners for Engineering and Surveying (NCEES).

Three of the predominant references used for the transportation depth exam are AASHTO's *A Policy on Geometric Design of Highways and Streets* (referred to as *GDHS* or the *Green Book*), the Transportation Research Board's *Highway Capacity Manual* (*HCM*), and the Federal Highway Administration's *Manual on Uniform Traffic Control Devices* (*MUTCD*). This book supplements the information given in the *HCM*, the *Green Book*, and the *MUTCD*, and condenses the most commonly used material so that you are able to quickly find the most relevant data and procedures.

To determine what to include, I reviewed each transportation depth exam design standard, the basic concepts most frequently covered in textbooks, and the specifications outlined by NCEES in order to develop a sense of what an engineer with moderate experience (i.e., four years after graduation) would most likely be expected to know for the exam. I tempered this level of knowledge with what a practicing engineer should know about traffic engineering, even if not directly involved in traffic analysis in the workplace. In other words, my approach for this book was twofold: 1) eliminate nonessential topics, and 2) comprehensively cover essential topics. By following this approach, I feel I've successfully found a midpoint to define the upper limit of useful material to include in this book.

This book reconfigures the *HCM*'s approach to highway capacity analysis in a streamlined manner, using a set of sequential steps that are common or similar to all facility types. The *HCM* is organized around a system of concepts and procedures to evaluate quality of service on freeways, multilane and two-lane highways, urban streets, intersections, pedestrian paths, bicycle paths, and transit facilities. The *HCM* approaches each of these facilities as a separate entity under one or several chapters, which include all procedures for each facility type. In order to study a particular facility type using the *HCM*'s approach, an engineer must review multiple chapters to understand the analysis procedure and techniques for a facility type. The *HCM* chapters also often repeat concepts or procedures that are similar for more than one facility type.

To make the material in the *HCM* more accessible, I used the approach of a single core logic path, developed the concepts of traffic flow, then proceeded through the methodologies, combining similar or parallel procedures for each facility type. Methodology unique to a facility branches out as needed, then returns to the main procedure and advances to the next step. Therefore, only one baseline sequence has to be followed, and variations in each step are applied as necessary to account for the highway type being studied. Some factors, such as the heavy vehicle factor, are evaluated the same way irrelevant of highway type, while other factors use concepts shared by only one or two types. For example, level of service (LOS) for freeways and multilane highways is based on flow density, while LOS for two-lane highways is found using either percent time spent following or average travel speed. I cover LOS under one topic heading, but subdivide it within a section to explain the differences for each highway type. All of the data from a particular topic are grouped together to avoid confusion, and table and figure titles indicate when the information applies to more than one traffic condition.

In many cases, data is given in customary U.S. units, with notations added for converting to SI units. While this may lead to rounding errors in a few cases, the omission of SI units keeps the information more readable than including duplicate data in SI units. Final calculations using conversion factors will be very close to values derived using the *HCM* SI values.

For the chapter on geometry, I reviewed many highway engineering and surveying books. Although more recent books have a tendency to treat geometry as a less important subject in light of modern electronic measuring devices, I chose to emphasize the fundamentals of layout geometry. In particular, I focused on areas where there is often confusion, such as circular curve terminology, proper baseline or centerline stationing, and alignment traverse closures using PIs as traverse control points. Alignment layout accuracy requires professional field procedure working hand-in-hand with good office procedure.

In this book, I have emphasized the classical applications of geometric layout to illustrate the basics of horizontal alignment control, which has its roots in the U.S. Transcontinental Railroad. The U.S. Transcontinental Railroad developed the basic standards for all railroad geometry still used today. For highway applications, the Pennsylvania Turnpike Commission devised the design standards and the principles of high-speed, high-volume traffic accommodation that were later applied to the Interstate Highway system, often called the largest U.S. public works project ever built.

My emphasis on geometry was propelled in part by the advent of electronic calculation and layout devices, which have led many engineers to rely on shortcuts to geometric design. A true understanding of the fundamentals of geometry is crucial to success both on the PE exam and as a transportation engineer. Technology cannot replace an engineer's logical thinking processes or the application of engineering principles to a finished product. For roadway layout, the proof is often not seen until after the finished product is complete, but truly knowledgeable engineers can visualize their layout through the eye of the driver.

The organization of this book is similar to the development of transportation systems. Chapter 1 covers the planning process, assessing the desires and needs of a transportation system. Chapter 2 addresses the size and capacity of the various vehicular transportation components. Chapter 3 illustrates the processes of evaluating pedestrian, bicycle, and mass transit facilities. Chapter 4 analyzes details of geometry and physical features of the road network. Chapter 5 covers construction activities relating to transportation. Chapter 6 details advances in cost-effective analysis of traffic safety features. Taken together, the chapters reflect the almost overwhelming impact transportation has on the daily activities of individual users and on communities.

It is my hope that the *Transportation Depth Reference Manual* will be your primary transportation analysis reference for the PE transportation depth exam. Furthermore, just as the *Civil Engineering Reference Manual* has been found on many practicing engineers' bookshelves, I hope you will also find the *Transportation Depth Reference Manual* useful as a quick reference for transportation work in your professional career.

Finally, I ask that you submit any suspected errors to PPI's errata website at **www.ppi2pass.com/errata**. All errors will be reviewed, and verified mistakes will be incorporated into future editions of this book. You can also check the latest errata postings to find errors submitted by other readers.

Norman R. Voigt, PE, PLS

Acknowledgments

For many pursuing an engineering career, working as a team becomes second nature. Most projects require the knowledge and capabilities of many, and the sense of accomplishment is shared by all who participate. I am privileged to have started my career with one of the finest bridge engineers of the mid-twentieth century, George S. Richardson, PE, at Richardson, Gordon & Associates (RG&A) in Pittsburgh, PA. Mr. Richardson and his partners assembled a group of leading structural transportation engineers to tackle some of the most difficult road and bridge design challenges in the mid-Atlantic region and beyond. With over half the staff being registered engineers, RG&A received more awards for structural and roadway design projects than any firm we knew. This was quite an accomplishment for a medium-sized firm of the day, which has no comparison to today's mega-size firms.

I am particularly indebted to one of my mentors, James K. Arentz, PE, who was one of the designers of the western end of the original Pennsylvania Turnpike. Mr. Arentz was fond of describing how he participated in route layout reconnaissance using a bombsight mounted in the belly of a small plane. The plane would fly the proposed route at low altitude, and Mr. Arentz would view the terrain through the bombsight, envisioning the roadway alignment hugging the sides of hills and following the contours of the valleys. Mr. Arentz taught me how to apply curve geometry and superelevation along challenging routes to make a smooth and comfortable roadway. Our goal was a highway that would ride much like the finest railroad alignments used by first-class passenger trains. The standards set by the Pennsylvania Turnpike were copied in Germany's Autobahns and became the basis for the U.S. Interstate Highway System. It took me many years to appreciate what I learned from Mr. Richardson, Mr. Arentz, and their talented colleagues. The kind of mentoring they gave is what true professional development is all about.

Through many years of coordinating and teaching portions of the intensive civil PE review course sponsored by The Pennsylvania State University, I came to know many engineers and engineering candidates. I have shared their experiences and struggles to learn about the practical side of engineering, applying theory to a workable finished product. And, in some cases, defending the basics of engineering design against the business and political nature of large public engineering projects. The emphasis of the Penn State review courses has always been that engineering licensure is about the process of analysis and application of solutions, not simply finding a quick answer to a problem. My colleagues in the program, David Morse, Tom Leech, Kashi Banerjee, Sam Shamsi, and the original energy behind the program, Joe DeSalvo, have all set the tone for proper engineering analysis where the application of realistic solutions is a prerequisite for a commendable finished project.

I would like to thank the individuals who worked to ensure the accuracy of this book, including Suleiman Ashur, PhD, Maher Murad, PhD, and Julie M. Vandenbossche, PhD, for technically reviewing this book, and Patrick Albrecht and Todd Fisher for performing the calculation check.

I have received outstanding support from PPI on this project. There are many current and former employees (a project of this size doesn't just happen overnight) that helped with the many drafts and my scribbled comments. Thank you to Sarah Hubbard, director of product development and implementation; Cathy Schrott, production services manager; Megan Synnestvedt and Jenny Lindeburg King, editorial project managers; Marjorie Roueche; Heather Kinser; Tyler Hayes, Chelsea Logan, Scott Marley, Magnolia Molcan, and Julia White, copy editors; Kate Hayes, production associate; Tom Bergstrom, technical illustrator; and Amy Schwertman, cover designer. Michael R. Lindeburg, PE, graciously provided use of sections, tables, and figures from the *Civil Engineering Reference Manual.*

The final and most important supporting person has been my wife, Mary Jean, a professional librarian with a background in technology. She has been able to keep me headed in the right direction, teaching me how the written word should be handled so that others can share in the knowledge being presented.

Norman R. Voigt, PE, PLS

Codes and References

The *Civil Engineering Reference Manual* and the *Transportation Depth Reference Manual* are the minimum recommended library for the civil PE transportation depth exam. The exam is based on the following codes, as noted by the NCEES transportation specifications.

CODES AND STANDARDS USED ON THE EXAM

AASHTO: *A Policy on Geometric Design of Highways and Streets*, 5th ed., 2004, American Association of State Highway and Transportation Officials, Washington, DC.

AASHTO: *AASHTO Guide for Design of Pavement Structures*, 4th ed., 1993 and 1998 supplement, American Association of State Highway and Transportation Officials, Washington, DC.

AASHTO: *Roadside Design Guide*, 3rd ed., 2002, American Association of State Highway and Transportation Officials, Washington, DC.

AI: *The Asphalt Handbook* (MS-4), 7th ed., 2007, Asphalt Institute, Lexington, KY.

HCM: *Highway Capacity Manual* (HCM 2000), 2000 ed., Transportation Research Board—National Research Council, Washington, DC.

ITE: *Traffic Engineering Handbook*, 6th ed., 2009, Institute of Transportation Engineers, Washington, DC.

MUTCD: *Manual on Uniform Traffic Control Devices*, 2009 ed., U.S. Department of Transportation—Federal Highway Administration, Washington, DC.

PCA: *Design and Control of Concrete Mixtures*, 14th ed., 2002 (rev. 2008), Portland Cement Association, Skokie, IL.

REFERENCES USED IN THIS BOOK

The following references were used to prepare this book. You may also find them useful references to bring with you to the exam.

Elementary Surveying. Russell C. Brinker and Paul R. Wolf. HarperCollins.

Engineering Economic Analysis. Michael R. Lindeburg. Professional Publications, Inc.

Fundamentals of Traffic Engineering. Wolfgang S. Homburger, Jerome W. Hall, Edward C. Sullivan, and William R. Reilly. University of California, Berkeley.

Highway Engineering. Paul H. Wright and Karen Dixon. John Wiley & Sons.

Introduction to Transportation Engineering. James H. Banks. McGraw-Hill.

Manual of Transportation Engineering Studies. Institute of Transportation Engineers.

Pedestrian Planning and Design. John J. Fruin. Metropolitan Association of Urban Designers and Environmental Planners.

Principles of Highway Engineering and Traffic Analysis. Fred L. Mannering, Scott S. Washburn, and Walter P. Kilareski. John Wiley & Sons.

Project Management for Engineering and Construction. Garold D. Oberlender. McGraw-Hill.

Railroad Curves and Earthwork. C. Frank Allen. Norwood Press.

Route Location and Design. Thomas F. Hickerson. McGraw-Hill.

Route Surveying and Design. Carl L. Meyer and David W. Gibson. Harper & Row.

Surveying. Francis H. Moffitt and Harry Bouchard. Harper & Row.

The Surveying Handbook. Russell C. Brinker and Roy Minnick, eds. Kluwer Academic Publishers.

Superpave Mix Design (SP-2). The Asphalt Institute.

Thickness Design—Highways & Streets (MS-1). The Asphalt Institute.

Traffic and Highway Engineering. Nicholas J. Garber and Lester A. Hoel. Cengage Learning.

Traffic Flow Theory and Control. Donald R. Drew. McGraw-Hill.

Traffic System Analysis for Engineers and Planners. Martin Wohl and Brian V. Martin. McGraw-Hill.

Transition Curves for Highways. Joseph Barnett. Public Roads Administration.

Transportation Engineering Basics. A. S. Narasimha Murthy and Henry R. Mohle. American Society of Civil Engineers.

Transportation Engineering and Planning. C. S. Papacostas and P. D. Prevedouros. Prentice-Hall.

Trip Generation. The Institute of Transportation Engineers.

Introduction

The *Transportation Depth Reference Manual* covers the transportation depth section of the civil PE exam administered by the National Council of Examiners for Engineering and Surveying (NCEES). The civil PE exam provides the qualifying test for candidates seeking registration as civil engineers. The transportation section of the exam is intended to assess your knowledge of transportation design principles and field practice.

This book is written with the exam in mind. Major topics, equations, and example problems are presented, and practice problems are given at the end of each chapter. National design standards are referenced throughout the chapters and used in examples and practice problems. Appropriate sections of the standards are succinctly explained and analyzed to help you gain familiarity with the codes before using them during the exam. This book provides a comprehensive guide and reference for self-study of transportation engineering.

This book is organized into six chapters, corresponding to the following exam specifications. Traffic analysis is divided into two chapters to provide separate analysis of vehicular traffic and pedestrian, bicycle, and mass transit.

1. Transportation Planning (7.5%)

   Optimization, cost analysis, traffic impact studies, capacity analysis

2. Traffic Analysis (22.5%)

   Traffic capacity studies, traffic signals, speed studies, intersection analysis, traffic volume studies, sight distance evaluation, traffic control devices, pedestrian facilities, driver behavior and performance

3. Geometric Design (30%)

   Horizontal and vertical curves, sight distance, superelevation, vertical and horizontal clearances, acceleration and deceleration, intersections and interchanges

4. Traffic Safety (15%)

   Roadside clearance analysis, conflict analysis, work zone safety, accident analysis

5. Other Topics (25%)

   Hydraulics, hydrology, soils and materials properties, soil mechanics analysis, engineering economics, construction operations and methods, pavement structures

HOW TO USE THIS BOOK

The *Transportation Depth Reference Manual* provides comprehensive coverage of the major topics on the transportation depth exam and is designed to be used in conjunction with the *Civil Engineering Reference Manual*, which should be your primary breadth exam review resource. Start by reviewing the exam topics (listed in this introduction) and familiarizing yourself with the content and format of this book. Review the table of contents and the index, and flip through the chapters. Each chapter begins with a nomenclature list of the chapter's variables and ends with practice problems covering the chapter's major topics. Every significant term and concept has been indexed to provide a method of finding topics and data quickly.

Create a study schedule based on your strengths and weaknesses, and on how much time you think you'll need to spend reviewing each chapter. While chapters can be reviewed and referenced independently, each chapter builds on the topics presented previously. As you read each chapter, work the example problems and review the presented solution. At the end of the chapter, assess your understanding by solving the end-of-chapter practice problems. The practice problems are designed to give you experience applying relevant equations, data, and design standards to a given problem. Restrain yourself from reviewing the solutions until after you've solved each problem, then compare your solving approach with that given in the solution. With practice, you will be able to quickly decide which design standards, data, and equations are applicable to the problem at hand.

1 Transportation Planning

Nomenclature

a	curve adjustment factor	–	–
A	annual amount	\$	\$
AW	annual worth	\$	\$
B	benefit	\$	\$
C	cost	\$	\$
CF	count expansion factor	–	–
d	delay	min	min
f	factor	–	–
i	effective annual interest rate	%	%
L	sound level	dB	dB
n	number of compounding periods	–	–
N	number	–	–
p	proportion	–	–
P	period	min	min
r	correlation coefficient	–	–
r^2	coefficient of determination	–	–
t	time	min	min
t	user time	p-min	p-min
T	number of trips	–	–
v	flow rate	vph	vph
V	actual volume	veh	veh
V'	adjusted volume	veh	veh
X	quantifiable parameter	various	various

Subscripts

a	adults
A	annual maintenance
B	break
C	counting
e	employed adults
eq	hourly equivalent
h	households
I	initial
l	licensed drivers
p	peak or person-time
s	school-age children
seg	segment
sig	signal
t	time period, total, or transit accessible locations
ts	transit stop
v	vehicles
vtr	vehicle reduction

1. INTRODUCTION

Transportation planning involves travel demand analysis and forecasting and utilizes mathematical models to predict the volume of trips between activity centers. The Federal Highway Administration (FHWA) has a standardized process for all regions with a population of greater than 50,000.[1] The FHWA works with a region's existing plans for land use, environmental, or economic development to establish a transportation system's goals (irreducible values) and objectives (attainable targets). From these goals and objectives, the movements of people and products are mathematically modeled and calibrated to historical patterns prevalent in the region in order to project future traffic patterns. Alternative strategies for managing travel demand are proposed, tested, and evaluated using the modeled projections and the goals and objectives. Changes are made as needed to produce an optimal strategy to implement the proposed transportation system. The costs of alternative strategies are compiled, and a draft report is prepared for public comment. Simultaneously, a *draft environmental impact statement* (DEIS), commonly known as an *environmental impact report* (EIR), is prepared using the possible strategies to determine the projected environmental effects of each alternative. After a period of public input and assessment by the appropriate decision makers, a strategy is selected, the funding stream is secured, and the system's final design begins. Traffic volumes along planned travel paths, which were developed from the travel demand modeling, are used to design the facility to meet the required capacity.

Transportation planning is a process that persists over the life of a project, rather than a single task completed at the start of a project. The process allows transportation planners to make necessary improvements as the project evolves. There are certain considerations, however, that transportation planners must always keep in mind when planning a new project. For example, federal legislation requires that transportation planners follow certain environmental guidelines in every new project. Several important acts pertaining to transportation planning include the 1991 *Intermodal Surface Transportation Efficiency Act* (ISTEA), the 1998–2003 *Transportation Equity Act*

[1]Local and state governments set standards for areas with populations of fewer than 50,000.

for the 21st Century (TEA-21), and the 2005–2009 *Safe, Accountable, Flexible, Efficient Transportation Equity Act: A Legacy for Users* (SAFETEA-LU). Many of the environmental considerations in these acts are based on the *Clean Air Act*,[2] which places responsibility on the states to set pollution limits. The challenge transportation planning often faces is decreasing pollution while devising strategies that will meet the capacity needs of future transport.

2. ORIGIN-DESTINATION STUDIES

Before the *Federal-Aid Highway Act* of 1956, the development of most urban areas did not rely on travel surveys. With the Act, however, funds were appropriated to extend primary and secondary highway systems. Because of the increasing complexity of urban street systems, the improvements for these extended systems would need more than collected traffic volume data. The study of origins and destinations of trips became imperative for projecting future traffic demand.

Origin-destination studies, often called *O-D studies* or *O-D surveys*, vary widely in scope, but are most often employed during the preparation of comprehensive transportation plans for a large area. The O-D study report is often combined with other information, such as economic and employment data, and is used as baseline information for recommending or justifying transportation development programs and projects.

O-D studies are expensive and time-consuming to perform. While this limits their frequency, they are necessary to conduct when updating or supplementing existing O-D data, particularly if a major transportation improvement program is being planned. The projection of an O-D survey is usually 15–25 years forward[3] and includes input on future economic trends and population growth.

O-D studies classify trips into three distinct categories. *External-external trips* pass through a study area, but have neither their origin nor their destination within the study zone. *Internal-internal trips* have both their origin and destination within the study area. *External trips* (*internal-external* or *external-internal*) have either their origin or their destination outside of the study area. Trip categories are determined from the data collected during an O-D study, which can be performed using a variety of methods.

To perform an O-D study, the study area must first be chosen. A *study area* can be a single zone or can be divided into several zones. *Study zones* are a portion of a study area and are chosen based on several factors, such as the size of the study area and the study's objective. A study area can be small, such as a business district or residential neighborhood, or large enough to encompass an entire city center, containing several zones within the study area. For instance, a large study area for a small city could have zones designated for major clusters of activity within the city boundary. Examples include a central business district (CBD), a university center, a dense residential community, a park district with low density housing, a manufacturing center, and so on. The boundaries of the zones are usually selected along streets or roads for ease of identification, but another logical barrier, such as a railroad, canal, or river, can be selected. It is important that the zone boundaries are easily identified by the traveling population, and boundaries and zone identifications must be unique to avoid misunderstandings by the interview subjects.

Regardless of the size of the area or zone, it is important to select boundary locations where counting stations can be conveniently located. Boundary locations must also ensure all traffic entering and leaving the study area or zone passes through a counting station. Adjustment factors are incorporated into the data evaluation process to adjust for counting errors, but careful selection of counting station locations and data collection techniques can greatly reduce the amount of error correction needed, improving data reliability.

Study Methods

One of the first O-D study methods developed was the *home-interview origin-designation survey*, which interviewed household members to obtain information on the number, purpose, mode, origin, and destination of all trips made on a certain day. However, this method is invasive and time-consuming and is not often used today. Instead, methods such as license plate studies, tag-on vehicle studies, cordon surveys, corridor surveys, and postcard surveys are used to collect data.

A *license plate study* establishes multiple stations throughout the designated study area to read passing vehicles' license plates. The license plate numbers are then tracked to establish travel paths. Home addresses can be checked through motor vehicle registration lists to determine where the vehicle most likely came from or was headed. License plate studies have the advantage of being less disruptive than home interviews. With improvements in electronic remote license plate reading systems, license plate studies blend well with automated data collection techniques.

A *tag-on vehicle study*, also known as the *vehicle intercept method*, involves placing a small tag on a vehicle or handing a colored card to the driver of a vehicle as it enters the study area. The tag or card is color-coded to identify the location where it was first assigned. Internal stations can monitor the vehicle's path by observing the color of the tag or card. As the vehicle leaves the study area, the driver is asked to return the card to an observer.

[2]The *Clean Air Act* was enacted by Congress in 1990 and defines the Environmental Protection Agency's (EPA's) responsibilities for protecting and improving the nation's air quality. The EPA sets air pollution limits for the United States, and states are prohibited from having less stringent pollution controls than those set for the nation. However, while the EPA sets the national standards, states are responsible for carrying out the Act, as combating pollution requires regional knowledge.

[3]Because O-D studies involve such a large investment of effort, they are not warranted for shorter-term projects.

A *cordon survey* encircles a study zone with counting stations located on all travel paths into and out of the zone. Cordon surveys are comprehensive and can include all modes of transportation. They are often used to determine traffic flow from one major activity center, such as a business center, shopping district, or other activity center that has many different arrival and departure paths, to another. Figure 1.1 is an example of a cordon counting station setup.

Figure 1.1 *Cordon Survey Counting Station Locations*

Adapted from GARBER/HOEL. *Traffic and Highway Engineering*, 3E. © 2002 Cengage Learning, a part of Cengage Learning, Inc. Reproduced by permission. www.cengage.com/permissions

Because cordon surveys involve all vehicles entering or leaving the activity center, the stations perform simultaneously for a period of 24 hours. These surveys require months of planning and involve extensive notification of the motoring public. To compensate for motorists' tendency to avoid the study area during a count period, adjustment factors are often inserted into the data. Cordon surveys are complex undertakings and are therefore performed infrequently, as many agencies feel reliable results can be obtained instead through a series of carefully planned corridor surveys on each leg of the activity center approach.

A *corridor survey* covers all major routes through a traffic corridor, that is, major traffic flows between significant nodes of a network. Like cordon surveys, corridor surveys are often used to determine traffic flow from one major activity center to another. They are typically used between towns and major industrial, shopping, and residential areas, and can yield significant data, such as vehicle occupancy and origin-destination information.

Roadside interviews require thorough training of interview personnel, careful selection of station sites, and significant advance publicity in order to encourage public cooperation. Sample size is often a function of traffic flow and the public's tolerance for traffic delays. The Institute of Transportation Engineers' (ITE's) *Manual of Transportation Engineering Studies* recommends interviewing the drivers of at least 20% of vehicles, although a 50% sample is recommended to obtain reliable results. Inevitably, there will be some degree of undercounting, which should be considered in the final tabulation of results.

Postcard studies stop motorists at survey stations and distribute survey questions on postcards with prepaid postage. The motorists can complete and return the surveys at their leisure, making postcard studies considerably less invasive than roadside or home interviews and cutting down on traffic delays. The *Manual of Transportation Engineering Studies* suggests a minimum response rate of 20% is needed to obtain reliable data. Typically, postcard studies have shown responses in the range of 25% to 35%.

Onboard transit surveys offer another important interviewing option that features trained interviewers, usually recently retired transit operators. These interviewers ride on transit vehicles, hand survey forms and pencils to passengers as they board, and then ask the passengers to check off applicable boxes while riding to their destination. This type of survey usually results in reliable data, especially if the interviewers are familiar with the transit route system. Transit surveys are important tools in O-D studies, especially in cities that have a significant proportion of transit usage, as they provide zone-to-zone travel patterns due to the fixed-route nature of transit.

In addition to the survey methods previously discussed, O-D studies may use other data sources (often termed *secondary sources*). One of the most-used secondary sources is census data, because they are updated at least once every 10 years and are uniform in quality. Other secondary sources include voter registration records, city directories, property tax records, and land use data.

Once a method's final data are collected, they are reviewed and checked for accuracy. Data checks include comparing the data against selected known high-accuracy samples, which can be obtained from two or three control points within the study area, such as a bridge or other well-known point of traffic restriction. Accuracy can be checked by comparing origin and destination points against census data to ascertain if driver populations closely match survey data, or by checking employment destinations against businesses' employee counts. After the data's accuracy is verified, the data are plotted in a variety of forms to illustrate the intent and purpose of the study. Links between origin and destination nodes are usually bundled along existing travel corridors and presented as a graphic illustration of the relative numbers of trips along each corridor. Histograms are often prepared to illustrate the daily, weekly, or monthly distribution of trips per unit of time. Trip plots can be further divided into the basic categories of external and internal trips, and by the nature of a trip's purpose, such as shopping, work, to and from home, and so on.

Cumulative statistics on *person-trips*, or one person taking one trip, are often shown as simple graphics. Person-trip ends can be illustrated as a series of three-dimensional blocks, with the height drawn in proportion to the number of trip ends, or by using a horizontal bar chart. The ITE defines a *trip* as a single movement between two locations with either the origin or the destination inside the study zone. The ITE uses the term *total trip ends* for the total number of trips in and out of a study zone. Therefore, for the purposes of traffic engineering, the terms "total trips" and "trip ends" can be used synonymously. Vehicular traffic volumes are often plotted using line graphs or as histograms. They are also frequently illustrated on a traffic flow map with the line weight proportional to the relative traffic volume on each segment.

Figure 1.2 is an example of the representation of trips by auto and trips by transit into and out of a city business center during the morning and afternoon peak periods. The number of trips are also compared to distance, which is illustrated by concentric rings that represent the approximate distance from the city center. In Fig. 1.2, the morning trips entering and leaving the city center are shown on the left, and the afternoon trips are shown on the right. The graphic arrow widths are proportional to the number of trips crossing each concentric distance ring and indicate the direction of trips into and out of the city center. For example, there are 75,000 trips by transit and 15,000 trips by auto crossing ring 1 in the afternoon outbound direction.

Figure 1.2 *Internal Trips by Auto and Transit Inbound and Outbound During Peak Periods*

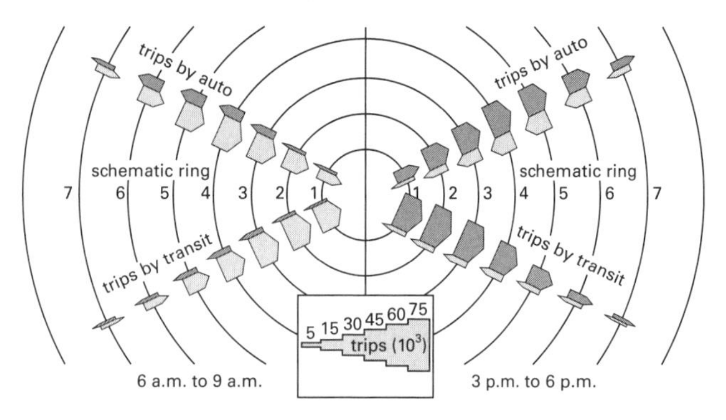

Figure 1.3 is a *traffic flow map*, which illustrates traffic volume using line weight. The line weight is proportional to the relative traffic volume on each segment. Numerical traffic volumes are provided along each segment. Two-directional volumes can be illustrated with separate line weights when the directional traffic volume varies significantly. Unlike the representation shown in Fig. 1.2, traffic flow maps provide the traffic volumes on individual roadway segments and allow comparison between segments. Engineers and statisticians can use these kinds of visualizations to assist in transportation policy decisions, using them to communicate with city officials, planners, and the public, who may not have a background in transportation.

Figure 1.3 *Traffic Flow Map*

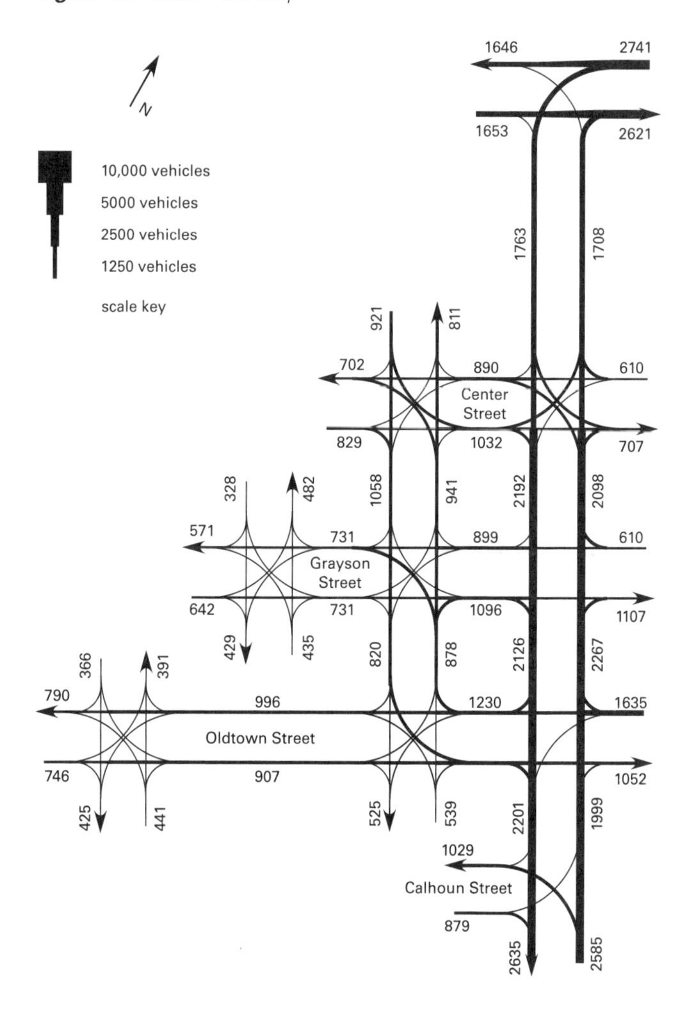

Due to the standardization of study formats driven by mandated formulas and study parameters, it is possible to compare interpolated data from region to region and from study period to study period.

3. TRAFFIC VOLUME STUDIES

Traffic volume studies determine the numbers, movements, and classifications of vehicles (and/or bicycles and pedestrians) at specific locations and times. They are essential for many types of traffic analysis, such as determining the influence of pedestrians on traffic flow or calculating annual average daily traffic (AADT, see Chap. 6). The length of a study, or *count period*, depends on the study method and the intended use of the recorded data. Volumes are usually counted over a set time period, which can range from a few minutes to more than one year. The count period should represent the typical conditions (such as time of day, day of month, and month) applicable to the desired analysis. For example, studies at schools should not usually be conducted on weekends.

The shortest count periods are used to determine the peak period flow rate. If the peak hour is to be determined, the count period should start one-half hour to one hour before the expected peak hour and extend one-half hour to one hour after the expected peak hour. The peak hour is defined and determined as the four contiguous 15-minute count periods that contribute to the highest one-hour count.

The tools and equipment used to determine the traffic volume can be as simple as tally sheets or manual count boards or as complex as electronic detectors and computer data generation programs. Regardless of the tools used, the purpose for a traffic volume study remains the same: to accurately establish the actual amount of traffic passing a given location during a unit of time.

Counting Methods

There are two counting methods for collecting traffic volume data: manual and automatic. Manual and automatic counting methods are virtually the same technique for any period being analyzed, except that manual counting methods are usually limited to periods shorter than a few days or hours. Both methods may be used to count vehicles (e.g., to determine the number of multi-axle or large vehicles at a certain location), pedestrians (e.g., to identify shopping, casual, or commuter traffic), and bicycles (e.g., to separate bicycle traffic by recreation or commuter purpose).

Manual Counting Methods

Manual counting methods require observers to go to a site and collect data for a specific time period using stopwatches or timers. They may be used to collect small data samples, or when automatic equipment is not available or is too expensive to use. Examples of data gathered by manual counting include vehicle classifications, direction of travel, vehicle occupancy, and pedestrian movements.

Manual methods are typically used when the count period is limited to a few hours or a few days (also known as *short period sampling*), as longer time periods tend to be fatiguing and generally may be handled more efficiently with automatic methods. In most cases, short period sampling adequately establishes baseline parameters for design purposes. For instance, intersections may be counted at intervals of 5 minutes or 15 minutes to establish peak hour flow rates for a period of 2 hours over the morning or afternoon peak. For roadway segments that lie between intersections, the peak 15 minute and 60 minute flow periods are selected to determine the peak hour flow rate and the peak hour factor. Midday counts at intervals between 9:00 a.m. and 3:00 p.m. may be used to determine base flow rates. Full 12 hour daytime counts (7:00 a.m. to 7:00 p.m.) may be used to determine typical design-day traffic flow. Except for intense evening shopping activity, the remaining 12 hours rarely accounts for more than 15% of the total daily traffic.

The number of observers needed to collect data will depend on the number of lanes or crosswalks being observed, the types of data collected, the traffic volume, and the length of the count period. There should be a sufficient number of observers at count stations to provide frequent intervals of relief. For example, on busy intersections with *full movement activity*,[4] two or more observers may use short-break and/or alternating counting procedures to help fight fatigue. Observers may count for alternating periods of 5 minutes each and use the intervening rest periods to review the count tabulations, reset their counters, and prepare for the next counting period. Rest periods should be alternated among the observers, and there will be little loss in count accuracy if observers carefully coordinate their break periods.

An *adjusted volume*, V', is calculated to account for break periods. To determine the adjusted volume for each counting period, P_C, the actual volume, V, is expanded in proportion to each break period, P_B, by a count expansion factor, CF. When the counting periods and break periods are of constant duration and are equal to each other, the number of cycles can be substituted for P_C and P_B to calculate the count expansion factor.

$$V' = V(\text{CF}) \qquad 1.1$$

$$\text{CF} = \frac{P_C + P_B}{P_C} \qquad 1.2$$

In order to perform a manual count, observers need portable recording devices. Three devices commonly used are tally sheets, mechanical counting boards, and electronic counting boards. *Tally sheets* are simple, inexpensive tools on which observers record data. The sheets may be prepared in advance, or observers may draw a diagram of the location and use labeled boxes to record tallies of the desired information for each period.

Mechanical counting boards have counters mounted to a board in various configurations to represent typical intersection layouts. Observers carry the board as they move to accurately observe and record vehicle flows. After the data have been collected, the totals are summarized on field forms and then entered into a computer to provide a convenient check on the count accuracy. *Electronic counting boards* are battery-operated, handheld devices that have been developed to bridge the gap from mechanical boards to computer input by providing electronic input of data. Electronic counting boards have counting buttons and an internal clock that automatically separates data by time intervals. End-of-day data are downloaded to a computer. Fully equipped offices may have weatherproof laptops or tablet computers to enter data directly at the field count station. Data reduction, such as accuracy checks and count comparisons, can be performed on the portable computer prior to leaving the count location.

[4]Full movement activity is categorized as traffic patterns involving through traffic in four directions, including turning traffic.

Example 1.1

Observers are stationed at an intersection with a traffic signal that operates on a 90 sec cycle. The observers each take counts for 7 complete cycles, then break for 2 cycles in order to review their count tabulations and prepare for the next series of counts. The total vehicle count averages 160 veh during the 7 cycle counting period for one of the observer's movements. What is the adjusted vehicle volume for this movement?

Solution

The total counting period duration is

$$P_C = \frac{\left(90\ \frac{\text{sec}}{\text{cycle}}\right)(7\ \text{cycles})}{60\ \frac{\text{sec}}{\text{min}}} = 10.5\ \text{min}$$

The total break period duration is

$$P_B = \frac{\left(90\ \frac{\text{sec}}{\text{cycle}}\right)(2\ \text{cycles})}{60\ \frac{\text{sec}}{\text{min}}} = 3\ \text{min}$$

Using Eq. 1.2, the count expansion factor is

$$\text{CF} = \frac{P_C + P_B}{P_C} = \frac{10.5\ \text{min} + 3\ \text{min}}{10.5\ \text{min}} = 1.29$$

Alternatively, since the counting and break periods are equal (i.e., 90 sec/cycle), the count expansion factor could also have been calculated using the number of cycles. There are 7 counting periods and 2 break periods. Therefore, the count expansion factor is

$$\text{CF} = \frac{N_C + N_B}{N_C} = \frac{7+2}{7} = 1.29$$

Using Eq. 1.1, the adjusted volume is

$$\begin{aligned} V' &= V(\text{CF}) = (160\ \text{veh})(1.29) \\ &= 206\ \text{veh} \end{aligned}$$

Automatic Counting Methods

Automatic counting methods gather large quantities of traffic data over an extended, often unattended, period of time. They are generally used to determine traffic patterns and trends, and observers may use permanent or portable automatic counters to collect data.

Permanent automatic counters consist of vehicle detection devices placed along a roadway, such as magnetic loops imbedded into the pavement surface. Placing a single loop in each lane will count the total vehicle flow along a roadway segment. Placing a pair of loops a set distance apart will detect the average traffic speed as well.

Portable automatic counters are generally used to collect data similar to that collected in manual counts, but for longer periods of time. They use pneumatic road tubes that are temporarily attached to the roadway surface. The air in the tubes is compressed each time an axle crosses the tubes, which actuates a switch in the control box. Timers may be set to determine the number of actuations in preset periods. In order to convert axle counts to vehicle counts, a manual observation count is taken over a sample period to determine the typical vehicle mix. Software is available to process data and provide detailed information based on the setting of the pneumatic tubes. The processed data from the sample observations are applied to the axle counts to determine the extracted vehicle counts. Pneumatic tubes may also be placed a set distance apart to determine average traffic speed. This is useful for setting the *amber interval* (i.e., the length of time the yellow light is on) for traffic signals.

Count Locations

Count locations are selected based on the purpose of the study. Observers are positioned to have a clear view of traffic, with two of the most common locations being intersections and roadway or pathway mid-block segments. *Intersection counts* are used to collect approach volumes and turning volumes for traffic signal timing, for high-crash intersection analysis, and for evaluating congestion. The intersection and/or the approach roadways are the most likely locations to perform an intersection study, and the upstream counts (or approach roadway counts) are used to establish arrival type.

Roadway or *pathway mid-block segment counts* are used to determine network volumes and traffic flow or capacity requirements. For mid-block or roadway segment counts, the count can be located anywhere convenient and accessible that has mostly uniform traffic flow. Mid-block count locations are selected at a point where intersection turning movements least affect the traffic flow. Roadway segment count locations are selected far enough from the survey end points that traffic represents uniform flow conditions of the segment.

4. TRAFFIC FLOW AND TRIP GENERATION MODELS

Traffic Flow

Traffic is the flow of people and vehicles on roads. It occurs for many reasons, such as the necessity to transport goods and the human desire to travel. Traffic involves interactions between vehicles, drivers, and the roadway system.

Traffic flow studies are performed when the effort is warranted, such as for estimating the needs of a new trip generator (e.g., a new shopping center or park) or for a planned improvement to an existing location. Broad traffic studies, such as regional or corridor studies, concentrate on traffic volumes, travel times, and capacities. Limited area studies, such as those performed at intersections, tend to focus on accidents, delays, gaps, and detailed physical conditions. Therefore, depending on their size and purpose, traffic flow studies usually cover

- *volume*, on each leg, link, or segment
- *accident experience*, especially at intersections
- *speed*, primarily in peak periods
- *capacity*, primarily in peak periods
- *physical conditions*, including effect on safety
- *travel time*, in relation to other factors
- *delays*, especially those caused by signals
- *gaps*, or the ability to enter the flow

The data gathered in traffic flow studies are used by traffic engineers who need specific data about study locations in order to create accurate model analyses. General information available in traffic standards and handbooks can then be used to compare data and help validate study conclusions.

Trip Generation and Attraction

The simple form of a trip has an origin and a destination and a link between the two. A complex form may have many incremental segments, with origins, destinations, and purposeful activities along the way. The *trip destination*, commonly called the *trip generator*, determines significant characteristics of the trip. *Trip generation studies* are used to determine the predominant trip generators in the study region, which are then used to estimate the number of trips generated. Examples of trip generators include factories, office buildings, shopping centers, resorts, state parks, airports, transit centers, county fairs, and sporting arenas. *Trip attraction studies* predict the number of trip ends to a nonhome destination (e.g., to a school or shopping center). Trip generation and attraction studies are closely related and often share methodologies.

Each type of trip purpose, such as for work, shopping, or recreation, has characteristics that transportation planners may use to project trip behavior. The ITE publishes *Trip Generation*, which defines a large number of widely accepted trip generator types and characteristics. Developed using current methodology of land-use-based traffic and transportation engineering, it describes procedures for quantifying the number of trips generated and attracted for various types of land uses. Trip generation studies are often used as *traffic impact studies* to determine the volume generated by a new trip generator and its effect on surrounding roadways. These data are then used to evaluate both the existing and new facilities to determine design improvements to accommodate the increase in trip volume. Because trip generation studies link trip volume and land use, trip generation data are often used in zoning and land use planning. Data determined using trip generation studies are often used in *site-impact studies* to estimate environmental impacts due to new development.

Trip Generation provides a collection of data for 12 major land use categories comprised from more than 4800 traffic studies conducted at over 550 sites in the U.S. and Canada. Figure 1.4 provides an example of data from *Trip Generation* for an office park and shows the data given for each land use. The equations and weighted average trip generation rate given in *Trip Generation* are usually sufficient to determine the number of trip ends, but they can also be determined from local data.

Figure 1.4 *Office Park Trip Generation Curve*

average vehicle trip ends vs: 1000 ft² gross floor area
on a: weekday,
peak hour of adjacent street traffic
one hour between 7 a.m. and 9 a.m.

number of studies: 27
average 1000 ft² GFA: 331
directional distribution: 89% entering, 11% exiting

trip ends generated per 1000 ft² gross floor area

average rate	range of rates	standard deviation
1.84	0.98–5.89	1.50

data plot equation

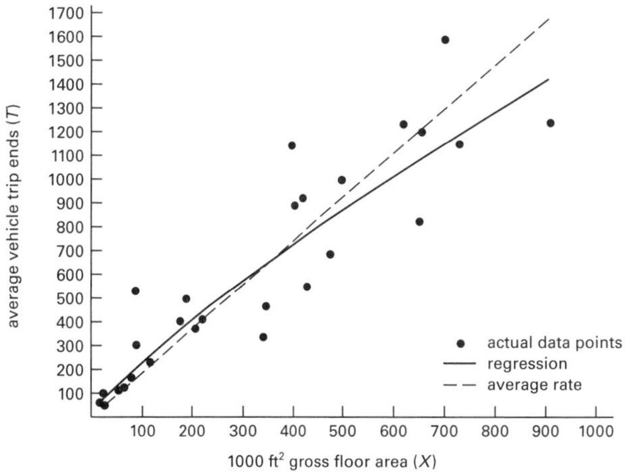

average vehicle trip ends (T)
1000 ft² gross floor area (X)
● actual data points
— regression
– – average rate

regression equation: $\ln T = 1.679 + 0.818 \ln X$
$r^2 = 0.89$

(Multiply ft² by 0.929 to obtain m².)

Reprinted with permission from the Institute of Transportation Engineers, *Trip Generation.*

For trip generation studies, the study area is divided into zones, or tracts, and data are obtained for each zone. Trip generation studies typically use socioeconomic data, such as household income or car ownership, while trip attraction studies most often use employment data, dividing employment by type, such as retail or service. Socioeconomic data can be obtained from sources such as the U.S. Census, while employment information is most often obtained from workplace or household surveys.

Regression Equations

The number of generated trips can be calculated using independent variables or estimated using traffic counts. Independent variables, commonly denoted as X, differ based on the zone and its land use. X must be a quantifiable parameter, such as the number of dwelling units or the number of beds in a hospital. For residential zones, independent variables include household income, car ownership, and family size. For retail zones, examples of independent variables include store type, gross leasable area, customer income level, and proximity to transit. Trip generation studies are often based on several independent variables to develop a zone's total number of trips in order to cover the anticipated behavior of the majority of trip purposes.

Once independent variables are selected and the necessary data are collected, the total number of trips, or total trip ends, can be calculated using either a linear model, Eq. 1.3, or a nonlinear model, Eq. 1.4. T is the total number of trips, bX and $b \ln X$ are the slopes of the plotted curve, and a is the curve's vertical intercept. The units of X depend on the variable being used. For example, if X were the gross leasable area, the units would be square feet.

$$T = a + bX \quad \text{[linear model]} \qquad 1.3$$

$$\ln T = a + b \ln X \quad \text{[nonlinear model]} \qquad 1.4$$

The ITE refers to both the linear and nonlinear models as *regression equations*, or *fitted curve equations*. A regression equation must be determined for each independent variable used. The choice of the linear model versus the nonlinear model is typically based on the model that has the highest coefficient of determination. The *coefficient of determination*, r^2, is a statistical measure that determines how well the least squares relate to the original data. The value of r^2 ranges from 0 to 1, with 1 indicating that all plotted data fit on the line (i.e., an ideal fit). 0 indicates that the data are scattered and does not indicate an adequate correlation between the data points and the regression equation.

The data given in *Trip Generation* include a regression equation for each land use, although care should be taken when selecting regression equations. Whether the regression equation is calculated or selected from *Trip Generation*, the suitability of the equation to the location should be verified against plotted data points to determine the suitability of fit. The standard deviation and the coefficient of determination for a land use's data are provided as an indication of the variation in trip generation data for the land use. Therefore, proper engineering judgment should be used when using the average rates and regression equations from *Trip Generation*.

Weighted Average Generation Rate

The generated trip ends can also be calculated using known vehicle counts and applying weighted averages. The *weighted average trip generation rate* is the sum of all trip ends during the study period divided by the sum of all independent variables. For example, if the independent variable, X, is the gross floor area, the weighted average rate is the sum of all vehicle trips divided by the sum of the square footage of the gross leasable area in the zone and would be reported as the weighted average rate per 1000 ft^2. The weighted average trip rate is then multiplied by X to find the number of generated trip ends. The weighted average rate can then be plotted as shown in Fig. 1.4, using the assumption that a linear relationship between trip ends and the independent variable exists.

Data Suitability

As outlined by the ITE, the suitability of the regression equation and the weighted average trip generation rate to the location should be verified as follows. *Trip Generation* suggests first comparing the number of trip ends calculated with the regression equation and the weighted average rate. If X is within the range of data points, the curve is not projecting beyond known information. If the difference between the two methods is less than 5%, then either method is acceptable. However, if there is more than a 5% spread, and if the fitted curve shows zero or close to zero trips when $X = 0$, then the equation for the fitted curve should be used. While it is recognized that ln(0) is undefined, a very small number, such as $X = 1$, can be used.

If the previous conditions are not met, then data points should be compared with the independent variable X to see if there are data points near that value to favor one plot over the other.

As a final check, the standard deviation and the *coefficient of determination*, r^2, the square of the *correlation coefficient*, r, can be compared to the data. As previously outlined, an r^2 value of 1 indicates that the data have a perfect fit, and a value close to 0 indicates that the data are scattered. For trip generation studies, an r^2 value less than 0.75 indicates that the weighted average rate should be used, and an r^2 value greater than or equal to 0.75 indicates that the regression equation should be used. The standard deviation should also be compared to the weighted average rate. If the standard deviation is greater than 110% of the weighted average rate, the regression equation should be used. Likewise, if the standard deviation is less than or equal to 110% of the weighted average rate, the weighted average rate should be used.

In problems where none of the previous conditions are acceptably met, the trip planner must make a selection using acceptable levels of data collected at or near the study site and at least two other sites of similar nature to confirm the data.

Example 1.2

Using Fig. 1.4, estimate the number of vehicle trip ends for the peak one hour between 7:00 a.m. and 9:00 a.m. for an office park with 625,000 ft^2 GFA, and determine which rate is more accurate.

Solution

The example may be solved either graphically or mathematically. To solve graphically, examine the average rate and the fitted curve plots. As shown in the following illustration, $X = 625$ on the horizontal scale intersects the average rate line at 1150 trip ends and the fitted curve at 1050 trip ends.

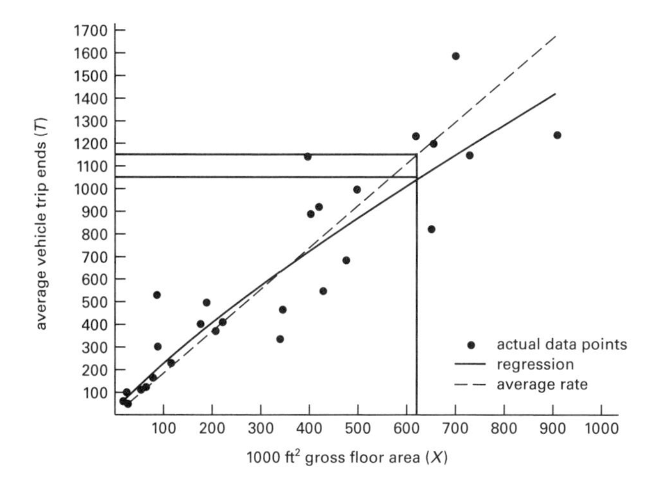

To solve mathematically, use the weighted average rate and the regression equation. For the weighted average rate,

$$T = (1.84 \text{ trip ends})\left(\frac{625{,}000 \text{ ft}^2}{1000 \text{ ft}^2}\right) = 1150 \text{ trip ends}$$

X is the gross floor area per 1000 ft^2. Using Eq. 1.4 and the values shown in Fig. 1.4,

$$\begin{aligned} \ln T &= 1.679 + 0.818 \ln X \\ &= 1.679 + 0.818 \ln \frac{625{,}000 \text{ ft}^2}{1000 \text{ ft}^2} \\ &= 6.945 \\ T &= e^{\ln T} = e^{6.945} \\ &= 1038 \text{ trip ends} \end{aligned}$$

The graphical procedure shows 1150 trip ends on the average curve and 1050 trip ends on the fitted curve, while the values calculated are 1150 trip ends for the weighted average rate and 1038 trip ends for the regression equation. Although mathematical results are preferable, either graphical or calculated results are acceptable, as 1050 trip ends is within 1% of 1038 trip ends.

Although the difference between the two methods is less than 5% (i.e., either result is acceptable), additional data checks can be performed. There are 27 data points, as shown in the illustration. Substituting $X = 0$ into the fitted curve equation yields 5 trips, which is close to 0 and indicates that the fitted curve is valid.

Perform the final check by evaluating the standard deviation and the correlation coefficient of determination. The standard deviation is $^{1.84 \text{ trip ends}}/_{1.50 \text{ trip ends}} \times 100\% = 123\%$, meaning it is greater than 110% and the regression equation rate should be used. The correlation coefficient of determination, r^2, is greater than 0.75, showing that the regression equation rate is the better choice.

5. CAPACITY PLANNING ANALYSIS

Capacity planning analysis is used to estimate the traffic carrying ability of transportation network segments at prescribed levels of operation. Criteria for operational levels are characterized by the level of service for each type of facility, which relates to the amount of traffic accommodated within a desired time period (i.e., the flow rate). The *Highway Capacity Manual* (*HCM*) lists six *levels of service* (LOS), which are described in Table 1.1.

Table 1.1 *Levels of Service*

LOS	description
A	*free flow:* traffic flows at or above the posted speed limit; all motorists have complete mobility between lanes
B	*reasonably free flow:* slightly more congested, with some impingement of maneuverability
C	*stable flow:* roads remain below but efficiently close to capacity; posted speed is maintained, but the ability to pass or change lanes is not always assured
D	*approaching unstable flow:* speeds are somewhat reduced; motorists are hemmed in by other cars and trucks
E	*unstable flow:* flow is irregular and speed varies rapidly, but rarely reaches the posted limit
F	*forced or breakdown flow:* flow is forced, with frequent drops in speed to nearly 0 mph (0 kph)

The criteria for the various LOS categories are set using base conditions (i.e., ideal conditions), such as 12 ft (3.7 m) wide lanes, at-level grades, and adequate sight distances, for each roadway type. The characteristics of a particular study segment are compared to the base conditions using adjustments to the flow rate made in passenger car equivalents (pce) for capacity and LOS predictions.

Capacity analysis is performed on two major network elements—intersections and uniform segments of extended roadway length. The analysis begins with volume and speed studies during the planning stage, which deal primarily with base conditions over broader segments of the network. This is in contrast to later operational analyses, which deal with precise definitions of roadway characteristics over shorter segments, spot conditions, and site-specific variations from the base conditions. The concepts of LOS and capacity flow of network segments form the fundamentals of network planning.

Volume and Speed Relationships

Once trip paths are defined by a regional study, a series of *desire lines* develop, which may be formed into a connecting link between pairs of origin and destination nodes. The accumulation of O-D nodes and links between nodes may be developed into a *traffic network diagram* to estimate expected volumes. In order for the planner to determine how much capacity to build into the connecting link, decisions must be made that weigh the link's capacity against a variety of factors, such as economic constraints, the available right-of-way, available funds for improvements, and the type of facility proposed.

Traffic link and node diagrams are developed from traffic network diagrams and traffic volume maps by codifying nodes at major intersections or activity centers. Figure 1.5 shows a traffic link network diagram, with the numbers shown representing the *average annual daily traffic*, or AADT (sometimes referred to as the *average daily traffic*, or ADT). The AADT is the total traffic for one year divided by 365 days. Traffic data for various links and nodes are recorded in a spreadsheet format for further computational analysis.

Figure 1.5 *Traffic Network Diagram*

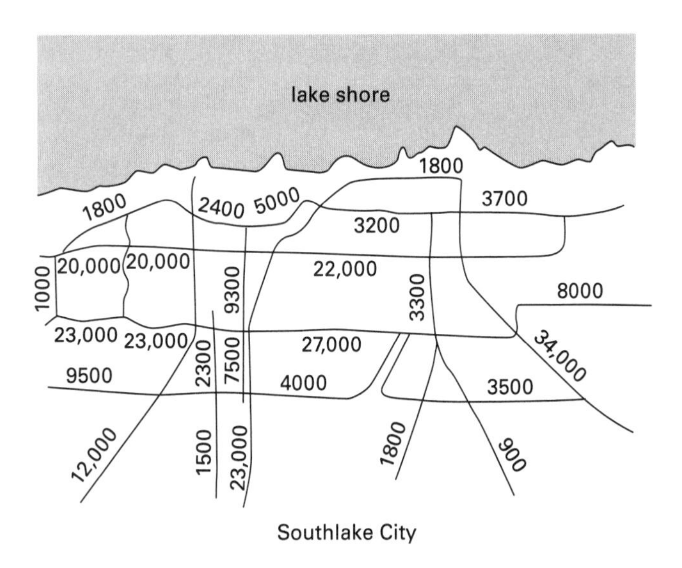

Roadway Types

Each roadway type has a capacity limit based on geometry, access point density, type of traffic control, and number of lanes. Generally, roadways are broken down into the categories of freeways, multilane highways, two-lane highways, arterials, and local streets. Hard-paved surfaces, such as concrete (i.e., rigid pavement) or a bituminous road mix (i.e., flexible pavement) capable of supporting travel at high speeds, will carry the most traffic. Gravel or earth surfaces are considered low-type pavements and are not capable of sustaining frequent traffic or high-speed travel, but are necessary to access many areas where very low-volume traffic prevails.

Freeways are designed for safe operating speeds ranging from 50 mph to 70 mph (80 kph to 110 kph) or greater. *Multilane* and *two-lane highways* generally have design speeds of 40 mph to 65 mph (65 kph to 105 kph). *Arterials* are usually designed for at least 30 mph (50 kph). When considering posted speeds, a design speed well above the actual posted speed will result in a greater safety factor, and to some degree a greater capacity, than a posted speed at or very near the design speed. The posted speed should not be greater than the design speed, but it may be difficult to enforce if the posted speed is unrealistically low.

Typical roadway analysis or design criteria develop a relationship between traffic characteristics such as volume, capacity, trucks, grades, number of lanes, peak hour factors, and speed. Alternatives combine economics with design criteria to make comparisons between selected corridors. The outcome may be different for differing types of economic analysis. For instance, an economic analysis based on delays in shopping center access may show a considerably different solution than an analysis that emphasizes minimal through-traffic delay. Furthermore, opinions on the monetary value of time vary considerably, making it difficult to assess the purely economic value of many incremental improvements, such as intersection widening or the addition of turning lanes.

Intersections

An *intersection* is the place where two or more vehicle pathways join, cross, or separate. With pathways that join or cross at unsignalized intersections, each traffic flow must seek gaps in the other flows to avoid collision. Should an arriving vehicle not find a gap of sufficient size to join or cross a path, a delay[5] will occur while that vehicle waits for a gap. To help create or force traffic gaps, a traffic control or signaling system, such as a signal roundabout, may be employed. Once a signaling system is installed, gaps become systematized and predictable. However, in order to create gaps, delays caused by the traffic system will inevitably occur.

[5]A *delay* can be considered the additional time needed to travel a route below a normal free-flow speed. Significant delays create driver discomfort and frustration, lost time, and a less smooth flow of transport.

Capacity and Other Types of Analysis

Many preliminary intersection studies address only the capacity and not the details of intersection design. These studies, called *planning analyses*, are broad and give an estimation of an intersection's level of use, but provide little information on delay. *Operational analysis*, on the other hand, is performed to evaluate the full details of four primary components: demand, or service flow; signalization; geometric design; and delay, or level of service. Operational analysis can be used to test various intersection configurations and the effect of detail adjustments on the four components.

6. OPTIMIZATION AND COST ANALYSIS

Optimization is the process of comparing alternative strategies in order to satisfy a project's objectives. In transportation engineering, optimization involves choosing among alternative plans based on costs and/or benefits in order to achieve the best possible approach to a project's objective. Optimization is rarely a stand-alone activity, but rather is performed in conjunction with other elements of project execution, and often requires weighting objectives when they are in conflict with each other. Optimization can also be applied to adjust varying quantities of factors to achieve a desired performance characteristic of a material, a design element, or a standard of compliance.

There are three basic characteristics of optimization—objective functions, variables, and constraints. The first step in optimization is to define a project's *objective function*, that is, something to be maximized or minimized. Objective functions can be parameters of cost, the accuracy of a point location, a defined spread of data values, traffic density thresholds, a durability function, or some other parameter. Objective functions are most often attainable values, but they also can be loosely defined or not defined at all in cases where there are no functions to optimize. Objective functions that are loosely or not defined are similar to the goals of feasibility studies, where a set of objective functions can be further defined for future study.

There can be multiple objective functions, in which a number of different objectives are to be optimized simultaneously. These objectives are often noncomplementary or noncompatible, such as minimizing initial cost while also minimizing operating and maintenance cost. For instance, a paving project can be constructed using a more-durable surface material, such as concrete or asphalt, or a less-durable material such as plain crushed-stone aggregate. A crushed-stone surface requires more maintenance over time to keep the surface free of wheel ruts and drainage erosion. For low traffic volumes, a crushed-stone road may at first seem economical, but the long-term operating and maintenance cost could be much greater than hard surface maintenance—even for low volume conditions. Most often, multiple objective functions will result in a less-than-optimal use of one or more resources in order to optimize the project as a whole. Having multiple objectives or less-than-optimal usage can lead to simplification by weighting the variables and reforming the functions into a single objective function. However, sublevel objective functions can also be studied in projects with multiple objectives when the sublevel objectives are not fixed in time or value. For example, optimizing a concrete mixture uses sublevel objectives of aggregate gradation, cement quality, and water quality. Usually, the sublevel objectives are optimized before the superior objectives are analyzed. However, the superior objectives can be evaluated using a "what if" scenario to set the performance level of a sublevel objective. For instance, in a concrete-mix design, if there are several available sources of aggregate that vary in cost and quality, the overall mix design can be evaluated using two or more aggregate quality objectives. After the aggregate is selected, the superior objective of concrete mixture design is approached. The optimum concrete mixture may be a slightly different formulation, depending on which aggregate is used.

The second optimization characteristic, a set of *variables*, or *unknowns*, determines the performance value of the objective function. The variables can include factors such as time, material, shape or dimension, and the manufacturing process.

The third characteristic, *constraints*, sets the condition of the variables. However, constraints are not essential to a project's optimization if there are no variables. With variables, the objective function is judged by the level of achievement attained. Usually constraints involve thresholds, such as greater than or less than values, acceptable ranges, and negative eliminators. For example, a transportation project may have a selection of aggregates from several sources. The aggregate requirements can be defined as having a minimum crushing strength and a maximum organic material content, a certain size gradation range with a maximum dust content, and a certain freeze-thaw resistance. Cost (such as excavation, hauling, and storage) and environmental limitations add to the broader range of constraints. These constraints may determine which source cannot be used due to negative consequences. In transportation projects, constraints are primarily specifications, codes or laws, design guides, standards of practice, costs, and human factors responses.

Optimization analysis can be performed as a one-time condition, such as selecting the size of structural members. The analysis can also be a continuous process, such as adjusting the proportions of a pavement mixture to compensate for properties of supplied materials that vary as the job progresses. In cases where the data are not accurate or the future predictions are not entirely known, the analysis must take into account these uncertainties. In either case, the output of the optimization analysis requires a *recourse action* to take place to adjust input parameters or product ingredients. After the recourse action, the optimization analysis is performed again, and the output is adjusted using the new data.

During the planning stage, engineers employ various models to determine not only construction costs, but also operating and replacement costs that will be necessary to attain certain benefits. The optimal solution may be the one with the least cost, or it may be the one that provides the most benefit to the expected user. When optimizing by cost analysis, it is important that the objectives of the finished product are clearly defined. The degrees of attainment of objectives become qualifying factors for ranking the alternative outcomes against the cost to build each alternative. Costs can be evaluated using not only direct costs of construction, but also financing and maintenance costs, liability costs, or other costs evaluated using standard engineering economic analysis. *Life-cycle cost analysis* (LCCA) involves predicting the expected useful life of a project and the costs of construction, operation and maintenance, and demolition and disposal, especially when environmentally sensitive materials are used in construction.

There are several different types of optimization models, including linear programming, integer programming, and dynamic programming models. *Linear programming* (LP) is used in optimization of a linear objective function. There are two types of LP models: *single objective*, in which there is one objective to optimize, and *multiple objective*, which tries to find solutions satisfying more than one objective. *Integer programming* restricts some or all of the variables in optimization to be integers. *Dynamic programming* (DP) breaks up a problem into smaller, simpler components. It is best used when the issue of optimization is very complex, as it may be easier to find solutions to each separate component, rather than try to solve the problem as a whole.

Example 1.3

An isolated signalized intersection with multiple approaches and movements is being studied for improvement. After selecting a lane configuration, the criterion for setting the cycle length is to minimize average delay of all approaching vehicles. Several cycle lengths have been tested with the following results.

cycle length (sec)	average delay (sec/veh)
90	95
105	83
120	69
135	68
150	78

When will the minimum delay occur?

Solution

The solution can be found graphically by plotting the data and fitting a curve through the data points.

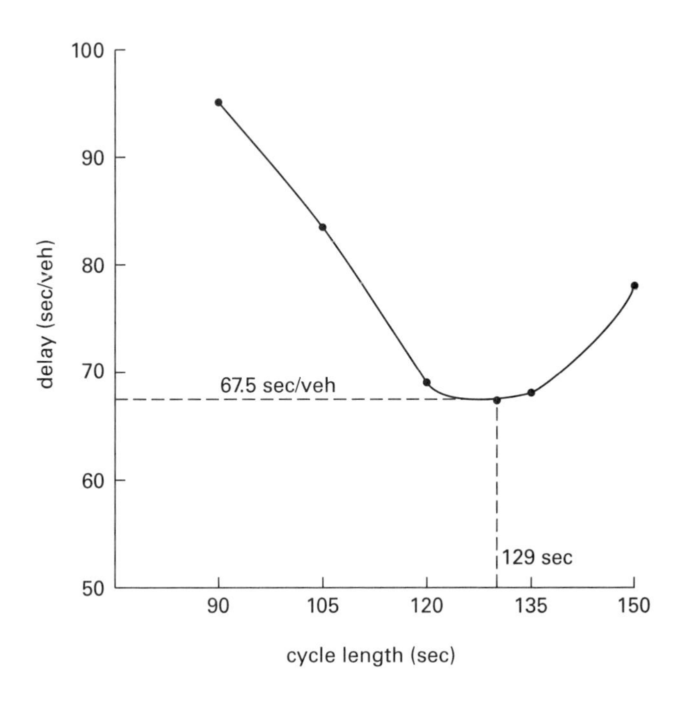

By inspection, the curve shows a minimum delay of 67.5 sec/veh occurring with a cycle length of 129 sec.

Setting Optimization Criteria

Optimization can be performed at many stages in a project. The criterion for optimization depends on the overall goals of the project and the objectives that are to be met at a particular level of execution. In the early stages, the criteria to optimize a project's major components are usually broad. Such criteria would be added to project-cost estimates and funding requirements, and include the following components: community needs, schedule and phasing, material selection, site restrictions, environmental controls, risk assessments, and neighborhood impacts.

Community needs often start with the desire to build or modify a particular transportation facility, such as a bridge, roadway, or intersection. From these needs, *planning studies* are used to determine optimization based on capacity analysis, access improvements, and crash mitigation.

Schedule and phasing criteria are set by both internal and external constraints. External events, such as coordination with transportation network connections, seasonal weather changes, and major community activities affected by the project, should be optimized during the project planning phase. Events internal to the project, such as material production, delivery schedules, staging, and environmental controls, usually have more precise criteria to optimize the cost and quality of the completed project.

Material selection focuses on the cost and availability of primary materials, such as earthwork, concrete, and steel, and secondary materials, such as coatings, accessory hardware, and landscape treatments. The cost of purchase, delivery, and placement, and the impact on the project schedule are criteria for optimization.

Site restriction optimization is primarily a function of topographic features and available property needed to construct a project. Optimization based on topographic features means selecting the most effective method to build over watercourses, hills, and valleys, in various soil conditions, and to account for climate factors such as ice and snow. Optimizing property usage aims to make efficient use of the space available in relation to the value of the property. Higher-priced land in well-developed locations calls for more compact design of structures and drainage facilities, while lower-cost land in less dense locations allows less costly alternatives to be considered.

Environmental controls include factors such as noise and dust mitigation, VOC emissions, and trash disposal from the jobsite. As more stringent engine emissions controls take place, contractors have been assisted greatly by equipment manufacturers in optimizing equipment leasing for specific job applications. Compliance with community regulations for dust, noise, and silt runoff is often handled by firms specializing in these tasks. Optimization of productivity, or the dollar value of work performed per unit of time, is compared to the cost of complying with applicable regulations and controls.

Risk assessment is becoming an increasing part of optimization as transportation improvement funding becomes more difficult to justify. The early days of interstate highway construction applied a broad brush of the safety improvement provided by limited-access divided highway construction. With the near-completion of the interstate highways, attention is now more focused on specific elements of design, such as guiderail location, paved shoulder treatment, median separation treatment, and crash-cost abatement versus the cost of mitigation measures. For example, transportation planners might discuss whether or not the cost of constructing a highway guiderail is more or less than the cost of a crash with a guiderail in place. Similarly, risk assessments may be made for intersections when optimizing the placement of a traffic signal, establishment of the width of a median, elimination of hazardous geometry conditions, and so on. While risk assessment is constantly changing, there are many established standards (e.g., AASHTO's *A Policy on Design Standards—Interstate System* or the FHWA's *Manual on Uniform Traffic Control Devices* (MUTCD)) that have been thoroughly tested and that represent the level of achievement to be attained during optimization analysis of risk.

Neighborhood impact plays an increasingly important role as quality of life issues enter project planning. Noise barriers, access controls, school bus safety, and public safety issues are traditional concerns, while a new emphasis is being placed on sustainable living and close-knit neighborhood development. Optimization of these values not only affects the initial cost of a project, but has long-term consequences as well.

Optimization During Later Phases of a Project

A transportation project begins the process of optimization with a transportation planner, an engineer, and/or an architect, who selects the initial solution based on the current understanding of an optimal solution. As the project progresses, optimization moves from a broad perspective of solutions to a narrower focus on the best solution for the objective at hand. For instance, transportation planners that are optimizing access to a commercial district might initially select a highway as the best solution. Once a highway is selected, the next phase is to select the type of highway—limited access, multilane arterial, two-lane, and so on. This phase involves careful analysis of benefits, user and construction cost, capacity, funding availability, and institutional support. Optimization of each of these elements involves cost and traffic projections and assurances of future use patterns. When the type of highway is selected, designs are tested and optimized using delay models, user-cost models, environmental cost models, and so on. These cost models can be very specific when the base data are highly reliable.

As a project moves into final design and construction, optimization moves toward focusing on material selection, source of supply, labor cost, contractor capability, and the specifics of scheduling and phasing the work. Through the progression of a transportation project's stages, it becomes increasingly difficult to return to earlier definitions of broad optimization. For instance, once a concrete bridge has been selected and is in production, changing to a steel superstructure would require including the cost of reconstructing already completed portions in the evaluation of optimizing the change to steel.

Benefit-Cost Analysis

In the simplest terms, optimization can be thought of as choosing among alternatives based on costs (i.e., dollar value) and/or benefits. *Benefit-cost analysis* was introduced at the federal level by the U.S. Army Corps of Engineers to evaluate flood control projects. The analysis method involved identifying measurable costs of a project during the planning stage and comparing the dollar-value of expected benefits. It was used to justify and prioritize many of the flood control, dam, and navigation lock public works programs carried out in the twentieth century. While effective in ensuring that tax-supported projects achieve the highest economic return, the method of benefit-cost analysis doesn't always accurately reflect less precise costs, such as the environmental pollution created by new project developments. In spite of this drawback, a variety of benefit-cost analysis is used on almost all projects when evaluating the design of subelements. For instance, pavement designs are usually analyzed for the optimal balance between required layer strength for a predicted service life (the benefit) and the cost of construction.

Benefit-cost (B/C) comparisons become more difficult when alternatives have differing lengths of service life, or when comparing funding from different sources, such as capital budgets versus operating and maintenance (OM) budgets.

The benefit-cost ratio can be calculated from Eq. 1.5.

$$B/C = \frac{\Delta \begin{array}{c}\text{user}\\ \text{benefits}\end{array}}{\Delta \begin{array}{c}\text{investment}\\ \text{cost}\end{array} + \Delta\,\text{maintenance} - \Delta \begin{array}{c}\text{residual}\\ \text{value}\end{array}} \qquad 1.5$$

Equation 1.5 is a conventional method of comparing the gross benefits of a project in the numerator to all of the costs in the denominator, which are the initial investment plus the operating and maintenance costs, less the residual value. The *residual value* (also known as the *salvage value*) is the market value at the end of the project life.

In some cases, part of the cost elements may not be known or may not be available. For instance, the residual value or market worth at the end of service life may not be known or readily predicted. Evaluation can proceed, however, recognizing that once the residual value is known, the ratio may change. Another approach is to assume that residual value of the new project is an inherent part of its replacement project at the end of its service life, just as residual value of the existing condition is included in the current project cost.

One of the biggest drawbacks of benefit-cost analysis is the inability to accurately predict the dollar value of future benefits. This is especially true for projects that have a long expected life (50 years or more), such as major water transportation projects. The closer the benefit-cost ratio moves to unity (1.00), the more the inaccuracies of predictions increase the opportunities for negative benefit conditions. The project then becomes more difficult to sell to the public based on the benefit-cost ratio alone.

Example 1.4

A proposal to dredge a shipping channel to a new depth has an estimated cost of $248 million over three years, plus $15 million per year to maintain the channel at its new depth. The port facility served by the shipping channel expects additional revenue of $70 million per year to occur due to the increased access by larger cargo ships. There is no residual value for the project. Using an effective interest rate of 5%, what is the benefit-cost ratio of the project over 10 years?

Solution

The total cost for 10 years, C_t, will be the initial construction cost, C_I, plus the annual maintenance cost, C_A, times the number of years, N_{yr}.

$$\begin{aligned} C_t &= C_I + C_A N_{\text{yr}} \\ &= \$248\text{M} + \left(\$15\ \frac{\text{M}}{\text{yr}}\right)(10\ \text{yr}) \\ &= \$398\text{M} \end{aligned}$$

Calculate the benefits, B, over 10 years.

$$B = \left(\$70\ \frac{\text{M}}{\text{yr}}\right)(10\ \text{yr}) = \$700\text{M}$$

Calculate the present worth of the costs.

$$\begin{aligned} C &= \$398\text{M}(P/F, 5\%, 10) \\ &= (\$398\text{M})(0.6139) \\ &= \$244\text{M} \end{aligned}$$

Calculate the present worth of the benefits.

$$\begin{aligned} B &= \$700\text{M}(P/F, 5\%, 10) \\ &= (\$700\text{M})(0.6139) \\ &= \$430\text{M} \end{aligned}$$

The benefit-cost ratio, B/C, is

$$B/C = \frac{\$430\text{M}}{\$244\text{M}} = 1.76$$

Therefore, the benefits over the 10 year life of the project will be 1.76 times more than the cost to install the channel and maintain it for 10 years.

7. SITE IMPACT ANALYSIS

Site impact analysis involves examining the environmental issues of transportation projects. Site impact analysis can include, but is not limited to, examining a project's potential effect on

- natural resources
- the relocation of residential areas and businesses
- air quality
- sound and noise
- wetlands and coastal zones
- society and the economy
- water quality

- flood hazards, floodplains, and flood liability
- emergency management

Sound, noise, and air quality are especially pertinent to civil engineers, who must monitor and control these factors during project construction.

Sound and Noise

Sound occurs when an ear senses pressure vibrations, while *noise* is unwanted sound. *Noise studies* are conducted to determine the nature and intensity of noise generated by a roadway. They determine existing background noise levels, called *ambient sound* levels, which may also include existing transportation facilities. Sound levels are measured at the L_{10}, L_{50}, and L_{90} levels, which means the sound levels are exceeded 10%, 50%, and 90% of the time, respectively. Traffic noise studies are most often conducted to project the noise levels for a new facility. However, noise levels for an existing facility can also be determined. Data from trip generation studies are often used to estimate the effect of a new facility on existing noise levels.

Sound levels are expressed in levels of pressure and measured in decibels (dB). The decibel scale is logarithmic at base 10, meaning a change of 1 dB reflects a ten-fold change in sound pressure level. The sound level *A-weighted scale*, which suppresses low frequency sounds, is often used to simulate human hearing response, in which case the sound level is given in A-weighted decibels (dBA). The *reference level* for sound in air is usually chosen as 0.02 MPa. This is the lowest level of normal sensitivity for the human ear found in those without hearing damage due to loud noise exposure. The *maximum audible field* (MAF) threshold varies with audible frequency, measured in hertz. The following four levels are often used for reference.

- 70 dB at 2 Hz
- 100 dB at 100 Hz
- 0 dB at 2000 Hz
- 20 dB at 16,000 Hz

The *threshold of discomfort* also varies with audible frequency, but is generally accepted at approximately 120 dB for all cases. The *threshold of pain* is approximately 140 dB, and the *threshold of hearing damage* occurs at approximately 145 dB.

Traffic noise studies are commonly required by the FHWA for federally funded transportation projects. The FHWA classifies transportation projects as either type I or type II. A *type I project* is one that involves constructing a new highway or altering an existing highway such that significant horizontal or vertical alignment changes are made or a new lane is added. The definition of "significant" changes to horizontal and vertical alignment is defined by a state's Department of Transportation. A noise study must be performed for all type I projects. *Type II projects*, also known as *retrofit projects*, must also include a noise study and are defined as projects undertaken on existing roadways to provide noise abatement.

During a noise study, sound levels are recorded over a period of hours, days, or months, depending on the study objective. Recorded noise levels are then compared to the noise abatement criteria to determine if traffic noise impacts exist. A *traffic noise impact* occurs when existing or projected noise levels approach or exceed the noise abatement criteria or when a substantial increase in the existing noise environment occurs. The FHWA defines a substantial increase as 10–15 dBA higher than the current noise levels, but this may vary depending on state policies. The *noise abatement criteria* (NAC) are the absolute noise levels for different land use categories and are outlined in 23 CFR 772 (*Code of Federal Regulations*, Title 23, Part 772) and given in Table 1.2. The FHWA defines *absolute noise levels* as those that are unacceptably high for a given land use. L_{10} is a time-varying traffic noise level that is measured over a one hour period and is calculated as the percentage of the study period, in this case 10% of the time, that a sound level is exceeded. L_{eq} is the hourly equivalent sound level, which is the equivalent constant sound level which contains the same amount of energy as the time-varying noise level (i.e., L_{10}), both measured for one hour. For most traffic conditions, L_{eq} is about 3 dBA less than L_{10} for the same study period.

If traffic noise impacts are determined to occur, abatement measures must be considered. These measures can include sound barrier walls, berms, dikes, earth mounds, vegetation, or other passive devices to deflect or soften the sound from entering the sensitive sites. Additional measures may be employed as needed, such as changing the grade line of a project, rerouting the configuration ramps, adjusting the location of truck-climbing lanes, and so forth. In most instances, small changes needed to accommodate noise mitigation measures have little effect on a project's cost. However, in some cases, expensive sound barriers may be needed when retrofitting sensitive neighborhoods, such as those near hospitals. This cost must be compared with the cost of relocating the roadway to a less sensitive area.

Air Quality

Transportation, by its very nature of propulsion energy, affects air quality, whether through power plant emissions for electrically powered transit and vehicles or exhaust pipe emissions for combustion systems. Therefore, when a transportation project is proposed, *air quality studies* must be conducted. Like noise studies, air quality studies for new facilities are often based on data obtained through trip generation studies.

The Environmental Protection Agency (EPA) is tasked with developing emission factors for all emission sources. Air quality impact studies must be directly related to the project at hand in order to have valid application. Studies that exist whether or not the transportation

Table 1.2 Noise Abatement Criteria (NAC), Hourly Sound Level in A-Weighted Decibels (dBA)[a]

activity category	L_{eq}[b] (dBA)	L_{10} (dBA)	description of activity category
A	57 (exterior)	60 (exterior)	lands on which serenity and quiet are of extraordinary significance and serve an important public need, and where the preservation of those qualities is essential if the area is to continue to serve its intended purpose
B	67 (exterior)	70 (exterior)	picnic areas, recreation areas, playgrounds, active sports areas, parks, residences, motels, hotels, schools, churches, libraries, and hospitals
C	72 (exterior)	75 (exterior)	developed lands, properties, or activities not included in categories A or B
D	–	–	undeveloped lands
E	52 (interior)	55 (interior)	residences, motels, hotels, public meeting rooms, schools, churches, libraries, hospitals, and auditoriums

[a]Either L_{eq} or L_{10} (but not both) may be used on a project.
[b]L_{eq} represents the hourly equivalent sound level.

Reprinted from *Code of Federal Regulations*, Title 23, Part 772, Table 1, U.S. Department of Transportation, Federal Highway Administration.

project is implemented are not valid uses of transportation funds (e.g., studying mitigation methods for the effects of acid rain would not be included in a transportation air quality impact study). Starting with a concise project description, the report should clearly show the nature of the project, the proposed alternatives, unusual topographic features, and ambient conditions. The data collection and analysis must be complete and sufficient to provide reliable projections based on current industry standards and government regulations or guidelines. The report should contain a complete air quality analysis of each alternative. Projections of total emissions for the entire project and analysis for the year of completion should show the operating emissions environment at the completion of construction.

Because transportation is one of the largest energy consumers, it is often targeted as a significant contributor of pollutants to *nonattainment areas* (areas that do not meet national primary or secondary air quality standards for a pollutant). The EPA establishes criteria pollutants and the *National Ambient Air Quality Standards* (NAAQS), which can be found at **www.ppi2pass.com/NAAQS**. Though not all polluting emissions are applicable to a transportation project, all motive power units emit some pollutants into the atmosphere. Depending on the project, air quality studies may test for carbon monoxide (CO), carbon dioxide (CO_2), hydrocarbons (HC), nitrous oxides (NOx), ozone (O_3), lead (Pb, where leaded fuel is still used), sulfurous compounds, mercury (Hg), nuclear radiation (from power plants), and, in some cases, heat energy. Pollutant abatement from sources not directly related to the transportation project at hand, such as noxious odors from a sewage treatment plant or coke oven, are generally not eligible for transportation project funding.

Air quality studies are generally either mesoscale or microscale studies. *Mesoscale* studies include the total emissions generated by various project alternatives and involve calculating the impacts of photochemical oxidants as related to local jurisdiction requirements. Mesoscale studies cover contributing sources for several miles outside the project location and include changes in traffic on nearby highways or other transportation segments resulting from the proposed project. Computer modeling is usually employed in these studies.

Microscale studies concentrate on one or a few pollutants, such as carbon monoxide or lead. Dispersion models are used to predict concentrations at critical locations. As with mesoscale studies, computer modeling is used extensively to analyze the data and reduce the outputs into an understandable presentation. Transportation construction projects can significantly affect air quality to such an extent that a separate air quality impact study may need to be specifically performed for construction activities.

Transportation construction activities usually impact air quality in three major ways. First, impact initially occurs when earth is excavated and the protective vegetative cover is removed. This generates a considerable amount of dust that is picked up by the surrounding atmosphere, even in areas where there is little or no vegetation covering the native soil. The U.S. Department of Agriculture's (USDA's) Natural Resources Conservation Service (NRCS), formerly the Soil Conservation Service, has determined the susceptibility of different soils to wind erosion. (See NRCS' website at www.nrcs.usda.gov for more information.) Once a soil type is known, dust suppression methods may be built into a transportation project's construction procedures. Second, transportation construction projects affect air quality through the concentrated collections of exhaust-gas emissions that are caused by traffic congestion. To offset these emissions, traffic routing and congestion relief measures may be specified for a project based on the amount of excess emissions permitted during periods of restricted traffic-flow rates. Third, emissions from the construction equipment affect air quality. In response, equipment manufacturers have steadily improved engine performance and emissions control (for instance, the large clouds of black diesel smoke pouring out from exhaust stacks have been reduced dramatically). EPA standards may also be utilized to prepare project impact reports from construction equipment operations.

PRACTICE PROBLEMS

1. A medium-sized, 20 ac, mixed-use commercial development is planned, consisting of offices, retail stores, and light manufacturing. The existing frontage road is a nearly level, straight, two-lane highway with unlimited sight distance and no median. The posted speed and average speed are both 45 mph. Traffic currently operates at LOS C, and there is no signal planned along the proposed frontage. Analysis of site access is being performed to determine design elements. For vehicular access, five common design groups from AASHTO design guides are being used.

P: passenger car

SU: single-unit truck or bus

WB-40 (WB-12): semitrailer combination, medium

WB-50 (WB-15): semitrailer combination, large

WB-62 (WB-19): interstate semitrailer

(a) What is the minimum design vehicle for this site?

(A) SU

(B) WB-40 (WB-12)

(C) WB-50 (WB-15)

(D) WB-62 (WB-19)

(b) Proposed alternatives to access the driveway from the existing frontage road include

I. three driveway entrances: one entrance for delivery trucks and two smaller entrances where delivery trucks would be prohibited

II. a continuous left-turn lane along the frontage road and multiple access driveways for the retail businesses

III. a single large driveway entrance for all vehicles

IV. a jug handle to the right for left turns

Which alternative would be the least confusing for unfamiliar drivers?

(A) I

(B) II

(C) III

(D) IV

(c) The local planning board suggested minimizing the traffic impact on the frontage road by designing the entrance driveways for quick entry into the site and restricting the exit flow rate with narrow exit lanes and speed control humps. What is the main reason restricting the exit flow rate is NOT a good idea?

(A) Drivers leaving the site may become frustrated with the exit restrictions and attempt to leave by traveling the wrong way on the entrance driveways.

(B) Designing exit driveways to restrict both truck and automobile movement is difficult, as trucks are much larger than automobiles.

(C) Drivers wanting to leave the site in a hurry may become discouraged and not return, opting to visit businesses at more accessible locations.

(D) Restricting egress from a site limits the speed of and capacity for emergency evacuation.

2. An interchange is being evaluated for construction cost versus user cost, with the lowest user cost being preferred. There are two major traffic movements through the interchange, with two alternative ramp configurations being considered. User cost is comprised of the vehicle cost and the person cost. Ramp lengths are measured from the same points on the approach roadways for each alternative. The difference in route lengths is due to ramp configuration. Use a 3% annual growth in traffic volume and a 3% annual increase in per-mile and per-hour user costs. A 20 yr projected user benefit-cost analysis is to be used.

	alternative A	alternative B
route 10 traffic volume	42,000 ADT	42,000 ADT
route 14 traffic volume	18,000 ADT	18,000 ADT
ramp length and design speed		
route 10	1.32 mi, 25 mph	0.65 mi, 45 mph
route 14	0.97 mi, 30 mph	0.97 mi, 30 mph
total bridge cost and length	none	$8,500,000; 0.1 mi
time per vehicle in interchange		
route 10	3.16 min	1.0 min
route 14	1.93 min	1.93 min
vehicle cost	$0.45/veh-mi	$0.45/veh-mi
person cost	$8.00/p-hr	$8.00/p-hr
vehicle occupancy	1.05 p/veh	1.05 p/veh
ramp construction cost (excluding bridge costs)	$6,000,000/mi	$6,000,000/mi

(a) How much more does alternative B cost to construct than alternative A?

(A) $4,020,000

(B) $4,480,000

(C) $5,870,000

(D) $7,920,000

(b) The current annual user time (in person-minutes) is, most nearly, how much larger for alternative A than alternative B?

(A) 36,500 p-min

(B) 80,600 p-min

(C) 90,700 p-min

(D) 95,300 p-min

(c) What is most nearly the current annual total user cost savings for alternative B?

(A) $10,800

(B) $21,500

(C) $25,400

(D) $52,200

(d) The present value of the user cost over the life of a project is compared with the construction cost using the formula $P = A(P/A,i\%,n)$. Conduct an incremental benefit-cost analysis to compare the additional construction cost for alternative B to the difference in user cost.

(A) 0.04:1

(B) 0.08:1

(C) 2.5:1

(D) 12:1

3. Travel behavior and population characteristics along a corridor are found to correspond to the following factors derived from surveying a sample population of 1000 people.

proportion of households in the population, p_h	0.35
number of adults per household, N_a	0.74
number of vehicles per household, N_v	1.55
number of licensed drivers per adult, N_l	0.61

The census population of the study zone is 25,000, and there are 14,000 average one-way, weekday trip ends generated from the study zone.

(a) The number of households in the study zone is most nearly

(A) 3500 households

(B) 4400 households

(C) 4900 households

(D) 8800 households

(b) The number of vehicles in the study zone is most nearly

(A) 1550 veh

(B) 13,600 veh

(C) 21,700 veh

(D) 38,800 veh

(c) How many one-way, weekday trip ends are generated within the study zone for each licensed driver?

(A) 1.12 trip ends/licensed driver

(B) 1.40 trip ends/licensed driver

(C) 3.54 trip ends/licensed driver

(D) 7.09 trip ends/licensed driver

4. A population zone of 1700 households has proximity to transit. Planners are making a projection of transit ridership potential. A survey of neighborhood characteristics resulted in the following data.

proportion of households in the population, p_h	0.35
number of employed adults per household, N_e	0.74
proportion of employed adults working at transit accessible locations, $p_{e,t}$	0.30
proportion of households within 0.3 mi of a transit stop, p_{ts}	0.40
number of vehicles per household, N_v	2.1
vehicle reduction factor for households within 0.3 mi of a transit stop, f_{vtr}	0.25
number of licensed drivers per adult, N_l	2.3

The survey found that of workers living within 0.3 mi of a transit stop, 85% would use transit if their workplace were accessible to it ($p_{<0.3} = 0.85$). Of workers living more than 0.3 mi from a transit stop, 15% would use transit if their workplace were accessible to it ($p_{>0.3} = 0.15$).

(a) The population of the study zone is most nearly

(A) 600

(B) 1700

(C) 3590

(D) 4860

(b) How many vehicles are in the study zone?

(A) 900 veh

(B) 2680 veh

(C) 3400 veh

(D) 3570 veh

(c) The number of potential employed adult transit riders that would be generated from this zone is most nearly

(A) 130 riders

(B) 160 riders

(C) 460 riders

(D) 540 riders

5. Two zones, A and B, are connected by three alternative highway travel routes. There are 10,000 vehicles traveling from zone A to zone B during a weekday. The distances and average travel speeds are shown. Route 1 has 2 signals, each with a 70 sec delay, route 2 has 15 signals with a 30 sec delay each, and route 3 has 2 signals with a 30 sec delay each.

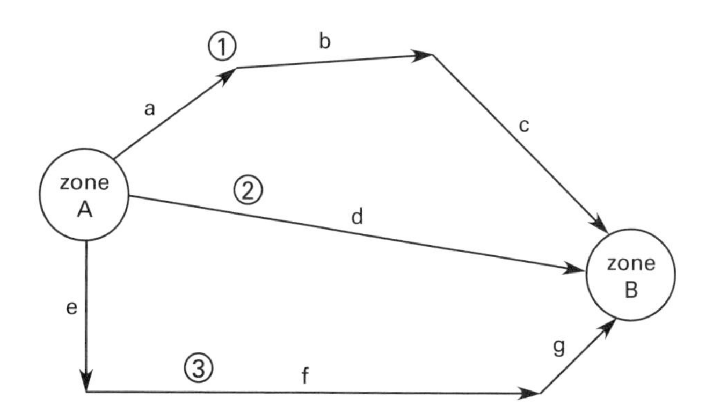

segment	segment length (mi)	average travel speed (mph)
a	4	25
b	5	35
c	7	30
d	15	40
e	4	30
f	8	65
g	3	30

Trips are to be assigned to each route based on the inverse of the proportion of the total travel time delay.

(a) What is most nearly the trip delay per vehicle on the quickest route?

(A) 22.4 min on route 3

(B) 23.0 min on route 2

(C) 30.0 min on route 2

(D) 34.5 min on route 1

(b) What percentage of vehicles uses the route with the least delay?

(A) 27.1%

(B) 31.2%

(C) 41.7%

(D) 72.9%

(c) The number of vehicles on the route with the least delay is most nearly

(A) 2320 veh on route 1

(B) 2710 veh on route 2

(C) 3120 veh on route 2

(D) 4170 veh on route 3

SOLUTIONS

1. (a) While retail commercial stores can usually be served with SU minimum design, light manufacturing more often needs access by a WB-40 (WB-12) medium semitrailer combination. This design can accommodate an occasional WB-50 (WB-15) large semitrailer combination if there are few close obstructions.

The answer is (B).

(b) Multiple access points with restrictions may cause confusion to unfamiliar drivers and lead to potential violations of the restrictions, even with careful signage or other control devices. A continuous left-turn lane should be avoided for the same reason and will become less effective under increased traffic volumes. A jug handle can be confusing because of the need to go to the right to turn left. A jug handle also increases the amount of traffic crossing the mainline flow. The most comprehensible access involves a single combined access point with adequate capacity for all vehicle types and the expected traffic volume. Option III provides a single access point for all drivers.

The answer is (C).

(c) Fire safety and life should be paramount over all other concerns when designing site access. The ability to exit the site quickly is important if there is a need to evacuate people in an emergency situation. Restricting exiting ability can be dangerous and increase the potential for panic.

The answer is (D).

2. (a) Determine the construction cost of both routes for each alternative.

The construction cost for alternative A, route 10, is

$$C_{\text{A},10} = (1.32 \text{ mi})\left(\frac{\$6{,}000{,}000}{\text{mi}}\right) = \$7{,}920{,}000$$

The construction cost for alternative A, route 14, is

$$C_{\text{A},14} = (0.97 \text{ mi})\left(\frac{\$6{,}000{,}000}{\text{mi}}\right) = \$5{,}820{,}000$$

The total cost of alternative A, $C_{\text{A},t}$, is

$$\begin{aligned} C_{\text{A},t} &= C_{\text{A},10} + C_{\text{A},14} \\ &= \$7{,}920{,}000 + \$5{,}820{,}000 \\ &= \$13{,}740{,}000 \end{aligned}$$

The construction cost for alternative B, route 10, is

$$\begin{aligned} C_{\text{B},10} &= (0.65\ \text{mi})\left(\frac{\$6{,}000{,}000}{\text{mi}}\right) \\ &= \$3{,}900{,}000 \end{aligned}$$

The construction cost for alternative B, route 14, is

$$C_{\text{B},14} = (0.97\ \text{mi})\left(\frac{\$6{,}000{,}000}{\text{mi}}\right) = \$5{,}820{,}000$$

The total cost of alternative B, $C_{\text{B},t}$, including the cost of the bridge, is

$$\begin{aligned} C_{\text{B},t} &= C_{\text{B},10} + C_{\text{B},14} + C_{\text{B,bridge}} \\ &= \$3{,}900{,}000 + \$5{,}820{,}000 + 8{,}500{,}000 \\ &= \$18{,}220{,}000 \end{aligned}$$

The cost difference is

$$\begin{aligned} C_{\text{B},t} - C_{\text{A},t} &= \$18{,}220{,}000 - \$13{,}740{,}000 \\ &= \$4{,}480{,}000 \end{aligned}$$

The answer is (B).

(b) The user time for alternative A is

$$\begin{aligned} t_{\text{A}} = t_{\text{A},10} + t_{\text{A},14} &= (42{,}000\ \text{veh})\left(1.05\ \frac{\text{p}}{\text{veh}}\right)(3.16\ \text{min}) \\ &\quad + (18{,}000\ \text{veh})\left(1.05\ \frac{\text{p}}{\text{veh}}\right)(1.93\ \text{min}) \\ &= 175{,}833\ \text{p-min} \end{aligned}$$

The user time for alternative B is

$$\begin{aligned} t_{\text{B}} = t_{\text{B},10} + t_{\text{B},14} &= (42{,}000\ \text{veh})\left(1.05\ \frac{\text{p}}{\text{veh}}\right)(1.0\ \text{min}) \\ &\quad + (18{,}000\ \text{veh})\left(1.05\ \frac{\text{p}}{\text{veh}}\right)(1.93\ \text{min}) \\ &= 80{,}577\ \text{p-min} \end{aligned}$$

The time difference is

$$\begin{aligned} t_{\text{A}} - t_{\text{B}} &= 175{,}833\ \text{p-min} - 80{,}577\ \text{p-min} \\ &= 95{,}256\ \text{p-min} \quad (95{,}300\ \text{p-min}) \end{aligned}$$

The user time for alternative A is more than the user time for alternative B.

The answer is (D).

(c) The total user cost is the per vehicle cost plus the per person cost.

The per vehicle cost for alternative A, $C_{\text{A},v}$, is

$$\begin{aligned} C_{\text{A},v} &= \left(0.45\ \frac{\$}{\text{veh-mi}}\right)\left(\begin{array}{r}(42{,}000\ \text{veh})(1.32\ \text{mi}) \\ + (18{,}000\ \text{veh})(0.97\ \text{mi})\end{array}\right) \\ &= \$32{,}805 \end{aligned}$$

The per person cost for alternative A, $C_{\text{A},p}$, is

$$\begin{aligned} C_{\text{A},p} &= \frac{(175{,}833\ \text{p-min})\left(8.00\ \frac{\$}{\text{p-hr}}\right)}{60\ \frac{\text{min}}{\text{hr}}} \\ &= \$23{,}444 \end{aligned}$$

The total user cost for alternative A, $C_{\text{A},t}$, is

$$\begin{aligned} C_{\text{A},t} &= C_{\text{A},v} + C_{\text{A},p} \\ &= \$32{,}805 + \$23{,}444 \\ &= \$56{,}249 \end{aligned}$$

The per vehicle cost for alternative B, $C_{\text{B},v}$, is

$$\begin{aligned} C_{\text{B},v} &= \left(0.45\ \frac{\$}{\text{veh-mi}}\right)\left(\begin{array}{r}(42{,}000\ \text{veh})(0.65\ \text{mi}) \\ + (18{,}000\ \text{veh})(0.97\ \text{mi})\end{array}\right) \\ &= \$20{,}142 \end{aligned}$$

The per person cost for alternative B, $C_{\text{B},p}$, is

$$C_{\text{B},p} = \frac{(80{,}577\ \text{p-min})\left(8.00\ \frac{\$}{\text{p-hr}}\right)}{60\ \frac{\text{min}}{\text{hr}}} = \$10{,}744$$

The total user cost for alternative B, $C_{\text{B},t}$, is

$$\begin{aligned} C_{\text{B},t} &= C_{\text{B},v} + C_{\text{B},p} \\ &= \$20{,}142 + \$10{,}744 \\ &= \$30{,}886 \end{aligned}$$

The current annual user benefit for alternative B is

$$\begin{aligned} B &= C_{\text{A},t} - C_{\text{B},t} \\ &= \$56{,}249 - \$30{,}886 \\ &= \$25{,}363 \quad (\$25{,}400) \end{aligned}$$

The answer is (C).

(d) The present value of the user cost over the life of a project is compared with the construction cost. Using the uniform series present worth formula, the difference in present worth of each alternative user cost is compared with the added construction cost of alternative B. Written in symbol form, the present worth formula is

$$P = A(P/A, i\%, n)$$

The P/A factor can be found using factor tables, or the formula can be calculated. Using an interest rate, i, of 3% and a number of compounding periods, n, of 20, the multiplying factor is

$$(P/A, 3\%, 20) = \frac{(1+i)^n - 1}{i(1+i)^n} = \frac{(1+0.03)^{20} - 1}{(0.03)(1+0.03)^{20}} = 14.8775$$

The total 20 yr user cost for alternative A is

$$C_{\text{A,20 yr}} = (14.8775)(\$56{,}249) = \$836{,}844$$

The total 20 yr user cost for alternative B is

$$C_{\text{B,20 yr}} = (14.8775)(\$30{,}886) = \$459{,}506$$

The user benefit of alternative B over 20 yr is

$$B_{\text{B,20 yr}} = \$836{,}844 - \$459{,}506 = \$337{,}338$$

The benefit-cost ratio is

$$\begin{aligned} B/C &= \frac{\Delta \begin{array}{c}\text{user}\\\text{benefits}\end{array}}{\Delta \begin{array}{c}\text{investment}\\\text{cost}\end{array} + \Delta\,\text{maintenance} - \Delta \begin{array}{c}\text{residual}\\\text{value}\end{array}} \\ &= \frac{\$377{,}338}{\$4{,}480{,}000} \\ &\approx 0.08{:}1 \end{aligned}$$

The 20 year user benefit is about 0.08 times the construction cost.

The answer is (B).

3. (a) The number of households is found by multiplying the proportion of households in the population by the population.

$$\begin{aligned} N_h &= p_h(\text{population}) \\ &= (0.35 \text{ households})(25{,}000) \\ &= 8750 \text{ households} \quad (8800 \text{ households}) \end{aligned}$$

The answer is (D).

(b) The number of vehicles is found by multiplying the number of vehicles per household by the number of households.

$$\begin{aligned} \text{no. of vehicles} &= N_v N_h \\ &= \left(1.55 \ \frac{\text{veh}}{\text{household}}\right)(8750 \text{ households}) \\ &= 13{,}563 \text{ veh} \quad (13{,}600 \text{ veh}) \end{aligned}$$

The answer is (B).

(c) The amount of one way weekday trip ends generated within the study zone for each licensed driver is

$$\begin{aligned} T &= \frac{T_t}{N_l N_a N_h} \\ &= \frac{14{,}000 \text{ trip ends}}{\left(0.61 \ \frac{\text{licensed driver}}{\text{adult}}\right)\left(0.74 \ \frac{\text{adults}}{\text{household}}\right) \times (8750 \text{ households})} \\ &= 3.54 \text{ trip ends/licensed driver} \end{aligned}$$

The answer is (C).

4. (a) The population is found by dividing the number of households, N_h, by the proportion of households in the population, p_h.

$$\begin{aligned} \text{population} &= \frac{N_h}{p_h} = \frac{1700 \text{ households}}{0.35 \text{ households}} \\ &= 4857 \quad (4860) \end{aligned}$$

The answer is (D).

(b) The number of vehicles is found by

$$\begin{aligned} \text{no. of vehicles} &= N_v N_h - f_{\text{vtr}} p_{\text{ts}} N_h \\ &= \left(2.1 \ \frac{\text{veh}}{\text{household}}\right)(1700 \text{ households}) \\ &\quad - \left(0.25 \ \frac{\text{veh}}{\text{household}}\right)(0.40) \times (1700 \text{ households}) \\ &= 3400 \text{ veh} \end{aligned}$$

The answer is (C).

(c) The number of potential transit riders is found by adding the proportion of employed adults that would use transit and have access to transit at their workplace from the two subzones. That is, the proportion of those households less than 0.3 mi from a transit stop, $p_{<0.3}$, and those households more than 0.3 mi from a stop, $p_{>0.3}$.

$$\text{no. of riders} = \begin{pmatrix} N_e p_{e,t} p_{\text{ts}} p_{<0.3} N_h \\ + N_e p_{e,t}(1 - p_{\text{ts}}) p_{>0.3} N_h \end{pmatrix} p_h^{-1}$$

$$= \begin{pmatrix} \left(0.74 \ \frac{\text{employed adults}}{\text{household}}\right) \\ \times \left(0.30 \ \frac{\text{transit accessible adults}}{\text{employed adults}}\right) \\ \times (0.40) \\ \times \left(0.85 \ \frac{\text{riders}}{\text{transit accessible adults}}\right) \\ \times (1700 \text{ households}) \\ + \left(0.74 \ \frac{\text{employed adults}}{\text{household}}\right) \\ \times \left(0.30 \ \frac{\text{transit accessible adults}}{\text{employed adults}}\right) \\ \times (1 - 0.40) \\ \times \left(0.15 \ \frac{\text{riders}}{\text{transit accessible adults}}\right) \\ \times (1700 \text{ households}) \end{pmatrix} \times (0.35 \text{ households})^{-1}$$

$$= 464 \text{ riders} \quad (460 \text{ riders})$$

The answer is (C).

5. (a) Trip delay times are found for each route by adding the travel delay for each segment together with the signal delay.

The trip delay time for route 1 is

$$\begin{aligned} d_{\text{R1}} &= \sum_{i=1}^{m} d_{\text{seg},i} + \sum_{i=1}^{n} d_{\text{sig},i} \\ &= d_{\text{seg a}} + d_{\text{seg b}} + d_{\text{seg c}} + N_{\text{sig}} d_{\text{sig}} \\ &= \left(\frac{4 \text{ mi}}{25 \ \frac{\text{mi}}{\text{hr}}} + \frac{5 \text{ mi}}{35 \ \frac{\text{mi}}{\text{hr}}} + \frac{7 \text{ mi}}{30 \ \frac{\text{mi}}{\text{hr}}} \right) \left(60 \ \frac{\text{min}}{\text{hr}}\right) \\ &\quad + \frac{(2 \text{ signals})\left(70 \ \frac{\text{sec}}{\text{signal}}\right)}{60 \ \frac{\text{sec}}{\text{min}}} \\ &= 34.5 \text{ min} \end{aligned}$$

The trip delay time for route 2 is

$$\begin{aligned} d_{\text{R2}} &= \sum_{i=1}^{m} d_{\text{seg},i} + \sum_{i=1}^{n} d_{\text{sig},i} = d_{\text{seg d}} + N_{\text{sig}} d_{\text{sig}} \\ &= \left(\frac{15 \text{ mi}}{40 \ \frac{\text{mi}}{\text{hr}}} \right) \left(60 \ \frac{\text{min}}{\text{hr}}\right) \\ &\quad + \frac{(15 \text{ signals})\left(30 \ \frac{\text{sec}}{\text{signal}}\right)}{60 \ \frac{\text{sec}}{\text{min}}} \\ &= 30.0 \text{ min} \end{aligned}$$

The trip delay time for route 3 is

$$\begin{aligned} d_{\text{R3}} &= \sum_{i=1}^{m} d_{\text{seg},i} + \sum_{i=1}^{n} d_{\text{sig},i} = d_{\text{seg e}} + d_{\text{seg f}} + d_{\text{seg g}} \\ &\quad + N_{\text{sig}} d_{\text{sig}} \\ &= \left(\frac{4 \text{ mi}}{30 \ \frac{\text{mi}}{\text{hr}}} + \frac{8 \text{ mi}}{65 \ \frac{\text{mi}}{\text{hr}}} + \frac{3 \text{ mi}}{30 \ \frac{\text{mi}}{\text{hr}}} \right) \left(60 \ \frac{\text{min}}{\text{hr}}\right) \\ &\quad + \frac{(2 \text{ signals})\left(30 \ \frac{\text{sec}}{\text{signal}}\right)}{60 \ \frac{\text{sec}}{\text{min}}} \\ &= 22.4 \text{ min} \end{aligned}$$

The smallest delay is 22.4 min on route 3.

The answer is (A).

(b) Proportioning the traffic according to the inverse of the time delay means the route with the least delay receives the largest proportion of traffic. To find the proportions, the inverse of delay for each route is divided by the total of inverse delays for all routes. The total of inverse delays is

$$\begin{aligned} \sum_{1}^{3} d &= \frac{1}{d_{\text{R1}}} + \frac{1}{d_{\text{R2}}} + \frac{1}{d_{\text{R3}}} \\ &= \frac{1}{34.5 \text{ min}} + \frac{1}{30.0 \text{ min}} + \frac{1}{22.4 \text{ min}} \\ &= 0.0289 \ \frac{1}{\text{min}} + 0.0333 \ \frac{1}{\text{min}} + 0.0446 \ \frac{1}{\text{min}} \\ &= 0.107 \text{ 1/min} \end{aligned}$$

The proportion for route 1 is

$$p_1 = \frac{0.0289 \ \frac{1}{\text{min}}}{0.107 \ \frac{1}{\text{min}}} = 0.0271$$

The proportion for route 2 is

$$p_2 = \frac{0.0333\ \frac{1}{\text{min}}}{0.107\ \frac{1}{\text{min}}} = 0.311$$

The proportion for route 3 is

$$p_3 = \frac{0.0446\ \frac{1}{\text{min}}}{0.107\ \frac{1}{\text{min}}} = 0.417$$

Check that the proportions total 1.00.

$$\begin{aligned} p_t &= 0.271 + 0.311 + 0.417 \\ &= 0.999 \quad \text{[OK]} \end{aligned}$$

The proportion of traffic for route 3 is 0.417 (41.7%).

The answer is (C).

(c) The vehicle volumes are proportioned by

$$V_i = p_i V_t$$

Route 3 has the least delay. The vehicle volume on route 3 is

$$V_3 = p_3 V_t = (0.417)(10{,}000 \text{ veh}) = 4170 \text{ veh}$$

The answer is (D).

Traffic and Capacity Analysis

Nomenclature

A	area	ft^2	m^2
A_{pbT}	permitted phase adjustment	–	–
ATS	average travel speed	mph	kph
b	length	ft	m
BFFS	base free-flow speed	mph	kph
BPTSF	base percent time spent following	%	%
c	capacity	vph	vph
C	confidence level	%	%
C	cycle length	sec	s
d	control delay	sec/veh	s/veh
d	distance	mi	km
d_1	uniform control delay	sec/veh	s/veh
d_2	incremental delay	sec/veh	s/veh
d_3	initial queue delay	sec/veh	s/veh
df	degrees of freedom	–	–
D	clear storage distance	ft	m
D	density	various	various
E	passenger car equivalent	–	–
f	adjustment factor	various	various
FFS	free-flow speed	mph	kph
g	effective green time	sec	s
g/C	green ratio	–	–
G	grade	%	%
HV	percentage of heavy vehicles	%	%
I	upstream filtering/metering adjustment factor	–	–
k	incremental delay calibration factor	–	–
L	limit of acceptable error	–	–
L	lost time	sec	s
L	segment length	mi	km
LC	lateral clearance	ft	m
M	vehicles met	veh	veh
N	number	–	–
O	overtaking vehicles	veh	veh
p	fraction	–	–
P	vehicles passed	veh	veh
PF	progression adjustment factor	–	–
PHF	peak hour factor	–	–
PTSF	percent time spent following	%	%
Q_b	initial queue at beginning of period T	veh	veh
R_p	platoon ratio	–	–
$\overline{R}$	average range in running speed	mph	kph
s	saturation flow rate	various	various
s	sample standard deviation	mph	kph
s_o	base saturation flow rate	pc/hr-ln	pc/h·ln
S	speed	mph	kph
t	Student's t-distribution variate	–	–
t	time	various	various
$\bar{t}$	adjusted average travel time	min	min
T	analysis period duration	hr	h
TLC	total lateral clearance	ft	m
u	delay parameter	–	–
v	rate of flow (service flow rate)	various	various
v_g	unadjusted demand flow rate for lane group	vph	vph
v_{g1}	unadjusted demand flow rate for single lane with highest volume	vph	vph
v/c	volume-capacity ratio	–	–
V	volume	vph	vph
W	width	ft	m
X	lane group v/c ratio or degree of saturation	–	–
Y	volume-capacity ratios of critical movements	–	–
z	standard normal variate confidence limit	mph	kph

Symbols

α	significance level	%	%
σ	standard deviation	mph	kph

Subscripts

a	area type
ave	average
A	access-point density or approach A
bb	bus blocking
B	buses
c	clearing or critical
d	directional traffic distribution

EB	eastbound
f	field-observed, free-flow, or percent time spent following
FM	field-measured
g, G	grade
HV	heavy vehicle
i	ith vehicle or lane group i
ID	interchange density
l	intersection l
L	lanes or left
LC	lateral clearance
Lpb	left turn for pedestrians and bicycles
LS	lane and shoulder
LT	left-turn
LTA	left turn under protected green
LU	lane utilization
LW	lane width
m	parking maneuvers
M	median or multilane
min	minimum
n	northbound
np	no passing zone
N	number of lanes
o	base or optimal
p	driver population, parking, passenger, passenger car, peak, or platoon
ped	pedestrian
PA	platoons arriving during green
r	receiving
R	recreational vehicle or right
Rpb	right turn for pedestrians and bicycles
RT	right-turn
RTA	right turn under protected green
s	average travel speed, saturated, southbound, or speed
SB	southbound
t	test run, time period, total, travel time, or turn
T	truck and bus
TLC	total lateral clearance
u	undersaturated
w	lane width
WB	westbound

1. TRAVEL TIME AND DELAY STUDIES

A *travel time and delay study* assesses the time it takes a vehicle to travel between two points along a given route (referred to as the *study segment*). The data collected from these studies are used by traffic engineers and analysts to identify a route's problem locations and to determine design or operational improvements that will better facilitate traffic flow.

Traffic studies are most frequently conducted on major arterials leading to and from a region's *central business district* (CBD), but a study segment can include any type of roadway.

Some traffic studies require the use of *test vehicles*, vehicles either manned by personnel in charge of gathering data for the studies or equipped with special instruments for traffic study data measurement. A travel time and delay study can be conducted with or without a test vehicle, depending on the purpose of the study, the type of roadway being studied, the length of the study segment, the time of day the study is conducted, and the personnel and equipment resources available. For more on studies conducted using test vehicles, see "Test Vehicle Methods."

Before any data are collected, precise starting and ending points are identified and marked so they will be visible to the observers measuring data for the study. The preferred method is to mark the starting and ending points using existing, easy-to-identify objects, such as road signs or the end of a median barrier, so that motorists are not aware that a traffic study is taking place and will not change speed or otherwise alter their behavior from the norm. Placement of temporary markers, such as traffic cones, is suboptimal, as temporary markers can cause motorists to change speed.

Once the starting and ending points are marked, the distance between the points is measured to find the study segment length. If a test vehicle is to be used, the study segment length should be measured along the vehicle's most probable path. For the most accurate results, the distance should be measured by a surveyor. The distance can also be measured on accurate maps or roadway plans or using the vehicle's odometer if a test vehicle is being used, but these methods are less accurate than surveying.

Test site data, such as the time and date, temperature, weather conditions, and the test run direction, should be recorded and verified before data collection begins. Any roadside observers should remain out of sight as much as possible, and in-vehicle observers should operate any equipment and take any measurements below the eye level of passing drivers.

The essential measurements of a traffic system's performance are travel time, delay, and vehicle speed. *Travel time* is the length of time it takes to traverse a study segment. Travel time varies inversely with travel speed and is comprised of the *running time* (the time the vehicle is in motion) and *delay* (the time the vehicle is stopped due to factors beyond the driver's control).

Delay can be classified as fixed delay, stopped delay, control delay, or operational delay. *Fixed delay* is caused by a traffic control device, such as a stop sign or traffic signal. *Stopped delay* is the amount of time a vehicle is stopped at a red light. *Control delay* is delay at a signalized intersection attributed to traffic signal operation and includes delays due to vehicle acceleration and deceleration in addition to the stopped delay. Control delay is a more precise measurement than stopped delay, but calculation is often complicated and time-consuming. Therefore, travel time and delay studies frequently use stopped delay, while signal timing and other studies that require highly accurate delay times use control delay. Calculation of control delay is described in more detail in Sec. 2.5. *Operational delay* is the delay caused by the traffic itself when there is no control device causing the traffic delay. Delay can also include the time spent traveling below a certain speed, which is determined by a study's purpose and objectives.

The sum of all delay types on a given study segment is referred to as the segment's *total delay*. Each type of delay can be measured, tabulated, and reported for individual test runs, as well as for the total of all test runs.

Test runs with unusual delay can be left out of the final data. For example, an accident would not be considered a typical traffic condition, and a test run delayed by an accident could be eliminated from the data set. Observers may also choose to omit test runs that occur during the time it takes for traffic to regain normal flow patterns after an unusual flow event.

Vehicle speed within a study segment can be measured and expressed in a variety of ways. The study method, study purpose, and available resources dictate which speed measurement is used. Average travel speed, space mean speed, and running speed are the measurements most commonly used in travel time and delay studies.

The *Highway Capacity Manual* (*HCM*) defines *average travel speed*, ATS, as the study segment length divided by the average vehicle travel time. ATS is considered a *space mean speed*, which is measured in respect to space (i.e., the segment length). Equation 2.1 is used to calculate the average travel speed, S_{ave}, where L is the study segment length (in miles or kilometers), N_t is the number of test runs observed, and t_i is the travel time (in hours) of the ith vehicle to traverse the segment.

$$S_{\text{ave}} = \frac{N_t L}{\sum_{i=1}^{N_t} t_{i,\text{hr}}} = \frac{L}{\frac{1}{N_t}\sum_{i=1}^{N_t} t_{i,\text{hr}}} \quad \text{[\textit{HCM} Eq. 7-4]} \qquad 2.1$$

If the average travel time (in hours) for the study segment length, t_{ave}, is known, Eq. 2.1 can be simplified to Eq. 2.2.

$$S_{\text{ave}} = \frac{L}{t_{\text{ave,hr}}} \quad \text{[\textit{HCM} Eq. 7-4]} \qquad 2.2$$

Equation 2.3 is used to calculate the running speed, S, for a single test run. *Running speed* is the distance a vehicle travels divided by the running time. L is the study segment length, and t is the travel time.

$$S = \frac{L}{t} \qquad 2.3$$

Confidence Level

Conclusions and predictions based on sampling studies are not 100% accurate 100% of the time, and travel time and delay studies are no exception. Studies should be repeated several times, and test runs with unusual measurements should be removed from the final data, in order to increase confidence in the results.

Confidence in the results of a study is measured in terms of the *confidence level*, C, a percentage expressing the likelihood that the results from a given study are accurate and correct. For example, results with a 95% confidence level are considered to have a 5% chance of being erroneous in some way.

The complement of the confidence level is the *significance level*, α. The significance level is a percentage expressing the chance that given results are incorrect. In other words, the significance level for given results is always equal to 100% minus the confidence level for those results; and conversely, the confidence level for given data is always equal to 100% minus the significance level for that data. The 95% confidence level, or 5% significance level, is the lowest level generally accepted as being statistically significant. Results with a confidence level of 99% are said to be *highly significant*.

Sample Size

A study's minimum sample size can be determined using the procedure outlined in the Institute of Transportation Engineers' (ITE's) *Manual of Transportation Engineering Studies*. The *minimum sample size* is the number of test runs required to achieve a certain permitted error. The *permitted error* is determined by the level of accuracy required in the study and is defined as the tolerance for error in a specific parameter. The ITE procedure applies the permitted error to the average range in running speed, $\overline{R}$, and has units of miles (kilometers) per hour. A transportation engineer typically determines the permitted error before data collection begins, and the value of permitted error may be increased or decreased based on an agency's reliability requirements, the availability of accurate data, and staff, time, and financial resources.

The ITE suggests the following ranges of permitted error for various types of studies. However, engineering judgment should be used when selecting a value of permitted error.

- transportation planning and highway capacity studies: ±3–5 mph (±5–8 kph)
- traffic operations, trend analysis, and economic evaluations: ±2–4 mph (±4–7 kph)
- before-and-after studies: ±1–3 mph (±2–5 kph)

Once four initial test runs are completed, the average range in running speed, $\overline{R}$, is determined using Eq. 2.4.

$$\overline{R} = \frac{|S_1 - S_2| + |S_2 - S_3| + \cdots + |S_{N_t-1} - S_{N_t}|}{N_t - 1} \qquad 2.4$$

The permitted error and the average range in running speed are compared to Table 2.1 to establish the

Traffic/Capacity Analysis

minimum sample size necessary to achieve a 95% confidence level. Table values should not be interpolated. If the average range in running speed falls between two listed ranges, the higher range is always used. Additional test runs are completed as necessary to satisfy the minimum sample size requirement.

Table 2.1 *Minimum Sample Size Requirements for Travel Time and Delay Studies with a Confidence Level of 95%*

average range in running speed, $\overline{R}$, (mph)	minimum number of test runs, N_t, for a permitted error of 1.0 mph	2.0 mph	3.0 mph	4.0 mph	5.0 mph
2.5	4	2	2	2	2
5.0	8	4	3	2	2
10.0	21	8	5	4	3
15.0	38	14	8	6	5
20.0	59	21	12	8	6

(Multiply mph by 1.609 to obtain kph.)

Reprinted with permission from the Institute of Transportation Engineers, *Manual of Transportation Engineering Studies.*

The sample size can also be calculated using either the normal distribution or the t-distribution. The *normal distribution*, also referred to as the *Gaussian distribution*, is a symmetrical distribution that is used extensively in statistical analysis. The normal distribution is based on known values of the standard deviation and mean. A *t-distribution* is used when the true population standard deviation and mean are unknown. *Student's t-distribution* is a standardized t-distribution that is commonly used in travel time and delay studies. The Federal Highway Administration (FHWA) recommends that a t-distribution be used if the estimated sample size, N, is less than 30, which applies to the majority of travel time and delay studies.

Equation 2.5 calculates a study's minimum sample size, N, using Student's t-distribution. L is the limit of acceptable error, t_α is the t-distribution parameter for a desired confidence level, and s is the sample's standard deviation in mph (kph).

$$N = \left(\frac{t_\alpha s}{L}\right)^2 \qquad 2.5$$

The t-distribution parameter is typically determined from Student's t-distribution tables using the degrees of freedom, df, which is one less than the sample size. However, the sample size is determined using the t-distribution parameter. Therefore, the one- and two-tail limits for a normal distribution, z, are substituted for t_α to calculate an estimated sample size. Values of z for various confidences levels on a standard normal curve are given in Table 2.2. The estimated sample size calculated using the values from Table 2.2 should be checked by calculating the degrees of freedom (i.e., the estimated sample size minus 1), determining the t-distribution parameter from Student's t-distribution tables, and recalculating the sample size. Both calculated sample sizes should be relatively close in value.

Table 2.2 *Confidence Levels for One- and Two-Tail Confidence Limits*

confidence level	one-tail limit, $\pm z$	two-tail limit, $\pm z$
90%	1.28	1.645
95%	1.645	1.96
97.5%	1.96	2.17
99%	2.33	2.575
99.5%	2.575	2.81
99.75%	2.81	3.00

Example 2.1

A major arterial through a shopping district is being evaluated for signal upgrades. The arterial is 2.3 mi long and has 14 intersections with varying spacing. Four test runs have been completed with the following results.

run number	elapsed time (sec)	speed (mph)
1	420	17.3
2	370	19.7
3	340	21.3
4	520	14.0

The permitted error is 1.0 mph with a confidence level of 95%. How many additional test runs are needed to complete the study?

Solution

Use Eq. 2.4 to find the average range in running speed.

$$\begin{aligned}\overline{R} &= \frac{|S_1 - S_2| + |S_2 - S_3| + |S_3 - S_4|}{N_t - 1}\\ &= \frac{\left|17.3\ \frac{\text{mi}}{\text{hr}} - 19.7\ \frac{\text{mi}}{\text{hr}}\right| + \left|19.7\ \frac{\text{mi}}{\text{hr}} - 21.3\ \frac{\text{mi}}{\text{hr}}\right| + \left|21.3\ \frac{\text{mi}}{\text{hr}} - 14.0\ \frac{\text{mi}}{\text{hr}}\right|}{4 - 1}\\ &= 3.77\ \text{mi/hr}\end{aligned}$$

Use Table 2.1 to find the minimum sample size requirement for the study. 3.77 mph is between 2.5 mph and 5.0 mph, so use the values for an average range in running speed of 5.0 mph. For a 5.0 mph average range and a permitted error of ±1.0 mph, a total of eight test runs are needed. Four test runs have already been completed, so four more test runs are needed to complete the study.

Data Collection

Data collection for travel time and delay studies consists of measuring the travel time and delay time. Multiple test runs are performed until the sample size is large enough to determine ATS, running speed, and/or space mean speed.

Manual data collection requires observers. In most tests, observers stay out of sight on the roadside. If a test vehicle is to be used, an observer can serve as the driver of the test vehicle or ride in the test vehicle as a passenger. Observers record data at predetermined points along the study segment using two stopwatches, one tracking total elapsed time and one tracking delay time.

Automatic data collection uses a variety of devices to measure total time and delay time. One common method of automatic data collection is installing sensors connected to a laptop computer in a test vehicle. An observer riding in or driving the test vehicle keys into the computer when the test vehicle passes the starting point, and again when the test vehicle passes the ending point. A computer program then uses the data collected from the sensors and the information keyed in by the observer to calculate the travel time and delay time. Another common method of automatic data collection is the use of sensors or other devices installed at various locations along a study segment, providing continuous data on an observed vehicle's speed and position. This method typically uses sensors installed for other primary purposes, such as real-time traffic monitoring or vehicles' global positioning systems (GPS). Ultimately, the advantages of automatic data collection are the reduction of the number of field personnel required for data collection and the decreased possibility of human error.

Regardless of the type of data collection, personnel involved in the study should take note of the type, location, duration, and cause of any traffic delays occurring during a given test run. This information is necessary for proper data analysis.

Test Vehicle Methods

Test vehicle methods require a test vehicle, a driver, and/or an observer. Depending on the data collection method used, the observer may be a passenger in the test vehicle, or the driver may serve as the observer. Test vehicle methods typically require study segments of no less than 1 mi (1.6 km), with end points distant enough from intersections to obtain accurate approach and recovery speeds. An intersection can be used as the starting point or end point for a study segment when using a test vehicle method, as long as the test vehicle can proceed uninhibited through the intersection and the speed at which vehicles travel through the intersection matches the travel flow speed.

Three common methods requiring a test vehicle are the floating car, average speed, and moving vehicle test methods. When using the *floating car method*, the driver "floats" with the traffic flow, attempting to safely pass as many vehicles as pass the test vehicle. The total elapsed time is recorded from the beginning to the end of the study segment. Trips are recorded several times in each direction to establish a reliable average travel time.

For the *average speed method*, the driver makes several practice runs, then drives the length of the study segment going a speed that, in the opinion of the driver, is the average speed of the traffic flow.

The *moving vehicle method* is used on two-directional roadways where opposing traffic is visible at all times, usually over a study segment with a convenient turnaround just beyond the segment. The moving vehicle method requires several round trips, and it produces results and data that can be used to calculate the average travel time, space mean speed, and hourly volume for the study segment. Observers count vehicles traveling in the opposite direction of the test vehicle (i.e., *vehicles met*), vehicles that pass the test vehicle (i.e., *vehicles overtaking*) and vehicles moving in the same direction that are passed by the test vehicle (i.e., *vehicles passed*). While simply recording the total travel time for a segment may be sufficient for a study's purpose, data gathered at intermediate points along the study segment, such as intersections, can be recorded to study speed variations along the route. Additional points should only be included if the observers are able to safely and accurately record the data. The additional data can be used to determine the location of the most severe delay or locations where speed improvements would have the greatest effect on the overall average travel time.

An example of a travel time and delay study field sheet for the moving vehicle method can be found in App. 2.A. Rows and columns can be arranged to meet the needs of the study and location so data can be readily entered by observers, allowing for faster recording and reducing errors caused by missing critical information.

Figure 2.1 shows a typical study segment for a study using the moving vehicle method, with end points at Haysville St. and Emsworth Ave. Table 2.3 shows the data for one round trip on the study segment. Data for each intersection include the arrival time, which is the time the vehicle enters the intersection, and the departure time, which is the time the vehicle leaves the intersection. The delay is found by subtracting the arrival time from the departure time. Because the only typical cause of delay along the study segment is delay from time spent waiting at traffic signals, all the delay for this study is classified as stopped delay. The travel time is determined by subtracting the departure time of the last intersection from the arrival time of the current intersection. For example, the travel time for Chateau Ave. for the southbound run is calculated as the time the test vehicle departed Belleview Ave., 92 sec, subtracted from the time the test vehicle arrived at Chateau Ave., 139 sec. Therefore, the travel time for Chateau Ave. is 47 sec.

The *hourly volume* is the number of vehicles that will pass the beginning of the study segment in the time it takes the test vehicle to make a round trip through the study segment. The calculations assume a number of vehicles met, M, will pass the starting point between the time the test vehicle passes the starting point and the time the test vehicle returns to the starting point. Vehicles overtaking the test vehicle, O, minus those passed by the test vehicle, P, compensate for the test

Figure 2.1 Beaver River Blvd. Study Segment for the Moving Vehicle Method

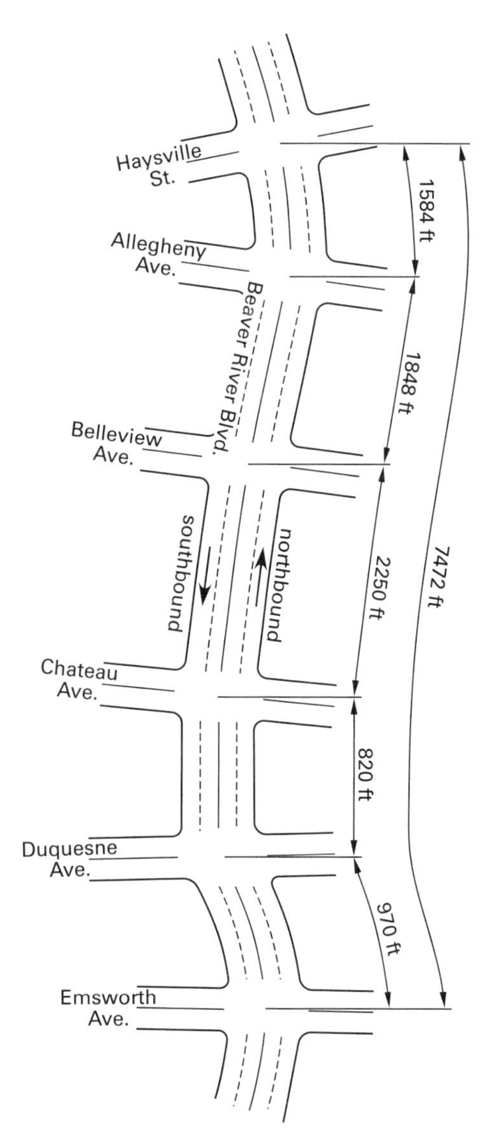

vehicle not traveling at the exact speed of traffic. The following moving vehicle method equations can be rewritten to determine the directional volume, average travel time, and space mean speed for eastbound and westbound segments. Equation 2.6 and Eq. 2.7 are used to determine directional volumes.

$$V_s = \frac{M_n + O_s - P_s}{t_n + t_s} \quad \text{[southbound]} \qquad 2.6$$

$$V_n = \frac{M_s + O_n - P_n}{t_n + t_s} \quad \text{[northbound]} \qquad 2.7$$

Volume in vehicles per minute is multiplied by 60 min/hr to obtain the volume in vehicles per hour, which is how volumes are usually reported.

Volumes are estimates, as errors in counting are inevitable. For example, some vehicles will not be counted during the time it takes to turn the test vehicle around at the end of the segment, opposing vehicles that exit or enter at cross streets are not accounted for, and so on.

The adjusted average travel time is determined from the average of all test run travel times adjusted by the vehicles passed and the overtaking vehicles, as shown in Eq. 2.8 and Eq. 2.9.

$$\bar{t}_s = t_s - \frac{O_s - P_s}{V_s} \quad \text{[southbound]} \qquad 2.8$$

$$\bar{t}_n = t_n - \frac{O_n - P_n}{V_n} \quad \text{[northbound]} \qquad 2.9$$

The space mean speed for the moving vehicle method is calculated using Eq. 2.10 or Eq. 2.11.

$$S_s = \frac{d}{\bar{t}_s} \quad \text{[southbound]} \qquad 2.10$$

$$S_n = \frac{d}{\bar{t}_n} \quad \text{[northbound]} \qquad 2.11$$

Example 2.2

Seven test trips over the Monongahela Blvd. study segment were conducted using the moving vehicle method. The total study segment distance is 8150 ft, and only the total travel time from end point to end point was recorded during the test. Using the test trip data in the following table, calculate (a) the directional volumes, (b) the directional adjusted average travel times, and (c) the directional space mean speeds using customary U.S. units.

test trip summary, Monongahela Blvd.

southbound trips	t_s (sec)	M_s (veh)	O_s (veh)	P_s (veh)
1	214	84	7	10
2	220	90	6	9
3	223	82	7	8
4	210	75	8	11
5	214	81	8	10
6	235	85	4	3
7	212	90	6	7
total	1528	587	46	58
average	218.3	83.9	6.6	8.3

test trip summary, Monongahela Blvd.

northbound trips	t_n (sec)	M_n (veh)	O_n (veh)	P_n (veh)
1	208	135	6	7
2	206	131	5	5
3	211	142	6	4
4	196	120	4	5
5	210	128	3	2
6	215	143	3	1
7	208	143	1	2
total	1454	942	28	26
average	207.7	134.6	4.0	3.7

Solution

(a) The directional volumes are calculated using Eq. 2.6 and Eq. 2.7 and the averages of all runs. For the southbound direction, use Eq. 2.6.

$$V_s = \frac{M_n + O_s - P_s}{t_n + t_s} = \left(\frac{134.6 \text{ veh} + 6.6 \text{ veh} - 8.3 \text{ veh}}{207.7 \text{ sec} + 218.3 \text{ sec}}\right) \times \left(60 \ \frac{\text{min}}{\text{hr}}\right)\left(60 \ \frac{\text{sec}}{\text{min}}\right)$$
$$= 1264 \text{ vph}$$

For the northbound direction, use Eq. 2.7.

$$V_n = \frac{M_s + O_n - P_n}{t_n + t_s} = \left(\frac{83.9 \text{ veh} + 4.0 \text{ veh} - 3.7 \text{ veh}}{207.7 \text{ sec} + 218.3 \text{ sec}}\right) \times \left(60 \ \frac{\text{min}}{\text{hr}}\right)\left(60 \ \frac{\text{sec}}{\text{min}}\right)$$
$$= 712 \text{ vph}$$

(b) For the adjusted average travel time in the southbound direction, use Eq. 2.8.

$$\bar{t}_s = t_s - \frac{O_s - P_s}{V_s} = \frac{218.3 \text{ sec}}{60 \ \frac{\text{sec}}{\text{min}}} - \frac{(6.6 \text{ veh} - 8.3 \text{ veh}) \times \left(60 \ \frac{\text{min}}{\text{hr}}\right)}{1265 \ \frac{\text{veh}}{\text{hr}}}$$
$$= 3.56 \text{ min}$$

For the northbound direction, use Eq. 2.9.

$$\bar{t}_n = t_n - \frac{O_n - P_n}{V_n}$$
$$= \frac{207.7 \text{ sec}}{60 \ \frac{\text{sec}}{\text{min}}} - \frac{(4.0 \text{ veh} - 3.7 \text{ veh})\left(60 \ \frac{\text{min}}{\text{hr}}\right)}{712 \ \frac{\text{veh}}{\text{hr}}}$$
$$= 3.44 \text{ min}$$

(c) The space mean speed in the southbound direction can be found using Eq. 2.10.

$$S_s = \frac{d}{\bar{t}_s} = \frac{\left(\frac{8150 \text{ ft}}{5280 \ \frac{\text{ft}}{\text{mi}}}\right)\left(60 \ \frac{\text{min}}{\text{hr}}\right)}{3.56 \text{ min}} = 26.0 \text{ mph}$$

For the northbound direction, use Eq. 2.11.

$$S_n = \frac{d}{\bar{t}_n} = \frac{\left(\frac{8150 \text{ ft}}{5280 \ \frac{\text{ft}}{\text{mi}}}\right)\left(60 \ \frac{\text{min}}{\text{hr}}\right)}{3.44 \text{ min}} = 26.9 \text{ mph}$$

Methods Without Test Vehicles

Many transportation studies do not require test vehicles to gather data. For instance, the *direct observation method* involves observers placed at an elevated vantage point who measure travel time between two points a known distance apart. This method requires good visibility and is suitable for study segments of less than 0.5 mi (0.8 km).

The *license plate method* requires observers, placed at checkpoints along the test segment, to record the last two, three, or four digits of vehicle license plates and the time at which the vehicle passes each checkpoint. The spacing between checkpoints will vary depending on the roadway's characteristics, the configuration of nearby roads, and the purpose of the study. The FHWA guidelines for checkpoint spacing are 1–3 mi (1.6–4.8 km) between checkpoints on freeways with high levels of access; 3–5 mi (4.8–8.0 km) on freeways with low levels of access; 0.5–1 mi (0.8–1.6 km) on arterials with high levels of cross street and driveway access; and 1–2 mi (1.6–3.2 km) on arterials with low levels of cross street and driveway access. Clocks are coordinated between checkpoints to ensure accuracy. Tape recorders can be used to record the precise time of day at periodic intervals and the reading of license plates at each checkpoint in order to reduce the number of missed vehicles. The plate readings are matched along with the observed times to obtain travel time between the checkpoints. According to the *Manual of Transportation Engineering Studies*, a sample size of 50 matched plate readings usually provides sufficient accuracy.

A variation of the license plate method uses *toll road cards* or automatic toll collection system data. The time a vehicle crosses the toll gateway is compared with the time the vehicle crosses the exit recorder, and the ATS is calculated using the distance between the two locations. This method is useful for comparing travel speeds at various times of the day, as toll information is recorded continuously. Toll road cards can only be used to obtain data on travel speed, not actual running speed, as delay cannot be obtained from toll collection data.

An *interview method* can be set up using selected individuals, such as truck, taxi, or delivery drivers. The drivers are asked to record their start and end times for designated routes or test segments. Volunteer participants need a short training session, a stop watch, and a log sheet to record their time. This method provides a lot of data in a short amount of time with little expense, although the reliability of the data may be less than data collected by fully trained observers.

Data Analysis

Data must be analyzed once it is collected. After the speeds, delays, and travel and running times have been tabulated, the data can be analyzed to determine whether delays exist. Once the types and severity of

Table 2.3 Moving Vehicle Method Data for Beaver River Blvd. Study Segment

Study segment: Beaver River Blvd. Date: May 12
Weather: 75°F, clear Time: 4:30–6:00 p.m. End points: Emsworth Ave. and Haysville St.

way point	distance (ft)	arrival time (sec)	departure time (sec)	vehicles met	vehicles passed	overtaking vehicles	travel time (sec)	stopped delay (sec)	delay reason	segment speed (mph)
run 1 southbound										
Haysville St.	0	–	0	–	–	–	–	–	–	–
Allegheny Ave.	1584	30	30	21	3	1	30	0	–	35.9
Belleview Ave.	1848	70	92	20	2	2	40	22	signal	31.4
Chateau Ave.	2250	139	142	23	3	1	47	3	signal	32.6
Duquesne Ave.	820	159	159	9	1	0	17	0	–	32.8
Emsworth Ave.	970	176	–	11	1	3	17	–	–	38.8
total	7472	176	–	84	10	7	151	25	–	–
average										28.8
run 1 northbound										
Emsworth Ave.	0	–	0	–	–	–	–	–	–	–
Duquesne Ave.	970	17	17	18	1	0	17	0	–	38.8
Chateau Ave.	820	31	39	16	0	1	14	8	signal	39.8
Belleview Ave.	2250	81	96	41	3	2	42	15	signal	36.4
Allegheny Ave.	1848	132	132	31	2	3	36	0	–	34.9
Haysville St.	1584	164	–	29	1	0	32	–	–	33.7
total	7472	164	–	135	7	6	141	23	–	–
average										31.0

(Multiply ft by 0.305 to obtain m.)
(Multiply mph by 1.609 to obtain kph.)

delays are established, operational improvements that may reduce delay and improve overall flow can be determined.

Data can be reported in tabular form but are often shown using graphical plots for enhanced visualization. This kind of plot is particularly useful when the study zone covers a very large area, such as an entire CBD. Figure 2.2 shows a typical travel time, speed, and delay diagram along a study route. Study results can also be displayed graphically as speed contours laid over a map of the study zone.

Figure 2.2 *Travel Time, Speed, and Delay Diagram Along a Route*

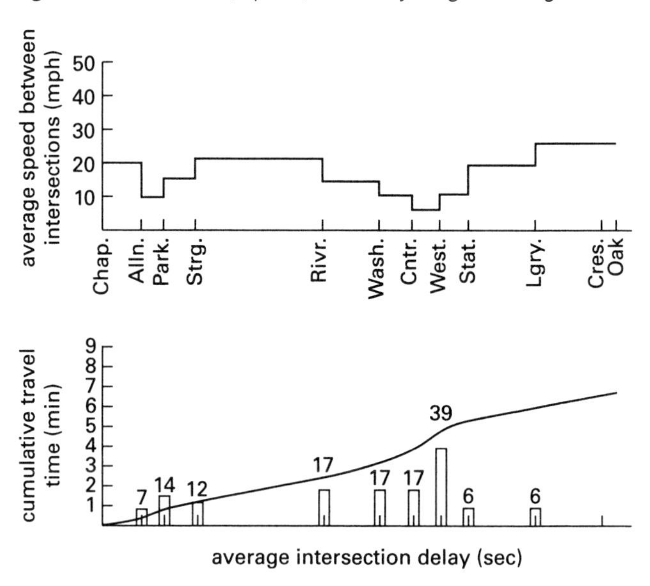

Reprinted with permission from the Institute of Transportation Engineers, *Manual of Transportation Engineering Studies.*

2. CAPACITY ANALYSIS PROCEDURES

Traffic capacity analysis estimates the ability of transportation facilities to sustain traffic under a range of defined conditions. The *HCM* is the primary authority on capacity analysis. The methods presented in the *HCM* are used to analyze capacity and predict traffic flow by accounting for a facility's specific characteristics, such as roadway type, traffic flow, and control systems. The *HCM* procedures use data accumulated over several decades. Due to improvements in computer-assisted analysis, the mathematical models the procedures are based on now closely represent actual traffic conditions. While there is no substitute for using field data for the specific location being evaluated, the *HCM* procedures give a highly probable, mathematical approximation of what may occur under actual field conditions.

Analytical Procedures and Applications

Traffic capacity analysis in the *HCM* divides a given roadway into facilities, segments, and points. A *facility* is a length of roadway composed of contiguous segments. A *segment* is a one-directional length of roadway, marked by two *points* that indicate where a given segment begins and ends. Points are usually locations where traffic enters, exits, or crosses a facility. Two-lane highway segments are unique in that segments can be analyzed as either two-way segments or as one-way segments with a passing lane.

The *HCM* divides transportation facilities into several categories. The two major facility categories are uninterrupted flow facilities and interrupted flow facilities. *Uninterrupted flow facilities* include freeways, multilane highways, and two-lane highways on segments that do not have intersections that affect the traffic flow. *Interrupted flow facilities* include unsignalized and signalized intersections, urban streets, roundabouts (sometimes called traffic circles), and transit, pedestrian, and bicycle facilities.

The *HCM* also classifies transportation facilities by mode of transportation. Transit, pedestrian, and bicycle modes are self-explanatory. *Highway modes* include freeways, multilane highways, two-lane highways, and urban streets. A *freeway* is a divided highway with two or more lanes in each direction, fully controlled access, and no fixed interruption points (e.g., signalized intersections). *Multilane highways* have at least two lanes in each direction, no or partial control of access, and signalized intersections spaced 2 mi (3 km) or more apart. *Two-lane highways* have one lane in each direction and may have signalized intersections and no or partial controlled access. *Urban streets* typically have a high density of access points and signalized intersections less than 2 mi (3 km) apart.

Capacity analysis is performed on segments where *demand* (expressed as vehicles or passenger cars per hour) and *capacity* (the maximum sustainable flow rate over a segment during a specific time period under set conditions) are relatively constant. Capacity analysis methods compare a facility's demand to its capacity. The procedure used for capacity analysis depends on whether or not the segment being analyzed includes signalized intersections. Capacity analysis for unsignalized intersections is covered in Sec. 2.3. Section 2.4 covers the fundamentals of traffic signals, which are the basis for signalized intersection capacity analysis outlined in Sec. 2.5.

Analytical methods are designed to be sensitive to roadway and traffic characteristics. However, the methods cannot predict the effects of posted speed limits, police enforcement, safety features, driver education, or unusual vehicle performance, as these are isolated conditions that do not represent the majority of traffic flow.

3. CAPACITY ANALYSIS FOR UNSIGNALIZED INTERSECTIONS

Unsignalized intersection capacity analysis follows a step-by-step procedure: (1) adjust the traffic volume to determine the demand flow rate, (2) calculate the free-flow speed (either from field measurement or by applying adjustment factors to a base free-flow speed), and (3) determine the LOS. While methods for determining the flow rate, speed, and LOS vary based on facility type and roadway characteristics, using this core procedure simplifies analysis.

Flow Rates

Flow rate is the number of vehicles (or bicycles or pedestrians) passing a given point in an hour. If the vehicle count (or bicycle or pedestrian count) applies to a time period other than one hour, the rate is converted to and expressed as vehicles (or bicycles or pedestrians) per hour.

Flow rate is closely related to density and speed, as shown in Eq. 2.12, and knowing any two values can determine the third. *Density* is the number of vehicles averaged over the roadway segment space and is typically given in units of vehicles per mile or vehicles per kilometer (veh/mi or veh/km), vehicles per mile per lane or vehicles per kilometer per lane (veh/mi-ln or veh/km·ln), or passenger cars per mile per lane or passenger cars per kilometer per lane (pc/mi-ln or pc/km·ln). Density is difficult to measure in the field and is typically calculated using the following speed-flow rate relationship.

$$D = \frac{v_p}{S} \quad [HCM \text{ Eq. 21-5}] \qquad 2.12$$

If the density is high, the speed drops, which corresponds to a decrease in flow rate. When density is low, the speed increases, which also increases the flow rate. However, increases in flow rate and speed under low density conditions are limited by geometry (see Chap. 4) or driver comfort, whichever is lower. For design and analysis purposes, the *HCM* uses a maximum speed of 75 mph (120 kph), and flow rates at greater speeds are not significant for analysis.

Flow rate is also affected by *flow friction*, which is caused by conditions deviating from the default (i.e., ideal) conditions. For example, the *HCM* sets the base lane width at 12 ft (3.7 m). Lane widths narrower than the base width cause drivers to decrease speed, which decreases the flow rate. The *HCM* uses a series of adjustment factors to correct the flow rate when roadway conditions fall outside the default values.

For unsaturated flow conditions, the terms flow rate, demand, demand flow rate, volume, and flow volume tend to be used interchangeably, which can be confusing. In general, the *demand* is the number of drivers desiring service on a given roadway, expressed in vehicles or passenger cars per hour. Demand is determined by actual vehicle counts and includes a mix of vehicle types (e.g., automobiles, trucks, and buses). For oversaturated conditions (i.e., when demand exceeds capacity), it is appropriate to refer to demand rather than flow rate, as the flow is often near zero. *Volume* is similar to flow rate and is based on the number of vehicles passing a point during a given interval, often one hour. Volume is typically given in vehicles per hour. Unlike flow rate, volume is not divided by the observation time. For example, a volume of 20 vph counted over a 15 minute period corresponds to a flow rate of 80 vph (i.e., the volume divided by the counting period length).

Because freeway and highway traffic are composed mainly of passenger cars, flow rate uses passenger cars as its unit of measure. Other types of vehicles, such as trucks, buses, or recreational vehicles (RVs), must be converted to *passenger car equivalents* (i.e., the number of passenger cars displaced by a heavy vehicle under set roadway, traffic, and control conditions) when calculating the flow rate.

Flow Rates on Multilane Highways and Freeways

The *demand flow rate*, v_p, also called the *passenger car flow rate*, is the adjusted hourly demand in passenger car equivalents for the peak 15 minutes. V is the hourly volume along the segment. The *HCM* gives three adjustments that must be made to hourly volume counts or estimates to arrive at the demand flow rate. These adjustments include the *peak hour factor*, PHF, the *heavy vehicle adjustment factor*, f_{HV}, and the *driver population adjustment factor*, f_p. The driver population adjustment factor ranges from 0.85 to 1.00, with 1.00 being the default value, which is used for most peak-period commuter traffic. Lower values may be necessary when the traffic stream is sufficiently influenced by a large proportion of weekend, recreational, or midday drivers. To express flow rate on a per-lane basis, the number of lanes, N_L, is used. The 15 minute passenger car equivalent flow rate, in passenger cars per hour per lane (pc/hr-ln or pc/h·ln), is calculated from Eq. 2.13.

$$v_p = \frac{V}{N_L(\text{PHF})f_{\text{HV}}f_p} \quad [\textit{HCM}\ \text{Eq. 21-3}] \qquad 2.13$$

Flow Rates on Two-Lane Highways

The demand flow rate for a two-lane highway is given in passenger cars per hour (pc/hr or pc/h). Flow rate is used to calculate both the ATS and the percent time spent following. *Percent time spent following*, PTSF, is the average percentage of the total travel time that a vehicle spends behind slower vehicles due to an inability to pass. In other words, PTSF represents a vehicle's freedom to maneuver. For clarity, $v_{p,s}$, $f_{G,s}$, and $f_{\text{HV},s}$ for ATS use are distinguished from their PTSF counterparts, $v_{p,f}$, $f_{G,f}$, and $f_{\text{HV},f}$, by the use of subscripts s and f, but have the same meanings otherwise.[1] ATS and PTSF are discussed further in "Travel Speed on Two-Lane Highways," later in this section. The demand flow rate for ATS is

$$v_{p,s} = \frac{v}{(\text{PHF})f_{G,s}f_{\text{HV},s}} \quad [\textit{HCM}\ \text{Eq. 20-3}] \qquad 2.14$$

Similarly, the demand flow rate for PTSF is

$$v_{p,f} = \frac{v}{(\text{PHF})f_{G,f}f_{\text{HV},f}} \quad [\textit{HCM}\ \text{Eq. 20-3}] \qquad 2.15$$

To calculate the flow rate for two-lane highways, the driver population factor used on multilane highways and freeways is replaced by a grade adjustment factor. The *grade adjustment factor*, $f_{G,f}$ or $f_{G,s}$, accounts for the the effect of terrain on PTSF and travel speeds. The grade adjustment factor ranges from 0.71 to 1.00, with 1.00 being the default value. The default factor is based on the range of two-way flow rates, the range of directional flow rates, and the type of terrain. Table 2.4 is used to find the grade adjustment factor, $f_{G,s}$, when calculating ATS, and Table 2.5 is used to find the grade adjustment factor, $f_{G,f}$, when calculating PTSF.

Peak Hour Factor

The peak hour factor, PHF, accounts for the variation in flow within the peak hour. The PHF is determined as the hourly volume of the maximum volume hour of the day divided by the peak flow rate within the same hour.

$$\text{PHF} = \frac{V_{\text{vph}}}{v_p} = \frac{V_{\text{vph}}}{\left(\frac{60\ \text{min}}{t}\right)V_t} = \frac{V_{\text{vph}}}{N_t V_t} \quad [\textit{HCM}\ \text{Eq. 7-1}] \qquad 2.16$$

The *peak flow rate*, v_p, is the maximum equivalent hourly volume during the peak hour. The peak flow rate is determined from the peak volume of vehicles, V_t, counted during a given time period, t, and the number of time periods, N_t, in one hour. Although the time period of the peak flow rate can be any length, standard practice in capacity analysis is to use a 15 minute peak flow period. The PHF for a 15 minute flow period is determined from Eq. 2.17.

$$\text{PHF}_{15\ \text{min}} = \frac{V_{\text{vph}}}{4V_{15\ \text{min,peak}}} \quad [\textit{HCM}\ \text{Eq. 7-2}] \qquad 2.17$$

The range of PHFs for 15 minute counting periods is 0.25–1.00. In order to achieve 0.25, there would need to be no traffic for 45 minutes during the peak hour, and all of the traffic would have to occur within the remaining 15 minute period (e.g., along a road outside a factory during a shift change). A PHF of approximately 1.00 indicates that traffic is uniformly distributed throughout the peak hour, which commonly occurs along highly congested corridors. On freeways, typical PHF values range from 0.80 to 0.95. Lower values are typical for rural and off-peak conditions,

[1]The *HCM* does not use s and f to differentiate the ATS and PTSF cases.

Table 2.4 Grade Adjustment Factor ($f_{G,s}$) for Two-Lane Roads to Determine ATS on Two-Way Directional Segments

range of two-way flow rates (pc/hr)	range of directional flow rates (pc/hr)	level terrain	rolling terrain
<600	<300	1.00	0.71
≥600 and <1200	≥300 and <600	1.00	0.93
≥1200	≥600	1.00	0.99

Highway Capacity Manual 2000. Copyright, National Academy of Sciences, Washington, D.C. Exhibit 20-7. Reproduced with permission of the Transportation Research Board.

Table 2.5 Grade Adjustment Factor ($f_{G,f}$) for Two-Lane Roads to Determine PTSF on Two-Way Directional Segments

range of two-way flow rates (pc/hr)	range of directional flow rates (pc/hr)	level terrain	rolling terrain
<600	<300	1.00	0.77
≥600 and <1200	≥300 and <600	1.00	0.94
≥1200	≥600	1.00	1.00

Highway Capacity Manual 2000. Copyright, National Academy of Sciences, Washington, D.C. Exhibit 20-8. Reproduced with permission of the Transportation Research Board.

while higher values are typical for urban and congested suburban conditions.[2]

In the absence of field data, the *HCM* recommends using the following approximations.

- 0.95 for congested conditions
- 0.92 for urban areas
- 0.88 for rural areas

Heavy Vehicle Factor

Heavy vehicles (defined by the *HCM* as vehicles with more than four tires touching the pavement) in the traffic stream can change the free-flow speed because trucks, buses, and RVs occupy more space and have different performance characteristics than automobiles. A heavy vehicle adjustment factor, f_{HV}, is needed so that the equivalent flow rate can be expressed in pc/hr-ln (pc/h·ln). The heavy vehicle adjustment factor is found from Eq. 2.18.

$$f_{HV} = \frac{1}{1 + p_T(E_T - 1) + p_R(E_R - 1)} \quad \text{[\textit{HCM} Eq. 21-4]}$$

2.18

[2]As with all traffic measuring estimates, field data values of PHFs take precedence over default or handbook values. In the absence of reliable field data, the default or handbook values are suitable for general conditions.

E_T and E_R are the passenger car equivalents for trucks and buses and for RVs, respectively. Truck and bus counts can be combined because trucks and buses share similar performance characteristics. When the number of trucks is more than five times the number of RVs for multilane highways and freeways, the RV count can be included with trucks rather than evaluated separately. The type of terrain dictates the E_T and E_R values, as shown in the following section. p_T and p_R are the proportion of trucks and buses and of RVs in the vehicle mix, respectively.

Roadway Terrain

Roadway terrain is classified into three general types. *Level terrain* is any combination of grades and vertical or horizontal alignment that permits heavy vehicles to operate at nearly the same speed as passenger cars. Grades are limited to 1–2% for short durations. *Rolling terrain* is any combination of grades and horizontal or vertical alignment that causes heavy vehicles to reduce their operating speed substantially below that of passenger cars, but does not cause heavy vehicles to operate at *crawl speeds* (the maximum sustained speed that can be maintained by a specific type of vehicle on a constant upgrade of a given percentage) for a significant length of time or extended distances. *Mountainous terrain* is any combination of grades and vertical or horizontal alignment that causes heavy vehicles to operate at crawl speeds for significant distances or at frequent intervals.

Extended freeway or highway segments can include a series of upgrades and downgrades that do not have a significant influence on speed. This can include grades of up to 3% for distances of 0.25 mi to 1 mi (0.4 km to 1.6 km). Grades greater than 3% or grades that are sustained for distances longer than 0.25 mi (0.4 km) should be considered as a specific grade and analyzed as a separate segment using the procedure given in *HCM* Chap. 20 and Chap. 21. Appendix 2.B gives passenger car equivalents relating the percentage of grade, the length of the grade segment, and the percentage of trucks and buses.

Table 2.6 gives the passenger car equivalents for multilane highways and freeways. Table 2.7 shows the passenger car equivalents for trucks and RVs on two-lane highways used to determine ATS on two-way and directional segments. Table 2.8 shows the passenger car equivalents for trucks and RVs on two-lane highways used to determine PTSF on two-way and directional segments.

Travel Speed on Multilane Highways and Freeways

For multilane highways and freeways, the average travel speed is determined from field data taken from the study segment or from a nearby segment with similar characteristics. In the absence of such data, or when projecting speed for a new facility, the free-flow speed, FFS, may be estimated from a base free-flow speed, BFFS. *Free-flow*

Table 2.6 *Passenger Car Equivalents on Extended General Highway and Freeway Segments*

factor	type of terrain		
	level	rolling	mountainous
trucks and buses, E_T	1.5	2.5	4.5
RVs, E_R	1.2	2.0	4.0

Highway Capacity Manual 2000. Copyright, National Academy of Sciences, Washington, D.C. Exhibit 21-8 and Exhibit 23-8. Reproduced with permission of the Transportation Research Board.

Table 2.7 *Passenger Car Equivalents for Trucks and Buses and RVs to Determine ATS on Two-Way and Directional Segments on Two-Lane Highways*

vehicle type	range of two-way flow rates (pc/hr)	range of directional flow rates (pc/hr)	type of terrain: level terrain	type of terrain: rolling terrain
trucks and buses, E_T	< 600	< 300	1.7	2.5
	≥ 600 and < 1200	≥300 and < 600	1.2	1.9
	≥1200	≥600	1.1	1.5
RVs, E_R	< 600	< 300	1.0	1.1
	≥600 and < 1200	≥300 and < 600	1.0	1.1
	≥1200	≥600	1.0	1.1

Highway Capacity Manual 2000. Copyright, National Academy of Sciences, Washington, D.C. Exhibit 20-9. Reproduced with permission of the Transportation Research Board.

Table 2.8 *Passenger Car Equivalents for Trucks and Buses and RVs to Determine PTSF on Two-Way and Directional Segments on Two-Lane Highways*

vehicle type	range of two-way flow rates (pc/hr)	range of directional flow rates (pc/hr)	type of terrain: level terrain	type of terrain: rolling terrain
trucks and buses, E_T	< 600	< 300	1.1	1.8
	≥600 and < 1200	≥300 and < 600	1.1	1.5
	≥1200	≥600	1.0	1.0
RVs, E_R	< 600	< 300	1.0	1.0
	≥600 and < 1200	≥300 and < 600	1.0	1.0
	≥1200	≥600	1.0	1.0

Highway Capacity Manual 2000. Copyright, National Academy of Sciences, Washington, D.C. Exhibit 20-10. Reproduced with permission of the Transportation Research Board.

speed is the theoretical speed of traffic under low-volume conditions. *Base free-flow speed* is based on an ideal (i.e., base) set of roadway conditions, such as 12 ft (3.7 m) lane width and at least 6 ft (1.8 m) clearance outside the lane. Drivers adjust speed according to the comfort level of the roadway in relation to these ideal conditions, which must be accounted for using adjustment factors when calculating free-flow speed. When a base free-flow speed is not available, the design speed can be used as a substitute.

To estimate free-flow speed from base free-flow speed, adjustments must be made for the width of lanes and available lateral clearance along the study segment. The *lane width adjustment factor*, f_{LW}, is based on the *HCM*'s ideal width standard of 12 ft (3.7 m), which provides the perception of a clear path for drivers of all skill levels without requiring undue attention to the precise positioning of the vehicle within the lane itself. As lane width is reduced from the ideal width, the average traffic speed reduces by a predictable amount. Lane widths narrower than 10 ft (3.1 m) are not practical for traffic with a mix of large trucks and buses as there is not enough lateral clearance for passing. Table 2.9 gives adjustments for lane widths of multilane highways and freeways.

Table 2.9 *Adjustment for Lane Widths (f_{LW}) for Multilane Highways and Freeways*

lane width (ft)	reduction in FFS (mph)
12	0.0
11	1.9
10	6.6

(Multiply ft by 0.305 to obtain m.)
(Multiply mph by 1.609 to obtain kph.)

Highway Capacity Manual 2000. Copyright, National Academy of Sciences, Washington, D.C. Exhibit 21-4. Reproduced with permission of the Transportation Research Board.

Reduced clearance from the edge of a lane reduces a driver's feeling of safety. Drivers will tend to reduce speed when shoulders are narrower than 6 ft (1.8 m). For freeways, the effect of reduced lateral clearance on free-flow speed is based on the right-shoulder lateral clearance. The left-shoulder lateral clearance is rarely less than 2 ft (0.6 m), so adjustments for that shoulder are not normally applied. Table 2.10 gives the *right-shoulder lateral clearance adjustment factor*, f_{LC}, for freeways.

Table 2.10 *Adjustment for Right-Shoulder Lateral Clearance (f_{LC}) on Freeways*

right-shoulder lateral clearance (ft)	reduction in FFS (mph)			
	lanes in one direction			
	2	3	4	≥ 5
≥ 6	0.0	0.0	0.0	0.0
5	0.6	0.4	0.2	0.1
4	1.2	0.8	0.4	0.2
3	1.8	1.2	0.6	0.3
2	2.4	1.6	0.8	0.4
1	3.0	2.0	1.0	0.5
0	3.6	2.4	1.2	0.6

(Multiply ft by 0.305 to obtain m.)
(Multiply mph by 1.609 to obtain kph.)

Highway Capacity Manual 2000. Copyright, National Academy of Sciences, Washington, D.C. Exhibit 23-5. Reproduced with permission of the Transportation Research Board.

For multilane highways, the effect of reduced lateral clearance on free-flow speed is based on the lateral clearance for fixed objects along both sides of the roadway. Fixed objects include signs, trees, bridge rails, traffic barriers, and retaining walls. Standard raised curbs are not considered fixed objects for the purpose of determining lateral clearance. For divided multilane highways, the lateral clearance for both sides of the roadway is added as shown in Eq. 2.19 to find the *total lateral clearance*, TLC. Once the total lateral clearance is calculated, the *total lateral clearance adjustment factor*, f_{TLC}, is found from Table 2.11.

$$\text{TLC} = \text{LC}_R + \text{LC}_L \quad [\textit{HCM}\ \text{Eq. 21-2}] \qquad 2.19$$

The left-shoulder lateral clearance is always taken as 6 ft (1.8 m) for undivided highways and highways with *two-way left-turn lanes*, TWLTLs. The total lateral clearance is calculated from Eq. 2.19, and the adjustment factor for total lateral clearance, f_{TLC}, is found from Table 2.11.

Table 2.11 *Adjustment for Total Lateral Clearance (f_{TLC}) on Multilane Highways*

four-lane highway		six-lane highway	
total lateral clearance* (ft)	reduction in FFS (mph)	total lateral clearance* (ft)	reduction in FFS (mph)
12	0	12	0
10	0.4	10	0.4
8	0.9	8	0.9
6	1.3	6	1.3
4	1.8	4	1.7
2	3.6	2	2.8
0	5.4	0	3.9

(Multiply ft by 0.305 to obtain m.)
(Multiply mph by 1.609 to obtain kph.)
*Total lateral clearance is the sum of the lateral clearances of the median and shoulder. Use a value of 6 ft (1.8 m) if the width of the median or shoulder is greater than 6 ft (1.8 m). For purposes of analysis, the total lateral clearance cannot exceed 12 ft (3.7 m).

The base free-flow speed is based on five lanes or more in one direction on a freeway. If there are fewer than five lanes, the *number of lanes adjustment factor*, f_N, given in Table 2.12 must be applied.

The base free-flow speed for a freeway is based on a freeway that has no more than one interchange every 2 mi (3 km), expressed as an interchange density of 0.5 interchanges per mile (0.3 interchanges per kilometer). An *interchange* is any area where there is at least one on-ramp. Off-ramps do not count as interchanges for the purpose of interchange density. Interchange density for a given freeway is found by evaluating a 6 mi (10 km) segment, covering 3 mi (4.8 km) in each direction (i.e., 3 mi (4.8 km) upstream and 3 mi (4.8 km) downstream) from the starting point. If the given 6 mi (10 km) segment has an interchange density greater than 0.5 interchanges per mile (0.3 interchanges per kilometer), an *interchange density adjustment factor*, f_{ID}, should be included in the calculation for free-flow speed. Adjustment factors for interchange density are given in Table 2.13.

Table 2.12 *Adjustment for Number of Lanes (f_N) on Freeways*

number of lanes (one direction)	reduction in FFS* (mph)
≥ 5	0.0
4	1.5
3	3.0
2	4.5

(Multiply mph by 1.609 to obtain kph.)
*For all rural freeway segments, f_N is 0.

Table 2.13 *Adjustment for Interchange Density (f_{ID}) on Freeways*

interchanges per mile	reduction in FFS (mph)
0.50	0.0
0.75	1.3
1.00	2.5
1.25	3.7
1.50	5.0
1.75	6.3
2.00	7.5

(Multiply mph by 1.609 to obtain kph.)

Multilane highways without medians (i.e., undivided highways) have opposing flow friction that requires a *median type adjustment factor*, f_M, to be applied to the base free-flow speed as given in Table 2.14. Divided highways and highways with two-way left-turn lanes do not require an adjustment.

Table 2.14 *Adjustment for Median Type (f_M) for Multilane Highways*

median type	reduction in FFS (mph)
undivided highways	1.6
divided highways (including TWLTLs*)	0.0

(Multiply mph by 1.609 to obtain kph.)
*Two-way left-turn lanes

Some multilane highways have *access points*, such as intersections or driveways, that can interrupt the flow of traffic. Base free-flow speed assumes there are no access points along the study segment. If there are right-side access points along the study segment in the direction of travel, the base free-flow speed must be adjusted to account for the effect on the traffic flow. Access points that do not noticeably affect traffic flow should not be included in access-point density calculations. The *access-point density* of a highway is found by dividing the number of right-side access points in the direction of travel by the length of the study segment in miles (kilometers). Once the access-point density is known, the *access-point density adjustment factor*, f_A, can be found from Table 2.15.

Table 2.15 *Adjustment for Access-Point Density (f_A) on Multilane Highways*

access points per mile	reduction in FFS (mph)
0	0.0
10	2.5
20	5.0
30	7.5
≥ 40	10.0

(Multiply mph by 1.609 to obtain kph.)

Highway Capacity Manual 2000. Copyright, National Academy of Sciences, Washington, D.C. Exhibit 21-7. Reproduced with permission of the Transportation Research Board.

The FFS used in capacity analysis is determined by applying the adjustments to the BFFS using Eq. 2.20 and Eq. 2.21.

For freeways, the FFS is

$$\text{FFS} = \text{BFFS} - f_{\text{LW}} - f_{\text{LC}} - f_N - f_{\text{ID}} \quad [\textit{HCM}\text{ Eq. 23-1}] \qquad 2.20$$

For multilane highways, the FFS is

$$\text{FFS} = \text{BFFS} - f_M - f_{\text{LW}} - f_{\text{TLC}} - f_A \quad [\textit{HCM}\text{ Eq. 21-1}] \qquad 2.21$$

Example 2.3

An undivided, four-lane highway has a total lateral clearance of 4 ft. The lanes are 11 ft wide, and there are 10 access points per mile. Assuming a base free-flow speed of 55 mph, what is the free-flow speed?

Solution

The BFFS is given as 55 mph. From Table 2.14, the median type adjustment factor, f_M, for an undivided highway is 1.6 mph. From Table 2.9, the lane width adjustment factor, f_{LW}, for 11 ft lanes is 1.9 mph. From Table 2.11, the total lateral clearance adjustment factor, f_{TLC}, for a four-lane highway with a total lateral clearance of 4 ft is 1.8 mph. From Table 2.15, the access-point density adjustment factor, f_A, for 10 access points per mile is 2.5 mph. Using Eq. 2.21,

$$\begin{aligned}\text{FFS} &= \text{BFFS} - f_M - f_{\text{LW}} - f_{\text{TLC}} - f_A \\ &= 55\ \frac{\text{mi}}{\text{hr}} - 1.6\ \frac{\text{mi}}{\text{hr}} - 1.9\ \frac{\text{mi}}{\text{hr}} - 1.8\ \frac{\text{mi}}{\text{hr}} - 2.5\ \frac{\text{mi}}{\text{hr}} \\ &= 47.2\ \text{mi/hr} \quad (47\ \text{mph})\end{aligned}$$

Travel Speed on Two-Lane Highways

When field data are available, the FFS for a two-lane highway is determined using Eq. 2.22.

$$\text{FFS} = S_{\text{FM}} + 0.0125\frac{v_f}{f_{\text{HV}}} \quad [\textit{HCM}\text{ Eq. 20-1}]\ [\text{SI}] \qquad 2.22(a)$$

$$\text{FFS} = S_{\text{FM}} + 0.00776\frac{v_f}{f_{\text{HV}}} \quad [\textit{HCM}\text{ Eq. 20-1}]\ [\text{U.S.}] \qquad 2.22(b)$$

S_{FM} is the field-measured speed, and v_f is the field-observed flow rate for the period corresponding to S_{FM}. f_{HV} is the heavy vehicle adjustment factor calculated using Eq. 2.18.

When field data are not available, the free-flow speed can be estimated from Eq. 2.23 by subtracting the *lane and shoulder width adjustment factor*, f_{LS}, and the access-point density adjustment factor, f_A, from the base free-flow speed.

$$\text{FFS} = \text{BFFS} - f_{\text{LS}} - f_A \quad [\textit{HCM}\text{ Eq. 20-2}] \qquad 2.23$$

Table 2.16 gives the adjustment factors for lane and shoulder width, and Table 2.17 gives adjustment factors for access-point density for two-lane highways.

Table 2.16 *Adjustment for Lane and Shoulder Width (f_{LS}) on Two-Lane Highways*

	reduction in FFS (mph)			
	shoulder width (ft)			
lane width (ft)	≥ 0 and < 2	≥ 2 and < 4	≥ 4 and < 6	≥ 6
≥ 9 and < 10	6.4	4.8	3.5	2.2
≥ 10 and < 11	5.3	3.7	2.4	1.1
≥ 11 and < 12	4.7	3.0	1.7	0.4
≥ 12	4.2	2.6	1.3	0.0

(Multiply ft by 0.305 to obtain m.)
(Multiply mph by 1.609 to obtain kph.)

Highway Capacity Manual 2000. Copyright, National Academy of Sciences, Washington, D.C. Exhibit 20-5. Reproduced with permission of the Transportation Research Board.

Table 2.17 Adjustment for Access-Point Density (f_A) on Two-Lane Highways

access points per mile	reduction in FFS (mph)
0	0.0
10	2.5
20	5.0
30	7.5
40	10.0

(Multiply mph by 1.609 to obtain kph.)

Highway Capacity Manual 2000. Copyright, National Academy of Sciences, Washington, D.C. Exhibit 20-6. Reproduced with permission of the Transportation Research Board.

The ATS on two-lane highways is influenced by both the demand flow rate, $v_{p,s}$, and the *no passing zone adjustment factor*, f_{np}, and is calculated from Eq. 2.24.

$$\text{ATS} = \text{FFS} - 0.0125 v_{p,s} - f_{np} \quad [\textit{HCM} \text{ Eq. 20-5}] \text{ [SI]} \qquad 2.24(a)$$

$$\text{ATS} = \text{FFS} - 0.00776 v_{p,s} - f_{np} \quad [\textit{HCM} \text{ Eq. 20-5}] \text{ [U.S.]} \qquad 2.24(b)$$

Table 2.18 shows the ATS adjustment factors for no passing zones, on a two-way segment of a two-lane highway.

Table 2.18 ATS Adjustment for Effect of No Passing Zones (f_{np}) on Two-Way Segments of Two-Lane Highways

two-way demand flow rate, $v_{p,s}$ (pc/hr)	reduction in ATS (mph)					
	no passing zones (%)					
	0	20	40	60	80	100
0	0.0	0.0	0.0	0.0	0.0	0.0
200	0.0	0.6	1.4	2.4	2.6	3.5
400	0.0	1.7	2.7	3.5	3.9	4.5
600	0.0	1.6	2.4	3.0	3.4	3.9
800	0.0	1.4	1.9	2.4	2.7	3.0
1000	0.0	1.1	1.6	2.0	2.2	2.6
1200	0.0	0.8	1.2	1.6	1.9	2.1
1400	0.0	0.6	0.9	1.2	1.4	1.7
1600	0.0	0.6	0.8	1.1	1.3	1.5
1800	0.0	0.5	0.7	1.0	1.1	1.3
2000	0.0	0.5	0.6	0.9	1.0	1.1
2200	0.0	0.5	0.6	0.9	0.9	1.1
2400	0.0	0.5	0.6	0.8	0.9	1.1
2600	0.0	0.5	0.6	0.8	0.9	1.0
2800	0.0	0.5	0.6	0.7	0.8	0.9
3000	0.0	0.5	0.6	0.7	0.7	0.8
3200	0.0	0.5	0.6	0.6	0.6	0.7

(Multiply mph by 1.609 to obtain kph.)

Highway Capacity Manual 2000. Copyright, National Academy of Sciences, Washington, D.C. Exhibit 20-11. Reproduced with permission of the Transportation Research Board.

The PTSF is used to assess a two-lane highway's level of service. The PTSF is estimated using Eq. 2.25 from the base percent time spent following, BPTSF, the two-way passenger car flow rate, $v_{p,f}$, and the *directional distribution and no passing zone adjustment factor*, $f_{d/np}$, given in Table 2.19.

$$\text{PTSF} = \text{BPTSF} + f_{d/np} \quad [\textit{HCM} \text{ Eq. 20-6}] \qquad 2.25$$

The *base percent time spent following*, BPTSF, is calculated from Eq. 2.26.

$$\text{BPTSF} = \left(1 - e^{-0.000879 v_{p,f}}\right) \times 100\% \quad [\textit{HCM} \text{ Eq. 20-7}] \qquad 2.26$$

Levels of Service (LOS)

The *level of service* (LOS) rating system outlined in the *HCM* was created to measure the quality of service on a roadway. While capacity quantitatively measures the demand a roadway can handle, the LOS illustrates the operational conditions of a roadway. Levels of service are based on speed, flow rate, delay, and a vehicle's freedom to maneuver in traffic, as well as the driver's perception of these conditions. LOS designations are arranged into six levels of traffic accommodations. The levels range from A, which is the most free-flowing condition, to F, which is *jam density* (density at which flow comes to a standstill) with very low flow rates. The *HCM* lists the levels as

- A: free flow
- B: reasonably free flow
- C: stable flow
- D: approaching unstable flow
- E: unstable flow
- F: forced or breakdown flow or oversaturation

The freedom to maneuver is inversely related to density, as there is more maneuverability and a higher speed and flow rate in low density conditions. The LOS is most often determined by a roadway's density, which is calculated using Eq. 2.12 and then compared to the density limits for each LOS. The LOS can also be based on the flow rate of the segment, which is then compared to the service flow rate. The *service flow rate*, in pc/hr-ln (pc/h·ln), is the hourly rate at which vehicles can reasonably be expected to cross a given point during a period of time, typically 15 minutes, under typical roadway conditions while maintaining a designated LOS. Because the service flow rate is the maximum flow rate for a given LOS, the service flow rate is the effective boundary between levels of service. The maximum service flow rate for LOS E is considered the maximum sustainable flow rate and is referred to as the *limit of capacity*. When the density or service flow rate for a given roadway is between the values for one LOS and the next, the higher level of service is always used (e.g., if the density on a

Table 2.19 *Adjustment for Directional Distribution and No Passing Zones ($f_{d/np}$) on PTSF on Two-Way Segments of Two-Lane Highways*

two-way flow rate, $v_{p,f}$ (pc/hr)	increase in PTSF (%)					
	no passing zones (%)					
	0	20	40	60	80	100
directional split = 50/50						
≤ 200	0.0	10.1	17.2	20.2	21.0	21.8
400	0.0	12.4	19.0	22.7	23.8	24.8
600	0.0	11.2	16.0	18.7	19.7	20.5
800	0.0	9.0	12.3	14.1	14.5	15.4
1400	0.0	3.6	5.5	6.7	7.3	7.9
2000	0.0	1.8	2.9	3.7	4.1	4.4
2600	0.0	1.1	1.6	2.0	2.3	2.4
3200	0.0	0.7	0.9	1.1	1.2	1.4
directional split = 60/40						
≤ 200	1.6	11.8	17.2	22.5	23.1	23.7
400	0.5	11.7	16.2	20.7	21.5	22.2
600	0.0	11.5	15.2	18.9	19.8	20.7
800	0.0	7.6	10.3	13.0	13.7	14.4
1400	0.0	3.7	5.4	7.1	7.6	8.1
2000	0.0	2.3	3.4	3.6	4.0	4.3
≥ 2600	0.0	0.9	1.4	1.9	2.1	2.2
directional split = 70/30						
≤ 200	2.8	13.4	19.1	24.8	25.2	25.5
400	1.1	12.5	17.3	22.0	22.6	23.2
600	0.0	11.6	15.4	19.1	20.0	20.9
800	0.0	7.7	10.5	13.3	14.0	14.6
1400	0.0	3.8	5.6	7.4	7.9	8.3
≥ 2000	0.0	1.4	4.9	3.5	3.9	4.2
directional split = 80/20						
≤ 200	5.1	17.5	24.3	31.0	31.3	31.6
400	2.5	15.8	21.5	27.1	27.6	28.0
600	0.0	14.0	18.6	23.2	23.9	24.5
800	0.0	9.3	12.7	16.0	16.5	17.0
1400	0.0	4.6	6.7	8.7	9.1	9.5
≥ 2000	0.0	2.4	3.4	4.5	4.7	4.9
directional split = 90/10						
≤ 200	5.6	21.6	29.4	37.2	37.4	37.6
400	2.4	19.0	25.6	32.2	32.5	32.8
600	0.0	16.3	21.8	27.2	27.6	28.0
800	0.0	10.9	14.8	18.6	19.0	19.4
≥ 1400	0.0	5.5	7.8	10.0	10.4	10.7

freeway segment is 40 pc/mi-ln, the LOS on the segment, using Table 2.20, is E, not D).

Multilane Highways and Freeways LOS

LOS criteria for freeways are shown in Table 2.20. The maximum service flow rate at the limit of capacity for a freeway is 2400 pc/hr-ln, the flow rate at LOS E for a free-flow speed of 75 mph (120 kph).

Table 2.21 shows LOS criteria for multilane highways. The maximum service flow rate at the limit of capacity for an undivided highway is 2200 pc/hr-ln at a free-flow speed of 60 mph (97 kph). This is slightly less than the maximum flow rate at the limit of capacity for a freeway at the same free-flow speed (2300 pc/hr-ln at 60 mph (2300 pc/h·ln at 97 kph)).

The density limits for each LOS are thresholds for each level. For instance, a density of 11 pc/mi-ln (6.8 pc/km·ln) or less is considered LOS A for both freeways and multilane highways for all FFS. The density at the limit of capacity (i.e., LOS E) is 45 pc/mi-ln (28 pc/km·ln) for freeways and 40 pc/mi-ln (25 pc/km·ln) for multilane highways.

Example 2.4

A multilane highway is known to have a peak flow rate of 1150 pc/hr-ln at a free-flow speed of 50 mph. What is the LOS for the highway?

Solution

Use Eq. 2.12 to find the density of the highway.

$$D = \frac{v_p}{S} = \frac{1150 \ \frac{\text{pc}}{\text{hr-ln}}}{50 \ \frac{\text{mi}}{\text{hr}}} = 23 \text{ pc/mi-ln}$$

From Table 2.21, a density of 23 pc/mi-ln with a 50 mph free-flow speed means that the density lies between LOS B and LOS C. Therefore, the LOS is C.

Two-Lane Highways LOS

Two-lane highways are categorized into two classes in the *HCM*. *Class I* includes two-lane highways that are major intercity routes, primary arterials, daily commuter routes, and primary links in state or national highway networks. On these highways, drivers expect to travel at relatively high speeds. There are few driveway connections and intersections, and the geometry is such that drivers are comfortable with high-speed driving. Class I LOS criteria are PTSF and ATS, as shown in Table 2.22. Both criteria must be met for each LOS in order to be classified at that level.

Class II includes two-lane highways that are access routes for class I facilities, scenic or recreational routes that are not primary arterials, and commercial strips with many driveway connections and turning movements. Speeds on class II facilities are hindered by limited sight distance, intersections, driveways, or general topographic conditions, such as mountainous terrain with steep grades or sharp turns. Drivers generally tolerate higher levels of PTSF on class II facilities, as they typically serve shorter trips and a greater variety of trip purposes than class I facilities. Driver perception may

Table 2.20 *LOS Criteria for Freeways**

criteria	LOS A	B	C	D	E
FFS = 75 mph					
maximum density (pc/mi-ln)	11	18	26	35	45
minimum speed (mph)	75.0	74.8	70.6	62.2	53.3
maximum v/c	0.34	0.56	0.76	0.90	1.00
maximum service flow rate (pc/hr-ln)	820	1350	1830	2170	2400
FFS = 70 mph					
maximum density (pc/mi-ln)	11	18	26	35	45
minimum speed (mph)	70.0	70.0	68.2	61.5	53.3
maximum v/c	0.32	0.53	0.74	0.90	1.00
maximum service flow rate (pc/hr-ln)	770	1260	1770	2150	2400
FFS = 65 mph					
maximum density (pc/mi-ln)	11	18	26	35	45
minimum speed (mph)	65.0	65.0	64.6	59.7	52.2
maximum v/c	0.30	0.50	0.71	0.89	1.00
maximum service flow rate (pc/hr-ln)	710	1170	1680	2090	2350
FFS = 60 mph					
maximum density (pc/mi-ln)	11	18	26	35	45
minimum speed (mph)	60.0	60.0	60.0	57.6	51.1
maximum v/c	0.29	0.47	0.68	0.88	1.00
maximum service flow rate (pc/hr-ln)	660	1080	1560	2020	2300
FFS = 55 mph					
maximum density (pc/mi-ln)	11	18	26	35	45
minimum speed (mph)	55.0	55.0	55.0	54.7	50.0
maximum v/c	0.27	0.44	0.64	0.85	1.00
maximum service flow rate (pc/hr-ln)	600	990	1430	1910	2250

(Multiply mph by 1.609 to obtain kph.)
(Multiply pc/mi-ln by 0.621 to obtain pc/km·ln.)
*The exact mathematical relationship between density and the volume-capacity ratio, v/c, is not always maintained at LOS boundaries because of the use of rounded values. Density is the primary determinant of LOS. The speed criterion is the speed at maximum density for a given LOS. For measurements and calculations related to v/c, see Sec. 2.5. LOS F is characterized by highly unstable and variable traffic flow. Prediction of accurate flow rate, density, and speed at LOS F is difficult.

Table 2.21 LOS Criteria for Multilane Highways*

criteria	LOS A	LOS B	LOS C	LOS D	LOS E
FFS = 60 mph					
maximum density (pc/mi-ln)	11	18	26	35	40
average speed (mph)	60.0	60.0	59.4	56.7	55.0
maximum v/c	0.30	0.49	0.70	0.90	1.00
maximum service flow rate (pc/hr-ln)	660	1080	1550	1980	2200
FFS = 55 mph					
maximum density (pc/mi-ln)	11	18	26	35	41
average speed (mph)	55.0	55.0	54.9	52.9	51.2
maximum v/c	0.29	0.47	0.68	0.88	1.00
maximum service flow rate (pc/hr-ln)	600	990	1430	1850	2100
FFS = 50 mph					
maximum density (pc/mi-ln)	11	18	26	35	43
average speed (mph)	50.0	50.0	50.0	48.9	47.5
maximum v/c	0.28	0.45	0.65	0.86	1.00
maximum service flow rate (pc/hr-ln)	550	900	1300	1710	2000
FFS = 45 mph					
maximum density (pc/mi-ln)	11	18	26	35	45
average speed (mph)	45.0	45.0	45.0	44.4	42.2
maximum v/c	0.26	0.43	0.62	0.82	1.00
maximum service flow rate (pc/hr-ln)	490	810	1170	1550	1900

(Multiply mph by 1.609 to obtain kph.)
(Multiply pc/mi-ln by 0.621 to obtain pc/km·ln.)
*The exact mathematical relationship between density and the volume-capacity ratio, v/c, is not always maintained at LOS boundaries because of the use of rounded values. Density is the primary determinant of LOS. For measurements and calculations related to v/c, see Sec. 2.5. LOS F is characterized by highly unstable and variable traffic flow. Prediction of accurate flow rate, density, and speed at LOS F is difficult.

Highway Capacity Manual 2000. Copyright, National Academy of Sciences, Washington, D.C. Exhibit 21-2. Reproduced with permission of the Transportation Research Board.

Table 2.22 LOS Criteria for Two-Lane Highways in Class I*

LOS	PTSF (%)	ATS (mph)
A	≤ 35	> 55
B	> 35 and ≤ 50	> 50 and ≤ 55
C	> 50 and ≤ 65	> 45 and ≤ 50
D	> 65 and ≤ 80	> 40 and ≤ 45
E	> 80	≤ 40

(Multiply mph by 1.609 to obtain kph.)
*LOS F applies whenever the flow rate exceeds the segment capacity.

Highway Capacity Manual 2000. Copyright, National Academy of Sciences, Washington, D.C. Exhibit 20-2. Reproduced with permission of the Transportation Research Board.

not necessarily coincide with the definitions given in the *HCM*, and consideration should always be given to driver expectations when classifying a two-lane highway. For example, a primary arterial through a mountainous area may be classified as a class II highway if drivers understand that high speeds are unrealistic along the route. Class II LOS is determined only by PTSF, as shown in Table 2.23.

Because LOS F is difficult to calculate, ranges for LOS F are not included in Table 2.22 and Table 2.23. In order to determine if a highway is operating at LOS F, calculate the two-way demand flow rate, v_p, and the directional flow rate. The *directional flow rate* is the flow rate in the most heavily traveled direction and is determined as the two-way demand flow rate multiplied by the larger value in the directional split, in decimals. For example, if v_p is 1100 pc/hr and the directional split is 70/30, the

*Table 2.23 LOS Criteria for Two-Lane Highways in Class II**

LOS	PTSF (%)
A	≤ 40
B	> 40 and ≤ 55
C	> 55 and ≤ 70
D	> 70 and ≤ 85
E	> 85

*LOS F applies whenever the flow rate exceeds the segment capacity.

directional flow rate is 1100 pc/hr × 0.70 = 770 pc/hr. The two-way demand flow rate and the directional flow rate are then compared to the limits of capacity. The two-way limit of capacity is 3200 pc/hr, and the directional limit of capacity is 1700 pc/hr. If either flow rate is larger than the limits of capacity, then one or both directions are oversaturated, and the LOS is F. If both values are less than the limits of capacity, the LOS is determined from Table 2.22 or Table 2.23.

Example 2.5

A two-lane highway used as a scenic and recreational route is evaluated to determine level of service. A traffic study provides the following data for the roadway.

- 950 pc/hr two-way flow rate (field observed)
- 4% trucks and buses
- 0% RVs
- 0.85 PHF
- level terrain
- 3 ft shoulder width
- 10 ft lane width
- 60/40 directional split
- 50 mph base speed (field measured)
- 5.0 mi road length
- 12 access points per mile
- 60% no passing zones

What is the LOS for this highway?

Solution

The highway is used as a scenic and recreational route, so it is a class II facility. Class II LOS is determined using PTSF. To calculate the BPTSF for the highway, the peak flow rate for the highway must be known.

From Table 2.5, the grade adjustment factor, $f_{G,f}$, is 1.00.

Calculate the heavy vehicle adjustment factor for PTSF using Eq. 2.18. From Table 2.8, the passenger car equivalent is 1.1 for trucks on level terrain and a 950 pc/hr two-way flow rate, and the passenger car equivalent for RVs under the same conditions is 1.0. Since there are no RVs in the traffic flow, p_R is zero.

$$\begin{aligned} f_{\text{HV}} &= \frac{1}{1 + p_T(E_T - 1) + p_R(E_R - 1)} \\ &= \frac{1}{1 + (0.04)(1.1 - 1) + (0)(1.0 - 1)} \\ &= 0.996 \end{aligned}$$

Calculate the demand flow rate for PTSF using Eq. 2.15. From Table 2.5, the grade adjustment factor for two-lane roads on level terrain used to find PTSF is 1.00.

$$\begin{aligned} v_{p,f} &= \frac{v}{(\text{PHF}) f_{G,f} f_{\text{HV},f}} = \frac{950 \ \frac{\text{pc}}{\text{hr}}}{(0.85)(1.00)(0.996)} \\ &= 1122 \text{ pc/hr} \quad (< 3200 \text{ pc/hr}) \end{aligned}$$

Determine the highest directional flow rate.

$$v_{p,f} = (0.60)\left(1122 \ \frac{\text{pc}}{\text{hr}}\right) = 673 \text{ pc/hr} \quad (< 1700 \text{ pc/hr})$$

The demand flow rate and the directional flow rate are less than 3200 pc/hr and 1700 pc/hr, respectively. Therefore, the LOS is greater than LOS F and can be determined from Table 2.23.

Calculate the BPTSF using Eq. 2.26.

$$\begin{aligned} \text{BPTSF} &= (1 - e^{-0.000879 v_{p,f}}) \times 100\% \\ &= \left(1 - e^{(-0.000879)(1122 \text{ pc/hr})}\right) \times 100\% \\ &= 62.7\% \end{aligned}$$

Calculate the PTSF using Eq. 2.25. Interpolating from Table 2.19, $f_{d/\text{np}}$ is 9.8% for a 60/40 directional split, 60% no passing zones, and a two-way demand flow rate of 1122 pc/hr.

$$\begin{aligned} \text{PTSF} &= \text{BPTSF} + f_{d/\text{np}} \\ &= 62.7\% + 9.8\% \\ &= 72.5\% \end{aligned}$$

From Table 2.23, the LOS is D.

Example 2.6

A two-lane highway has a 70/30 directional split with 20% no passing zones. The two-way flow rate is 800 pc/hr. What is the PTSF?

Solution

From Table 2.19, for a two-lane highway with a 70/30 directional split and 20% no passing zones, the adjustment factor, $f_{d/\text{np}}$, is 7.7%. Using Eq. 2.25 and Eq. 2.26,

$$\begin{aligned}\text{PTSF} &= \text{BPTSF} + f_{d/\text{np}} \\ &= (1 - e^{-0.000879 v_{p,f}}) \times 100\% + f_{d/\text{np}} \\ &= \left(1 - e^{(-0.000879)(800 \text{ pc/hr})}\right) \times 100\% + 7.7\% \\ &= 58.2\%\end{aligned}$$

4. TRAFFIC SIGNALS

The *Manual on Uniform Traffic Control Devices* (*MUTCD*) defines a *traffic signal* as any device that alternately directs traffic, including pedestrians, bicyclists, and vehicles, to stop and then proceed. Traffic signals are used to reduce traffic conflict, crashes, and delays; to reduce police involvement in traffic management; and to improve the overall safety of the road system.

All traffic signal systems have at least three round signals arranged in a vertical configuration. Each signal (also called an *aspect*) is a different color. The top signal is always red, the center signal is always amber (also called yellow), and the bottom signal is always green.

In special, low-clearance cases, a traffic signal system may be arranged horizontally instead of vertically. In these instances, the leftmost signal is red, the center signal is amber, and the rightmost signal is green. A green arrow is often added to the bottom of a signal system to indicate permitted left or right movements (see Fig. 2.3). Arrows may also be added at complex intersections to indicate movements at various angles other than straight ahead.

The red aspect indicates all approaching traffic must stop. The green aspect is turned on (and the red aspect is turned off) when one or more approaching traffic streams is permitted to enter the intersection. At the end of the permitted approach time, the amber aspect turns on and the green aspect turns off to indicate that the green interval is about to end. After the red aspect turns on again, the approaching traffic must stop and allow a different traffic stream to enter the intersection.

Traffic Signal Warrants

The design, installation, operation, and maintenance of a signal system involve considerable expense, and much of the liability for any accidents or delays the signal system might cause is placed on the people responsible for designing, installing, and maintaining the signal system. Therefore, before a signal system is designed and installed, an engineering study is conducted to determine

Figure 2.3 Common Signal Arrangement

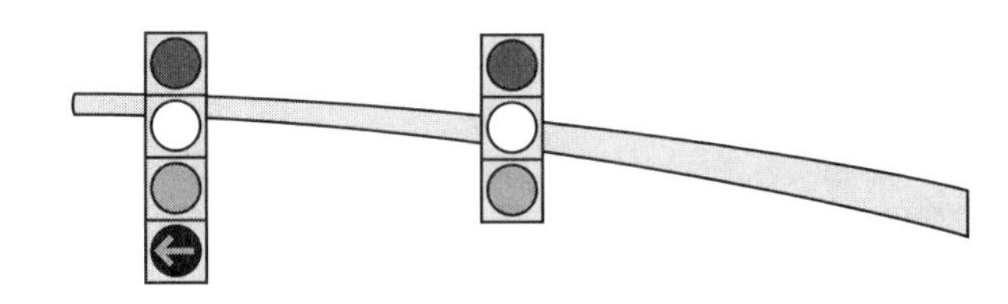

if installing a signal will improve the overall safety and/or operation of the traffic flow. The study should also describe intersection improvements that would reduce the impact of the signal, such as realigning the horizontal and vertical geometry of the intersection or reconfiguring the approach lanes. In addition, the study should evaluate improvements that could reduce the complexity of the signal or eliminate the need for a signal altogether.

When a signal is warranted, the usual goal is to minimize delay through the intersection for all traffic, including pedestrians and bicyclists. The following *MUTCD* warrants are used to determine if a new traffic signal or the removal of a traffic signal is warranted.

Warrant 1, eight-hour vehicular volume: Meeting either condition A or condition B, or a combination of both A and B, will achieve warrant 1. For condition A, the volume of traffic flowing in both directions on a major street during a day's highest volume eight-hour period is at least 500 vph for one-lane approaches or 600 vph for approaches with two or more lanes. A minor street must have at least 150 vph on the highest volume one-lane approach or 200 vph on an approach with two or more lanes during the same eight hours. Condition B is used when continuous traffic is interrupted (e.g., when major street flow is so heavy that the minor street flow cannot enter without significant delay). For condition B, the volume of traffic flowing in both directions on a major street is at least 750 vph for one-lane approaches or 900 vph for approaches with two or more lanes. The minor street's highest entering flow must have a volume of at least 75 vph for one-lane approaches or 100 vph for approaches with two or more lanes. To satisfy requirements for a combination of condition A and B, 80% of the minimum volumes can be used if alternative measures have been unsuccessful. When the major-street approach speed is greater than 40 mph (70 kph) or the road is in an isolated community of 10,000 or fewer, 70% of the minimum volumes can be used for conditions A or B or 56% for combinations of conditions A and B. Appendix 2.C gives the required traffic volumes for both conditions.

Warrant 2, four-hour vehicular volume: Warrant 2 is evaluated using the highest volume four-hour period. The highest volumes for the major street and for the minor street do not have to occur during the same four-hour period. To achieve warrant 2, the intersection of the four-hour major street volume (total of both approaches) and the highest four-hour volume on the minor street (one approach) must fall above the applicable curve in App. 2.D. If the community population is fewer than 10,000 or the major street speed exceeds 40 mph (70 kph), 70% of the minimum volumes may be used, as shown in App. 2.D.

Warrant 3, peak hour: Meeting either condition A or B will achieve warrant 3. Warrant 3 applies to intersections whose minor approach traffic experiences undue delay for at least one hour during the average day. These intersections are typically located near office complexes, industrial parks, manufacturing plants, or other vehicle-intensive sites that produce a high volume of vehicle traffic over a short period of time. For condition A, all of the following conditions must exist for the same one-hour period: 1) vehicle delay on a minor approach controlled by a stop sign is a minimum of 4 veh-hr for one-lane approaches and 5 veh-hr for two-lane approaches; 2) the four-hour volume is at least 100 vph for one-lane approaches 150 vph for two-lane approaches; and 3) the total volume is at least 650 vph for a three-approach intersection or 800 vph for intersections with four or more approaches. For condition B, the plotted vehicles per hour on a two-directional, major street and the corresponding vehicles per hour on a minor street (one direction) fall above the curve in App. 2.E.

Warrant 4, pedestrian volume: Meeting either condition A or B will achieve warrant 4. This warrant applies to both mid-block (i.e., between intersections) and pedestrian crossings at intersections, as long as the nearest existing signal is at least 300 ft (90 m) away. Condition A states that, for any four-hour period, the plotted vehicles per hour on a two-directional, major street and pedestrians per hour crossing the major street for all crossings should fall above the curve given in App. 2.F. Condition B states that, for any one-hour period, the plotted vehicles per hour and pedestrians per hour for a major-street approach must fall above the curve given in App. 2.F.

Warrant 5, school crossing: This warrant applies to locations with traffic streams that have fewer adequate crossing gaps than minutes required for children to cross the street. There must be at least 20 students during the highest crossing hour. The signal will not be warranted if within 300 ft (90 m) of another signal, unless the school crossing signal is coordinated to not restrict the progressive movement of traffic.

Warrant 6, coordinated signal system: Meeting either condition A or condition B will achieve warrant 6. Condition A applies to one-way streets, or streets that have predominantly one-way traffic flow, and where existing signals are too far apart to maintain efficient platooning. (A *platoon* is a group of vehicles or pedestrians traveling together.) Condition B applies to two-way streets where existing traffic signals impede platooning, and where the proposed signal will provide progressive operation. New signals should be spaced at least 1000 ft (300 m) from existing signals.

Warrant 7, crash experience: Meeting conditions A, B, and C will achieve warrant 7. Condition A applies when previous measures to reduce crash frequency have failed. Condition B applies when five or more reportable crashes that involve property damage or personal injury occur within reporting limits. These crashes must occur within a 12 month period and must be able to be corrected by a traffic signal. Condition C applies when the intersection experiences 80% of traffic volumes shown in warrant 1 for the major and minor streets during the same eight-hour period.

Warrant 8, roadway network: This warrant is achieved when either condition A or condition B is met. Condition A requires the intersection to have a volume of at least 1000 vph during the peak hour of a typical weekday and projected five-year volumes that meet at least one of warrants 1, 2, and 3. Condition B requires the intersection to have a five-hour volume of at least 1000 vph on a Saturday or Sunday. The intersection should also have at least one of the following characteristics: the intersection is a part of a principal network for through traffic, the intersection includes a rural or suburban highway near or through a city, or the intersection appears as a major route on an official transportation plan.

Warrant 9, intersection near a grade crossing: This warrant is achieved where an intersection controlled by a stop or yield sign is near a railroad grade crossing, none of the other warrants are met, and both condition A and condition B are met. Condition A is that the stop line or yield line on the approach is within 140 ft(43 m) of the center of the nearest railroad track. Condition B is a one- or two-lane approach meeting or exceeding, during the highest volume one-hour period, the plotted values of approach vehicles per hour. To determine if condition B is met, the clear storage distance is measured, and minor-street volumes are adjusted using the factors given in App. 2.G. The *clear storage distance*, *D*, is the available space for vehicle storage and is measured between 6 ft (2 m) from the rail nearest the intersection to the intersection stop line or normal stopping point on the highway. The intersection of the major street volume and the adjusted minor street volume is established, and this point is compared to the applicable graph in App. 2.G. If this point falls above the appropriate clear storage distance curve, the approach meets condition B.

The *MUTCD* provides detailed guidance on the application of these warrants. Note that meeting the criteria of these nine warrants does not mean the installation (or removal) of a traffic signal is mandatory. Traffic signals, even when they are warranted, can be ineffectively placed, incorrectly operated, or improperly maintained. Such signals create excessive delays, signal disobediences, increased use of alternate roads to avoid the signal, and/or escalated collision rates. Before a signal is installed or removed, the engineering study must determine improvements that will reduce the impact of the signal. Ultimately, a traffic signal's goal is to minimize delay through an intersection. Therefore, even if a warrant is met, a signal should not be installed or removed if it will disrupt the progressive traffic flow.

Traffic Signal Design

Properly designed traffic signals can be used to maintain the orderly flow of traffic, increase the traffic-handling capacity of the intersection, reduce collision rates, and/or interrupt heavy traffic flow to allow other vehicles or pedestrians to cross. To ensure a traffic signal is properly designed, use the following steps.

step 1: Assemble and organize geometric data. Include area type (e.g., CBD), number of lanes, lane widths, grades, and parking and turning conditions.

step 2: Assemble and organize traffic volume data, usually taken in 15 min periods. Include approach volumes, the PHF, heavy vehicle factors, pedestrian data, bus blocking rates, approach speeds, and turning volumes.

step 3: Organize signal phasing groups and determine minimum cycle times. Include yellow and red intervals, minimum pedestrian times, actuated phases, and the analysis period.

step 4: Calculate passenger car equivalents. Adjust for buses, trucks, lane utilization, turn movements, and parking.

step 5: Identify critical lane volumes.

step 6: Check level of service (usually LOS D or better).

step 7: Determine cycle length.

step 8: Determine phase lengths.

step 9: Check pedestrian intervals.

step 10: Convert yellow and green intervals to percentages of cycle lengths.

step 11: Prepare a sequence diagram and timing chart.

A new signal system should be installed soon after it is designed, as local conditions tend to change with time. After a system is installed, the traffic engineer who designed the system should visit the site to check that the signal is working properly. The signal should be observed during a critical time period (e.g., during the peak hour for an intersection) at a minimum, but further observation is advised. In some cases, additional data may need to be collected to determine how well the signal is operating.

Signal timing is an important aspect of the design process. Signal phases should be timed in proportion to the movements required at the intersection, such that all traffic flows can complete all necessary movements in the intersection safely and with a minimum of delay. Some movements, such as left turns, require a larger proportion of time than other movements, such as simple through-movements or right turns. Individual movements are factored in by estimating the delay time required for a given traffic flow to complete the movement and the conditions under which the movement must be completed.

The sum of all phase lengths, or cycle lengths, should be long enough to accommodate all required phases but not so long as to increase driver frustration. Generally, a phase should not be longer than 90–120 sec or shorter than 12–15 sec. In isolated cases of saturated flow, cycle lengths of 3 min or more have been found to be effective, but these are not considered usual conditions.

Within the general limits, the formulas, calculations, and adjustments shown in the *HCM* are for purposes of comparing to other ideal default conditions or for fine-tuning the signal for a particular situation, such as increasing the approach timc for vehicles making turns as opposed to those continuing straight. These detailed calculations closely approximate prevailing (i.e., local) conditions, thereby reducing the time and cost necessary to make field adjustments for optimizing the signal.

Traffic Signal Timing

In previous sections, free-flow analysis was described for freeways, multilane highways, and two-lane highways. The basis of signalized intersection analysis is measurement of the approaching flow conditions and volumes at the intersection, and allocation of signal timing in proportion to the time required for each phase to complete all necessary movements.

Traffic signal timing is described in terms of intervals, cycles, and phases. An *interval* is the period of time during which a particular signal (red, amber, or green) is lit. A *phase* is the portion of the cycle length during which a given set of approaching traffic streams receive a green light. For example, if the signals for both northbound and southbound traffic turn green at the same time, that is one phase, and another phase begins when the eastbound and westbound signals turn green while the northbound and southbound signals are red. The time it takes for the traffic during a given phase to enter the intersection and clear the intersection again, including any amber time or green time, is the *phase length.* The *signal cycle* is a complete set of phases, from green to red. The *cycle length* is the total time required for a signal system to complete one cycle of all phases, from red to green, including lost time.

Red time is the portion of the cycle time during which the signal is red for a given phase. The *effective red time* is the time during which a given traffic movement is actually required to stop (e.g., the red time, plus any green time during which a movement must remain stopped due to traffic streams that have not cleared the intersection). The effective red time is equal to the difference between the cycle length and the effective green time. The *all-red time* is the time during which all phases on all approaches are red (i.e., when no traffic movements can enter the intersection at all).

The *green time* is the time, in seconds, that the signal is green for a given phase. The *effective green time* is the time during which a given traffic movement or set of movements may proceed (i.e., the portion of the green time when there is a large enough gap in the intersection for the movements to proceed safely). Effective green time is equal to the difference between the cycle length and the effective red time and includes the first part of the amber time, during which traffic already in the intersection clears the intersection.

Amber time is the time, in seconds, that the signal is amber for a given phase. While the red time and green time are widely variable, the amber time is always approximately 3–5 sec, regardless of intersection.

Movement lost time is the green time lost due to delays, such as time to begin acceleration or time waiting for other vehicles to enter or clear the intersection. *Lost time* includes the movement lost time and the portion of the amber time during which all vehicles have cleared the intersection. Lost time can be estimated by summing the movement lost time and the amber time, which is a conservative estimate since a portion of the amber time is typically used by vehicles to clear the intersection.

Phases in a cycle can be numbered 1, 2, 3; I, II, III; or a, b, c. There can be as few as two phases in a cycle, or as many as are needed for all traffic streams to successfully and safely complete all necessary movements through the intersection. Phases can also have subphases, such as a green arrow that is lit for a few seconds during the green time to allow vehicles making left turns to do so safely. In most circumstances, the number of phases, including subphases, is limited to six or seven.

Phases can either be actuated by a signal controller that keeps each phase limited to a certain number of seconds or by sensors called *traffic detectors* that monitor approaching traffic and adjust phase lengths accordingly. *Fixed-time signals*, also known as *pretimed signals*, have set lengths for every phase, and traffic detectors are not used on any approach. *Semiactuated signals* are signals that are actuated by traffic detectors, but also have at least one phase that is timed by a controller. *Fully actuated signals* are signals that are actuated only by traffic detectors.

Turn movements can be described as protected, permitted, or protected-plus-permitted. A turn is *protected* if the movement can be made without interfering with other movements (either because of lane separation, traffic signals, or traffic signs). A turn is *permitted* if the movement is not expressly forbidden (either by a red light or by traffic signs) but must yield to or merge with traffic from another movement. *Protected-plus-permitted* turns are protected during the given phase, but also permitted after the amber time or red time has begun. Left turns that must yield to oncoming traffic are considered permitted. Right turns that occur simultaneously with permitted left turns onto the same approach are also considered permitted if there are not separate lanes for left and right turns. Right or left turns that must yield to pedestrian or bicycle flows are also considered permitted turns. Figure 2.4 illustrates signal phasing for different types of turns.

Figure 2.4 *Signal Phasing and Movement Diagrams*

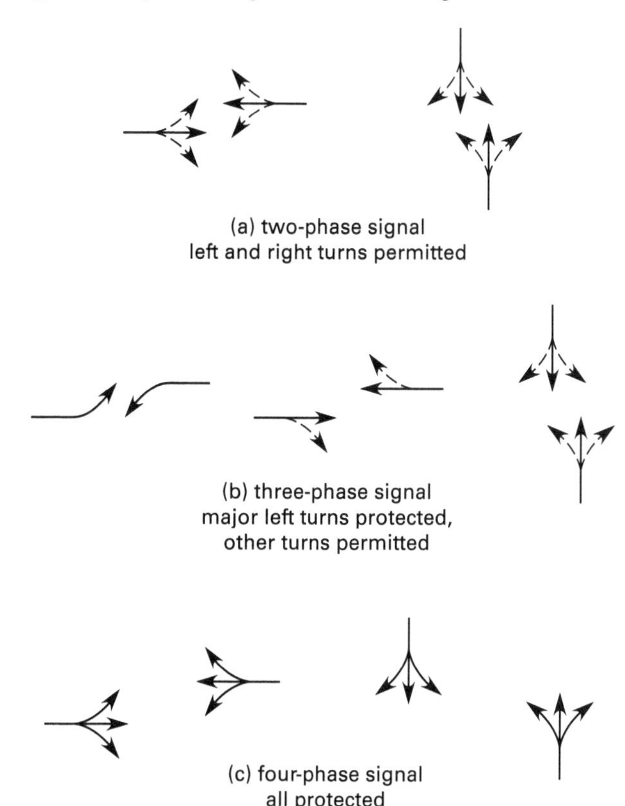

5. CAPACITY ANALYSIS FOR SIGNALIZED INTERSECTIONS

Saturation Flow Rate

Capacity analysis of signalized intersections is based on the saturation flow rate, s, which is considered the capacity of a signalized intersection approach. The *saturation flow rate* is the hourly rate, in vph or veh/hr-ln (veh/hr·ln), at which previously stopped traffic can move through an intersection under prevailing conditions. The saturation flow rate is found from the *base saturation flow rate*, which is the maximum steady flow rate, in passenger cars per hour per lane, at which previously stopped passenger cars can cross a signalized intersection under base conditions. Both saturation flow rate and base saturation flow rate assume that the green signal is available and no lost times are experienced. The base saturation flow rate used in the *HCM* is 1900 pc/hr-ln.

The base saturation flow rate, s_o, is modified by twelve factors to determine the saturation flow rate. These factors relate to a set of ideal conditions, such as zero grade, 12 ft (3.7 m) wide lanes, no parking, no bus stops, and no heavy vehicles in the traffic stream. Variations from the base conditions cause more or less resistance to the base flow rate. The equations for these factors are given as follows and explained in further detail in the *HCM*. For each of the factors, if there is no information available, the *HCM* default value is 1.0, meaning there is no effect on the flow. The saturation flow rate, s, for a subject lane group is expressed as a total for all lanes in the lane group and is calculated from Eq. 2.27.

$$s = s_o N_L f_w f_{HV} f_g f_p f_{bb} f_a f_{LU} f_{RT} f_{LT} f_{Rpb} f_{Lpb} \quad \text{[HCM Eq. 16-4]} \qquad 2.27$$

N_L is the number of lanes in the lane group. The *lane width adjustment factor*, f_w, accounts for the negative impact of narrow lanes on the saturation flow rate. It also allows for an increased flow rate on wide lanes. Standard lane widths, W, are 12 ft (3.6 m), and lane widths must be greater than 8 ft (2.4 m). If the lane width is greater than 16 ft (4.8 m), a two-lane analysis may be considered. The lane width adjustment factor is calculated using Eq. 2.28.

$$f_w = 1 + \frac{W_{\text{m}} - 3.6 \text{ m}}{9 \text{ m}} \quad \text{[HCM Exh. 16-7] [SI]} \qquad 2.28(a)$$

$$f_w = 1 + \frac{W_{\text{ft}} - 12 \text{ ft}}{30 \text{ ft}} \quad \text{[HCM Exh. 16-7] [U.S.]} \qquad 2.28(b)$$

The heavy vehicle adjustment factor, f_{HV}, is found from Eq. 2.29 and accounts for the additional space occupied by heavy vehicles as well as for the differing operating capabilities compared to passenger cars. The passenger car equivalent, E_T, used for heavy vehicles is equal to 2.0 passenger-car units (pc/HV). $HV_{\%}$ is the percentage of heavy vehicles for a lane group volume.

$$f_{HV} = \frac{100\%}{100\% + HV_{\%}(E_T - 1)} \quad \text{[HCM Exh. 16-7]} \qquad 2.29$$

The *grade adjustment factor*, f_g, accounts for the effects of grades on the operation of all vehicles and is calculated from Eq. 2.30. $G_{\%}$ is the percentage of grade on a lane group approach, which must be greater than -6 and less than 10 (negative for downhill, positive for uphill).

$$f_g = 1 - \frac{G_{\%}}{200} \quad \text{[HCM Exh. 16-7]} \qquad 2.30$$

The *parking adjustment factor*, f_p, accounts for the decrease in flow rate caused by a parking lane adjacent to a lane group, as well as for the occasional blockage of a lane as cars maneuver in and out of parking spaces. Each maneuver in or out is assumed by the *HCM* to block traffic for an average of 18 sec. f_p is calculated using Eq. 2.31 and must be greater than 0.05. Where no parking is permitted, it is equal to 1.0. N_L is the number of lanes in the lane group, and N_m is the number of parking maneuvers per hour, which has a practical limit of 180.

$$f_p = \frac{N_L - 0.1 - \dfrac{(18 \text{ sec})N_m}{3600 \dfrac{\text{sec}}{\text{hr}}}}{N_L} \quad \text{[HCM Exh. 16-7]} \qquad 2.31$$

The *bus blocking adjustment factor*, f_{bb}, accounts for the effect of buses that stop to let out or pick up passengers at near- or far-side bus stops within 250 ft (75 m) upstream or downstream of the intersection. A *near-side stop* is located before an intersection, and a *far-side stop* is located after an intersection. A 14.4 sec average blocking time is assumed during a green interval. The number of buses stopping per hour, N_B, is set at a practical limit of 250. The bus blocking adjustment factor is found from Eq. 2.32 and must be greater than 0.05.

$$f_{bb} = \frac{N_L - \dfrac{(14.4 \text{ sec})N_B}{3600 \dfrac{\text{sec}}{\text{hr}}}}{N_L} \quad \text{[HCM Exh. 16-7]} \qquad 2.32$$

The *area type adjustment factor*, f_a, accounts for the inefficiencies of intersections in business districts compared to other locations. A factor of 0.9 should be used in areas exhibiting CBD characteristics, including narrow street right-of-way, frequent parking maneuvers, vehicle blockages, taxi and bus activity, and high pedestrian activity. For all other locations, use a factor of 1.0.

For lane groups with more than one lane, the *lane utilization adjustment factor*, f_{LU}, is used to account for unequal distribution of flow among the lanes. The f_{LU} value is based on the flow in the lane with the highest volume and is given by Eq. 2.33. v_g is the unadjusted demand flow rate for the lane group, and v_{g1} is the unadjusted demand flow rate for the single lane that carries the highest volume, both in vehicles per hour.

$$f_{LU} = \frac{v_g}{v_{g1} N_L} \quad \text{[HCM Eq. 16-5]} \qquad 2.33$$

At volumes close to capacity, lane utilization is near 1.0. At lower demand, drivers tend to utilize one lane more than another or shy away from lanes closest to parked cars or other disturbances. In the absence of field data, default values may be used as given in Table 2.24.

The *right-turn adjustment factor*, f_{RT}, depends on whether the right turn is made from an exclusive or shared lane and on the proportion of right-turning vehicles in the lane group. The right-turn factor is 1.0 if the lane group does not include any right turns. For exclusive right-turn lanes, the factor is 0.85.

For a shared lane, the right-turn adjustment factor is found from Eq. 2.34. p_{RT} is the proportion of right turns

Table 2.24 Default Lane Utilization Adjustment Factors

lane group movements	no. of lanes in lane group, N_L	traffic in most heavily traveled lane (%)	lane utilization adjustment factor, f_{LU}
through or shared	1	100.0	1.000
	2	52.5	0.952
	3*	36.7	0.908
exclusive left turn	1	100.0	1.000
	2*	51.5	0.971
exclusive right turn	1	100.0	1.000
	2*	56.5	0.885

*If a lane group has more lanes than shown, it is recommended that data on the lane group be gathered or else that the smallest f_{LU} shown for that type of lane group be used.

in the lane group and is determined by dividing the turning volume by the total lane group volume.

$$f_{RT} = 1.0 - 0.15p_{RT} \quad \text{[shared]} \qquad 2.34$$

For a single lane, the right-turn adjustment factor is found from Eq. 2.35.

$$f_{RT} = 1.0 - 0.135p_{RT} \quad \text{[single]} \qquad 2.35$$

The *left-turn adjustment factor*, f_{LT}, is based on whether left turns are made from exclusive or shared lanes, the type of phasing (protected, permitted, or protected-plus-permitted) used for the lane group, the proportion of left-turning vehicles in the lane group, and the opposing flow rate when permitted left turns are made. There are six cases where the left-turn adjustment factor is used. (See App. 2.H for adjustment factors for nonprotected phasing alternatives.)

case 1: exclusive lane with protected phasing

case 2: exclusive lane with permitted phasing

case 3: exclusive lane with protected-plus-permitted phasing

case 4: shared lane with protected phasing

case 5: shared lane with permitted phasing

case 6: shared lane with protected-plus-permitted phasing

The left-turn adjustment factor is 1.0 if the lane group does not include any left turns. Otherwise, for an exclusive lane, the factor is 0.95. For a shared lane, the left-turn adjustment factor is determined using Eq. 2.36, where p_{LT} is the proportion of left turns in the lane group, determined by dividing the turning volume by the total lane group volume.

$$f_{LT} = \frac{1}{1.0 + 0.05p_{LT}} \quad \text{[shared]} \qquad 2.36$$

Adjustment factors for pedestrians and bicycles in the intersection during protected turns use formulas specific to right turns and left turns. The procedure to determine the factors is as follows. (Refer to *HCM* Chap. 16, App. D for a detailed step-by-step analysis.)

step 1: Determine the average pedestrian occupancy of the intersection.

step 2: Determine the relevant conflict zone occupancy.

step 3: Determine the permitted phase pedestrian-bicycle adjustment factors for turning movements.

step 4: Determine the saturation flow adjustment factors for left- and right-turning movements.

The *right-turn pedestrian and bicycle adjustment factor*, f_{Rpb}, is found using Eq. 2.37. p_{RT} is the proportion of right turns in a lane group, which is found by dividing the turning volume by the total lane group volume. A_{pbT} is the *permitted phase adjustment*, and p_{RTA} is the proportion of right turns during a protected phase to total right turns made by the lane group during green intervals.

$$f_{Rpb} = 1.0 - p_{RT}(1 - A_{pbT})(1 - p_{RTA}) \quad \text{[}HCM\text{ Eq. D16-11]} \qquad 2.37$$

The *left-turn pedestrian and bicycle adjustment factor*, f_{Lpb}, is found using Eq. 2.38. p_{LT} is the proportion of left turns in a lane group, which is found by dividing the turning volume by the total lane group volume. A_{pbT} is the permitted phase adjustment, and p_{LTA} is the proportion of left turns during a protected phase to the total left turns made by the lane group during green intervals.

$$f_{Lpb} = 1.0 - p_{LT}(1 - A_{pbT})(1 - p_{LTA}) \quad \text{[}HCM\text{ Eq. D16-12]} \qquad 2.38$$

The permitted phase adjustment is calculated from Eq. 2.39. Use Eq. 2.39(a) if the number of turning lanes, $N_{L,t}$, is less than the number of receiving lanes, $N_{L,r}$. If there is an equal number of turning and receiving lanes, use Eq. 2.39(b). V_{ped} is the volume of pedestrians in ped/hr.

$$A_{pbT} = 1 - \frac{V_{ped}}{2000} \quad [N_{L,t} < N_{L,r}] \qquad 2.39(a)$$

$$A_{pbT} = 1 - \frac{(0.6)V_{ped}}{2000} \quad [N_{L,t} = N_{L,r}] \qquad 2.39(b)$$

A summary of all the adjustment factors for saturation flow rate is given in Table 2.25.

***Table 2.25** Adjustment Factor Summary for Saturation Flow Rate*[a,b]

factor	formula	definition of variables	notes
lane width	$f_w = 1 + \frac{W_{\text{m}} - 3.6 \text{ m}}{9 \text{ m}}$ $f_w = 1 + \frac{W_{\text{ft}} - 12 \text{ ft}}{30 \text{ ft}}$	W = lane width (ft or m)	$W \geq 8.0$ ft (2.4 m) If $W > 16$ ft (4.8 m), a two-lane analysis may be considered.
heavy vehicle	$f_{\text{HV}} = \frac{100\%}{100\% + \text{HV}_{\%}(E_T - 1)}$	$\text{HV}_{\%}$ = percentage of heavy vehicles for lane group volume	$E_T = 2.0$ pc/HV
grade	$f_g = 1 - \frac{G_{\%}}{200}$	$G_{\%}$ = percentage of grade on a lane group approach	$-6\% \leq G_{\%} \leq 10\%$ Negative is downhill.
parking	$f_p = \frac{N_L - 0.1 - \frac{(18 \text{ sec})N_m}{3600 \frac{\text{sec}}{\text{hr}}}}{N_L}$	N_L = number of lanes in lane group N_m = number of parking maneuvers per hour	$0 \leq N_m \leq 180$ $f_p \geq 0.05$ $f_p = 1.0$ for no parking
bus blocking	$f_{\text{bb}} = \frac{N_L - \frac{(14.4 \text{ sec})N_B}{3600 \frac{\text{sec}}{\text{hr}}}}{N_L}$	N_L = number of lanes in lane group N_B = number of buses stopping per hour	$0 \leq N_B \leq 250$ $f_{\text{bb}} \geq 0.05$
area type	$f_a = 0.9$ in CBDs $f_a = 1.0$ in all other areas		
lane utilization	$f_{\text{LU}} = \frac{v_g}{v_{g1} N_L}$	v_g = unadjusted demand flow rate for the lane group (vph) v_{g1} = unadjusted demand flow rate on the single lane in the lane group with the highest volume (vph) N_L = number of lanes in lane group	
left-turn	protected phasing: exclusive lane: $f_{\text{LT}} = 0.95$ shared lane: $f_{\text{LT}} = \frac{1}{1.0 + 0.05 p_{\text{LT}}}$	p_{LT} = proportion of left turns (LTs) in lane group	See App. 2.H for nonprotected phasing alternatives.
right-turn	exclusive lane: $f_{\text{RT}} = 0.85$ shared lane: $f_{\text{RT}} = 1.0 - 0.15 p_{\text{RT}}$ single lane: $f_{\text{RT}} = 1.0 - 0.135 p_{\text{RT}}$	p_{RT} = proportion of right turns (RTs) in lane group	$f_{\text{RT}} \geq 0.05$
pedestrian and bicycle	LT adjustment: $f_{\text{Lpb}} = 1.0 - p_{\text{LT}}(1 - A_{\text{pbT}}) \times (1 - p_{\text{LTA}})$ RT adjustment: $f_{\text{Rpb}} = 1.0 - p_{\text{RT}}(1 - A_{\text{pbT}}) \times (1 - p_{\text{RTA}})$	p_{LT} = proportion of LTs in lane group A_{pbT} = permitted phase adjustment p_{LTA} = proportion of LT protected green over total LT green p_{RT} = proportion of RTs in lane group p_{RTA} = proportion of RT protected green over total RT green	Refer to *HCM* Chap. 16, App. D for step-by-step procedure.

(Multiply ft by 0.305 to obtain m.)

[a]See *HCM* Exh. 10-12 for default values of base saturation flow rates and variables used to derive adjustment factors.

[b]The table contains formulas for all adjustment factors. However, for situations in which permitted phasing is involved, either by itself or in combination with protected phasing, separate tables are provided, as indicated in the "notes" column.

Intersection Flow Rate

The flow rate, v_p, can be determined for an intersection, an approach, or a single movement during the peak 15 min period using the total hourly volume and the peak hour factor, as shown in Eq. 2.40.

$$v_p = \frac{V}{\text{PHF}} \quad [\textit{HCM}\ \text{Eq. 16-3}] \qquad 2.40$$

Capacity and the *v/c* Ratio

The capacity of a lane group i, c_i, at a signalized intersection is determined from Eq. 2.41. s_i is the lane group's saturation flow rate, g_i is the the effective green time, and C is the cycle length. The ratio g/C is known as the *green ratio* and represents the proportion of available green time.[3]

$$c_i = s_i\left(\frac{g_i}{C}\right) \quad [\textit{HCM}\ \text{Eq. 16-6}] \qquad 2.41$$

The *volume-capacity ratio*, v/c, also referred to as the *degree of saturation*, compares the flow rate through an intersection to the capacity the intersection was designed to handle. In intersection analysis, the symbol X is used to refer to the v/c ratio. For any lane group i, X_i may be estimated using Eq. 2.42.

$$X_i = \left(\frac{v}{c}\right)_i = \frac{v_i}{s_i\left(\frac{g_i}{C}\right)} = \frac{v_iC}{s_ig_i} \quad [\textit{HCM}\ \text{Eq. 16-7}] \qquad 2.42$$

The value of X_i normally ranges from 0–1.0. Values greater than 1.0 indicate an excess of demand over capacity (i.e., oversaturation), and multiperiod analyses should be conducted. Basic changes such as altering the number or usage of lanes and increasing the cycle or phase lengths will improve intersection capacity.

Intersections without overlapping phases can be evaluated by considering critical lane groups for each phase. The critical v/c ratio for the intersection is determined from Eq. 2.43. $\sum(v/s)_{c,i}$ is the sum of the critical lane group ratios, C is the cycle length, and L is the lost time.

$$X_c = \sum\left(\frac{v}{s}\right)_{c,i}\left(\frac{C}{C-L}\right) \quad [\textit{HCM}\ \text{Eq. 16-8}] \qquad 2.43$$

The *critical lane group* for each approach is determined by calculating the v/c ratio for each lane group in the approach and choosing the lane group with the largest v/c ratio. One-way streets have only one lane group, which is always the critical lane group for the approach. The critical lane group determines the capacity of the intersection, as the critical lane group requires the largest proportion of green time. Therefore, a phase time that accommodates the most saturated approach will also accommodate all other approaches. For peak periods, evaluating X_c can help assess signal timing problems, as shown by the following if-then statements.

- If $X_i \le 1$ and $X_c \le 1$ for all lane groups, then the signal timings are adequate, but not necessarily optimal.
- If $X_i > 1$ and $X_c \le 1$ for all lane groups, then the cycle length is adequate, but the green allocation is incorrect.
- If X_i equals any value and $X_c > 1$, then the cycle length is too short.
- If $X_i > 1$ and $X_c > 1$ for several lane groups, then another, possibly simpler, phasing may yield improvement. If this fails, and $C > 150$ sec, then the signal's ability to handle traffic demand has been exhausted. Measures such as lane addition and left-turn-movement deletion are needed.

Cycle Length Estimation

Cycle lengths are selected by considering the type of signal controller that will be used, the phasing plan that will be adopted, and the allocation of green time among the various phases. State and local policies have considerable influence on the method used to select the cycle length. Some jurisdictions have policy restrictions on minimum or maximum cycle lengths or on subphases of cycles (e.g., the all-red time). In general, cycles that are too long will cause driver or pedestrian frustration, while those that are too short will not allow one set of movements to adequately clear the intersection before the next set of movements begins.

For an isolated intersection, an optimal phasing plan, and its cycle lengths, will equalize the saturation of the critical lane groups without unnecessarily delaying any of the lane groups. *HCM* Chap. 16, App. B has several methods of estimating the cycle length from an analytical standpoint. These methods, though intensive, can yield estimates of phase and cycle timings if adequate data are available during the design process.

A less intensive procedure to determine the optimal cycle length, C_o, is to use *Webster's formula*, Eq. 2.44. L_t is the total lost time during the cycle. The *total lost time* is the portion of the green signal that is unused and is the sum of the start-up and clearance times. The *start-up time* is the total time required for the initial drivers in the queue to react to the signal turning green and accelerate through the intersection. The *clearance time* is caused by drivers decelerating when the signal turns yellow. The total lost time usually amounts to 3–4 sec, and this range can be used as a default value if data do not exist or cannot be obtained. $\sum Y_i$ is the sum of the volume-capacity ratios of the critical movements.

$$C_o = \frac{1.5L_t + 5}{1 - \sum\limits_{\text{critical phases}} Y_i} = \frac{1.5L_t + 5}{1 - \sum\limits_{\text{critical phases}}\left(\frac{v}{c}\right)_i} \qquad 2.44$$

[3]Estimation of the green ratio for semiactuated or fully acutated signals is covered in more detail in *HCM* Chap. 16, App. B.

Traffic/Capacity Analysis

Intersection Delay

The *control delay*, d, measured in seconds per vehicle, is the delay caused by a signal forcing a lane group to reduce speed or stop, including deceleration on a signal approach, queuing time, and acceleration after a stop. The control delay is comprised of three types of delay: uniform delay, incremental delay, and initial queue delay. The control delay per vehicle is used to determine an intersection's LOS and is calculated using Eq. 2.45.

$$d = d_1(\text{PF}) + d_2 + d_3 \quad [\textit{HCM}\ \text{Eq. 16-9}] \qquad 2.45$$

d_1 is the *uniform control delay*, d_2 is the *incremental delay*, and d_3 is the *initial queue delay*. These three values are estimates, based on other data and results found for the intersection being evaluated. The methods and values used to calculate these three delay estimations varies based on the applicable initial queue delay situation.

The initial queue delay, d_3, is used to estimate delay caused by initial queues. An *initial queue* occurs when there is a queue from the previous cycle (i.e., an unmet demand) at the beginning of an analysis period. When an initial queue exists, vehicles experience an additional delay while they wait for the initial queue to leave the intersection. The presence of a queue at the beginning of the analysis period indicates that the previous cycle time was not sufficient to permit all of the demand to clear the intersection before the end of the cycle.

As outlined in the *HCM*, there are five situations used to estimate delay. These situations are determined by combinations of the initial queue at the beginning of the analysis period and the nature of the queue remaining at the end of the analysis period. Situation I and situation II occur when no initial queue is present at the beginning of the analysis period. For situation I, the queue is served and there is no remaining queue at the end of the period, indicating that the flow is undersaturated. For situation II, there is no initial queue, but a queue develops during and remains at the end of the analysis period. Situation II indicates the flow is oversaturated on one or more of the approach lane groups. For both situation I and situation II, d_3 is 0.

Situations III, IV, and V occur when an initial queue is present at the beginning of the analysis period. In situation III, the initial queue is fully served, and no queue remains at the end of the analysis period. In situation IV, a queue remains at the end of the analysis period, but it is smaller than the initial queue, while in situation V, the queue remains and is larger than the initial queue. The five situations are summarized in Table 2.26, and the applicable equations and default values for all five situations are shown in Table 2.27.

PF is the *progression adjustment factor*, which represents the quality of signal coordination at the intersection. A well-timed signal has a large proportion of vehicles arriving during a green phase. PF can be determined using Table 2.28 or calculated using Eq. 2.46. g is the effective green time, in seconds, for a lane group. The effective green time is either the green time used in pretimed signal control or the average lane group effective green time used for actuated control. C is the *intersection cycle time*. g/C is referred to as the green ratio and represents the proportion of green time available. f_{PA} is the *supplemental adjustment factor for platoons arriving during the green phase*, determined based on the arrival type (see Table 2.29) and Table 2.30. The quality of progression is quantified using a set of six *arrival types* (ATs), which are given in Table 2.29. Values of PF are limited to 1.0 for arrival types 3, 4, 5, and 6. Because progression primarily affects the uniform control delay, PF is applied only to the uniform control delay.

Table 2.26 *Condition of Initial and Final Queue for Delay Models*

situation	initial queue	queue at end of analysis period?
I	no	no
II	no	yes
III	yes	no
IV	yes	yes, but smaller
V	yes	yes, but larger

$$\text{PF} = \frac{(1-p)f_{\text{PA}}}{1-\frac{g}{C}} \quad [\textit{HCM}\ \text{Eq. 16-10}] \qquad 2.46$$

p is the *proportion of vehicles arriving during the green phase*, and g/C is the green time. R_p is the *platoon ratio*, found from Table 2.30 and the arrival type given by Table 2.29. The value of p cannot exceed 1.0.

$$p = R_p\left(\frac{g}{C}\right) \quad [\textit{HCM}\ \text{Exh. 16-12}] \qquad 2.47$$

The uniform control delay, d_1, assumes uniform arrivals, stable flow, and no initial queue at the beginning of a green phase. C is the cycle length, and g/C is the green ratio. X is the degree of saturation for the lane group. Uniform control delay is calculated from Eq. 2.48 for situations I and II.

$$d_1 = \frac{0.5C\left(1-\frac{g}{C}\right)^2}{1-\left(\min(1, X)\frac{g}{C}\right)} \quad [\textit{HCM}\ \text{Eq. 16-11}] \qquad 2.48$$

For situations III, IV, and V, the degree of saturation, X, is assumed to be 1.0 for the period of time the initial queue exists and equal to the actual degree of saturation for the remainder of the analysis period. In these situations, a time-weighted value of d_1 can be found from Eq. 2.49. d_s is the *saturated delay*, found using Eq. 2.48 for a degree of saturation of 1.0. d_u is the *undersaturated delay*, found using Eq. 2.48 and the actual value of the

Table 2.27 Summary of Delay Model Estimation by Situation

situation	X	Q_b	d_1	d_2	t_{unmet}	u	d_3	t_c
I	≤ 1.0	0	Eq. 2.48	Eq. 2.51	0	0	0	T
II	>1.0	0	Eq. 2.48	Eq. 2.51	0	0	0	TX
III	≤ 1.0	>0	Eq. 2.49	Eq. 2.51	Eq. 2.50	0	Eq. 2.52	T
IV	≤ 1.0	>0	Eq. 2.49	Eq. 2.51	T	Eq. 2.53	Eq. 2.52	Eq. 2.54
V	>1.0	>0	Eq. 2.49	Eq. 2.51	T	1	Eq. 2.52	Eq. 2.54

Highway Capacity Manual 2000. Copyright, National Academy of Sciences, Washington, D.C. Exhibit F16-4. Reproduced with permission of the Transportation Research Board.

Table 2.28 Progression Adjustment Factor, PF, for Uniform Delay Calculation[a,b]

green ratio, g/C	AT 1	AT 2	AT 3	AT 4	AT 5	AT 6
0.2	1.167	1.007	1.000	1.000	0.833	0.750
0.3	1.286	1.063	1.000	0.986	0.714	0.571
0.4	1.445	1.136	1.000	0.895	0.555	0.333
0.5	1.667	1.240	1.000	0.767	0.333	0
0.6	2.001	1.395	1.000	0.576	0	0
0.7	2.556	1.653	1.000	0.256	0	0

[a]Tabulation is based on default values of f_{PA} and R_p.
[b]PF may not exceed 1.0 for AT 3 through AT 6.

Highway Capacity Manual 2000. Copyright, National Academy of Sciences, Washington, D.C. Exhibit 16-12. Reproduced with permission of the Transportation Research Board.

degree of saturation. T is the *analysis period duration* in hours. t_{unmet} is the time required for the unmet demand to clear the intersection, found from Eq. 2.50. For situations IV and V, T and t_{unmet} are equal, as shown in Table 2.27, so d_1 will be equal to d_s. Equation 2.48 is not applicable to intersections with compound left-turn protection. Calculations of uniform control delay for these intersections are covered in *HCM* Chap. 16, App. F.

$$d_1 = d_s\left(\frac{t_{\text{unmet}}}{T}\right) + d_u(\text{PF})\left(\frac{T - t_{\text{unmet}}}{T}\right)$$

[*HCM* Eq. F16-5] *2.49*

$$t_{\text{unmet}} = \min\left(T, \frac{Q_b}{c\left(1 - \min(1, X)\right)}\right)$$

[*HCM* Eq. F16-2] *2.50*

In Eq. 2.50, T is the analysis period duration, X is the degree of saturation, c is the lane group capacity, and Q_b is the quantity of vehicles in the initial queue at the beginning of the analysis period.

The incremental delay, d_2, is used to estimate delay caused by nonuniform arrivals (random delay) and delay caused by prolonged periods of oversaturation (oversaturation delay) and is found using Eq. 2.51 for all situations. Equation 2.51 assumes that there is no

Table 2.29 Arrival Types

arrival type, AT	description
1	Dense platoon containing over 80% of the lane group volume, arriving at the start of the red phase. This AT is representative of network links that may experience very poor progression quality as a result of conditions such as overall network signal optimization.
2	Moderately dense platoon arriving in the middle of the red phase, or dispersed platoon containing 40%–80% of the lane group volume arriving throughout the red phase. This AT is representative of unfavorable progression on a two-way street.
3	Random arrivals in which the main platoon contains less than 40% of the lane group volume. This AT is representative of operations at isolated and noninterconnected signalized intersections characterized by highly dispersed platoons. It may also be used to represent coordinated operation in which the benefits of progression are minimal.
4	Moderately dense platoon arriving in the middle of the green phase or dispersed platoon containing 40%–80% of the lane group volume, arriving throughout the green phase. This AT is representative of favorable progression on a two-way street.
5	Dense to moderately dense platoon containing over 80% of the lane group volume, arriving at the start of the green phase. This AT is representative of highly favorable progression quality, which may occur on routes with low to moderate side-street entries and which received high-priority treatment in the signal timing plan.
6	This AT is reserved for exceptional progression quality on routes with near-ideal progression characteristics. It is representative of very dense platoons progressing over a number of closely spaced intersections with minimal or negligible side-street entries.

Highway Capacity Manual 2000. Copyright, National Academy of Sciences, Washington, D.C. Exhibit 16-4. Reproduced with permission of the Transportation Research Board.

Traffic/Capacity Analysis

Table 2.30 Relationship Between Arrival Type and Platoon Ratio

arrival type, AT	range of platoon ratio, R_p	default value of R_p	default value of f_{PA}	progression quality
1	≤ 0.50	0.333	1.00	very poor
2	> 0.50 and ≤ 0.85	0.667	0.93	unfavorable
3	> 0.85 and ≤ 1.15	1.000	1.00	random arrivals
4	> 1.15 and ≤ 1.50	1.333	1.15	favorable
5	> 1.50 and ≤ 2.00	1.667	1.00	highly favorable
6	> 2.00	2.000	1.00	exceptional

Highway Capacity Manual 2000. Copyright, National Academy of Sciences, Washington, D.C. Exhibit 16-11 and Exhibit 16-12. Reproduced with permission of the Transportation Research Board.

unmet demand that causes an initial queue at the start of the analysis period.

$$d_2 = 900T\left(X - 1 + \sqrt{(X-1)^2 + \frac{8kIX}{cT}}\right) \quad [HCM \text{ Eq. 16-12}] \qquad 2.51$$

T is the analysis period duration and X is the degree of saturation. I is the *upstream filtering/metering adjustment factor*, which is 1.0 for isolated intersections. Incremental delay calculations for nonisolated intersections (i.e., intersections affected by upstream signals) are beyond the scope of this book and are covered in depth in *HCM* Chap. 15.

k is the *incremental delay calibration factor*. The value of k is dependent on the controller type, the degree of saturation for the lane group, and the unit extension of the controller, and can be determined from Table 2.31. The *unit extension* for a given actuated signal is the minimum gap between vehicles on the approach needed to trigger the end of the approach's green time. For pretimed signals, the incremental delay calibration factor is always 0.50, because pretimed signals do not adjust green time based on traffic demand, and therefore can neither reduce nor increase incremental delay.

For situations I and II, the initial queue delay, d_3, is 0. For situations III, IV, and V, the initial queue delay is found from Eq. 2.52. Q_b is the quantity of vehicles in the initial queue at the beginning of the analysis period, T is the duration of the analysis period, c is the lane group capacity, and t_{unmet} is the time required for the unmet demand to clear the intersection. u is the *delay parameter*, which is equal to 0 for situations I, II, and III and is found from Eq. 2.53 for situations IV and V.

$$d_3 = \frac{1800 Q_b (1+u) t_{\text{unmet}}}{cT} \quad [HCM \text{ Eq. F16-1}] \qquad 2.52$$

$$u = 1 - \frac{cT}{Q_b}\big(1 - \min(1, X)\big) \quad [HCM \text{ Eq. F16-3}] \qquad 2.53$$

For situations IV and IV, a queue remains at the end of the analysis period. Equation 2.54 can be used to estimate the *initial queue clearing time*, t_c, which is the time required for the last vehicle arriving during analysis period T to clear the intersection. t_c is measured from the beginning of the analysis period. While the initial queue clearing time is not necessary to calculate the control delay, it may be useful for signal timing adjustments.

$$t_c = \max\left(T, \frac{Q_b}{c} + TX\right) \quad [HCM \text{ Eq. F16-4}] \qquad 2.54$$

Table 2.31 k-values Accounting for Controller Type

	degree of saturation, X^a					
unit extension (sec)	≤ 0.50	0.60	0.70	0.80	0.90	≥ 1.0
≤ 2.0	0.04	0.13	0.22	0.32	0.41	0.50
2.5	0.08	0.16	0.25	0.33	0.42	0.50
3.0	0.11	0.19	0.27	0.34	0.42	0.50
3.5	0.13	0.20	0.28	0.35	0.43	0.50
4.0	0.15	0.22	0.29	0.36	0.43	0.50
4.5	0.19	0.25	0.31	0.38	0.44	0.50
5.0^b	0.23	0.28	0.34	0.39	0.45	0.50
pretimed or nonactuated movement	0.50	0.50	0.50	0.50	0.50	0.50

[a]For a given unit extension and its $k_{\min}$ value at $X = 0.5$: $k = (1 - 2k_{\min})(X - 0.5) + k_{\min}$, $k \geq k_{\min}$, and $k \leq 0.5$.
[b]For unit extension > 5.0 sec, extrapolate to find k, keeping $k \leq 0.5$.

Highway Capacity Manual 2000. Copyright, National Academy of Sciences, Washington, D.C. Exhibit 16-13. Reproduced with permission of the Transportation Research Board.

Aggregated Delay

Lane groups for an approach can be aggregated to provide an estimation of control delay for the entire approach or for the intersection as a whole. Lane group aggregation uses weighted averages, where the lane group delays are weighted by the adjusted flows in the lane groups. Use Eq. 2.55 for an individual approach, where d_A is the delay for approach A, d_i is the delay for lane group i, and v_i is the adjusted flow for lane group i.

$$d_A = \frac{\sum d_i v_i}{\sum v_i} \quad \text{[HCM Eq. 16-13]} \qquad 2.55$$

The average delay per vehicle for the entire intersection, d_I, can be estimated using Eq. 2.56.

$$d_I = \frac{\sum d_A v_A}{\sum v_A} \quad \text{[HCM Eq. 16-14]} \qquad 2.56$$

Intersection LOS

Intersection LOS measures the delay a vehicle experiences while waiting at an intersection. The LOS is determined by the average control delay per vehicle. Table 2.32 gives the intersection LOS designations according to control delay.

***Table 2.32** LOS and Control Delay*

LOS	control delay (sec/veh)
A	≤ 10
B	>10–20
C	>20–35
D	>35–55
E	>55–80
F	>80

Highway Capacity Manual 2000. Copyright, National Academy of Sciences, Washington, D.C. Exhibit 16-2. Reproduced with permission of the Transportation Research Board.

Example 2.7

A traffic signal is being designed for the intersection of Sierra Blvd., running east-west, and Tarentum St., running north-south. Sierra Blvd. has two 11 ft wide lanes running in each direction and no turning lanes. Tarentum St. has two 15 ft wide lanes running in the southbound direction and no turning lanes. The area where the intersection is located shows no characteristics of a central business district, and nearby intersections have no effect on the traffic flow.

The planned signal will have a two-phase, pretimed cycle, as shown in the illustration.

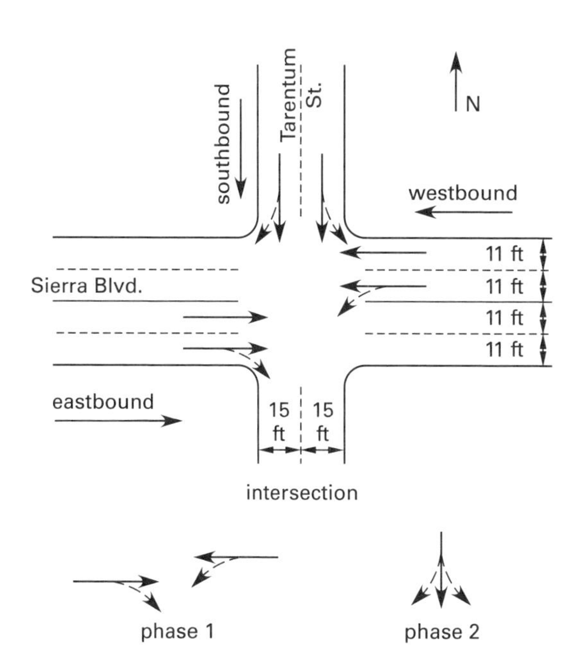

To ensure the design will improve traffic conditions at the intersection, a traffic study and a one-hour intersection analysis are performed. The traffic study gives the following data and results.

	traffic volumes (vph)		
	left-turning	through	right-turning
Sierra Blvd. eastbound	0	470	18
Sierra Blvd. westbound	25	630	0
Tarentum St. southbound	15	160	25

- level terrain, no grades
- arrival type: 3 for all approaches
- green time for Sierra Blvd. westbound and eastbound: 46 sec
- green time for Tarentum St. southbound: 26 sec
- amber time on all approaches: 4 sec
- movement lost time: 4 sec
- PHF: 0.90
- trucks in traffic flows: 5% in all movements on Sierra Blvd.; 8% in all movements on Tarentum St.
- no parking on any approach
- no pedestrians on any approach
- no bicycles on any approach
- no bus stops on any approach
- no initial queue

What will be the LOS at the intersection of Sierra Blvd. and Tarentum St. once the traffic signal is installed?

Solution

To find the LOS for the intersection, the average intersection delay must be found. Calculation of average intersection delay requires the control delay for the critical lane group of both Sierra Blvd. and Tarentum St.

Find the saturation flow rates for all approaches. To find the saturation flow rate, first find the proportion of right- and left-turning volumes for each street. The proportions are found by dividing the turning volumes by the related total lane group volumes.

For Sierra Blvd. eastbound, the proportion of right turns, p_{RT}, is

$$p_{\text{RT}} = \frac{18\ \frac{\text{veh}}{\text{hr}}}{0\ \frac{\text{veh}}{\text{hr}} + 470\ \frac{\text{veh}}{\text{hr}} + 18\ \frac{\text{veh}}{\text{hr}}} = 0.0369$$

For Sierra Blvd. westbound, the proportion of left turns, p_{LT}, is

$$p_{\text{LT}} = \frac{25\ \frac{\text{veh}}{\text{hr}}}{25\ \frac{\text{veh}}{\text{hr}} + 630\ \frac{\text{veh}}{\text{hr}} + 0\ \frac{\text{veh}}{\text{hr}}} = 0.0382$$

For Tarentum St. southbound, the proportion of right turns, p_{RT}, is

$$p_{\text{RT}} = \frac{25\ \frac{\text{veh}}{\text{hr}}}{15\ \frac{\text{veh}}{\text{hr}} + 160\ \frac{\text{veh}}{\text{hr}} + 25\ \frac{\text{veh}}{\text{hr}}} = 0.125$$

For Tarentum St. southbound, the proportion of left turns, p_{LT}, is

$$p_{\text{LT}} = \frac{15\ \frac{\text{veh}}{\text{hr}}}{15\ \frac{\text{veh}}{\text{hr}} + 160\ \frac{\text{veh}}{\text{hr}} + 25\ \frac{\text{veh}}{\text{hr}}} = 0.0750$$

Find the right-turn adjustment factor from Eq. 2.34, as all approaches are shared-lane.

For Sierra Blvd. eastbound,

$$f_{\text{RT}} = 1.0 - 0.15p_{\text{RT}} = 1.0 - (0.15)(0.0369) = 0.994$$

For Tarentum St. southbound,

$$f_{\text{RT}} = 1.0 - 0.15p_{\text{RT}} = 1.0 - (0.15)(0.125) = 0.981$$

Find the left-turn adjustment factor from Eq. 2.36.

For Sierra Blvd. westbound,

$$f_{\text{LT}} = \frac{1}{1.0 + 0.05p_{\text{LT}}} = \frac{1}{1 + (0.05)(0.0382)} = 0.998$$

Because there are no left turns permitted on Sierra Blvd. eastbound, f_{LT} is 1.0.

For Tarentum St. southbound,

$$f_{\text{LT}} = \frac{1}{1 + 0.05p_{\text{LT}}} = \frac{1}{1 + (0.05)(0.0750)} = 0.996$$

Find the lane width adjustment factor using Eq. 2.28(b).

For Sierra Blvd.,

$$f_w = 1 + \frac{W_{\text{ft}} - 12\ \text{ft}}{30\ \text{ft}} = 1 + \frac{11\ \text{ft} - 12\ \text{ft}}{30\ \text{ft}} = 0.967$$

For Tarentum St.,

$$f_w = 1 + \frac{W_{\text{ft}} - 12\ \text{ft}}{30\ \text{ft}} = 1 + \frac{15\ \text{ft} - 12\ \text{ft}}{30\ \text{ft}} = 1.10$$

Find the heavy vehicle adjustment factor using Eq. 2.29. The passenger car equivalent, E_T, is 2.0 pc/HV.

For Sierra Blvd.,

$$f_{\text{HV}} = \frac{100\%}{100\% + \text{HV}_{\%}(E_T - 1)} = \frac{100\%}{100\% + (5\%)(2.0 - 1)} = 0.952$$

For Tarentum St.,

$$f_{\text{HV}} = \frac{100\%}{100\% + \text{HV}_{\%}(E_T - 1)} = \frac{100\%}{100\% + (8\%)(2.0 - 1)} = 0.926$$

Find the grade adjustment factor from Eq. 2.30.

$$f_g = 1 - \frac{G_{\%}}{200} = 1 - \frac{0\%}{200} = 1.0$$

The parking adjustment factor, f_p, is 1.0 since there are no parking maneuvers on any approach.

The bus blocking adjustment factor, f_{bb}, is 1.0 since there are no bus stops at the intersection.

The area type adjustment factor, f_a, is 1.0, since there is no CBD-like activity near the intersection.

Since no field data are available for the lane utilization adjustment factor, use default values. From Table 2.24, f_{LU} is 0.952 for all approaches.

Since there are no pedestrians or bicycles, the default value of 1.0 is used for the right-turn adjustment factor, f_{Rpb}, and left-turn adjustment factor, f_{Lpb}, for pedestrians and bicycles.

Calculate the saturation flow from Eq. 2.27.

For Sierra Blvd. eastbound,

$$\begin{aligned} s_{\text{Sierra EB}} &= s_o N_L f_w f_{\text{HV}} f_g f_p f_{\text{bb}} f_a f_{\text{LU}} f_{\text{RT}} f_{\text{LT}} f_{\text{Rpb}} f_{\text{Lpb}} \\ &= \left(1900\ \frac{\text{pc}}{\text{hr-ln}}\right)(2\text{ lanes})(0.967)(0.952)(1.0)(1.0) \\ &\quad \times (1.0)(1.0)(0.952)(0.994)(1.0)(1.0)(1.0) \\ &= 3310\text{ vph} \end{aligned}$$

For Sierra Blvd. westbound,

$$\begin{aligned} s_{\text{Sierra WB}} &= s_o N_L f_w f_{\text{HV}} f_g f_p f_{\text{bb}} f_a f_{\text{LU}} f_{\text{RT}} f_{\text{LT}} f_{\text{Rpb}} f_{\text{Lpb}} \\ &= \left(1900\ \frac{\text{pc}}{\text{hr-ln}}\right)(2\text{ lanes})(0.967)(0.952)(1.0)(1.0) \\ &\quad \times (1.0)(1.0)(0.952)(1.0)(0.998)(1.0)(1.0) \\ &= 3324\text{ vph} \end{aligned}$$

For Tarentum St. southbound,

$$\begin{aligned} s_{\text{Tarentum SB}} &= s_o N_L f_w f_{\text{HV}} f_g f_p f_{\text{bb}} f_a f_{\text{LU}} f_{\text{RT}} f_{\text{LT}} f_{\text{Rpb}} f_{\text{Lpb}} \\ &= \left(1900\ \frac{\text{pc}}{\text{hr-ln}}\right)(2\text{ lanes})(1.10)(0.926)(1.0) \\ &\quad \times (1.0)(1.0)(1.0)(0.952)(0.981)(0.996) \\ &\quad \times (1.0)(1.0) \\ &= 3600\text{ vph} \end{aligned}$$

Determine the total cycle length by adding the green time for both phases, the amber time for both phases, and the movement lost time.

$$\begin{aligned} C &= 46\text{ sec} + 26\text{ sec} + 4\text{ sec} + 4\text{ sec} \\ &= 80\text{ sec} \end{aligned}$$

Find the peak flow rate for each approach using Eq. 2.40.

$$\begin{aligned} v_{p,\text{Sierra EB}} &= \frac{V}{\text{PHF}} = \frac{0\ \frac{\text{veh}}{\text{hr}} + 470\ \frac{\text{veh}}{\text{hr}} + 18\ \frac{\text{veh}}{\text{hr}}}{0.90} \\ &= 542\text{ vph} \end{aligned}$$

$$\begin{aligned} v_{p,\text{Sierra WB}} &= \frac{V}{\text{PHF}} = \frac{25\ \frac{\text{veh}}{\text{hr}} + 630\ \frac{\text{veh}}{\text{hr}} + 0\ \frac{\text{veh}}{\text{hr}}}{0.90} \\ &= 728\text{ vph} \end{aligned}$$

$$\begin{aligned} v_{p,\text{Tarentum SB}} &= \frac{V}{\text{PHF}} = \frac{15\ \frac{\text{veh}}{\text{hr}} + 160\ \frac{\text{veh}}{\text{hr}} + 25\ \frac{\text{veh}}{\text{hr}}}{0.90} \\ &= 222\text{ vph} \end{aligned}$$

Find the lane group capacity for each approach using Eq. 2.41.

For Sierra Blvd. eastbound,

$$\begin{aligned} c_{\text{Sierra EB}} &= s_{\text{Sierra EB}}\left(\frac{g_{\text{Sierra EB}}}{C}\right) = \left(3310\ \frac{\text{veh}}{\text{hr}}\right)\left(\frac{46\text{ sec}}{80\text{ sec}}\right) \\ &= 1903\text{ vph} \end{aligned}$$

For Sierra Blvd. westbound,

$$\begin{aligned} c_{\text{Sierra WB}} &= s_{\text{Sierra WB}}\left(\frac{g_{\text{Sierra WB}}}{C}\right) = \left(3324\ \frac{\text{veh}}{\text{hr}}\right)\left(\frac{46\text{ sec}}{80\text{ sec}}\right) \\ &= 1911\text{ vph} \end{aligned}$$

For Tarentum St. southbound,

$$\begin{aligned} c_{\text{Tarentum SB}} &= s_{\text{Tarentum SB}}\left(\frac{g_{\text{Tarentum SB}}}{C}\right) \\ &= \left(3600\ \frac{\text{veh}}{\text{hr}}\right)\left(\frac{26\text{ sec}}{80\text{ sec}}\right) \\ &= 1170\text{ vph} \end{aligned}$$

From Eq. 2.42, the v/c ratios are

$$X_{\text{Sierra EB}} = \frac{v}{c} = \frac{542\ \frac{\text{veh}}{\text{hr}}}{1903\ \frac{\text{veh}}{\text{hr}}} = 0.285$$

$$X_{\text{Sierra WB}} = \frac{v}{c} = \frac{728\ \frac{\text{veh}}{\text{hr}}}{1911\ \frac{\text{veh}}{\text{hr}}} = 0.381$$

$$X_{\text{Tarentum SB}} = \frac{v}{c} = \frac{222\ \frac{\text{veh}}{\text{hr}}}{1170\ \frac{\text{veh}}{\text{hr}}} = 0.190$$

Find the critical lane group for each phase.

For phase 1, the westbound v/c ratio is greater than the eastbound ratio. Therefore, Sierra Blvd. westbound is the critical lane group.

For phase 2, Tarentum St. is one-way, so the only lane group is the critical lane group by default.

The lost time, L, is the sum of the amber time and the movement lost time.

$$L = 4\text{ sec} + 4\text{ sec} = 8\text{ sec}$$

The critical volume-capacity ratio for the intersection is calculated using Eq. 2.43 and the sum of the critical lane group v/c ratios.

$$\begin{aligned} X_c &= \sum\left(\frac{v}{s}\right)_{c,i}\left(\frac{C}{C-L}\right) \\ &= (0.381 + 0.190)\left(\frac{80\text{ sec}}{80\text{ sec} - 8\text{ sec}}\right) \\ &= 0.634 \end{aligned}$$

The uniform control delay, d_1, is calculated for the critical lane groups using Eq. 2.48.

For Sierra Blvd.,

$$d_1 = \frac{0.5C\left(1-\frac{g}{C}\right)^2}{1-\left(\min(1,\,X)\frac{g}{C}\right)} = \frac{(0.5)(80\text{ sec})\left(1-\frac{46\text{ sec}}{80\text{ sec}}\right)^2}{1-(0.381)\left(\frac{46\text{ sec}}{80\text{ sec}}\right)}$$
$$= 9.25\text{ sec/veh}$$

For Tarentum St.,

$$d_1 = \frac{0.5C\left(1-\frac{g}{C}\right)^2}{1-\left(\min(1,\,X)\frac{g}{C}\right)} = \frac{(0.5)(80\text{ sec})\left(1-\frac{26\text{ sec}}{80\text{ sec}}\right)^2}{1-(0.190)\left(\frac{26\text{ sec}}{80\text{ sec}}\right)}$$
$$= 19.4\text{ sec/veh}$$

The incremental delay, d_2, is calculated using Eq. 2.51. From Table 2.31, the default value of k for a pretimed signal is 0.50. Use a value of 1.0 for I.

For Sierra Blvd.,

$$d_2 = 900T\left(X-1+\sqrt{(X-1)^2+\frac{8kIX}{cT}}\right)$$
$$= (900)(1.0\text{ hr})$$
$$\times\left((0.381-1)+\sqrt{\begin{array}{l}(0.381-1)^2\\ +\dfrac{(8)(0.50)(1.0)(0.381)}{\left(1911\ \frac{\text{veh}}{\text{hr}}\right)(1.0\text{ hr})}\end{array}}\right)$$
$$= 0.579\text{ sec/veh}$$

For Tarentum St.,

$$d_2 = 900T\left(X-1+\sqrt{(X-1)^2+\frac{8kIX}{cT}}\right)$$
$$= (900)(1.0\text{ hr})$$
$$\times\left((0.190-1)+\sqrt{\begin{array}{l}(0.190-1)^2\\ +\dfrac{(8)(0.50)(1.0)(0.190)}{\left(1170\ \frac{\text{veh}}{\text{hr}}\right)(1.0\text{ hr})}\end{array}}\right)$$
$$= 0.361\text{ sec/veh}$$

Since there is no initial queue, the initial queue delay, d_3, is zero.

Find the control delay, d, for each critical lane group using Eq. 2.45. From Table 2.28, the progression adjustment factor, PF, is 1.000 for arrival type 3.

For Sierra Blvd.,

$$d = d_1(\text{PF}) + d_2 + d_3$$
$$= \left(9.25\ \frac{\text{sec}}{\text{veh}}\right)(1.000) + 0.579\ \frac{\text{sec}}{\text{veh}} + 0\ \frac{\text{sec}}{\text{veh}}$$
$$= 9.82\text{ sec/veh}$$

For Tarentum St.,

$$d = d_1(\text{PF}) + d_2 + d_3$$
$$= \left(19.4\ \frac{\text{sec}}{\text{veh}}\right)(1.000) + 0.361\ \frac{\text{sec}}{\text{veh}} + 0\ \frac{\text{sec}}{\text{veh}}$$
$$= 19.8\text{ sec/veh}$$

The average intersection delay per vehicle is found using Eq. 2.56.

$$d_I = \frac{\sum d_A v_A}{\sum v_A}$$
$$= \frac{\left(9.82\ \frac{\text{sec}}{\text{veh}}\right)\left(728\ \frac{\text{veh}}{\text{hr}}\right) + \left(19.8\ \frac{\text{sec}}{\text{veh}}\right)\left(222\ \frac{\text{veh}}{\text{hr}}\right)}{728\ \frac{\text{veh}}{\text{hr}} + 222\ \frac{\text{veh}}{\text{hr}}}$$
$$= 12.2\text{ sec/veh}$$

Referring to Table 2.32, the overall LOS for the intersection is B.

6. PARKING FACILITIES

Introduction

Parking is an integral part of any transportation system. Parking provides storage for vehicles, and it serves as a place to change transportation modes (e.g., park and ride lots). *Parking facilities* are structures containing multiple parking spaces (e.g., parking lots or parking garages). Parking facilities are often difficult to design and costly to construct, and design of parking facilities is affected by many factors unique to these facilities.

Requirements for parking design are dictated both by local codes and ordinances and by more universal codes, such as the *International Building Code* (IBC). The Americans with Disabilities Act (ADA) *Standards for Accessible Design* governs design of *accessible parking spaces* (more commonly known as *handicapped spaces*), which are an essential design consideration for any parking facility. Other common design requirements are maneuvering room, total capacity, stall size, vertical clearance, and site dimensions.

Users have varying parking needs that must be considered and accommodated when designing a parking facility. For example, commuters who frequently use a given facility will adapt to the facility's design constraints better than tourists or shoppers who use the facility infrequently or only once. If a facility will primarily be used by drivers who are not familiar with the facility, measures such as additional signage may be needed to better accommodate the needs of unfamiliar users. Parking is sometimes segregated by user type (e.g., separate employee parking at a shopping mall) so the design can be tailored more easily to meet the needs of different types of users.

Because parking facilities often have large capital and maintenance costs, user patterns must be studied to ensure that the facility will be adequately used. Like roadway design, parking facility design is concerned with the volume of drivers and often uses the same peak hour analysis method. Trip generation for parking facilities is covered in more detail in the ITE's *Trip Generation*.

Considerations unique to parking facilities include peak arrival rates, peak departure rates, and turnover rates. Because the exact volume of parking facilities is highly variable, peak rates are measured in terms of the percentage of the capacity of a parking facility that arrives or departs over the course of a given hour (e.g., a peak arrival rate of 70 %/hr for a facility with a capacity of 100 vehicles means that 70 vehicles arrive at the facility and park during the peak hour). The *peak arrival rate* is the percentage of the total facility capacity that arrives and parks during the peak hour, and the *peak departure rate* is the percentage of the capacity that leaves the facility during the peak hour. Peak arrival and departure rates vary according to the facility type, as shown in Table 2.33.

Turnover is based on the duration for which vehicles park and varies by the user types and trip purposes associated with the facility. Parking at residential or office parking facilities tends to be long-term (i.e., 6–8 hr), and retail parking tends to be short-term (i.e., less than 3 hr). A high turnover rate means that the parking space can be shared among multiple users throughout the day. Areas with high parking demand, such as busy downtown areas, often promote high turnover by limiting the parking duration allowed at nearby parking facilities.

Capacity of Internal Layouts

The capacity of a parking facility is the based on the dimensions of the *parking stalls* (commonly referred to as *parking spaces*). Parking spaces are typically between 8.5 ft and 9.5 ft (2.6 m and 2.9 m) wide and between 17 ft and 18 ft (5.2 m and 5.5 m) long. These dimensions are based on *design vehicles*, which are determined from typical vehicle specifications for a wide range of U.S. vehicles. The American Association of State Highway and Transportation Officials (AASHTO) uses a passenger car design vehicle 7 ft (2.1 m) wide and 19 ft (5.8 m) long. (Design vehicles are covered in more detail in Chap. 4.) However, the selection of stall dimensions should consider land use, turnover, vehicle size, and user needs. For example, the use of shopping carts in a parking facility with high turnover may necessitate greater stall widths to allow quick loading and unloading of packages.

Designs for a parking facility may include the dimensions of a *parking module*, a section of a parking facility that includes one or two rows of parking spaces, as well as an aisle where vehicles can maneuver into and out of spaces. When module dimensions are given instead of parking dimensions, the area required for two facing spaces (the width of a parking space multiplied by the

Table 2.33 *Typical Peak Hour Parking Rates*

	percentage of total facility capacity arriving or departing in one hour[a]			
	peak a.m. hour		peak p.m. hour	
land use	arrival (%/hr)	departure (%/hr)	arrival (%/hr)	departure (%/hr)
residential	5–10	30–50	30–50	10–30
hotel/motel	30–50	50–80	30–60	10–30
office	40–70	5–15	5–20	40–70
general retail/restaurant	20–50	30–60	30–60	30–60
convenience retail/banking	80–150	80–150	80–150	80–150
central business district[b]	20–60	10–60	10–50	20–60
medical office	40–60	50–80	60–80	60–90
hospital				
visitor spaces	30–40	40–50	40–60	50–75
employee spaces	60–75	5–10	10–15	60–75
airport				
short-term (0–3 hr)	50–75	80–100	90–100	90–100
mid-term (3–24 hr)	10–30	5–10	10–30	10–30
long-term (24+ hr)	5–10	5–10	5–10	5–10
special event[c]	80–100	85–200		

[a]As a general rule, the larger the facility and/or the more diverse the tenants of the generating land uses, the lower the peak hour volume as a percentage of the capacity.
[b]It is generally more accurate to determine what fraction of the spaces are allotted to retail, office, and other uses than to use these typical rates.
[c]Assuming 100% of the capacity leaves over the course of 30 min. The equivalent volume in a full hour is 200% of capacity.

Reprinted with permission from Anthony P. Chrest, Mary S. Smith, and Sam Bhuyan, *Parking Structures: Planning Design, Construction, Maintenance, and Repair*, copyright © 2001, by Springer.

length of a parking module) can be used to calculate capacity; this will ensure that the capacity includes an allowance for the aisles between spaces.

Once the parking stall and module dimensions are selected, the capacity of the parking facility can be determined from the total available area. Before the capacity is estimated, a portion of the total area of the facility must be subtracted to allow for end row turning, exit and entrance ramps, turning loops between floors of a multistory facility, ticket booths and gates at ticketed facilities, and any other similar components of a facility that subtract from the total space available to park. This allowance can either be calculated as an estimate of the area required for all nonparking components of the facility (e.g., if a ticket booth is known to have an area of 1500 ft^2 (1350 m^2), 1500 ft^2 (1350 m^2) can be subtracted directly from the available area) or estimated as a percentage of the total available area (e.g., 5% of the total facility area will be required for nonparking construction). Once the available area is adjusted to account for these parts of the facility, the remaining available area is divided by the area of a single parking space to find the capacity of the parking facility. This procedure only produces an estimate, but is much faster than preparing a detailed layout.

Example 2.8

A parking facility is to be built on a 120 ft wide × 180 ft long rectangular site. The dimensions of a parking stall are designed as shown in the following illustration. Allowing 5% of the available area for maneuvering space beyond the aisles and parking spaces, how many parking spaces can the parking facility accommodate?

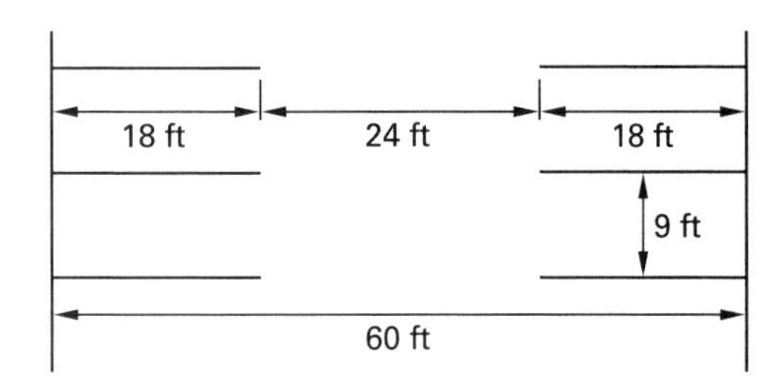

Solution

Parking spaces are 9 ft wide, and one parking module is 60 ft long, so the area required for two facing parking spaces and an aisle is

$$A = bW = (60 \text{ ft})(9 \text{ ft}) = 540 \text{ ft}^2$$

Determine the area of the site.

$$A = bW = (180 \text{ ft})(120 \text{ ft}) = 21{,}600 \text{ ft}^2$$

Subtract the maneuvering areas from the site area to determine the total area available for parking.

$$A_{\text{parking}} = 21{,}600 \text{ ft}^2 - (0.05)(21{,}600 \text{ ft}^2) = 20{,}520 \text{ ft}^2$$

The number of parking spaces available is

$$N_{\text{parking spaces}} = (2)\left(\frac{20{,}520 \text{ ft}^2}{540 \text{ ft}^2}\right) = 76$$

LOS Determination

LOS for parking facilities is based on the delay when attempting to use a given facility, measured in seconds per vehicle. User delay can be caused by a variety of factors that are dependent on actual site conditions and cannot be easily modeled. When a parking facility is designed, LOS is often approached in terms of the delays users will tolerate in the facility rather than designing for an acceptable level of delay.

For simplicity, delay tolerance is examined from the standpoint of users who are visitors to a given facility and users who regularly use the facility (e.g., employees). Table 2.34 shows the levels of service different kinds of users will accept for different design considerations. In general, visitors are more tolerant of longer travel distances and larger numbers of turns, while employees and other frequent users are more tolerant of lower turning radii and lower flow capacities.

Table 2.34 *Level of Service Tolerance for Parking Facilities*

design consideration	chief factor	acceptable LOS: A B	acceptable LOS: C D
turning radii, ramp slopes, etc.	freedom to maneuver	visitors	employees
travel distance, number of turns, etc.	travel time	employees	visitors
geometrics	freedom to maneuver	visitors	employees
flow capacity	v/c ratio	visitors	employees
entries/exits	average wait	employees	visitors

Reprinted with permission from Anthony P. Chrest, Mary S. Smith, and Sam Bhuyan, *Parking Structures: Planning Design, Construction, Maintenance, and Repair*, copyright © 2001, by Springer.

Table 2.35 shows sample design parameters that are commonly found in facilities meeting LOS A through LOS D. LOS E is generally considered unacceptable for parking facilities, as near-maximum flows will not typically be tolerated by users. LOS D is only minimally acceptable but will tend to be tolerated in high-traffic areas like CBDs or parking facilities for major events. Designers tend to design for LOS B or LOS C when designing parking facilities for suburban locations.

Table 2.35 *Typical Recommended Design Parameters to Meet Level of Service Criteria*

design standard	LOS A	LOS B	LOS C	LOS D
turning radius, ft[a]	42	36	30	24
driving lane width[b]				
straight				
single lane, ft[c]	12.5	12	11.5	11
multiple lane, ft	11.5	11	10.5	10
24 ft radius				
single lane, ft[d]	14.5	14	13.5	13
multiple lane, ft[e]	14	13.5	13	12.5
clearance from lane, ft[f]	3	2.5	2	1.5
turning bay[g]				
one-way, ft	20.5	19	17.5	16
two-way, concentric, ft[h]	33	31	29	27
two-way, nonconcentric, ft	36	34	32	30
circular helix				
single-threaded[i]				
outside diameter, ft[j]	99	86	73	60
inside diameter, ft[k]	52	42	32	22
slope[l]	4.2%	5.0%	6.1%	7.8%
double-threaded				
outside diameter, ft[j]	112	99	86	73
inside diameter, ft[k]	65	55	45	35
slope[l]	7.2%	8.3%	9.7%	11.8%
parking ramp slope	3.0%	4.0%	5.0%	6.0%
express ramp slope				
covered	8%	10%	12%	14%
uncovered	6%	8%	10%	12%
speed ramp slope	10%	12%	14%	16%
transition length, ft	13	12	11	10
360° turns to top	2.5	4.0	5.5	7.0
short circuit in long run, ft	250	300	350	400
travel "up" to crossover, ft	300	450	600	750
spaces passed				
angled	400	750	1100	N.R.[m]
perpendicular	250	500	750	N.R.[m]
volume-capacity ratio, v/c[n]	0.6	0.7	0.8	N.R.[m]
walking distance to elevator, ft	150	200	250	300
overhead clearance, ft[o]	10	9	8	7

(Multiply ft by 0.305 to obtain m.)
[a]Measured from center to edge of circle traced by outside front wheel of design vehicle.
[b]Based on AASHTO 1984.
[c]Add 5 ft (1.5 m) to provide enough space to pass a broken down vehicle.
[d]At less than 10 mph (16 kph); add 2 ft (0.6 m) for higher speeds such as in circular helix; add 10 ft (3 m) to allow vehicle to pass breakdown.
[e]At less than 10 mph (16 kph); add 2 ft (0.6 m) to each lane for higher speeds.
[f]To wall, column, or other obstruction per AASHTO 1984.
[g]Clear between face of columns; check clearance at parked vehicles with turning template.
[h]To be adequate for two-way traffic. If flow is predominately one-way, can reduce 3 ft (0.9 m).
[i]For left-hand turns. Add 5 ft (1.5 m) to diameter for right-hand turns. (Not applicable to double-threaded designs.)
[j]Outside-to-outside outer bumper wall.
[k]Inside-to-inside inner bumper wall using recommended lane widths and clearances above and 6 ft (2 m) walls. Reduce 10 ft (3 m) to provide space to pass breakdown.
[l]At centerline of driving lane with 10 ft (3 m) floor-to-floor height.
[m]N.R. = not recommended.
[n]Volume-capacity ratio; based on the *HCM*.
[o]Straight vertical clearance from top of floor surface to underside of overhead obstruction.

Reprinted with permission from Anthony P. Chrest, Mary S. Smith, and Sam Bhuyan, *Parking Structures: Planning Design, Construction, Maintenance, and Repair,* copyright © 2001, by Springer.

PRACTICE PROBLEMS

1. A six-lane major arterial through a city's central business district is being studied for possible development and improvement. If the development takes place, the anticipated increase in commercial traffic may require additional traffic management measures to minimize delay along the arterial. Therefore, a speed study is being performed along a 2.5 mi long segment to evaluate current speeds and delays along the arterial. If development takes place, the study data will be used as the "before" condition in a before-and-after study. The study segment operates near capacity during peak commuter times and has 14 signalized intersections to allow access to the businesses in the district. Four initial test runs are completed with the following results.

run no.	elapsed time
1	613 sec
2	484 sec
3	662 sec
4	570 sec

(a) What is most nearly the average running speed of the initial runs?

(A) 15 mph

(B) 16 mph

(C) 19 mph

(D) 23 mph

(b) What is most nearly the average range in running speed?

(A) 3.7 mph

(B) 5.1 mph

(C) 6.5 mph

(D) 7.4 mph

(c) How many additional runs are required for a confidence level of 95%?

(A) 0 runs

(B) 2 runs

(C) 3 runs

(D) 4 runs

2. A major approach to an intersection is analyzed during a 1 hr period. The data gathered from the analysis is as follows.

approach lanes	2 lanes
lane width	12 ft
approach volume	2970 vph
PHF	0.95
trucks	3%
buses	10 buses/hr
CBD characteristics	none
pedestrians crossing	none
bicycles	none
parking	none
grade	0%
left turns	none
right turns	50 vph, shared
green phase	18 sec
cycle length	60 sec
initial queue	none
final queue	none
arrival type	random

(a) Assuming the traffic is evenly distributed between the two lanes, what is most nearly the saturation flow rate?

(A) 1700 vph

(B) 3400 vph

(C) 3600 vph

(D) 4400 vph

(b) What is the level of service (LOS) of this approach?

(A) LOS A

(B) LOS B

(C) LOS C

(D) LOS D

SOLUTIONS

1. (a) Determine the running speed of each test run using Eq. 2.3.

$$S_1 = \frac{L}{t_{\text{sec},1}} = \frac{(2.5\text{ mi})\left(3600\ \frac{\text{sec}}{\text{hr}}\right)}{613\text{ sec}} = 14.7\text{ mph}$$

$$S_2 = \frac{L}{t_{\text{sec},2}} = \frac{(2.5\text{ mi})\left(3600\ \frac{\text{sec}}{\text{hr}}\right)}{484\text{ sec}} = 18.6\text{ mph}$$

$$S_3 = \frac{L}{t_{\text{sec},3}} = \frac{(2.5\text{ mi})\left(3600\ \frac{\text{sec}}{\text{hr}}\right)}{662\text{ sec}} = 13.6\text{ mph}$$

$$S_4 = \frac{L}{t_{\text{sec},4}} = \frac{(2.5\text{ mi})\left(3600\ \frac{\text{sec}}{\text{hr}}\right)}{570\text{ sec}} = 15.8\text{ mph}$$

Determine the average speed of all test runs.

$$S_{\text{ave}} = \frac{14.7\ \frac{\text{mi}}{\text{hr}} + 18.6\ \frac{\text{mi}}{\text{hr}} + 13.6\ \frac{\text{mi}}{\text{hr}} + 15.8\ \frac{\text{mi}}{\text{hr}}}{4} = 15.7\text{ mi/hr}\quad(16\text{ mph})$$

The answer is (B).

(b) Determine the average range in running speed using Eq. 2.4.

$$\overline{R} = \frac{|S_1 - S_2| + |S_2 - S_3| + |S_3 - S_4|}{N_t - 1} = \frac{\left|14.7\ \frac{\text{mi}}{\text{hr}} - 18.6\ \frac{\text{mi}}{\text{hr}}\right| + \left|18.6\ \frac{\text{mi}}{\text{hr}} - 13.6\ \frac{\text{mi}}{\text{hr}}\right| + \left|13.6\ \frac{\text{mi}}{\text{hr}} - 15.8\ \frac{\text{mi}}{\text{hr}}\right|}{4-1} = 3.70\text{ mi/hr}\quad(3.7\text{ mph})$$

The answer is (A).

(c) Refer to Table 2.1. Since the average range in running speed is 3.7 mph, choose the next highest average range in running speed from Table 2.1, which is 5.0 mph. For economic evaluations, a permitted error of 2.0–4.0 mph is recommended, with 2.0 mph providing greater accuracy. From Table 2.1, for an average range in running speed of 5.0 mph and a permitted error of 2.0 mph, a sample size of four is required. However, should the development take place, the speed study will be used as a "before" condition for any before-and-after study. Therefore, a permitted error of ±1.0–3.0 mph will be required based on the ITE's recommendations. Using a permitted error of 1.0 mph will yield the most accurate results. A total of eight runs are needed to establish a 95% confidence level with a 1.0 mph permitted error. Therefore, four additional test runs are required.

The answer is (D).

2. (a) Determine the adjustment factors.

Use 2.0 pc/HV as the passenger car equivalent for heavy vehicles. From Eq. 2.29, the heavy vehicle factor is

$$f_{\text{HV}} = \frac{100\%}{100\% + \text{HV}_{\%}(E_T - 1)} = \frac{100\%}{100\% + (3\%)(2.0 - 1)} = 0.971$$

Calculate the bus blocking adjustment factor using Eq. 2.32.

$$f_{\text{bb}} = \frac{N_L - \dfrac{(14.4\text{ sec})N_B}{3600\ \frac{\text{sec}}{\text{hr}}}}{N_L} = \frac{2\text{ lanes} - \dfrac{(14.4\text{ sec})\left(10\ \frac{\text{buses}}{\text{hr}}\right)}{3600\ \frac{\text{sec}}{\text{hr}}}}{2\text{ lanes}} = 0.980$$

Calculate the proportion of right turns by dividing the right turn volume by the total approach volume.

$$p_{\text{RT}} = \frac{50\ \frac{\text{veh}}{\text{hr}}}{2970\ \frac{\text{veh}}{\text{hr}}} = 0.0168$$

Traffic/Capacity Analysis

Using Eq. 2.35, the right-turn adjustment factor for a shared lane is

$$\begin{aligned} f_{\text{RT}} &= 1.0 - 0.15p_{\text{RT}} = 1.0 - (0.15)(0.0168) \\ &= 0.997 \end{aligned}$$

Use Table 2.24 to find the lane utilization adjustment factor, f_{LU}. The lane group has two lanes, a shared right-turn lane, and evenly distributed traffic. Therefore, from Table 2.24, the lane utilization adjustment factor is 0.952. Since there is no CBD activity, the area type adjustment factor, f_a, is 1.0. The remaining factors default to 1.0. Using Eq. 2.27, the saturation flow rate is

$$\begin{aligned} s &= s_o N_L f_w f_{\text{HV}} f_g f_p f_{\text{bb}} f_a f_{\text{LU}} f_{\text{RT}} f_{\text{LT}} f_{\text{Rpb}} f_{\text{Lpb}} \\ &= \left(1900\ \frac{\text{pc}}{\text{hr-ln}}\right)(2\text{ lanes})(1.0)(0.971)(1.0)(1.0)(0.980) \\ &\quad \times (1.0)(0.952)(0.997)(1.0)(1.0)(1.0) \\ &= 3432\text{ veh/hr} \quad (3400\text{ vph}) \end{aligned}$$

The answer is (B).

(b) Level of service for intersections is determined by control delay (in seconds per vehicle), as shown in Table 2.32. Find the degree of saturation, X, using Eq. 2.42.

$$\begin{aligned} X &= \frac{v}{c} = \frac{2970\ \frac{\text{veh}}{\text{hr}}}{3432\ \frac{\text{veh}}{\text{hr}}} \\ &= 0.865 \end{aligned}$$

Since no initial or final queue is present, the intersection is classified as situation I according to Table 2.26. Therefore, the uniform control delay, d_1, is found from Eq. 2.48.

$$\begin{aligned} d_1 &= \frac{0.5C\left(1 - \frac{g}{C}\right)^2}{1 - \left(\min(1, X)\frac{g}{C}\right)} \\ &= \frac{(0.5)(60\text{ sec})\left(1 - \frac{18\text{ sec}}{60\text{ sec}}\right)^2}{1 - (0.865)\left(\frac{18\text{ sec}}{60\text{ sec}}\right)} \\ &= 19.9\text{ sec} \end{aligned}$$

From Table 2.29, the arrival type for random arrivals is arrival type 3 (AT 3). From Table 2.28, the PF is 1.0 for AT 3.

The incremental delay, d_2, is found from Eq. 2.51. For pretimed signals, k is 0.50, and I defaults to 1.0.

$$\begin{aligned} d_2 &= 900T\left(X - 1 + \sqrt{(X-1)^2 + \frac{8kIX}{cT}}\right) \\ &= (900)(1.0\text{ hr}) \\ &\quad \times \left(\begin{array}{l} (0.865 - 1) \\ + \sqrt{\begin{array}{l} (0.865-1)^2 \\ + \frac{(8)(0.50)(1.0)(0.865)}{\left(3432\ \frac{\text{veh}}{\text{hr}}\right)(1.0\text{ hr})} \end{array}} \end{array} \right) \\ &= 3.32\text{ sec} \end{aligned}$$

There is no initial queue delay. Therefore, d_3 is 0 sec. From Eq. 2.45, the total control delay is

$$\begin{aligned} d &= d_1(\text{PF}) + d_2 + d_3 \\ &= (19.9\text{ sec})(1.0) + 3.32\text{ sec} + 0\text{ sec} \\ &= 23.2\text{ sec} \end{aligned}$$

From Table 2.32, the LOS of the intersection is C.

The answer is (C).

3 Pedestrian and Mass Transit Analysis

Nomenclature

a	acceleration	ft/sec^2	m/s^2
A	area	ft^2	m^2
c	capacity	various	various
C	signal cycle length	sec	s
d	deceleration	ft/sec^2	m/s^2
d	delay	sec	s
d_b	control delay	sec/bicycle	s/bicycle
D	density	ped/ft^2	ped/m^2
F	number of events	bicycles/hr	bicycles/h
FDW	flashing DON'T WALK time	sec	s
g	effective green time	sec	s
g	gravitational acceleration, 32.2 (9.81)	ft/sec^2	m/s^2
G	grade rise/run	%	%
h	headway	min/bus, min/train	min/bus, min/train
L	length	ft	m
M	pedestrian space	ft^2/ped	m^2/ped
N	number	–	–
p	proportion	–	–
Q_{tco}	holding area waiting time, minor street	ped-sec, sec	ped·s, s
Q_{tdo}	holding area waiting time, major street	ped-sec, sec	ped·s, s
r	radius of curb	ft	m
R	red phase time	sec	s
s	distance	ft	m
s_b	base saturation flow rate	bicycles/hr	bicycles/h
S	speed	various	various
S_j	intersection delay	sec	s
t	time	sec	s
T	total crosswalk occupancy time	ped-sec	ped·s
TS	total time-space	ft^2-sec	m^2·s
v	pedestrian volume	ped/cycle	ped/cycle
v	rate of flow (service flow rate)	various	various
v_b	one-directional flow rate	bicycles/hr	bicycles/h
$v_{p,\text{ci}}$	number of pedestrians waiting to cross minor street, inbound	ped/cycle	ped/cycle
$v_{p,\text{co}}$	number of pedestrians waiting to cross minor street, outbound	ped/cycle	ped/cycle
$v_{p,\text{di}}$	number of pedestrians waiting to cross major street, inbound	ped/cycle	ped/cycle
$v_{p,\text{do}}$	number of pedestrians waiting to cross major street, outbound	ped/cycle	ped/cycle
v_t	total number of circulating pedestrians	ped, ped/cycle	ped, ped/cycle
v/c	volume-capacity ratio	–	–
v	velocity	mph	kph
W	width	ft	m
W_o	sum of obstruction widths and shy distances	ft	m
WALK	WALK time in crosswalk	sec	s

Subscripts

0	at rest
a	acceleration
at	average travel
ave	average
b	bicycle or bike lane
c	circulating pedestrian, critical, or crossing platoon
d	crossing major street or deceleration
dw	dwell
e	equivalent or exclusive facility
E	effective
g	gap
G	group
i	segment i or inbound direction
j	intersection j
LT	left-turn
max	maximum
mi	minor street
min	minimum
mj	major street
o	opposing or outbound direction
p	passenger, peak, pedestrian, perception, or platoon
ps	passenger service
r	reaction
RT	right-turn
s	shared facility, start-up and end clearance, or subject direction
ST	straight
t	total or train
tv	turning vehicle
T	urban street
v	vehicle

1. PEDESTRIAN FACILITIES

Pedestrian facilities are areas or structures designed to accommodate pedestrians, either exclusively or in conjunction with nonmotorized and motorized vehicles. Pedestrian facilities include sidewalks, stairways, queuing, off-street paths, and crosswalks. The *Highway Capacity Manual* provides more information on types of pedestrian facilities.

Design of pedestrian facilities is based on optimizing *pedestrian flow rate*, the number of pedestrians that can pass through a given point in a given amount of time (typically measured as either pedestrians per minute or pedestrians per 15 minutes). *Pedestrian flow* is similar to vehicle flow, but is subject to more random and difficult-to-measure variables. This section discusses methods of analyzing pedestrian facilities and determining pedestrian flow for those facilities, as well as those aspects of facility design that must be accounted for by transportation engineers.

Analysis Input Data and Default Values

When analyzing a pedestrian facility, the most reliable data are field-measured values of the pedestrian facility and of the pedestrian flow at the study location. If the facility is not yet constructed, a nearby location that has similar characteristics to the proposed facility can be used for field data. The field study must take into account the flow traffic demographics, regional behavior of pedestrian traffic, geometric conditions, walkway surface, ambient weather and lighting, and other factors that influence traffic flow characteristics. In the absence of field data, default values can be considered. Table 3.1 shows default values that may apply to general situations without known field data.

Walking Speed

All evaluations of pedestrian flow and derivative calculations are in part a function of the average walking speed of pedestrians in the flow. *Average walking speed* (also referred to as *design speed*) is normally measured in feet per minute or feet per second (meters per minute or meters per second) and can be affected by a variety of factors, such as the grade of the walkway or the number of elderly pedestrians or slowly walking children in the pedestrian flow. *Pedestrian trip purpose* can also affect walking speed; studies have found that commuters, students, and people who frequently travel to and from a given location have greater walking speeds than shoppers, children, and older pedestrians.

On normal, level sidewalks, average walking speed can be assumed to be 5.0 ft/sec (1.5 m/s). Under other conditions where the walkway is still level, flow populations where less than 20% of the pedestrians are elderly can be assumed to have an average speed of 4.0 ft/sec (1.2 m/s), and flow populations where 20% or more of the pedestrians are elderly can be assumed to have an average walking speed of 3.0 ft/sec (0.9 m/s). If the

Table 3.1 *Input Data Default Values for Pedestrian Flow*

item	default	definition
	geometry	
length of sidewalk	–	approximately equal to the length of an urban street; varies from city to city
walkway width	5.0 ft	with a buffer zone between sidewalk and curb
	7.0 ft	with no buffer zone between sidewalk and curb
effective width	5.0 ft	
street corner radius	45 ft	with trucks and buses in turning volume
	25 ft	without trucks and buses in turning volume
crosswalk length	–	curb-to-curb dimension of traveled roadway[a]
	demand data	
analysis period	–	15 min[b]
number of pedestrians in a platoon	Eq. 3.24	
pedestrian walking speed	4.0 ft/sec	–[c]
pedestrian start-up time	1.8 sec	younger male
	2.0 sec	younger female
	2.4 sec	older male 50th percentile
	3.0 sec	default value
	3.7 sec	older male 85th percentile
	2.6 sec	older female 50th percentile
	4.0 sec	older female 85th percentile
	intersection data	
LOS E for pedestrian cross flows	≥ 13 ft^2/ped	space
	≤ 23 ped/min-ft	flow
	≥ 3.28 ft/sec	speed
	≤ 0.07 ped/ft^2	density

(Multiply ft by 0.305 to obtain m.)
(Multiply ft^2 by 0.093 to obtain m^2.)
[a]Signal design procedures may use other values.
[b]subject to code or policy
[c]see *MUTCD* for values

walkway has an upgrade of 10% or more, the average walking speed should be reduced by 0.5 ft/sec (0.15 m/s). Other, less predictable conditions can also affect the average walking speed of a pedestrian flow, such as an abundance of obstacles in the path of the flow. Good engineering judgment must be used when determining the average speed of a flow.

For pedestrian facilities where pedestrians will form a line, such as on a mass transit platform or at a signalized intersection, average walking speed must take into account the pedestrian start-up time. *Pedestrian start-up time* is different for each pedestrian, as the pedestrians in the back cannot step forward without interference until the front pedestrians begin to walk. Queues move forward in a *wave pattern* so that as each pedestrian walks, the entire queue moves forward in a fairly ordered progression. The time it takes for the wave progression to move from the front to the back of the queue depends largely on the length of the queue, though an average 3 sec start-up time is used for signal analyses at crosswalks.

The average walking speed is greater for terminals than for crosswalks or street sidewalks. A *terminal* is a transportation facility that serves as a loading and unloading area for passengers, such as a bus depot, train station, or airport, and is often used to transfer between transit modes and vehicles. Terminals are subject to more directed flow with little cross-traffic interference. Figure 3.1 shows that the majority of pedestrians in a terminal have a free-flow walking speed of 240–260 ft/min or 4.0–4.3 ft/sec (73–79 m/min or 1.2–1.3 m/s).

Figure 3.1 *Typical Free-Flow Walking Speed Distributions*

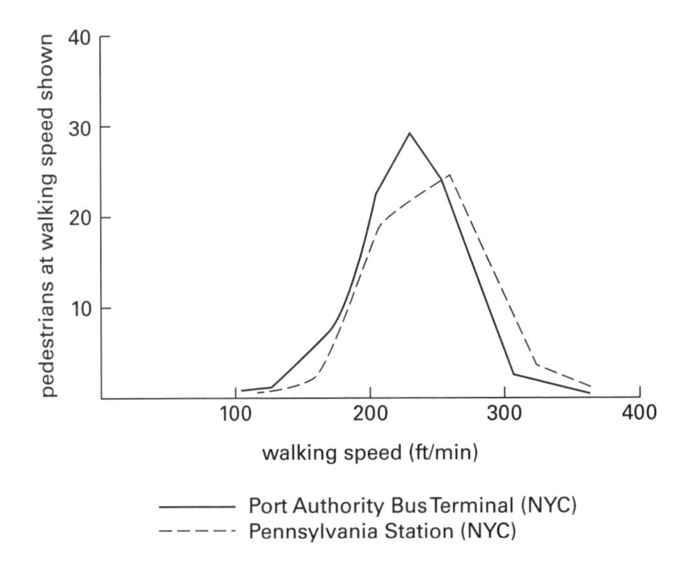

(Multiply ft/min by 0.305 to obtain m/min.)

Highway Capacity Manual 2000. Copyright, National Academy of Sciences, Washington, D.C. Exhibit 11-6. Reproduced with permission of the Transportation Research Board.

Pedestrian Space Requirements

Average walking speed of pedestrians is directly affected by the amount of space available for each pedestrian to walk in. Design of pedestrian facilities must account for a certain minimum amount of space for each pedestrian.

For the sake of simplicity, a pedestrian is assumed to occupy an oval-shaped body ellipse that represents the minimum space required for the pedestrian to stand and move around without impediment. The elliptical space occupied by a pedestrian standing upright is assumed to be 2.0 ft (0.6 m) wide at the shoulders and 1.5 ft (0.5 m) deep front to back. (See Fig. 3.2(a).) This represents a minimum standard for standing pedestrians.

Figure 3.2 *Pedestrian Ellipse for Standing Areas and Pedestrian Walking Space Requirements*

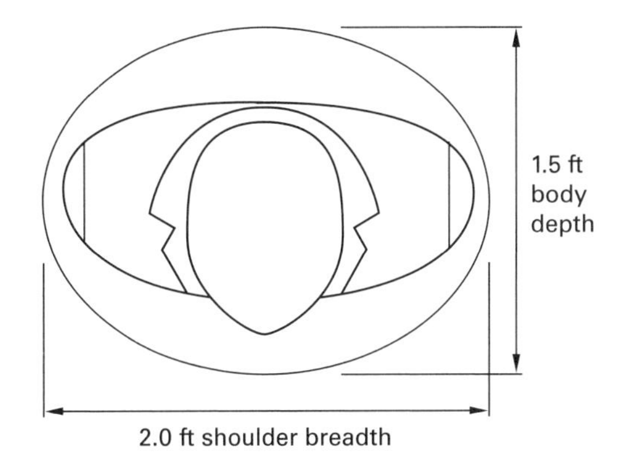

(a) pedestrian body ellipse

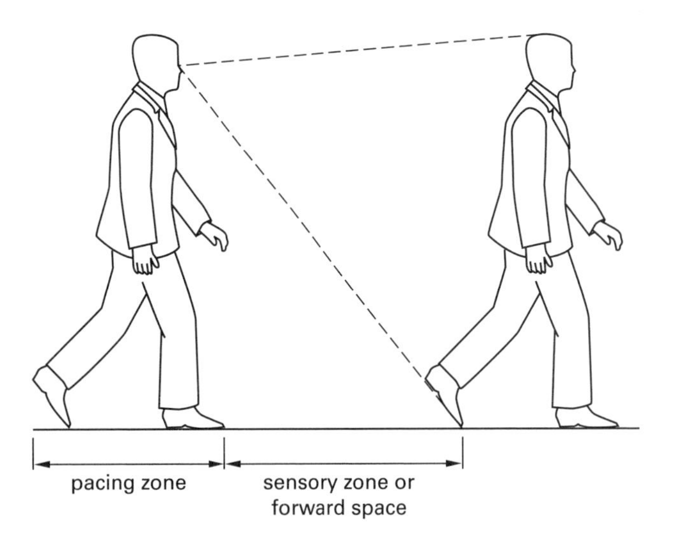

(b) pedestrian walking space requirements

(Multiply ft by 0.305 to obtain m.)

Highway Capacity Manual 2000. Copyright, National Academy of Sciences, Washington, D.C. Exhibit 11-5. Reproduced with permission of the Transportation Research Board.

Pedestrian/ Transit Analysis

When walking, normal arm and leg movements will increase the required space both in front and to the sides of a pedestrian. Increasing the width of the ellipse to 2.5 ft (0.8 m) allows for 3 in (8 cm) of extra distance on each side to account for bodily sway, extra berth from carrying packages or avoiding strangers, and other modifications to the pedestrian flow. Pedestrian walkways must be designed to accommodate this 2.5 ft (0.8 m) *minimum walking path*, so that pedestrians can use the space without significant risk of running into obstacles or being forced to slow their walking speed. The minimum walking path is different from the *effective walkway width*, the portion of the walkway that can be used effectively for pedestrian movement. Walkway design must also take the minimum clear width into account, ensuring that proper space is allotted for buffer zones between pedestrians and vehicular traffic or similar environmental concerns.

Design for forward space in a pedestrian facility is not as straightforward as design for lateral space. The amount of forward space a pedestrian requires can be divided into the *pacing zone*, the area required for a pedestrian to place his or her feet during forward movement, and the *sensory zone*, the area required for a pedestrian to comfortably perceive and react to obstacles and changes in the environment while walking. (See Fig. 3.2(b).) The length of a pacing zone is directly related to the pedestrian's travel speed, but is also dependent on the pedestrian's age, sex, physical condition, and other factors. The sensory zone varies considerably from person to person, depending on a given pedestrian's familiarity with the walkway, the amount of clutter or obstacles in the walkway, the level of lighting, the number of other pedestrians in the flow, and a variety of other factors that could cause a pedestrian to require a greater amount of space in which to perceive and react to changes in the walking environment. For this reason, pedestrian flow must be carefully observed to ensure that the body ellipse used for design does not require modification.

Pedestrian Unit Flow Rate

The *pedestrian unit flow rate* is the number of pedestrians passing a point per unit of time, typically expressed as pedestrians per minute. The pedestrian unit flow rate, v_p, is given by Eq. 3.1. S_p is the average walking speed of pedestrians (*pedestrian speed*) in a given area in feet or meters per minute (found from empirical observation or estimated as explained in the subsection, "Walking Speed"), and D_p is *pedestrian density* in the area in pedestrians per square foot or square meter.

$$v_p = S_p D_p \quad [\textit{HCM} \text{ Eq. 11-1}] \qquad 3.1$$

While an increase in pedestrian density will mean an increase in the flow rate, an increase in pedestrian density will also mean a decrease in the physical space available to each pedestrian in the flow, referred to as the *pedestrian space*. As available pedestrian space decreases, pedestrian speed also decreases, due to the decreased room to maneuver around obstacles or slower-moving pedestrians. As pedestrian density increases, average speed decreases, until eventually the average speed of the pedestrian flow is equal to the speed of the slowest person in the flow. Figure 3.3 shows the relationship between pedestrian speed and pedestrian density, Fig. 3.4 shows the relationship between pedestrian flow rate and pedestrian space, and Fig. 3.5 shows the relationship between pedestrian speed and pedestrian space. As shown in Fig. 3.4, the *maximum flow rate* occurs when the available pedestrian space is between 5 ft²/ped and 9 ft²/ped (0.5 m²/ped and 0.8 m²/ped). When available pedestrian space is 5 ft²/ped (0.5 m²/ped) or less, the flow rate drops dramatically, and when the pedestrian space is between 2 ft²/ped and 4 ft²/ped (0.2 m²/ped and 0.4 m²/ped), all flow stops.

Figure 3.3 *Relationship Between Pedestrian Speed and Density*

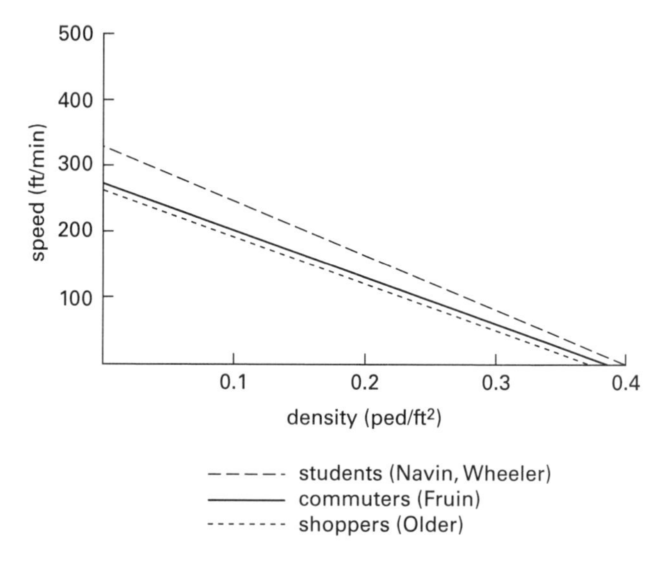

(Multiply ft/min by 0.305 to obtain m/min.)
(Multiply ped/ft² by 0.093 to obtain ped/m².)

Highway Capacity Manual 2000. Copyright, National Academy of Sciences, Washington, D.C. Exhibit 11-1. Reproduced with permission of the Transportation Research Board.

Given these correlations, the *HCM* recommends Eq. 3.2 be used to find the pedestrian flow rate rather than Eq. 3.1. M is the available pedestrian space, the reciprocal of the pedestrian density, as shown in Eq. 3.3.

$$v_p = \frac{S_p}{M} \quad [\textit{HCM} \text{ Eq. 11-1}] \qquad 3.2$$

$$M = \frac{1}{D_p} \quad [\textit{HCM} \text{ Eq. 11-2}] \qquad 3.3$$

Effective Walkway Width

Effective walkway width is the portion of the walkway that can be used effectively for pedestrian travel. Portions of walkway that cannot be used effectively must be

Figure 3.4 *Relationships Between Pedestrian Flow and Space*

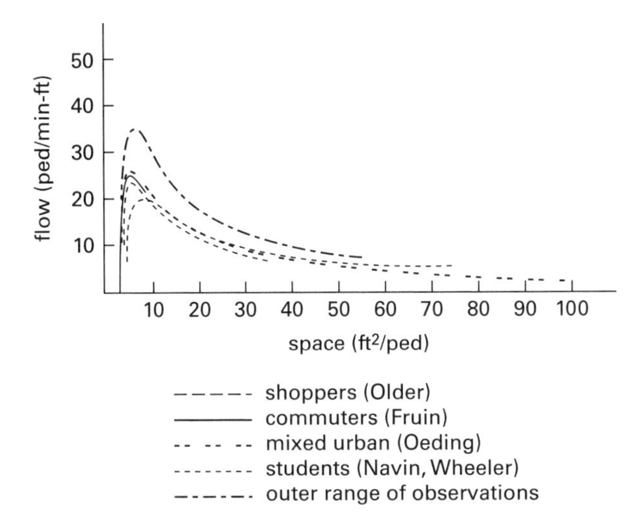

(Multiply ped/min-ft by 0.305 to obtain ped/min·m.)
(Multiply ft^2/ped by 0.093 to obtain m^2/ped.)

Highway Capacity Manual 2000. Copyright, National Academy of Sciences, Washington, D.C. Exhibit 11-2. Reproduced with permission of the Transportation Research Board.

Figure 3.5 *Relationship Between Pedestrian Speed and Space*

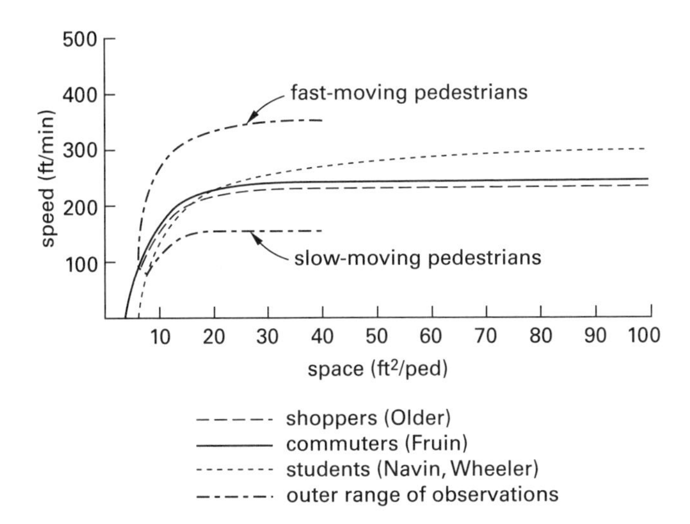

(Multiply ft/min by 0.305 to obtain m/min.)
(Multiply ft^2/ped by 0.093 to obtain m^2/ped.)

Highway Capacity Manual 2000. Copyright, National Academy of Sciences, Washington, D.C. Exhibit 11-4. Reproduced with permission of the Transportation Research Board.

deducted when determining if a walkway meets minimum walking path requirements. For instance, curb edges and vertical surfaces adjacent to walkways, such as building faces, storefronts, walls, and tall vegetation, are not generally used by pedestrians. Sidewalk fixtures and landscaping, such as trash containers, utility poles, signs, fire hydrants, mailboxes, parking meters, or trees, can impede the free movement of pedestrians. Storefront window recesses and column pillars along the sidewalk edge are often used as standing spaces by pedestrians.

Although a single pole or obstruction has little effect on walkway capacity under light to moderate flow conditions, for flows near the capacity limit, the presence of a series of obstructions has an effect similar to that of a *continuous obstruction.* The effective length of an obstruction is assumed to be five times its effective width. Therefore, *occasional obstructions* can be evaluated by multiplying their effective width by the ratio of their effective length to the average distance between them. Equation 3.4 is used to calculate the effective walkway width, W_E. W_t is the total walkway width, and W_o is the sum of obstruction widths and shy distances.

$$W_E = W_t - W_o \quad [\textit{HCM}\ \text{Eq. 18-1}] \qquad 3.4$$

Effective width reductions are called *shy distances* (i.e., the distance pedestrians "shy away from" an obstacle), or *preemptive widths.* The shy distances are summed using the most prevalent feature on each side of the walkway. Figure 3.6 illustrates typical walkway width adjustments caused by objects placed in the total walkway width, and Table 3.2 provides a list of shy distances.

Figure 3.6 *Width Adjustments for Fixed Obstacles*

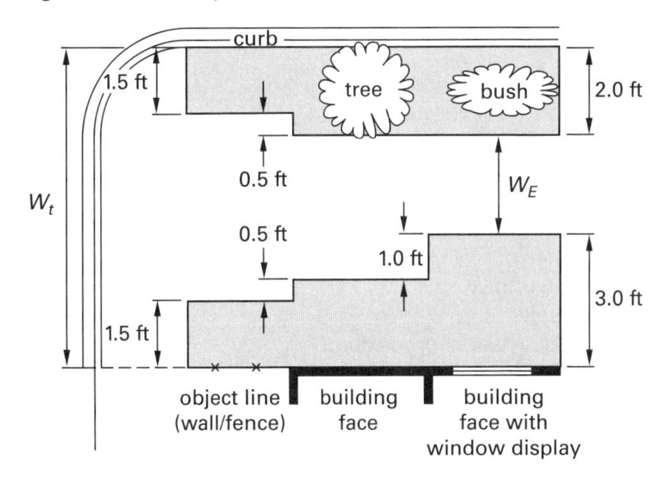

(Multiply ft by 0.305 to obtain m.)

Highway Capacity Manual 2000. Copyright, National Academy of Sciences, Washington, D.C. Exhibit 18-1. Reproduced with permission of the Transportation Research Board.

Flow-Conflict Performance

A walkway's greatest capacity occurs when 100% of the flow travels in the same direction (*unidirectional flow*). When *two-directional flow* occurs, the capacity is nearly the same as unidirectional flow, provided the flow is balanced (i.e., the directional split is 50/50). When the directional split is considerably unbalanced under high flows, such as 90/10, the minor directional flow will have considerable difficulty using the walkway, as the major flow often spills over into the minor flow direction. This is

Table 3.2 Shy Distances*

obstacle	approximate width preempted (ft)
street furniture	
light poles	2.5–3.5
traffic signal poles and boxes	3.0–4.0
fire alarm boxes	2.5–3.5
fire hydrants	2.5–3.0
traffic signs	2.0–2.5
parking meters	2.0
mailboxes (1.7 ft × 1.7 ft)	3.2–3.7
telephone booths (2.7 ft × 2.7 ft)	4.0
wastebaskets	3.0
benches	5.0
public underground access	
subway stairs	5.5–7.0
subway ventilation gratings (raised)	6.0+
transformer vault ventilation gratings (raised)	5.0+
landscaping	
trees	2.0–4.0
planter boxes	5.0
commercial uses	
newsstands	4.0–13.0
vending stands	variable
advertising displays	variable
store displays	variable
sidewalk cafes (two rows of tables)	7.0
building protrusions	
columns	2.5–3.0
stoops	2.0–6.0
cellar doors	5.0–7.0
standpipe connections	1.0
awning poles	2.5
truck docks (trucks protruding)	variable
garage entrances/exits	variable
driveways	variable

(Multiply ft by 0.305 to obtain m.)

*To account for the avoidance distance between pedestrians and obstacles, 1.0 ft to 1.5 ft (0.3 m to 0.5 m) must be added to the preemption width for individual obstacles. Widths are from curb to edge of object or from building face to edge of object.

Highway Capacity Manual 2000. Copyright, National Academy of Sciences, Washington, D.C. Exhibit 18-2. Reproduced with permission of the Transportation Research Board.

called *cross-flow walkway interference* and reduces the capacity of both walkways. The probability of contact between two pedestrians increases when the space per person becomes less than 35 ft^2/ped (3.3 m^2/ped). When there is 35 ft^2/ped (3.3 m^2/ped) or greater, the probability of conflict is effectively zero, and as the space approaches 18 ft^2/ped (5.5 m^2/ped) or less, passing becomes virtually impossible. Between these extremes, conflict probability follows a nonlinear relationship, as shown in Fig. 3.7.

Figure 3.7 Cross-Flow Traffic Probability of Conflict

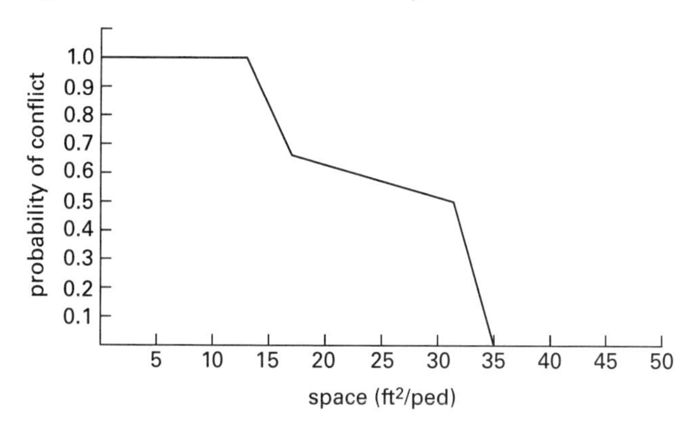

(Multiply ft^2/ped by 0.093 to obtain m^2/ped.)

Highway Capacity Manual 2000. Copyright, National Academy of Sciences, Washington, D.C. Exhibit 11-7. Reproduced with permission of the Transportation Research Board.

Whether there is a cross flow or an opposite direction flow, an average of at least 60 ft^2 to 100 ft^2 (5.6 m^2 to 9.3 m^2) of space per pedestrian is necessary to avoid interference.

Level of Service and Pedestrian Flow

Level of service (LOS) criteria measure the quality of pedestrian flow by comparing flow to defined ranges of available space per pedestrian, flow rates, and speeds. LOS for pedestrians is similar to LOS for highway traffic, using relative freedom to maneuver in traffic flow as a guideline. LOS is divided into five levels of flow, with LOS A describing the least dense flow and LOS F describing the most dense flow that is still moving. LOS E is considered the *nominal walkway capacity*, with flow that is tightly compressed, allowing for very little ability to pass slower pedestrians. As for vehicular flow, LOS F for pedestrians is a condition when all flow is severely restricted and flow is sporadic and unstable. The *HCM* provides descriptions of each LOS for pedestrian walkways, shown in Fig. 3.8, and for queuing areas, shown in Fig. 3.9.

Assigning a quality of flow index based on pedestrian space establishes a common parameter for flow analysis. Table 3.3 shows LOS criteria for pedestrian walkways and for queuing areas based on the pedestrian space, in square feet (square meters) per pedestrian in the study zone.

The usual analysis method is to consider a time period (i.e., 5 min, 10 min, or 15 min) during which the flow tends to be average and is more representative of continuous flow. The selection of a time period is a matter of pedestrian convenience, available space, economics, policy, code requirements, and the type of design conditions being considered. For instance, a fire or panic escape walkway requires a greater level of short-term capacity in order to evacuate a space quickly and must be capable of handling pedestrian

Figure 3.8 *Pedestrian Walkway LOS*

LOS A
pedestrian space > 60 ft^2/ped
flow rate ≤ 5 ped/min-ft
At LOS A, pedestrians move in desired paths without altering their movements in response to other pedestrians. Walking speeds are freely selected, and conflicts between pedestrians are unlikely.

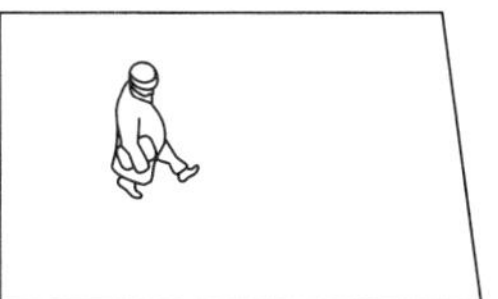

LOS B
pedestrian space > 40–60 ft^2/ped
flow rate > 5–7 ped/min-ft
At LOS B, there is sufficient area for pedestrians to select walking speeds freely, to bypass other pedestrians, and to avoid crossing conflicts. At this level, pedestrians begin to be aware of other pedestrians, and to respond to their presence when selecting a walking path.

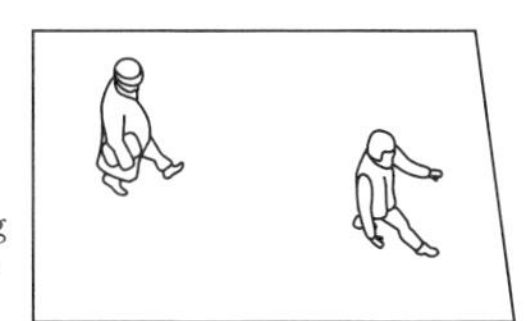

LOS C
pedestrian space > 24–40 ft^2/ped
flow rate > 7–10 ped/min-ft
At LOS C, space is sufficient for normal walking speeds, and for bypassing other pedestrians in primarily unidirectional streams. Reverse-direction or crossing movements can cause minor conflicts, and speeds and flow rates are somewhat lower.

LOS D
pedestrian space > 15–24 ft^2/ped
flow rate > 10–15 ped/min-ft
At LOS D, freedom to select walking speed and to bypass is restricted. Crossing or reverse-flow movements face a high probability of conflict, requiring frequent changes in speed and position. LOS D provides reasonably fluid flow, but friction and interaction between pedestrians is likely.

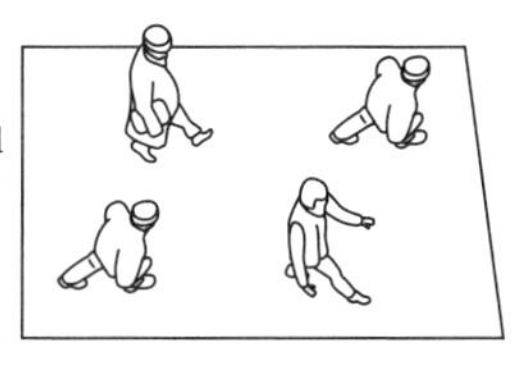

LOS E
pedestrian space > 8–15 ft^2/ped
flow rate > 15–23 ped/min-ft
At LOS E, virtually all pedestrians restrict their normal walking speed, frequently adjusting their gait. At the lower range, forward movement is possible only by a shuffle. Space is not sufficient for passing slower pedestrians. Cross-flow or reverse-flow movements are possible only with extreme difficulties. Design volumes approach the limit of walkway capacity with stoppages and interruptions to flow.

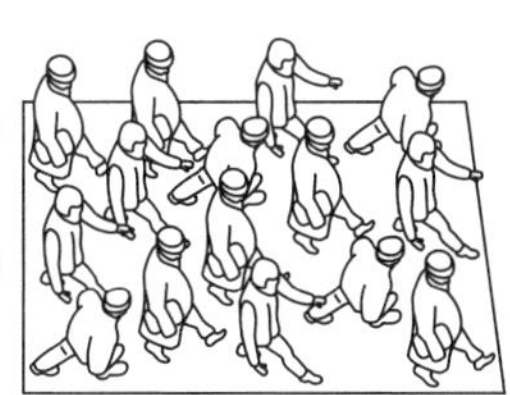

LOS F
pedestrian space ≤ 8 ft^2/ped
flow rate varies
At LOS F, all walking speeds are severely restricted, and forward progress is only made by shuffling. There is frequent unavoidable contact with other pedestrians. Cross-flow and reverse-flow movements are virtually impossible. Flow is sporadic and unstable. Space is more characteristic of queued pedestrians than of moving pedestrian streams.

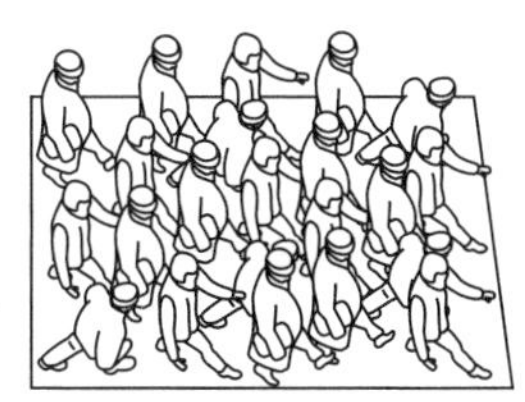

(Multiply ft^2/ped by 0.093 to obtain m^2/ped.)
(Multiply ped/min-ft by 0.305 to obtain ped/min·m.)

Highway Capacity Manual 2000. Copyright, National Academy of Sciences, Washington, D.C. Exhibit 11-8. Reproduced with permission of the Transportation Research Board.

Figure 3.9 *Queuing Area LOS*

LOS A
average pedestrian space > 13 ft^2/ped
Standing and free circulation through the queuing area is possible without disturbing others within the queue.

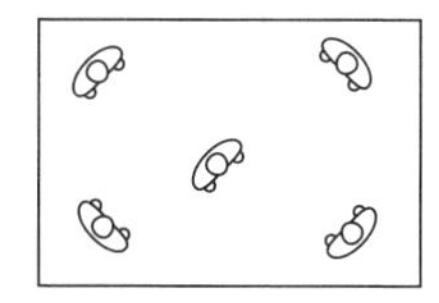

LOS B
average pedestrian space > 10–13 ft^2/ped
Standing and partially restricted circulation to avoid disturbing others in the queue is possible.

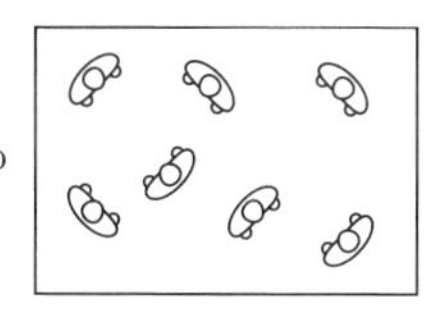

LOS C
average pedestrian space > 6–10 ft^2/ped
Standing and restricted circulation through the queuing area by disturbing others in the queue is possible. This density is within the range of personal comfort.

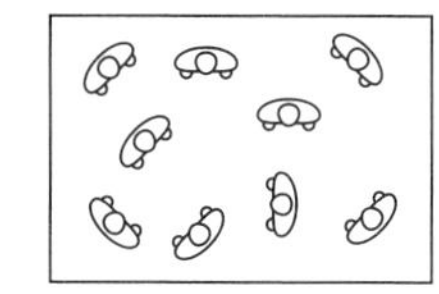

LOS D
average pedestrian space > 3–6 ft^2/ped
Standing without touching is possible. Circulation is severely restricted within the queue and forward movement is only possible as a group. Long-term waiting at this density is uncomfortable.

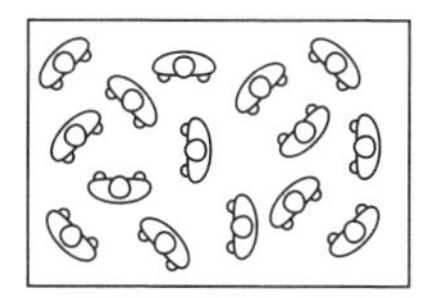

LOS E
average pedestrian space > 2–3 ft^2/ped
Standing in physical contact with others is unavoidable. Circulation in the queue is not possible. Queuing can only be sustained for a short period without serious discomfort.

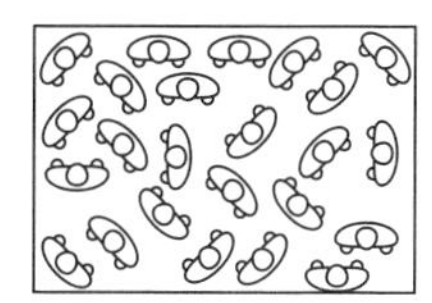

LOS F
average pedestrian space ≤ 2 ft^2/ped
Virtually all persons within the queue are standing in direct physical contact with others. This density is extremely uncomfortable. No movement is possible in the queue. There is potential for panic in large crowds at this density.

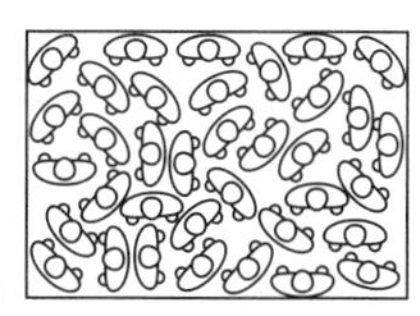

(Multiply ft^2/ped by 0.093 to obtain m^2/ped.)
Highway Capacity Manual 2000. Copyright, National Academy of Sciences, Washington, D.C. Exhibit 11-9. Reproduced with permission of the Transportation Research Board.

flows that are faster and have greater purpose than flows on a terminal concourse or sidewalk.

Service Volumes

The *HCM* defines *service volume* as the maximum hourly rate at which pedestrians can be reasonably expected to traverse a point along a walkway while maintaining a designated LOS during an hour where assumed conditions exist. The service volume takes into account the physical conditions of the walkway, such as the grade, effective width, average pedestrian speed, and

Table 3.3 *Average Flow LOS Criteria for Walkways and Sidewalks*

LOS	space (ft^2/ped)	flow rate (ped/min-ft)	speed (ft/sec)	v/c ratio
A	>60	≤5	>4.25	≤0.21
B	>40–60	>5–7	>4.17–4.25	>0.21–0.31
C	>24–40	>7–10	>4.00–4.17	>0.31–0.44
D	>15–24	>10–15	>3.75–4.00	>0.44–0.65
E	>8–15	>15–23	>2.50–3.75	>0.65–1.0
F	≤8	variable	≤ 2.50	variable

(Multiply ft^2/ped by 0.093 to obtain m^2/ped.)
(Multiply ped/min-ft by 0.305 to obtain ped/min·m.)
(Multiply ft/sec by 0.305 to obtain m/s.)

Highway Capacity Manual 2000. Copyright, National Academy of Sciences, Washington, D.C. Exhibit 18-3. Reproduced with permission of the Transportation Research Board.

other features that have an effect on the total rate of traffic flow.

Service volumes for pedestrian flow are dependent on a variety of assumptions. Service volumes for one location may not be entirely similar to another location with the same or similar physical parameters. This is why field data are needed to obtain accurate results from the analysis. The *HCM* provides a sample of service volumes for each LOS based on an effective sidewalk width of 5 ft (1.5 m) (see Table 3.4). This sample gives a general range of service volumes that may occur on a walkway of similar nature. Operational analysis and final design should not be based on these values, but these values can be used as a starting point to verify the actual conditions in the field.

Table 3.4 *Example of Service Volumes for a 5 ft Wide Pedestrian Sidewalk**

LOS	15 min pedestrian volume
A	375
B	525
C	750
D	1125
E	1725

*Assumes effective sidewalk width of 5.0 ft (1.5 m).

Highway Capacity Manual 2000. Copyright, National Academy of Sciences, Washington, D.C. Exhibit 11-16. Reproduced with permission of the Transportation Research Board.

Number of Pedestrians in a Platoon

Pedestrian flow tends to occur in a series of *platoons*, a group of pedestrians walking together. Platooning is generally involuntary and is often caused by an accumulation of faster walkers behind slower ones or a time-dependent event, such as a traffic signal or a bus arrival at a terminal. In transportation terminals, platooning usually occurs along walkways and concourses when large volumes of pedestrians emerge from arriving vehicles.

Because platooning reduces a pedestrian's freedom to move, flow parameters are altered. Table 3.5 gives LOS criteria for pedestrian flows not subject to platooning, while Table 3.6 shows the LOS for pedestrian flows with platooning.

Table 3.5 *LOS Criteria for Pedestrians on Walkways and in Queues*

LOS	space, pedestrian walkway (ft^2/ped)	space, queuing area (ft^2/ped)
A	>60	>13
B	>40–60	>10–13
C	>24–60	>6–10
D	>15–24	>3–6
E	>8–15	>2–3
F	≤8	≤2

(Multiply ft^2/ped by 0.093 to obtain m^2/ped.)

Highway Capacity Manual 2000. Copyright, National Academy of Sciences, Washington, D.C. Exhibit 18-3 and Exhibit 18-7. Reproduced with permission of the Transportation Research Board.

Table 3.6 *Platoon-Adjusted LOS Criteria for Walkways and Sidewalks*

LOS	space (ft^2/ped)	flow rate* (ped/min-ft)
A	>530	≤0.5
B	>90–530	>0.5–3
C	>40–90	>3–6
D	>23–40	>6–11
E	>11–23	>11–18
F	≤11	>18

(Multiply ft^2/ped by 0.093 to obtain m^2/ped.)
(Multiply ped/min-ft by 0.305 to obtain ped/min·m.)
*Rates in the table represent average flow rates over a 5 min to 6 min period.

Highway Capacity Manual 2000. Copyright, National Academy of Sciences, Washington, D.C. Exhibit 18-4. Reproduced with permission of the Transportation Research Board.

Uninterrupted Flow Pedestrian Facilities

Uninterrupted flow pedestrian facilities are established either exclusively for pedestrian use or for shared use with nonmotorized forms of transportation. Because the facilities have no disruptions except for interactions with other pedestrians (or with nonmotorized forms of transportation on shared facilities), pedestrian flow after start-up is relatively uniform over extended distances. Therefore, it can be analyzed using the pedestrian walking speed and pedestrian space requirements given in Table 3.3. Note that Table 3.3 does not consider platoon flow, but assumes average flow throughout the effective walkway width.

Sidewalks, concourses, and walkways that separate pedestrians from vehicular traffic and on which bicycles or other nonpedestrian uses are prohibited are often established along wide city streets, through park-like settings, and in airport, bus, or train terminals. These facilities are generally constructed to accommodate high volumes of pedestrian traffic and are designed along straight or gently curving routes with minimal cross-flow interference.

Performance measures can be determined by available space per person along the facility, M, found from Eq. 3.3. Field observations of space are made by counting the number of pedestrians in a given area per unit of time. When calculating the *pedestrian unit flow rate*, v_p, a 15 min unit of time is often used to compensate for normal flow variations, as shown in Eq. 3.5.

$$v_p = \frac{v_{15}}{15 W_E} \quad \text{[}HCM\text{ Eq. 18-2]} \qquad 3.5$$

Interrupted Flow Pedestrian Facilities

Pedestrians experience interrupted flow at signalized intersections. The delay, d_p, experienced at the intersection is calculated from Eq. 3.6.

$$d_p = \frac{0.5(C - g)^2}{C} \quad \text{[}HCM\text{ Eq. 18-5]} \qquad 3.6$$

C is the cycle length in seconds, and g is the effective green time in seconds. Pedestrians use both the "*WALK*" *interval* and the first few seconds of the "*Flashing DON'T WALK*" (FDW) *interval* to enter the intersection. For delay calculations, the *effective green time interval* is equal to the walk interval plus the first 4 sec of the FDW interval.

LOS criteria for pedestrians at signalized intersections are shown in Table 3.7. When delay is greater than 30 sec, pedestrians become impatient and will take more risks, such as disregarding signal indicators (i.e., "WALK" and/or "DON'T WALK" intervals). *Pedestrian noncompliance* with signal indicators is reduced by high traffic volumes, as pedestrians have little choice but to wait for the walk signal.

Table 3.7 *LOS Criteria for Pedestrians at Signalized Intersections*

LOS	pedestrian delay (sec/ped)	likelihood of noncompliance
A	< 10	low
B	≥ 10–20	↑
C	> 20–30	↕
D	> 30–40	↕
E	> 40–60	↓
F	60	very high

Highway Capacity Manual 2000. Copyright, National Academy of Sciences, Washington, D.C. Exhibit 18-9. Reproduced with permission of the Transportation Research Board.

Pedestrian Sidewalks on Urban Streets

Pedestrian sidewalks on urban streets include both interrupted and uninterrupted flows and are comprised of segments and intersections. *Segments* are defined from a signalized intersection and an upstream portion of pedestrian sidewalk that begins immediately after a signalized or unsignalized intersection. *Average pedestrian travel speed*, S_{at}, is the service measure for sidewalks. It is based on the time required, including stops, to traverse between two points and is found using Eq. 3.7. L_t is the total length of the street, L_i is the length of segment i, S_i is the pedestrian walking speed over segment i, and d_j is the pedestrian delay at intersection j.

$$S_{\text{at}} = \frac{L_t}{\sum \frac{L_i}{S_i} + \sum d_j} \quad \text{[}HCM\text{ Eq. 18-22]} \qquad 3.7$$

Pedestrians may experience delays on sidewalks due to commercial and residential driveways, upgrades and downgrades, pavement surface conditions, and other local features. As delay and sidewalk density increase, pedestrian risk-taking also increases, such as accepting shorter traffic gaps in order to enter or cross the traffic flow, dodging around lateral obstructions, or bumping into other pedestrians.

Determining sidewalk LOS is similar to determining vehicular LOS and uses thresholds of percentage of base speed. The LOS criteria for pedestrians on urban street sidewalks are presented in Table 3.8.

Table 3.8 *LOS Criteria for Pedestrians on Urban Street Sidewalks*

LOS	travel speed (ft/sec)
A	>4.36
B	>3.84–4.36
C	>3.28–3.84
D	>2.72–3.28
E	≥1.90–2.72
F	<1.90

(Multiply ft/sec by 0.305 to obtain m/s.)

Highway Capacity Manual 2000. Copyright, National Academy of Sciences, Washington, D.C. Exhibit 18-14. Reproduced with permission of the Transportation Research Board.

Pedestrian Area Requirements at Street Corners

A certain amount of *pedestrian area* is required at all signalized street corners. A *circulation area* is needed, providing space for pedestrians crossing the street during the green signal phase, pedestrians joining the queue of people waiting to cross during the red signal phase (the *red-phase queue*), and pedestrians moving between the sidewalks adjoining the corner who are not crossing the street. A *temporary holding area* is also required to accommodate the red-phase queue while those pedestrians wait to cross.

Figure 3.10 and Fig. 3.11 show the two *signal phase conditions* assessed when calculating values for crosswalks and street corners. *Condition one*, shown in Fig. 3.10, applies when pedestrians are crossing the minor street and the red-phase queue is waiting to cross the major street. *Condition two*, shown in Fig. 3.11, applies when pedestrians are crossing the major street and the red-phase queue is waiting to cross the minor street.

Figure 3.10 *Condition One: Minor Street Crossing*

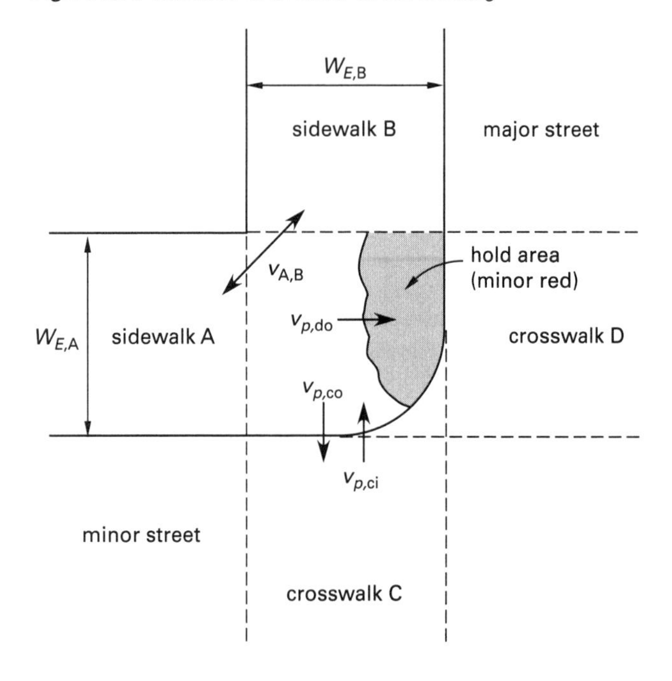

key condition 1

$v_{A,B}$ = sidewalk flow
$v_{p,do}$ = peds joining queue
$v_{p,co}$ = outbound crossing peds
$v_{p,ci}$ = inbound crossing platoon
$W_{E,A\,or\,B}$ = effective width of sidewalks

Highway Capacity Manual 2000. Copyright, National Academy of Sciences, Washington, D.C. Exhibit 18-11. Reproduced with permission of the Transportation Research Board.

Assessment of pedestrian area requirements at a given street corner compares the time and space available on the corner to the pedestrian arrival demand at that corner. The space available is constrained by sidewalk geometry at the corner, and the time available is determined by the signal timing at the corner.

In order to calculate the circulation area, the total number of circulating pedestrians and the net corner time-space available for circulating pedestrians must be known. The total number of circulating pedestrians, v_t, can be found from Eq. 3.8. The variables required for this equation are illustrated in Fig. 3.10 and Fig. 3.11.

$$v_t = v_{p,\text{ci}} + v_{p,\text{co}} + v_{p,\text{di}} + v_{p,\text{do}} + v_{\text{A,B}} \quad 3.8$$

Figure 3.11 *Condition Two: Major Street Crossing*

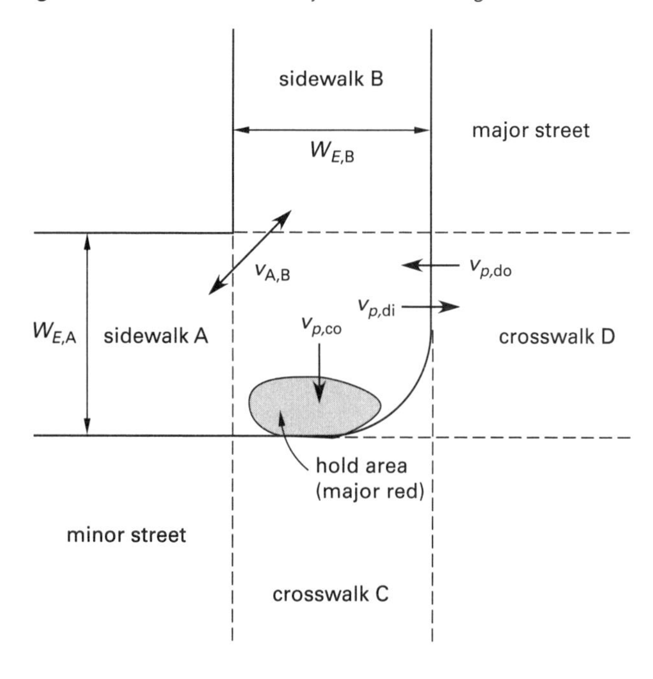

key condition 2

$v_{A,B}$ = sidewalk flow
$v_{p,co}$ = peds joining queue
$v_{p,do}$ = outbound crossing peds
$v_{p,di}$ = inbound crossing platoon
$W_{E,A\,or\,B}$ = effective width of sidewalks

Highway Capacity Manual 2000. Copyright, National Academy of Sciences, Washington, D.C. Exhibit 18-12. Reproduced with permission of the Transportation Research Board.

If the number of pedestrians waiting to cross in any direction on either street during one cycle is not known, it can be found from Eq. 3.9.

$$v = \left(\frac{v_p}{15\text{ min}}\right)C \quad 3.9$$

Finding the net corner time-space available requires the total time-space available and the time-space occupied by pedestrians waiting to cross. The *total time-space available*, TS, at a given corner can be found from Eq. 3.10. r is the radius of the curb, and $W_{E,\text{A}}$ and $W_{E,\text{B}}$ are the effective widths of sidewalk A and sidewalk B, respectively, as shown in Fig. 3.10 and Fig. 3.11.

$$\text{TS} = C(W_{E,\text{A}}W_{E,\text{B}} - 0.215r^2) \quad [\textit{HCM}\text{ Eq. 18-6}] \quad 3.10$$

The time-space occupied by pedestrians waiting to cross is a function of the *holding area waiting time* under condition one, Q_{tdo}, given by Eq. 3.11, and the holding area waiting time under condition two, Q_{tco}, given by Eq. 3.12. $v_{p,\text{do}}$ and $v_{p,\text{co}}$ are the number of pedestrians waiting to cross the major and minor streets, respectively, during one cycle. R_{mi} is the length of the minor-street red phase, and R_{mj} is the length of the major-street red phase. If pedestrian signals are installed, the red phase of both streets is equal to the length of the "DON'T WALK" phase.

$$Q_{\text{tdo}} = \frac{v_{p,\text{do}} R_{\text{mi}}^2}{2C} \quad \text{[HCM Eq. 18-7]} \qquad 3.11$$

$$Q_{\text{tco}} = \frac{v_{p,\text{co}} R_{\text{mj}}^2}{2C} \quad \text{[HCM Eq. 18-8]} \qquad 3.12$$

When all these values are known, the *net corner time-space available*, TS_c, can be found using Eq. 3.13.

$$\text{TS}_{c,\text{m}^2\cdot\text{s}} = \text{TS}_{\text{m}^2\cdot\text{s}} - 0.5(Q_{\text{tdo}} + Q_{\text{tco}})$$
[*HCM* Eq. 18-9] [SI] *3.13(a)*

$$\text{TS}_{c,\text{ft}^2\text{-sec}} = \text{TS}_{\text{ft}^2\text{-sec}} - 5(Q_{\text{tdo}} + Q_{\text{tco}})$$
[*HCM* Eq. 18-9] [U.S.] *3.13(b)*

Circulating space for each pedestrian, M, can be found from Eq. 3.14. The total circulation volume, v_t, is multiplied by 4 sec, the assumed average circulation time. The calculated space for each pedestrian can be used with Table 3.3 to determine the LOS for the walkway being studied.

$$M = \frac{\text{TS}_c}{4v_t} \quad \text{[HCM Eq. 18-10]} \qquad 3.14$$

In addition to adequate circulating area at a street corner, adequate circulating area must also be provided in a crosswalk to ensure pedestrians are able to clear the crossing during the green phase. Required circulating area for a crosswalk is determined similarly to required circulating area for a street corner, with some variations to address the safety concerns and time constraints placed on pedestrians crossing a street.

To find the circulation space available per pedestrian in the crosswalk, the crosswalk time-space available and the total crosswalk occupancy time must be known. The *crosswalk time-space available* can be found from Eq. 3.15. L is the length of the crosswalk, W_E is the effective width of the crosswalk, WALK + FDW is the effective green time used by a crossing pedestrian, S_p is the average speed of pedestrians in the crosswalk, and g is the green time when a pedestrian signal is not installed. These variations of the same equation reflect the fact that where pedestrian signals are installed, pedestrians use both the "WALK" interval and the first few seconds of the FDW interval to cross an intersection, whereas when no pedestrian signal is installed, pedestrians will tend to use only the green time available.

$$\text{TS} = LW_E\left(\text{WALK} + \text{FDW} - \frac{L}{2S_p}\right)$$
[*HCM* Eq. 18-11] [when signal is installed] *3.15(a)*

$$\text{TS} = LW_E\left(g - \frac{L}{2S_p}\right)$$
[*HCM* Eq. 18-11] [when signal is not installed] *3.15(b)*

Equation 3.16 can be used to calculate the number of pedestrians crossing during a single cycle. v is the volume of pedestrians on the walkway that feeds into the crossing.

$$N_p = \frac{v(C - g)}{C} \quad \text{[HCM Eq. 18-12]} \qquad 3.16$$

To find the total crosswalk occupancy time, the *total crossing time* (also called the *effective green time*), the time it takes for a crosswalk to clear once pedestrians enter the crosswalk, must be known. For calculations of total crossing time, pedestrian start-up time is assumed to be 3.2 sec. Equation 3.17 gives the total crossing time for crosswalks more than 10 ft (3 m) wide, and Eq. 3.18 gives the total crossing time for crosswalks 10 ft (3 m) wide or less.

$$t = 3.2 + \frac{L_{\text{m}}}{S_p} + 0.81\left(\frac{N_p}{W_{\text{m}}}\right)$$
[*HCM* Eq. 18-13] [SI] *3.17(a)*

$$t = 3.2 + \frac{L_{\text{ft}}}{S_p} + 2.7\left(\frac{N_p}{W_{\text{ft}}}\right)$$
[*HCM* Eq. 18-13] [U.S.] *3.17(b)*

$$t = 3.2 + \frac{L}{S_p} + 0.27N_p \quad \text{[HCM Eq. 18-13]} \qquad 3.18$$

Pedestrian Area Requirements in Crosswalks

The *total crosswalk occupancy time*, T, is given by Eq. 3.19. t is the total crossing time given from Eq. 3.17 or Eq. 3.18, v_i is the volume of inbound pedestrians crossing the intersection, and v_o is the volume of outbound pedestrians crossing the intersection.

$$T = (v_i + v_o)t \quad \text{[HCM Eq. 18-14]} \qquad 3.19$$

The circulation space per pedestrian in the crosswalk is found from Eq. 3.20. Like the circulation space at a street corner, this value can be used with Table 3.3 to determine the LOS for the walkway.

$$M = \frac{\text{TS}}{T} \quad [\textit{HCM} \text{ Eq. 18-15}] \qquad 3.20$$

The effect of vehicles turning into the crosswalk on the LOS for pedestrians crossing the crosswalk can be estimated by assuming an 8 ft (2.4 m) swept path for the vehicle. The time the vehicle will occupy the crosswalk is assumed to be 5 sec. Equation 3.21 is used to determine the time-space of turning vehicle interference, TS_{tv}. N_{tv} is the number of turning vehicles during the green phase. The turning-vehicle time-space is subtracted from the crosswalk time-space calculated in Eq. 3.15 to determine the effective time-space, TS_E, available for pedestrians.

$$\text{TS}_{\text{tv}} = 12N_{\text{tv}}W_E \quad [\textit{HCM} \text{ Eq. 18-16}] \quad [\text{SI}] \qquad 3.21(a)$$

$$\text{TS}_{\text{tv}} = 40N_{\text{tv}}W_E \quad [\textit{HCM} \text{ Eq. 18-16}] \quad [\text{U.S.}] \qquad 3.21(b)$$

$$\text{TS}_E = \text{TS} - \text{TS}_{\text{tv}} \quad [\textit{HCM} \text{ Exh. 18-16}] \qquad 3.22$$

Signalized Intersection LOS

LOS for the major components of a signalized intersection are determined using the following steps. The calculated values and corresponding LOS can be recorded in tabular form to allow comparisons between LOS on the different sections of the intersection to determine which areas could benefit from improvement.

step 1: Calculate the average delay, d_p, for the major and minor street using Eq. 3.6. Determine the LOS for both streets using Table 3.7.

step 2: Determine the total time-space, TS, available for crossing the major and minor streets using Eq. 3.15. Use Eq. 3.15(a) if a pedestrian signal is installed and Eq. 3.15(b) if a signal is not installed. If right-turning vehicles move through the crosswalk during the crossing period, calculate the time-space occupied by turning vehicles, TS_{tv}, from Eq. 3.21, then use Eq. 3.22 to determine the effective time-space, TS_E, available for pedestrians.

step 3: For the major and minor street, multiply the pedestrian inbound and outbound counts per 15 min period, $N_{p,15\,\text{min}}$, by the cycle length, C, to determine the flow rates in pedestrians per cycle. Using the same method, convert the number of pedestrians waiting on the corner during a 15 min period to pedestrians per cycle.

step 4: Calculate the total circulating pedestrian flow, v_t, using Eq. 3.8 and the total time-space, TS, available on the corner from Eq. 3.10.

step 5: Find the holding-area waiting time for the major street, Q_{tdo}, and the minor street, Q_{tco}, using Eq. 3.11 and Eq. 3.12, respectively.

step 6: Determine the net corner time-space available, TS_c, using Eq. 3.13.

step 7: Find the space per circulating pedestrian at the corner, M, using Eq. 3.14, then determine the corner LOS using Table 3.3.

step 8: Calculate the number of pedestrians that will cross the major and minor streets during one cycle, N_p, using Eq. 3.16.

step 9: Find the crossing time needed for the major and minor street, t, using Eq. 3.17 for crosswalk widths, W, more than 10 ft (3 m) wide and Eq. 3.18 for crosswalk widths 10 ft (3 m) wide or less.

step 10: Calculate the total crosswalk occupancy time, T, using Eq. 3.19 and the circulation space, M, using Eq. 3.20 for the major and minor streets. Using the circulation space, determine the LOS for the major and minor streets.

step 11: Complete the table and compare the LOS on the various components of the intersection to determine where incremental improvements would most likely have the greatest effect.

Unsignalized Intersections

When a pedestrian needs to cross an unsignalized intersection, the pedestrian will only cross if he or she can find a gap in vehicle traffic that he or she considers acceptable for crossing. The *critical gap*[1] is the time which a pedestrian considers the minimum time acceptable for crossing. If the available gap in traffic is less than the critical gap, the pedestrian will not cross.

The critical gap for a single pedestrian, t_c, is calculated by Eq. 3.23. The pedestrian start-up time and the end clearance time (the time between the pedestrian clearing the traveled lane and the arrival of the next vehicle at the pedestrian crossing) are combined into one term, t_s. L is the length of the crosswalk, and S_p is the walking speed of the pedestrian.

$$t_c = \frac{L}{S_p} + t_s \quad [\textit{HCM} \text{ Eq. 18-17}] \qquad 3.23$$

Most pedestrians are comfortable with an end clearance time of at least 0.5 sec, but for calculation of critical gap, end clearance time can be assumed to be zero unless study data suggest otherwise.

Calculation of a *group critical gap* depends on whether or not platooning is observed among pedestrians crossing the intersection. If platooning is observed, the spatial distribution of the platoon must be found in order

[1]The critical gap does not pertain to pedestrians using a crosswalk marked with zebra stripes, as pedestrians have the right-of-way, even though they may still hesitate to make sure vehicular traffic stops. Instead, pedestrian delay analysis is performed using the method for two-way stop-controlled intersections covered in the *HCM*.

to calculate the group critical gap. If no platooning is observed, the spatial distribution of the platoon can be assumed to be 1.

Calculation of the platoon spatial distribution requires knowing the total number of pedestrians in the platoon. If the total number of pedestrians in the platoon, N_c, is not known from field study, it can be found using Eq. 3.24. v_p is the pedestrian flow rate, v_v is the vehicle flow rate, and t_c is the critical gap for a single pedestrian.

$$N_c = \frac{v_p e^{v_p t_c} + v_v e^{-v_v t_c}}{(v_p + v_v) e^{(v_p - v_v) t_c}} \quad [HCM \text{ Eq. 18-19}] \qquad 3.24$$

Where platooning is not observed (i.e., arrivals are random), the average number of pedestrians waiting to cross the intersection is found from Eq. 3.25. v_p is the pedestrian flow rate in the intersection, and N_g is the total number of expected gaps that are greater than or equal to the minimum acceptable gap. However, the number of crossing pedestrians is not required to calculate the group critical gap for random arrivals, which is calculated using a platoon spatial distribution of 1.

$$N_c = \frac{v_p}{N_g} \qquad 3.25$$

The platoon spatial distribution can be found using Eq. 3.26. The value calculated in the first part of the expression must be rounded to the nearest integer.

$$N_p = \text{INT}\left(\frac{(0.75)(N_c - 1)}{W_{E,\text{m}}}\right) + 1 \quad [HCM \text{ Eq. 18-18] [SI]} \qquad 3.26(a)$$

$$N_p = \text{INT}\left(\frac{(8.0)(N_c - 1)}{W_{E,\text{ft}}}\right) + 1 \quad [HCM \text{ Eq. 18-18] [U.S.]} \qquad 3.26(b)$$

Once the platoon spatial distribution is known, the group critical gap, t_G, can be found using Eq. 3.27.

$$t_G = t_c + 2(N_p - 1) \quad [HCM \text{ Eq. 18-20}] \qquad 3.27$$

LOS for pedestrians crossing at an unsignalized intersection is determined based on the delay experienced by pedestrians waiting to cross the intersection. The delay experienced by a pedestrian, d_p, can be found using Eq. 3.28. Table 3.9 shows LOS criteria for pedestrians at unsignalized intersections.

$$d_p = \frac{1}{v_v}\left(e^{v_v t_G} - v_v t_G - 1\right) \quad [HCM \text{ Eq. 18-21}] \qquad 3.28$$

Table 3.9 *LOS Criteria for Pedestrians at Unsignalized Intersections*

LOS	average delay/pedestrian (sec)	likelihood of risk-taking behavior*
A	< 5	low
B	≥ 5–10	↑
C	> 10–20	↕
D	> 20–30	↕
E	> 30–45	↓
F	> 45	high

*Likelihood of acceptance of short gaps.

Highway Capacity Manual 2000. Copyright, National Academy of Sciences, Washington, D.C. Exhibit 18-13. Reproduced with permission of the Transportation Research Board.

LOS thresholds for unsignalized intersections have much smaller tolerances for delay than LOS thresholds for signalized intersections, as there is a higher likelihood of risk-taking behavior in these situations.

Example 3.1

A four-way stop intersection on a major street has no signal on the major street and stop control on the minor street. Pedestrian crossing information for the major street is as follows.

walking speed	4 ft/sec
start-up and clearance time	3 sec
crosswalk length	42 ft
effective width	8 ft
vehicle flow rate	375 vph
pedestrian flow rate	85 ped/hr

What is the pedestrian LOS for this intersection?

Solution

Determine the vehicle flow rate in vehicles per second and the pedestrian flow rate in pedestrians per second.

$$v_v = \frac{375 \frac{\text{veh}}{\text{hr}}}{3600 \frac{\text{sec}}{\text{hr}}} = 0.104 \text{ veh/sec}$$

$$v_p = \frac{85 \frac{\text{ped}}{\text{hr}}}{3600 \frac{\text{sec}}{\text{hr}}} = 0.0236 \text{ ped/sec}$$

Find the critical gap for one pedestrian using Eq. 3.23.

$$t_c = \frac{L}{S_p} + t_s = \frac{42 \text{ ft}}{4 \frac{\text{ft}}{\text{sec}}} + 3 \text{ sec} = 13.5 \text{ sec}$$

Pedestrian/ Transit Analysis

From Eq. 3.24, the number of pedestrians in a crossing platoon is

$$N_c = \frac{v_p e^{v_p t_c} + v_v e^{-v_v t_c}}{(v_p + v_v) e^{(v_p - v_v) t_c}}$$

$$= \frac{\left(0.0236\ \frac{\text{ped}}{\text{sec}}\right) e^{(0.0236\ \text{ped/sec})(13.5\ \text{sec})} + \left(0.104\ \frac{\text{veh}}{\text{sec}}\right) e^{(-0.104\ \text{veh/sec})(13.5\ \text{sec})}}{\left(0.0236\ \frac{\text{ped}}{\text{sec}} + 0.104\ \frac{\text{veh}}{\text{sec}}\right) \times e^{(0.0236\ \text{ped/sec} - 0.104\ \text{veh/sec})(13.5\ \text{sec})}}$$

$$= 1.35\ \text{ped}$$

From Eq. 3.26, the group spatial distribution is

$$N_p = \text{INT}\left(\frac{(8.0)(N_c - 1)}{W_{E,\text{ft}}}\right) + 1$$

$$= \text{INT}\left(\frac{(8.0\ \text{ft})(1.35\ \text{ped} - 1)}{8\ \text{ft}}\right) + 1$$

$$= 1\ \text{ped}$$

From Eq. 3.27, the group critical gap is

$$t_G = t_c + 2(N_p - 1)$$

$$= 13.5\ \text{sec} + (2)(1\ \text{ped} - 1)$$

$$= 13.5\ \text{sec}$$

From Eq. 3.28, the average pedestrian delay for the crosswalk is

$$d_p = \frac{1}{v_v}(e^{v_v t_G} - v_v t_G - 1)$$

$$= \left(\frac{1}{0.104\ \frac{\text{veh}}{\text{sec}}}\right)\left(e^{(0.104\ \text{veh/sec})(13.5\ \text{sec})} - \left(0.104\ \frac{\text{veh}}{\text{sec}}\right)(13.5\ \text{sec}) - 1\right)$$

$$= 16.03\ \text{sec}$$

From Table 3.9, the LOS for the intersection is C.

Stairways

Stairways can be designed using two approaches—either by fitting within structural limits of a facility or by pedestrian travel flow requirements. Usually, the structural approach results in stairs that fit the steepest angle, the shortest tread, and the highest riser permitted by codes or local ordinances, or by the facility's architectural aspects when meeting the minimum code requirements. For example, codes may set widths in multiples of 22 in (56 cm), with a vertical angle of 32.5°, using a 7 in (18 cm) riser and an 11 in (28 cm) tread.

The pedestrian travel flow approach views horizontal travel speed, service volume, and population agility as significant design parameters. When pedestrians walk up or down stairs, horizontal speed is reduced from that on level walkways. Agility and locomotion studies show that a greater horizontal speed is attained in the up-direction on a vertical angle of 26.5° using a 6 in (15 cm) riser and a 12 in (30 cm) tread. Often a compromise of a 28.6° vertical angle will be used, with a 6 in (15 cm) riser and an 11 in (28 cm) tread, because it is closer to a pedestrian's average 22 in (56 cm) pace length. The average pedestrian's down speed does not significantly change when the vertical angle changes from 32.5° to 28.6°. However, older pedestrians or pedestrians with disabilities may be more confident using stairs with less steep angles, particularly on outdoor stairs in inclement weather. In locations where snow and ice accumulate, vertical angles of 28.6° or 26.5° are safer than steeper angles.

Stair width should never be less than the effective width of approach sidewalks, and it should be increased by the equivalent width of central handrails when handrails are installed. Widths determined by multiples of 30 in (76 cm) are more comparable to pedestrian body motion than the 22 in (56 cm) code requirements. When the opposing flow is a small proportion of the prevailing flow, stair width should be increased by 30 in (76 cm), which is the equivalent of adding another lane.

Architectural details that interfere with snow and ice clearing reduce the ability of maintenance crews to clear the steps, and therefore, reduce stair capacity, during inclement weather. In high-volume locations, this may require additional stair width in order to maintain fair weather flow rates.

Landings must be the same width as the stairs. They can be as short as 40 in (1 m) according to some building codes. However, a length of 44 in (1.1 m) fits better with the average pedestrian's gait and is more comfortable for pedestrians with mobility challenges.

Adequate landing space should be provided at the top and bottom of the stairway to allow for pedestrian queues. The total elevation change from floor to floor should be divided into approximately equal sections, with landings placed at no more than every 12 ft (3.7 m) of elevation change. Mobility challenges arise when the elevation change is great, so landings may be placed at more frequent intervals.

Straight runs of stairs have the highest service capacity compared to stairs that jog to the left or right at landings, or require pedestrians to spiral to a reverse direction at landings. LOS criteria is given for stairways in Table 3.10. The volume-capacity ratios given in Table 3.10 are based on a stairway capacity of 530 ped/min-ft (162 ped/min·m). The LOS of a stairway should match the LOS of the adjoining walkway.

Table 3.10 LOS Criteria for Stairways

LOS	space (ft^2/ped)	flow rate (ped/min-ft)	average horizontal speed (ft/sec)	v/c ratio
A	>20	≤5	>1.74	≤0.33
B	>17–20	>5–6	>1.74	>0.33–0.41
C	>12–17	>6–8	>1.57–1.74	>0.41–0.53
D	>8–12	>8–11	>1.38–1.57	>0.53–0.73
E	>5–8	>11–15	>1.31–1.38	>0.73–1.00
F	≤5	variable	≤1.31	variable

(Multiply ft^2/ped by 0.093 to obtain m^2/ped.)
(Multiply ped/min-ft by 0.305 to obtain ped/min·m.)
(Multiply ft/sec by 0.305 to obtain m/s.)

Highway Capacity Manual 2000. Copyright, National Academy of Sciences, Washington, D.C. Exhibit 18-5. Reproduced with permission of the Transportation Research Board.

Summary of Analysis Procedures

The most accurate and reliable data are obtained through field observations of the pedestrian population at or near the design location. However, in the absence of field data, default values may be used (see Fig. 3.6, Table 3.2, and Table 3.1). The following are summaries of the input parameters and applicable formulas for the previously described pedestrian facilities.

Walkways

- length of the walkway study segment
- width of the walkway or crosswalk
- grade of the walkway (use a value of zero if the grade is less than 10% for short distances)
- type of pedestrian traffic (e.g., commuters, shoppers, elderly, school-age children)
- pedestrian travel speed (see Eq. 3.7)

Walkways and Sidewalk Pedestrian Facilities

- sum of obstruction and/or shy widths
- peak 15 min flow rate for both directions
- pedestrian unit flow rate (see Eq. 3.5)

Crossings at Signalized Intersections, Unsignalized Intersections, and Urban Street Facilities

- cycle length
- effective green time (see Eq. 3.17)
- average delay (see Eq. 3.6)
- single pedestrian critical gap (see Eq. 3.23)
- typical pedestrian number in platoon (see Eq. 3.24)
- spatial pedestrian distribution (see Eq. 3.26)
- group critical gap (see Eq. 3.27)
- vehicular flow rate
- average pedestrian delay (see Eq. 3.28)

Street Corner Time-Space Analysis (for Signalized Intersections)

- number of pedestrians waiting to cross the major street (see Eq. 3.12)
- number of pedestrians waiting to cross the minor street (see Eq. 3.11)
- total time-space available (see Eq. 3.10)
- circulation space per pedestrian (see Eq. 3.14)

Crosswalk Time-Space Analysis (for Signalized Intersections)

- average delay (see Eq. 3.6)
- number of pedestrians arriving during "DON'T WALK" or red indication
- total crossing time (see Eq. 3.17)
- total crosswalk time-space (see Eq. 3.15)
- total crosswalk occupancy time (see Eq. 3.19)
- number of conflicting right-turn movements
- time-space of turning vehicles (see Eq. 3.21)
- effective time-space (see Eq. 3.22)
- circulation space per pedestrian (see Eq. 3.20)

Pedestrian Level of Service Criteria

The LOS criteria for pedestrians are given in the following tables.

- Table 3.3 (Average Flow LOS Criteria for Walkways and Sidewalks)
- Table 3.5 (LOS Criteria for Pedestrians on Walkways and in Queues)
- Table 3.6 (Platoon-Adjusted LOS Criteria for Walkways and Sidewalks)
- Table 3.9 (LOS Criteria for Pedestrians at Unsignalized Intersections)
- Table 3.7 (LOS Criteria for Pedestrians at Signalized Intersections)
- Table 3.10 (LOS Criteria for Stairways)
- Table 3.8 (LOS Criteria for Pedestrians on Urban Street Sidewalks)

Example 3.2

A pedestrian crossing is located at a signalized intersection with a two-phase, 75 sec cycle, 5 sec clearance interval signal. There are no pedestrian signals. The green time is 40 sec for the major street and 30 sec for the minor street. Data collected at the intersection during field studies are shown. Assuming a pedestrian crossing speed of 4 ft/sec and no pedestrian lost time, create a summary table showing the pedestrian LOS at the intersection based on delay and available space, and determine the location where improvements would most likely have the most impact on the intersection.

major street	
crosswalk length, L	40 ft
effective crosswalk width, $W_{E,\mathrm{A}}$	14 ft
inbound pedestrian count, $v_{p,\mathrm{di}}$	300 ped/15 min
outbound pedestrian count, $v_{p,\mathrm{do}}$	150 ped/15 min
phase green time, g_d	40 sec
minor street	
crosswalk length, L	20 ft
effective crosswalk width, $W_{E,\mathrm{B}}$	14 ft
inbound pedestrian count, $v_{p,\mathrm{ci}}$	500 ped/15 min
outbound pedestrian count, $v_{p,\mathrm{co}}$	200 ped/15 min
phase green time, g_c	30 sec
corner	
radius	20 ft
sidewalk flow, $v_{\mathrm{A,B}}$	225 ped/15 min
effective sidewalk width, $W_{E,\mathrm{A}}$	14 ft
effective sidewalk width, $W_{E,\mathrm{B}}$	14 ft

Solution

step 1: Find the average delay and LOS for both the major and minor street.

For the major street, from Eq. 3.6,

$$d_p = \frac{0.5(C-g_c)^2}{C} = \frac{(0.5)(75 \text{ sec} - 30 \text{ sec})^2}{75 \text{ sec}}$$
$$= 13.5 \text{ sec}$$

From Table 3.7, the LOS is B.

For the minor street, from Eq. 3.6,

$$d_p = \frac{0.5(C-g_d)^2}{C} = \frac{(0.5)(75 \text{ sec} - 40 \text{ sec})^2}{75 \text{ sec}}$$
$$= 8.17 \text{ sec}$$

From Table 3.7, the LOS is A.

step 2: Determine the net time-space available for crossing the major and minor streets. Use Eq. 3.15(b) since there is not a pedestrian signal.

For the major street, from Eq. 3.15(b),

$$\text{TS} = LW_E\left(g_c - \frac{L}{2S_p}\right)$$
$$= (40 \text{ ft})(14 \text{ ft})\left(30 \text{ sec} - \frac{40 \text{ ft}}{(2)\left(4\ \frac{\text{ft}}{\text{sec}}\right)}\right)$$
$$= 14{,}000 \text{ ft}^2\text{-sec}$$

For the minor street, from Eq. 3.15(b),

$$\text{TS} = LW_E\left(g_d - \frac{L}{2S_p}\right)$$
$$= (20 \text{ ft})(14 \text{ ft})\left(40 \text{ sec} - \frac{20 \text{ ft}}{(2)\left(4\ \frac{\text{ft}}{\text{sec}}\right)}\right)$$
$$= 10{,}500 \text{ ft}^2\text{-sec}$$

step 3: Convert flows to pedestrians per cycle. For the major street,

$$v_{p,\mathrm{di}} = \left(\frac{N_{p,15\text{ min}}}{60\ \frac{\text{sec}}{\text{min}}}\right)C$$
$$= \left(\frac{\frac{300 \text{ ped}}{15 \text{ min}}}{60\ \frac{\text{sec}}{\text{min}}}\right)\left(75\ \frac{\text{sec}}{\text{cycle}}\right)$$
$$= 25.0 \text{ ped/cycle}$$

$$v_{p,\mathrm{do}} = \left(\frac{N_{p,15\text{ min}}}{60\ \frac{\text{sec}}{\text{min}}}\right)C$$
$$= \left(\frac{\frac{150 \text{ ped}}{15 \text{ min}}}{60\ \frac{\text{sec}}{\text{min}}}\right)\left(75\ \frac{\text{sec}}{\text{cycle}}\right)$$
$$= 12.5 \text{ ped/cycle}$$

For the minor street,

$$v_{p,\mathrm{ci}} = \left(\frac{N_{p,15\text{ min}}}{60\ \frac{\text{sec}}{\text{min}}}\right)C$$
$$= \left(\frac{\frac{500 \text{ ped}}{15 \text{ min}}}{60\ \frac{\text{sec}}{\text{min}}}\right)\left(75\ \frac{\text{sec}}{\text{cycle}}\right)$$
$$= 41.7 \text{ ped/cycle}$$

$$v_{p,\text{co}} = \left(\frac{N_{p,15\,\text{min}}}{60\ \frac{\text{sec}}{\text{min}}}\right)C = \left(\frac{\frac{200\ \text{ped}}{15\ \text{min}}}{60\ \frac{\text{sec}}{\text{min}}}\right)\left(75\ \frac{\text{sec}}{\text{cycle}}\right) = 16.7\ \text{ped/cycle}$$

For the corner,

$$v_{\text{A,B}} = \left(\frac{N_{p,15\,\text{min}}}{60\ \frac{\text{sec}}{\text{min}}}\right)C = \left(\frac{\frac{225\ \text{ped}}{15\ \text{min}}}{60\ \text{min}}\right)\left(75\ \frac{\text{sec}}{\text{cycle}}\right) = 18.8\ \text{ped/cycle}$$

step 4: Analyze the street corner by finding the total circulating pedestrian flow and the available time-space. From Eq. 3.8, the total pedestrian flow is

$$\begin{aligned} v_t &= v_{p,\text{ci}} + v_{p,\text{co}} + v_{p,\text{di}} + v_{p,\text{do}} + v_{\text{A,B}} \\ &= 41.7\ \frac{\text{ped}}{\text{cycle}} + 16.7\ \frac{\text{ped}}{\text{cycle}} + 25.0\ \frac{\text{ped}}{\text{cycle}} \\ &\quad + 12.5\ \frac{\text{ped}}{\text{cycle}} + 18.8\ \frac{\text{ped}}{\text{cycle}} \\ &= 115\ \text{ped/cycle} \end{aligned}$$

From Eq. 3.10, the available time-space on the corner is

$$\begin{aligned} \text{TS} &= C(W_{E,\text{A}}W_{E,\text{B}} - 0.215r^2) \\ &= (75\ \text{sec})\Big((14\ \text{ft})(14\ \text{ft}) - (0.215)(20\ \text{ft})^2\Big) \\ &= 8250\ \text{ft}^2\text{-sec} \end{aligned}$$

step 5: Use Eq. 3.11 and Eq. 3.12 to find the holding-area waiting time for pedestrians waiting to cross the major and minor street. The minor-street red phase, R_{mi}, and the major-street red phase, R_{mj}, are equal to the green time plus one clearance interval.

For the major street, from Eq. 3.11,

$$Q_{\text{tdo}} = \frac{v_{p,\text{do}}R_{\text{mi}}^2}{2C} = \frac{\left(12.5\ \frac{\text{ped}}{\text{cycle}}\right)(40\ \text{sec} + 5\ \text{sec})^2}{(2)\left(75\ \frac{\text{sec}}{\text{cycle}}\right)} = 169\ \text{ped-sec}$$

For the minor street, from Eq. 3.12,

$$Q_{\text{tco}} = \frac{v_{p,\text{co}}R_{\text{mj}}^2}{2C} = \frac{\left(16.7\ \frac{\text{ped}}{\text{cycle}}\right)(30\ \text{sec} + 5\ \text{sec})^2}{(2)\left(75\ \frac{\text{sec}}{\text{cycle}}\right)} = 136\ \text{ped-sec}$$

step 6: Find the time-space available at the corner using Eq. 3.13(b).

$$\begin{aligned} \text{TS}_{c,\text{ft}^2\text{-sec}} &= \text{TS}_{\text{ft}^2\text{-sec}} - 5(Q_{\text{tdo}} + Q_{\text{tco}}) \\ &= 8250\ \text{ft}^2\text{-sec} - \left(5\ \frac{\text{ft}^2}{\text{ped}}\right)\left(\begin{array}{r}169\ \text{ped-sec} \\ +\ 136\ \text{ped-sec}\end{array}\right) \\ &= 6725\ \text{ft}^2\text{-sec} \end{aligned}$$

step 7: Find the space per circulating pedestrian at the corner, then find the LOS using Table 3.3.

From Eq. 3.14,

$$M = \frac{\text{TS}_c}{4v_t} = \frac{6725\ \text{ft}^2\text{-sec}}{(4\ \text{sec})\left(115\ \frac{\text{ped}}{\text{cycle}}\right)} = 14.6\ \text{ft}^2/\text{ped}$$

From Table 3.3, the LOS is E.

step 8: Calculate the number of pedestrians that will cross the major and minor streets during the cycle length interval using Eq. 3.16.

For the major street, from Eq. 3.16,

$$N_p = \frac{v_{p,\text{do}}(C - g_c)}{C} = \frac{\left(12.5\ \frac{\text{ped}}{\text{cycle}}\right)\left(75\ \frac{\text{sec}}{\text{cycle}} - 30\ \frac{\text{sec}}{\text{cycle}}\right)}{75\ \frac{\text{sec}}{\text{cycle}}} = 7.50\ \text{ped/cycle}$$

For the minor street, from Eq. 3.16,

$$N_p = \frac{v_{p,\text{co}}(C - g_d)}{C} = \frac{\left(16.7\ \frac{\text{ped}}{\text{cycle}}\right)\left(75\ \frac{\text{sec}}{\text{cycle}} - 40\ \frac{\text{sec}}{\text{cycle}}\right)}{75\ \frac{\text{sec}}{\text{cycle}}} = 7.79\ \text{ped/cycle}$$

step 9: Find the crossing time needed for both the major and minor street.

Since the crosswalk width is greater than 10 ft for both the major and minor street, use Eq. 3.17(b). For the major street,

$$t = 3.2 + \frac{L_{\text{ft}}}{S_p} + 2.7\left(\frac{N_p}{W_{\text{ft}}}\right) = 3.2\ \text{sec} + \frac{40\ \text{ft}}{4\ \frac{\text{ft}}{\text{sec}}} + (2.7)\left(\frac{7.50\ \frac{\text{ped}}{\text{cycle}}}{14\ \text{ft}}\right) = 14.6\ \text{sec}$$

For the minor street,

$$t = 3.2 + \frac{L_{\text{ft}}}{S_p} + 2.7\left(\frac{N_p}{W_{\text{ft}}}\right) = 3.2\ \text{sec} + \frac{20\ \text{ft}}{4\ \frac{\text{ft}}{\text{sec}}} + (2.7)\left(\frac{7.79\ \frac{\text{ped}}{\text{cycle}}}{14\ \text{ft}}\right) = 9.70\ \text{sec}$$

step 10: Find the total crosswalk occupancy time required for crossing and the space per pedestrian at crossing for both the major and minor streets.

For the major street, from Eq. 3.19,

$$T = (v_{p,\text{di}} + v_{p,\text{do}})t = \left(25.0\ \frac{\text{ped}}{\text{cycle}} + 12.5\ \frac{\text{ped}}{\text{cycle}}\right)(14.6\ \text{sec}) = 548\ \text{ped-sec}$$

From Eq. 3.20, the circulation space for the major street is

$$M = \frac{\text{TS}}{T} = \frac{14{,}000\ \text{ft}^2\text{-sec}}{548\ \text{ped-sec}} = 25.5\ \text{ft}^2/\text{ped}$$

From Table 3.3, the LOS is C.

For the minor street, from Eq. 3.19,

$$T = (v_{p,\text{ci}} + v_{p,\text{co}})t = \left(41.7\ \frac{\text{ped}}{\text{cycle}} + 16.7\ \frac{\text{ped}}{\text{cycle}}\right)(9.70\ \text{sec}) = 566\ \text{ped-sec}$$

From Eq. 3.20, the circulation space for the minor street is

$$M = \frac{\text{TS}}{T} = \frac{10{,}500\ \text{ft}^2\text{-sec}}{566\ \text{ped-sec}} = 18.6\ \text{ft}^2/\text{ped}$$

From Table 3.3, the LOS is D.

step 11: Complete the table. Comparing LOS, improving the corner for waiting pedestrians, such as enlarging the corner to provide more circulation space, would most likely provide the greatest improvement for the intersection.

facility and activity	LOS criterion	value	LOS
corner, waiting time, crossing major street	delay	13.5 sec	B
corner, waiting time, crossing minor street	delay	8.17 sec	A
corner, circulation space	space	14.6 ft^2/ped	E
crosswalk space, major street	space	25.5 ft^2/ped	C
crosswalk space, minor street	space	18.6 ft^2/ped	D

2. BICYCLE FACILITIES

Bicycle facility evaluations assess the effects of pedestrians, traffic signals, and interactions between bicyclists on the LOS of the bicycle facility. Bicycle facilities tend to operate well below maximum capacity for the facility. As such, performance measures used for vehicle facilities or exclusive pedestrian facilities do not generally work well with bicycles. With the exception of bicycles operating within the normal flow of vehicle traffic, bicycle facilities require unique LOS calculations.

Bicycle facilities are separated into three classes by the Federal Highway Administration (FHWA) and into five facility types by the *HCM*, as shown in Table 3.11. The type of bicycle facility dictates the LOS criteria for the facility and the methods used to find the criteria values. Table 3.12 gives bicycle facility LOS for all bicycle facility types. The two most basic designations are *off-street bicycle facilities* (*bikeways*), facilities which are separated from all nonbicycle traffic or at least from motorized vehicle traffic, and *on-street bicycle facilities* (*bike lanes*), lanes designated for bicycles alongside normal vehicle traffic and separated from the general traffic only by pavement markings and/or pavement surface colorings.

Table 3.11 *Bicycle Facility Designations*

design facility	FHWA	*HCM*
off-street, bicycle use only	class I	bicycles only
off-street, shared use	class I	bicycles, pedestrians, skaters
on-street, designated lane	class II	uninterrupted flow—urban streets
on-street, designated lane	class III	interrupted flow—urban streets
on-street, shared lane with vehicles	class III	assign passenger car equivalents

AASHTO recommends off-street bicycle facilities be 10 ft (3 m) wide, and on-street bicycle facilities be 4 ft (1.2 m) wide, with a minimum width of 8 ft (2.4 m) for two-way, two-lane on-street bicycle traffic and a minimum width of 10 ft (3 m) for two-way, three-lane on-street traffic. However, in situations where adjoining motor vehicle traffic speeds and/or bicycle volume are high, wider bike lanes should be implemented to help ensure safety.

Whatever the actual width of the lanes, because bicycle facilities operate well below maximum capacity, LOS criteria for bicycle facilities are evaluated in terms of the number of *effective lanes* bicyclists use in a given facility, rather than the actual total width of the facility.

Calculations for LOS criteria of the different bicycle facility types are as follows.

Uninterrupted Flow Facilities

Uninterrupted flow facilities include all bikeways designed either exclusively for bicycles or for shared use with pedestrians and/or other nonmotorized vehicles. All uninterrupted flow bicycle facilities must be completely separated from general vehicular traffic and have no fixed interruption points.

The LOS of an uninterrupted flow bicycle facility is based on the frequency of *flow events* in the facility, encounters with other bicyclists that hinder the ability of a bicyclist to operate his or her bicycle in an optimal manner. The *HCM* divides flow events into *passing events*, for bicyclists traveling in the same direction, and *opposing* (*meeting*) *events*, for bicyclists traveling in opposing directions. For bikeways shared with pedestrians or other nonmotorized vehicles, flow events include passing and opposing other facility users as well as other bicyclists.

Table 3.12 *LOS Criteria for Defined Bicycle Facility Conditions*

	frequency of events (events/hr)			
LOS	two-way, two-lane paths exclusive and shared[a]	two-way, three-lane paths exclusive and shared[b]	signal control delay, d_b (sec/bicycle)	urban street bicycle lane travel speed (mph)
A	≤ 40	≤ 90	< 10	> 14
B	> 40–60	> 90–140	≤ 10–20	> 9–14
C	> 60–100	> 140–210	> 20–30	> 7–9
D	> 100–150	> 210–300	> 30–40	> 5–7
E	> 150–195	> 300–375	> 40–60	≥ 4–5
F	> 195	> 375	> 60	< 4

(Multiply mph by 1.61 to obtain kph.)
[a]8 ft wide paths
[b]10 ft wide paths

Highway Capacity Manual 2000. Copyright, National Academy of Sciences, Washington, D.C. Exhibit 19-1, Exhibit 19-2, Exhibit 19-4, and Exhibit 19-5. Reproduced with permission of the Transportation Research Board.

Pedestrian/ Transit Analysis

For shared-use facilities, the number of passing events, $F_{p,s}$, is found using Eq. 3.29. $v_{b,s}$ is the bicycle flow rate in the subject direction, S_p is the pedestrian speed, and S_b is the bicycle speed.

$$F_{p,s} = v_{b,s}\left(1 - \frac{S_p}{S_b}\right) \quad \text{[}HCM\text{ Eq. 18-3]} \qquad 3.29$$

For exclusive bicycle facilities, the number of passing events, $F_{p,e}$, is found using Eq. 3.30.

$$F_{p,e} = 0.188v_{b,s} \quad \text{[}HCM\text{ Eq. 19-1]} \qquad 3.30$$

For shared facilities, the number of meeting events, $F_{m,s}$, is found by Eq. 3.31. $v_{b,o}$ is the bicycle flow rate in the opposing direction. S_p is the pedestrian speed, and S_b is the bicycle speed.

$$F_{m,s} = v_{b,o}\left(1 - \frac{S_p}{S_b}\right) \quad \text{[}HCM\text{ Eq. 18-3]} \qquad 3.31$$

Meeting events on exclusive bicycle facilities, $F_{m,e}$, are estimated by Eq. 3.32.

$$F_{m,e} = 2v_{b,o} \quad \text{[}HCM\text{ Eq. 19-2]} \qquad 3.32$$

The equations for exclusive and shared facility events can be combined in Eq. 3.33 and Eq. 3.34, respectively. $v_{p,o}$ is the pedestrian flow rate in the opposing direction, and $v_{p,s}$ is the flow rate in the subject direction.

$$F_e = v_o + 0.188v_s \qquad 3.33$$

$$F_s = 2.5v_{p,o} + v_{b,o} + 3v_{p,s} + 0.18v_{b,s} \qquad 3.34$$

The total number of events occurring in a facility, F_t, with a weighting factor of 0.5 for meeting events, is found using Eq. 3.35.

$$F_t = 0.5F_m + F_p \quad \text{[}HCM\text{ Eq. 19-3]} \qquad 3.35$$

Once the total frequency of events is known, the number of events per hour can be found. The number of events per hour gives the LOS for exclusive and shared bicycle facilities from Table 3.12.

The total frequency of events can also be used to find the total bicycle flow rate for a bikeway at LOS F, as shown in Eq. 3.36. Proportions of the total flow rate traveling in the subject direction, p_s, and the corresponding bicycle flow rates are shown in Table 3.13.

$$v_{b,t} = \frac{F_t}{1 - 0.812p_s} \quad \text{[}HCM\text{ Eq. 19-4]} \qquad 3.36$$

Table 3.13 *Total Bicycle Flow Rate for Proportion of Flow in Subject Direction*

proportion of total flow in subject direction, p_s	total bicycle flow rate, $v_{b,t}$ (bicycles/hr)
0.1	212
0.2	233
0.3	257
0.4	288
0.5	328
0.6	380
0.7	452
0.8	557
0.9	724
1.0	1037

Interrupted Flow Facilities

Interrupted flow bicycle facilities are bike lanes that must pass through signalized intersections. LOS criteria for interrupted flow bicycle facilities are based on *control delay*, the delay resulting from a signal that causes the bicycle to reduce speed or stop. Control delay is calculated using Eq. 3.37. C is the cycle length, g is the green time, and v_b is the volume of bicycles at the intersection.

$$d_b = \frac{0.5C\left(1 - \frac{g}{C}\right)^2}{1 - \left(\frac{g}{C}\min\left(\frac{v_b}{c_b}, 1.0\right)\right)} \quad \text{[}HCM\text{ Eq. 19-10]} \qquad 3.37$$

Assuming there is no *overflow delay*, delay caused by an arrival rate that exceeds the service flow rate of the intersection, a value of 1.0 can be used in the denominator. This is more common with bicycles than with motorized vehicles, as bicyclists will generally find an alternate route (or engage in risk-taking behavior) rather than tolerate overflow delay. If there is overflow delay at the intersection, calculation of control delay requires the capacity of the bike lane, c_b, be known. Equation 3.38 gives the capacity of a bike lane at a signalized intersection. g is the green time at the intersection, C is the signal cycle length, and s_b is the *base saturation flow rate* of the intersection, in bicycles/hr. Base saturation flow rates can be as high as 2600 bicycles/hr, but this value assumes right-turning vehicle traffic will yield to bicycles or that bicycles will experience minimal interference with straight-through flow. Given that very few bicyclists experience such minimal interference, the *HCM* suggests base saturation flow rate of 2000 bicycles/hr be used unless study data suggests a different value be used.

$$c_b = s_b\frac{g}{C} = 2000\frac{g}{C} \quad \text{[}HCM\text{ Eq. 19-9]} \qquad 3.38$$

Once the control delay for an intersection is known, the LOS for bicycles at the intersection can be found from Table 3.14.

Table 3.14 *LOS Criteria for Bicycles at Signalized Intersections*

LOS	control delay (sec/bicycle)
A	<10
B	≥10–20
C	>20–30
D	>30–40
E	>40–560
F	>60

Highway Capacity Manual 2000. Copyright, National Academy of Sciences, Washington, D.C. Exhibit 19-4. Reproduced with permission of the Transportation Research Board.

Bicycles at Unsignalized Intersections

The *HCM* does not cover LOS for bicycles at unsignalized intersections, even in situations where one or more approaches have a bike lane. This is the one situation where performance measures used for motorized vehicles or pedestrians are considered to be applicable to bicycles. Low to moderate volumes of bicycle traffic often mix with motor vehicle traffic streams for cross-movements or left turns, and right turns are frequently accomplished without affecting the flow of motor vehicles, so the methodology used for motor vehicles at unsignalized intersections can also be applied to bicycles at unsignalized intersections. If a bicyclist dismounts from a bicycle and walks it through an intersection, methodologies used for pedestrians can be applied instead.

On-Street Bicycle Facilities

Designated on-street bicycle facilities, or bicycle lanes, are generally provided on streets where bicycle use is moderate to high and the separation of bicycles and vehicles is warranted. Paved highway shoulders can also serve as bicycle lanes.

LOS for on-street bicycle facilities is determined based on the frequency of events, just as with uninterrupted flow facilities. However, events in bike lanes are not limited to interactions with other bicyclists in the flow. The bicycle flow rate in bike lanes can also be affected by the quality of pavement in the bike lane, the amount of debris in the bike lane, lateral obstructions or hazards in the bike lane (e.g., grate openings), the number of driveways intersecting the bike lane, sight distance or other geometric constraints, and interference from drifting vehicle traffic or the opening of vehicle doors.

The number of events for bicycles in a bike lane can be found from Table 3.15, which shows the number of events given the mean speed of bicycles in the facility and a standard deviation in the speed. If insufficient data is available, a mean speed of 11.2 mph (18 kph) and a standard deviation as described in the footnote of Table 3.15 can be used. Using the number of events from Table 3.15, the LOS for the bike lane can be selected from Table 3.12.

LOS for on-street bicycle lanes can be influenced by the bicycle traffic conditions and the number of events. Table 3.15 illustrates the effect of on-street events for a range of bicycle flow rates and standard deviations of bicycle speeds. Using the number of events from Table 3.15, the LOS can be selected from Table 3.12. Should insufficient information be available, a mean speed of 11.2 mph (18 kph) and a standard deviation as described in the footnote of Table 3.15 can be used.

Pedestrian/ Transit Analysis

Table 3.15 *Effect of Mean and Standard Deviation for Speeds on Events for One-Way On-Street Bicycle Facilities*

		number of events								
		bicycle mean speed (mph)								
bicycle flow rate (bicycles/hr)	standard deviation* (mph)	7.5	8.1	8.7	9.3	9.9	10.6	11.2	11.8	12.4
100	0.9	14	13	12	11	10	10	9	9	8
	1.0	29	26	25	23	22	20	19	18	17
	2.8	42	39	36	34	32	30	28	27	25
200	0.9	27	25	23	22	21	19	18	17	16
	1.9	57	53	49	46	43	40	38	36	35
	2.8	84	78	73	68	64	60	56	54	51
300	0.9	41	38	35	33	31	29	27	26	25
	1.9	86	79	74	69	65	61	57	55	52
	2.8	126	117	109	102	96	89	85	80	76

(Multiply mph by 1.61 to obtain kph.)

*Standard deviation of bicycle speeds. If standard deviation data are unavailable, use the following default values:

0.9 mph for facilities used primarily by commuters

1.9 mph for facilities used by various user types

2.8 mph for facilities used primarily for recreational purposes

Highway Capacity Manual 2000. Copyright, National Academy of Sciences, Washington, D.C. Exhibit 19-3. Reproduced with permission of the Transportation Research Board.

Bike Lanes on Urban Streets

Bike lanes on urban streets affect the general vehicle traffic flow. The degree to which bike lanes affect vehicle traffic depends on the design conditions under which bicycles and vehicles interact. The *HCM* provides adjustment factors that consider the percentage of bicycles in left-turn or right-turn movements, as well as the relative interference of pedestrians for right turns. These procedures should be applied at intersections where there are extended bicycle lanes, lanes with both interrupted and uninterrupted flow elements, and where bicycle traffic causes observable delays. For bicycle lanes, uninterrupted flow segments begin immediately after the nearest upstream signal, and interrupted flow elements include the immediate approach to the next signalized intersection, as well as the intersection itself.

For extended designated on-street bicycle lanes, the LOS criteria are based on the average speed of bicycles in the section being analyzed. To analyze the effect of lane sharing between bicycles and motor vehicles on traffic flow, the bicycle flow rate at a given intersection must be adjusted using passenger car equivalents, as shown in Table 3.16.

Table 3.16 *Passenger Car Equivalents for Bicycles*

	lane width (ft)		
bicycle movement	<11	11–14	>14
opposed	1.2	0.5	0.0
unopposed	1.0	0.2	0.0

(Multiply ft by 0.305 to obtain m.)

Bicycles turning left are considered opposing movements on two-way streets. Bicycles traveling straight are considered unopposed. Bicycles turning right can be either opposed or unopposed, depending on the amount of pedestrian interference. After determining whether a bicycle movement is opposed or unopposed, each movement's flow rate is multiplied by the passenger car equivalent, pce, from Table 3.16 and then added to the vehicle flow rate, v_v, to determine the *equivalent traffic volume*, v_e, using Eq. 3.39.

$$v_e = v_v + \sum v_{b,\mathrm{LT}}(\mathrm{pce}) + \sum v_{b,\mathrm{ST}}(\mathrm{pce}) + \sum v_{b,\mathrm{RT}}(\mathrm{pce}) \qquad 3.39$$

The *average travel speed*, S_{at}, over the entire section is calculated using Eq. 3.40. L_t is the total length of the street under analysis, L_i is the length of a segment i, S_i is the bicycle running speed over segment i, and d_j is the average bicycle delay at intersection j that intersects segment i. It is very difficult to calculate average bicycle running speed, given the large number of factors that affect the speed of bicycles, so the *HCM* recommends using an average running speed of 15 mph (24 kph) unless existing data suggests a different value be used.

$$S_{\mathrm{at}} = \frac{L_t}{\sum \frac{L_i}{S_i} + \sum d_j} \quad [\textit{HCM}\ \text{Eq. 19-11}] \qquad 3.40$$

Table 3.17 gives LOS criteria based on bicycle average travel speed.

Table 3.17 *LOS for Bicycle Lanes on Urban Streets with Extended Bike Lanes*

LOS	bicycle travel speed (mph)	bicycle travel speed (kph)
A	> 14	> 23
B	> 9–14	> 15–23
C	> 7–9	> 11–15
D	> 5–7	> 8–11
E	≥ 4–5	≥ 6–8
F	< 4	< 6

Example 3.3

An intersection with a 10.5 ft lane width on one approach has a traffic volume of 500 vph and shares the approach with 20 bicycles/hr that turn left unopposed, 70 bicycles/hr that move straight ahead unopposed, and 30 bicycles/hr that turn right and are opposed by pedestrians in the crosswalk. What is the equivalent traffic volume?

Solution

Use Eq. 3.39 to find the equivalent traffic volume.

$$\begin{aligned} v_e &= v_v + v_{b,\mathrm{LT}}(\mathrm{pce}) + v_{b,\mathrm{ST}}(\mathrm{pce}) + v_{b,\mathrm{RT}}(\mathrm{pce}) \\ &= 500\ \frac{\mathrm{veh}}{\mathrm{hr}} + \left(20\ \frac{\mathrm{bicycles}}{\mathrm{hr}}\right)(1.0) + \left(70\ \frac{\mathrm{bicycles}}{\mathrm{hr}}\right)(1.0) \\ &\quad + \left(30\ \frac{\mathrm{bicycles}}{\mathrm{hr}}\right)(1.2) \\ &= 626\ \mathrm{veh/hr} \end{aligned}$$

Example 3.4

A bicycle lane along an urban street travels through four block-long segments and three signalized intersections. Using the following data, what are the average speed and LOS of this section?

segment	segment length (mi)	average speed (mph)	average delay (sec)
1	0.35	13	45
2	0.25	11	39
3	0.28	12.5	22

Solution

The average travel speed using Eq. 3.40 is

$$S_{\text{at}} = \frac{L_t}{\sum \frac{L_i}{S_i} + \sum d_j}$$

$$= \frac{0.35\ \text{mi} + 0.25\ \text{mi} + 0.28\ \text{mi}}{\frac{0.35\ \text{mi}}{13\ \frac{\text{mi}}{\text{hr}}} + \frac{0.25\ \text{mi}}{11\ \frac{\text{mi}}{\text{hr}}} + \frac{0.28\ \text{mi}}{12.5\ \frac{\text{mi}}{\text{hr}}} + \frac{45\ \text{sec} + 39\ \text{sec} + 22\ \text{sec}}{3600\ \frac{\text{sec}}{\text{hr}}}}$$

$$= 8.7\ \text{mph}$$

From Table 3.17, the LOS is C for this section.

3. MASS TRANSIT STUDIES

Mass transit is all large-scale systems of public transportation that serve a city or other metropolitan area, as well as a variety of intercity systems. *Public transportation* is any network of passenger vehicles that the general public can use. It runs on set routes, has set hours of operation, and charges set fares. Public transportation includes rail passenger cars, buses, or other fixed guideway vehicles, such as monorails, that run on underground, elevated, or other grade-separated right-of-ways.

Transit systems can be operated by either public entities, such as municipal governments, county or regional government transit authorities, or multistate authorities, or by private companies. Private companies rarely can operate public transit profitably, as transit is considered part of the fabric of urban community life, and as such, service takes precedence over profit. Public authorities sometimes contract with private companies to operate significant parts of the transit system, primarily intercity rail and bus systems. However, government subsidies are often received for parts of their operation.

Transit systems, whether private or government operated, are heavily regulated by both state and federal programs. Service quality standards vary considerably from region to region, depending on the primary emphasis of the controlling government unit. Analyzing vehicle traffic and analyzing transit service is different in that vehicle traffic is concerned with moving vehicles and transit is concerned with moving people.

Transit systems have further constraints of availability compared to pedestrian or vehicle facilities. A highway system, once constructed, is open for traffic and is available 24 hours a day, whereas transit service is available only when the transit vehicles are operating. Furthermore, fixed-route transit movements through an intersection or line segment place nonoptional demands on the transit network. Automobile drivers who find an intersection or freeway blocked may choose another route with less congestion, but a transit vehicle usually has no other option but to work through the congestion, either because of physical restrictions (e.g., a light rail system cannot operate off the light rail track) or because the vehicle cannot avoid the congestion without deviating from its route.

Transit ridership is made up of two primary groups: riders of choice and captive riders. *Riders of choice* are those that voluntarily use transit for their commute or to avoid the hassles and expense of driving. *Captive riders* are those that are unable to use other forms of transportation because of age, income, or physical or mental limitations.

Transit passengers become users of other forms of transportation at the ends of their transit trips. These forms include, but may not be limited to, walking, bicycles, automobiles, carpools, or motorcycles. How much these other forms play a role in the transit trip depends on the proximity of the origin and destination to the access points of the transit trip. For instance, if a desirable goal for planning is to have transit available within 0.5 mi (0.8 km) walking distance from every residence in an urban area, then to be successful there must be a network of routes that covers the residential areas at all times when transit service is desired. This coverage may not be economically feasible in lower density housing or practically feasible in hilly or mountainous communities. One approach is to encourage development of housing close to existing transit through land use policies.

A contrasting approach is to allow housing to develop, then fit transit service into the land use after development occurs. The latter is mostly what has happened in U.S. suburban development since the advent of the interstate highway system, so much so that providing reasonably frequent transit service in low density housing areas results in insufficient ridership to be economically justified based on available seat-mile or passengers per hour per transit vehicle. As a result, the form of transportation used by a great majority of riders at the ends of transit trips has become the automobile, driving between their homes and a park and ride lot at the transit terminal. In this way, the transit vehicle is used for the more densely patronized portion of the trip, allowing the automobile to be used for collection and distribution of riders to and from lower density housing areas.

Other forms of transit, such as walking, become feasible when the destination is within the walking distance capability of the passenger. For this reason, increased development of apartments and condominiums has occurred within walking distance of outlying stations when transit service is stable and reliable. This is particularly true around rapid transit lines and busways, but it can also occur along major arterials and around freeway interchanges that have a high volume of reliable transit service available.

From the urban planning standpoint, mass transit systems are a way of moving large volumes of people without increasing the burden of auto traffic, parking, and the highway infrastructure needed to support auto travel.

Measures of transit operating effectiveness or efficiency are dependent on the view of the observers or the participants. For instance, efficiency from the transit operator's viewpoint is related to a high percentage of usage for the hours that service is offered, efficient use of vehicle capacity, and the adequacy of station facilities and manpower needs. The transit rider views efficiency as having the needed frequency, reliability, and location of service, as well as a sense of overall safety and fares kept as low as possible. The overall operating authority is concerned with annual ridership, vehicle operating expense, the overall economics of passengers per revenue-mile and revenue-hour, and the justification of grant or revenue streams for capital and operating expenses.

Transit vehicle capacity used for planning analysis can be determined in several ways. For seated loads, vehicle capacity is the number of seats provided on the vehicle. For routes or systems that accommodate standees, the number of standees can be limited by a percentage of seated load, by an absolute number, or by the square footage of the vehicle floor area. Limits based on percentage of seated load or absolute number of standees may be in place for routes that operate across state lines, while limits determined by floor area are typical for extremely crowded city lines. To calculate the maximum vehicle capacity determined by floor area, the load capacity limit, in passengers per unit area, is multiplied by the *floor area* (i.e., the length times the width of the bus) to determine the maximum capacity during the peak period, or the *crush load*.

Example 3.5

Determine the maximum capacity for a 40 ft (12 m) long × 8.5 ft (2.6 m) wide city transit bus with a load capacity limit of 1 passenger per 2.5 ft^2 (0.23 m^2).

SI Solution

The maximum capacity of the bus during the peak period is

$$c_{\text{max}} = \left(\frac{1 \text{ passenger}}{0.23 \text{ m}^2}\right)(12 \text{ m})(2.6 \text{ m}) = 136 \text{ passengers}$$

Customary U.S. Solution

The maximum capacity of the bus during the peak period is

$$c_{\text{max}} = \left(\frac{1 \text{ passenger}}{2.5 \text{ ft}^2}\right)(40 \text{ ft})(8.5 \text{ ft}) = 136 \text{ passengers}$$

System Capacity Determination

Capacity of a transit system is described in terms of *linehaul capacity*, the maximum number of vehicles or passengers that can travel in one direction past one point in the system during a given time period (usually an hour). The actual number of passengers or vehicles traveling past a given point in the system are referred to as the *linehaul throughput* or *linehaul flow*. Several factors influence the linehaul capacity of a transit line. These factors include the capacity at station platforms; station access capacity to and from surrounding streets; terminal, merge point, or intersection capacity; signal control capacity; control delays; surface traffic interference at grade crossings (mostly a concern with fixed-guideway systems like monorails); and traffic mingling (for buses or light rail systems on city streets). While linehaul capacity is the theoretical maximum capacity of a transit line, the maximum capacity is more often restricted by station delay or signal delay.

The time a transit vehicle spends stopped at a platform or station while passengers alight from and board onto the vehicle is called the *dwell time*. A station's dwell time plus the additional time it takes the vehicle or train to decelerate and accelerate is the *station delay* for a transit line stopping at the station. *Signal delay* is the time added to a trip due to waiting at traffic signals or rail crossing signals.

Station delay controls capacity when the station delay exceeds the *vehicle arrival spacing*, the amount of time that passes between vehicles arriving at the station. Signal delay controls capacity when the signal delay and interference from traffic or surface streets are greater than any other delay on the line.

When analyzing flow and capacity for a transit system, the most frequent results needed are travel time between stations, top speed, average speed, and vehicle headway. While vehicle (or train) headway spacing is controlled by a schedule, travel time and speeds are controlled by the vehicle characteristics as designed or as actually operated.

For the simplest case, in which curve geometry and traffic restrictions are not constraints, four elements make up a station-to-station segment: acceleration, constant running at peak speed, deceleration, and dwell time.

For the standard case, in which station spacing is great enough to allow vehicles to accelerate to top speed, the simplified and time diagram would look like Fig. 3.12. The diagram assumes acceleration and deceleration are constant, as are travel speeds, and

Figure 3.12 *Time-Velocity Diagram of Vehicle Operations*

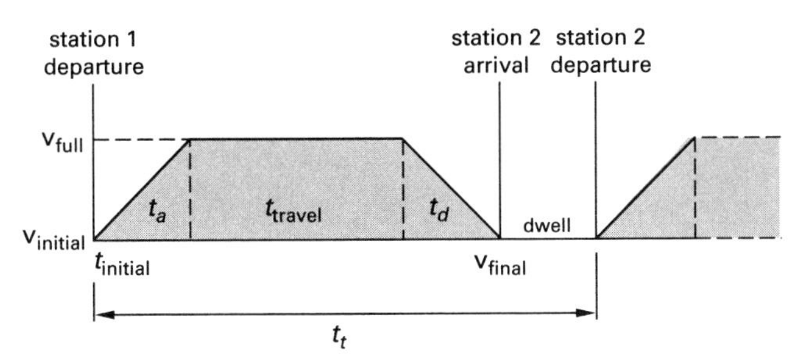

curve or grade restrictions are not taken into consideration. The velocity plotted as a function of distance shown in Fig. 3.13 has a similar shape as Fig. 3.12, except that the dwell time at the station has no change in distance.

Figure 3.13 *Distance-Velocity Diagram of Vehicle Operations*

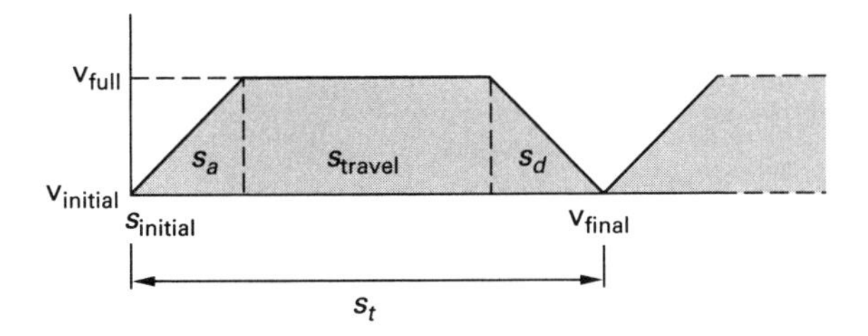

The travel time between stations is the sum of the time needed for acceleration, t_a, the time needed for travel, t_{travel}, and the time needed for deceleration, t_d, and is determined by applying Eq. 3.41, Eq. 3.42, and Eq. 3.43 to each element of Fig. 3.12. v_{full} is the design velocity of a vehicle traveling between the stations, v_{final} is the final velocity of a vehicle when it arrives at the station, and s_{travel} is the distance traveled at constant design speed.

$$t_a = \frac{v_{full} - v_{initial}}{a} \quad 3.41$$

$$t_{travel} = \frac{s_{travel}}{v_{initial}} \quad 3.42$$

$$t_d = \frac{v_{full} - v_{final}}{d} \quad 3.43$$

The average travel speed between stations is not affected by station dwell time, although the average travel speed of the line does include the station dwell time.

Applying Eq. 3.44 and Eq. 3.45 to the acceleration and deceleration rates results in the acceleration distance, s_a, and the deceleration distance, s_d. Subtracting s_a and s_d from the total distance, s_t, yields the travel distance, s_{travel}, traversed at a constant speed.

$$s_a = \frac{v_{full}^2 - v_{initial}^2}{2a} \quad 3.44$$

$$s_d = \frac{v_{full}^2 - v_{final}^2}{2d} \quad 3.45$$

If the transit vehicle is unable to accelerate to full design speed before it needs to decelerate for the next station, the distance-velocity diagram would look like a triangle, as shown in Fig. 3.14. For this case, the running velocity, v, acceleration, a, deceleration, d, and distance, s, are related by Eq. 3.46.

$$s = \frac{v^2}{2a} + \frac{v^2}{2d} = v^2\left(\frac{1}{2a} + \frac{1}{2d}\right) \quad 3.46$$

Figure 3.14 *Distance-Velocity Diagram When the Distance Between Stations Does Not Allow Acceleration to Full Design Speed*

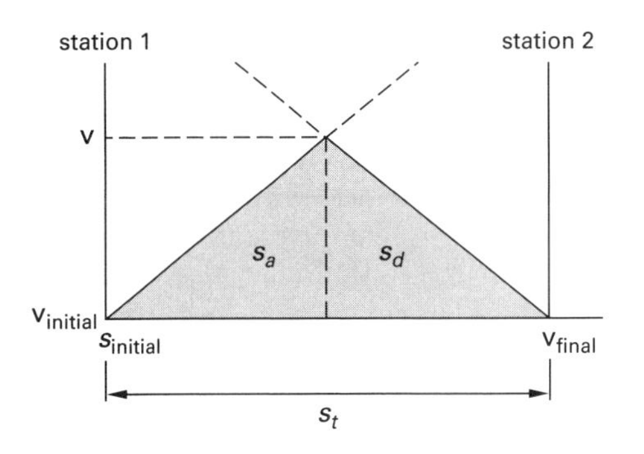

Equation 3.47 is used to solve for maximum velocity, v_{max}.

$$v_{max} = \sqrt{\frac{s}{\frac{1}{2a} + \frac{1}{2d}}} \quad 3.47$$

The maximum passenger flow rate of a given segment of a bus transit system can be found from Eq. 3.48. N_v is the number of transit vehicles arriving per hour at the station being studied, and c_{max} is the maximum capacity of each bus servicing the station.

$$v_{p,max} = N_v c_{max} \quad \text{[bus]} \quad 3.48$$

For rail transit systems, the maximum passenger flow rate of a segment is determined using Eq. 3.49. N_t is the number of trains arriving per hour at the station being studied, c_{car} is the capacity of one train car, and N_{cars} is the number of cars per train.

$$v_{p,max} = N_t c_{car} N_{cars} \quad \text{[rail]} \quad 3.49$$

The capacity of a transit vehicle can vary widely from city to city and from route to route, depending on local regulations and other requirements placed on the transit system. Capacity values should always be found from local transit information or from study data, and studies from one city should not be applied to studies from another city and assumed to fit.

Once the maximum passenger flow rate is known, the number of trips per hour required to accommodate that maximum flow rate can be found from Eq. 3.50 for bus transit systems and from Eq. 3.51 for rail transit systems. If one transit system must accommodate the flow from another transit system (e.g., if the passengers from a train must transfer to a bus for a portion of the trip from one destination to another), the flow rate from the first transit system can be inserted into the equation for the second transit system to find the required number of trips.

$$N_{\text{trips/hr}} = \frac{v_{p,\max}}{c_{\max}} \quad \text{[bus]} \qquad 3.50$$

$$N_{\text{trips/hr}} = \frac{v_{p,\max}}{c_{\text{car}} N_{\text{cars}}} \quad \text{[rail]} \qquad 3.51$$

Some transit systems, particularly bus systems and light rail systems, will have a maximum flow rate at a given stop that is much greater than the passenger service volume. In these situations, the most common solution is to build transit stations with multiple *berths* (areas where transit vehicles can arrive, load and unload passengers, and depart), so that multiple vehicles can arrive at once. The number of berths required to accommodate the service volume at a given transit stop can be found using Eq. 3.52.

$$N_{\text{berths}} = \frac{v_{p,\max}}{v_{\text{ps}}} \qquad 3.52$$

Example 3.6

A two-car light rail train has a capacity of 125 passengers per car. The scheduled arrival at a certain point on the line is at 6 min intervals. What is the maximum passenger flow rate at this location?

Solution

The number of trains per hour is found by dividing one hour by the train arrival interval.

$$N_t = \frac{60 \ \frac{\text{min}}{\text{hr}}}{6 \ \frac{\text{min}}{\text{train}}} = 10 \text{ trains/hr}$$

The maximum passenger flow rate is found from Eq. 3.49.

$$\begin{aligned} v_{p,\max} &= N_t c_{\text{car}} N_{\text{cars}} \\ &= \left(10 \ \frac{\text{trains}}{\text{hr}}\right)\left(125 \ \frac{\text{passengers}}{\text{car}}\right)\left(2 \ \frac{\text{cars}}{\text{train}}\right) \\ &= 2500 \text{ passengers/hr} \end{aligned}$$

Example 3.7

A train with a 200 mph design speed is to run on a perfectly straight track between two cities that are 25 mi apart. The acceleration of the train is limited to $0.03g$, and the deceleration is limited to $0.02g$. (a) What is the maximum distance the train will be able to run at 200 mph? (b) If a station were to be added at 10 mi from one city, what would be the maximum speed the train would attain on the 10 mi stretch of track?

Solution

(a) Convert miles per hour to feet per second.

$$\frac{\left(200 \ \frac{\text{mi}}{\text{hr}}\right)\left(5280 \ \frac{\text{ft}}{\text{mi}}\right)}{3600 \ \frac{\text{sec}}{\text{hr}}} = 293.3 \text{ ft/sec}$$

Find the acceleration distance from Eq. 3.44.

$$\begin{aligned} s_a &= \frac{\mathrm{v}_{\text{full}}^2 - \mathrm{v}_{\text{initial}}^2}{2a} = \frac{\mathrm{v}_{\text{full}}^2 - \mathrm{v}_{\text{initial}}^2}{(2)(0.03g)} \\ &= \frac{\left(293.3 \ \frac{\text{ft}}{\text{sec}}\right)^2 - \left(0 \ \frac{\text{ft}}{\text{sec}}\right)^2}{(2)(0.03)\left(32.2 \ \frac{\text{ft}}{\text{sec}^2}\right)\left(5280 \ \frac{\text{ft}}{\text{mi}}\right)} \\ &= 8.43 \text{ mi} \end{aligned}$$

Find the deceleration distance from Eq. 3.45.

$$\begin{aligned} s_d &= \frac{\mathrm{v}_{\text{full}}^2 - \mathrm{v}_{\text{final}}^2}{2d} = \frac{\mathrm{v}_{\text{full}}^2 - \mathrm{v}_{\text{final}}^2}{(2)(0.02g)} \\ &= \frac{\left(293.3 \ \frac{\text{ft}}{\text{sec}}\right)^2 - \left(0 \ \frac{\text{ft}}{\text{sec}}\right)^2}{(2)(0.02)\left(32.2 \ \frac{\text{ft}}{\text{sec}^2}\right)\left(5280 \ \frac{\text{ft}}{\text{mi}}\right)} \\ &= 12.6 \text{ mi} \end{aligned}$$

Subtract the acceleration and deceleration distances from the total distance to find the distance traveled at full design speed, 200 mph.

$$\begin{aligned} s_{\text{full}} &= s_t - s_a - s_d \\ &= 25 \text{ mi} - 8.43 \text{ mi} - 12.6 \text{ mi} \\ &= 3.97 \text{ mi} \end{aligned}$$

The train will run at 200 mph for a maximum distance of 3.97 mi.

(b) For a station at the 10 mi location, the maximum speed is found using Eq. 3.47.

$$
\begin{aligned}
v_{max} &= \sqrt{\frac{s}{\frac{1}{2a}+\frac{1}{2d}}} = \sqrt{\frac{s}{\frac{1}{(2)(0.03g)}+\frac{1}{(2)(0.02g)}}} \\
&= \frac{\sqrt{\frac{(10\text{ mi})\left(5280\ \frac{\text{ft}}{\text{mi}}\right)}{\frac{1}{(2)(0.03)\left(32.2\ \frac{\text{ft}}{\text{sec}^2}\right)}+\frac{1}{(2)(0.02)\left(32.2\ \frac{\text{ft}}{\text{sec}^2}\right)}}}}{5280\ \frac{\text{ft}}{\text{mi}}} \\
&\quad\times\left(3600\ \frac{\text{sec}}{\text{hr}}\right) \\
&= 138\text{ mph}
\end{aligned}
$$

The maximum speed the train will attain is 138 mph.

Urban Scheduled Transit

Urban scheduled transit is any and all scheduled transit service between locations within a single city, or scheduled intercity transit between cities in a larger metropolitan area. This includes transit service such as deviated-route bus service that operates on an overall schedule with slight variances in the scheduling of each individual stop.

Urban scheduled transit operates efficiently by grouping the maximum number of riders into each vehicle, allowing the transit authority to maintain a minimum number of vehicles while providing a high frequency of service. LOS criteria for urban scheduled transit systems are based on two factors: the service frequency along a scheduled transit route and the *headway*, a measurement of the time or distance between vehicles in the transit system. Table 3.18 shows LOS criteria for urban scheduled transit.

Table 3.18 *Service Frequency LOS for Urban Scheduled Transit Service*

LOS	headway (min)	frequency (vph)	comments
A	<10	>6	passengers don't need schedules
B	≥10–14	5–6	frequent service; passengers consult schedules
C	>14–20	3–4	maximum desirable time to wait if bus/train missed
D	>20–30	2	service unattractive to choice riders
E	>30–60	1	service available during hour
F	>60	<1	service unattractive to all riders

For passengers, LOS is determined based on the service to a given destination from a given transit stop, in which case multiple routes servicing the same destination are grouped into a single LOS classification. When transfers are involved to get from the stop to the destination, wait times at transfer stops are included in the headway. For trips where a rider must transfer to a route with less frequent service, the LOS of the transit system is the LOS of the less frequent service, regardless of the frequency or headway from the earlier leg of the trip.

The minimum service headway required for a system is found from Eq. 3.53. t_{dw} is the dwell time for the system, and t_{ave} is the average spacing interval between transit vehicles.

$$h_{min} = t_{dw} + t_{ave} \qquad 3.53$$

If the minimum headway and maximum capacity of a transit system are known, Eq. 3.54 can be used to find the passenger service volume of the system, the maximum number of passengers per hour the system should be designed to accommodate.

$$v_{ps} = \left(\frac{3600_{\text{sec/hr}}}{h_{min}}\right)c_{max} \qquad 3.54$$

If headway for a system is unknown, it can be found using Eq. 3.55. N_v is the number of transit vehicles per hour scheduled to arrive at a given stop.

$$h = \frac{60_{\text{min/hr}}}{N_v} \qquad 3.55$$

Paratransit

Paratransit, or *demand responsive transit*, is transit service that arrives at a given stop only after a rider contacts the service provider to schedule a pickup. The term paratransit is inclusive of all unscheduled transit service that makes stops only according to customer demand. The term is sometimes used, however, to refer to the the demand responsive service developed for senior citizens and persons with disabilities. The service may be operated by the local transit authority or a special division of the authority, usually with private contractors providing the service under contract with the transit authority. This type of paratransit accounts for less than 0.1% of all U.S. transit passengers, but performs a vital service for those that otherwise would have no regular means to travel to shopping locations, doctor visits, and other necessary life activities. This service is primarily point-to-point, and the service uses smaller vehicles, such as wheelchair accessible vans with seating capacity limited to fewer than 15 persons. Using smaller vehicles allows access to narrow streets and driveways not accessible to large buses and reduces the need for drivers to have a commercial driver's license.

LOS criteria for paratransit are based on the *access time*, the time that the rider must wait for pickup after requesting service. Table 3.19 shows LOS criteria for paratransit.

In addition to the advance notice requirement, paratransit may have a *scheduled arrival tolerance*, such that the rider must be ready a certain amount of time before the scheduled arrival in order to account for the demand and capacity of the service. For instance, a standard 20 min window may be established by the service provider, meaning the transit vehicle may arrive as much as 20 min earlier than the time scheduled by the rider, and the rider should be ready to be picked up at that time to avoid delays in service.

Table 3.19 *Service Access Frequency LOS for Paratransit Service*

LOS	access time (hr)	comments
A	0.0–0.5	fairly prompt response
B	>0.5–1.0	acceptable response
C	>1.0–2.0	tolerable response
D	>2.0–4.0	poor response, may require advance planning
E	>4.0–24.0	requires advance planning
F	>24.0	service not offered every weekday or at all

Highway Capacity Manual 2000. Copyright, National Academy of Sciences, Washington, D.C. Exhibit 27-2. Reproduced with permission of the Transportation Research Board.

Intercity Transit

Intercity transit service includes all transit going from one city to another, including bus, train, and ferry service. LOS criteria for intercity transit is based on the number of vehicles, trips, or trains per day, as shown in Table 3.20. If a service provides only one trip per day, there is not sufficient service to provide riders with a round trip that day. Similarly, if a return trip begins shortly after arrival at the destination, those two trips combine for a service of only one trip per day at the most, unless there is sufficient time to perform useful activities at the destination. Sufficient time is defined relative to what the destination is (e.g., if a bus travels to a theater but leaves before riders would have time to see a play, this would not count as a true round trip).

Table 3.20 *Service Access Frequency LOS for Intercity Scheduled Transit Service*

LOS	veh/day	comments
A	>15	numerous trips throughout the day
B	12–15	midday and frequent peak hour service
C	8–11	midday or frequent peak hour service
D	4–7	minimum service to provide choice travel times
E	2–3	round trip in one day is possible
F	0–1	round trip in one day is not possible

Highway Capacity Manual 2000. Copyright, National Academy of Sciences, Washington, D.C. Exhibit 27-3. Reproduced with permission of the Transportation Research Board.

Railroad Transit

Railroad operations include both longer distance suburban commuter service and intercity passenger service. For traffic planning analysis, rail transit is commonly evaluated in terms of speed, acceleration, braking, and capacity.

Bus Transit

Bus rapid transit, or BRT, includes all bus-only roadways created as part of a transit system. Bus-only roadways, either constructed as part of a major arterial street system or on a separated right-of-way similar to rail systems, are both common BRT designs. Transit planners consider BRT costly to construct. From a systems standpoint, it may be less costly to operate buses in an environment that is already oriented toward buses, rather than construct BRT facilities.

The capacity of buses varies depending on the number of entrance doors, the provision of wheelchair positions, and the details of the seating arrangement. Some systems prefer fewer seats and more standing room to allow for quicker boarding, while other systems that limit standing loads try to maximize the number of seats. Before buses were required to install wheelchair restraints, some U.S. city bus designs had as many as 45 seats in a 35 ft (11 m) long bus and 53 seats in a 40 ft (12 m) long bus. Now, a 40 ft (12 m) city bus seating arrangement has 28 seats or fewer when there are positions for four wheelchairs.

Buses that are 45 ft (14 m) long and 8.5 ft (2.6 m) wide can be operated in most states. Some manufacturers have introduced double-deck versions of the 45 ft (14 m) long bus that seat as many as 90 persons for commuter service. This large of a bus is practical only for longer runs with few stops, as the boarding times cause too much delay for typical city service with frequent stops. Articulated buses of 55 ft (17 m) or 60 ft (18 m) lengths have been used in some cities with mixed results. Although the capacity is increased for a single unit vehicle that can turn in less space than a 40 ft (12 m) or 45 ft (14 m) long bus, many older bus garages are not constructed to handle the longer articulated vehicles. Those systems face storage problems when the buses are not in service. Articulated buses handle the most heavily patronized routes, while 40 ft (12 m) or smaller buses handle the remaining service. Even with the many sizes available, the most predominant transit bus in U.S. cities continues to be 40 ft (12 m) long and 8.5 ft (2.6 m) wide. Additionally, low floor city bus designs are increasingly popular due to better handicapped accessibility and quicker boarding times.

If the bus system utilizes onboard fare collection, some additional delay is encountered at the platforms while passengers pay the fare (fare collection delay has been somewhat reduced in newer bus designs that have larger or better-designed boarding areas). Buses can operate at a closer spacing than other transit systems, and without the delay caused by signal systems. Furthermore, buses can operate as both the linehaul vehicle and the neighborhood distribution vehicle for a transit line, eliminating the delays caused by transfer between modes of transit.

If there is reason to calculate LOS for buses as a mass transit system rather than as part of normal traffic (e.g., BRT systems), LOS for buses is evaluated using the methodology for all fixed-transit systems. Buses can typically operate at a headway of 20–30 sec for extended periods. Passenger access to terminals and the alighting/boarding capacity of platforms are the controlling factors of system capacity, though linehaul capacity is still a controlling factor in all situations where it is greater than station capacity or platform capacity.

Bus Operations on Highways

For transit planning analysis, most situations involving buses on highways are handled as part of normal highway traffic, using bus blocking adjustment factors and passenger car equivalents. (See Chap. 2 and Chap. 4.)

Bus Operations on Streets

When buses are operated on streets, they should be analyzed with the flow of normal traffic, as with buses on highways. The *HCM* contains tables to convert bus traffic to passenger car equivalents for various conditions. Uninterrupted bus flow along open roadway is similar to truck flow. A bus can be considered to have the same passenger car equivalent as a truck if the bus density is low and stops are infrequent. The *HCM* generally uses the same passenger car equivalents for buses as for trucks. Unlike trucks, urban buses impede traffic flow by making frequent stops that must be accounted for when designing intersection signal timing.

Signal preemption, which extends the green time to allow an approaching transit vehicle to clear the intersection, can often improve flow capacity.

Bus stops can be located at an intersection corner before the bus enters the intersection—these stops are called *near-side stops*. Stops located after the bus crosses the intersection are known as *far-side stops* or *mid-block stops*. The decision to place stops on the near side or on the far side is largely a local one, and depends on what local drivers are used to encountering. While far-side bus stops may theoretically appear safer, local experience with the transit agency may show more rear-end collisions with far-side bus stops. Mid-block stops may not be effective in heavy traffic flows, as the transit vehicle may have difficulty re-entering the traffic lane after making a service stop. Many bus drivers tend to force their way into an unyielding traffic flow, which increases motorists' aggravation and generates complaints to municipal authorities about bus driver behavior.

Constraints of Fixed Routing

Urban transit buses and, for the most part, intercity buses, must stay on their fixed route. While a fixed route creates traffic demand predictability, it also can create increased passenger delay when sudden congestion occurs. The constraint of a fixed route must be accounted for when developing a traffic management system.

Effects of Signalization

In urbanized areas, signals can help bus operations by breaking up traffic flow and allowing gaps for pedestrian movements. Signals can also help slow or stop traffic at intervals to reduce conflicts with stopped buses or to increase the flow of buses to decrease acceleration delays.

When timing signal phases, time must be allocated for bus movements, particularly if the bus is making a turn through crowded foot traffic. Accounting for bus stop delays along a signal sequenced arterial can minimize delays experienced by the general traffic. It can also avoid unnecessary bus arrivals just as the given phase has terminated. A typical 2 mi (3 km) arterial street section with heavy retail activity can have travel time delays of 12–15 min for bus operations, added to normal traffic delays when signals are improperly timed.

PRACTICE PROBLEMS

1. A pedestrian crossing of a major street is at an unsignalized intersection with a minor street. The major street is 30 ft wide between curbs, and there is a 15 ft radius curb return at each corner of the intersection. The crosswalk is 10 ft wide, has a crossing flow rate of 250 ped/hr during the peak 10 min flow period, and has a pedestrian start-up time of 3.2 sec. Arrivals are random, and the average walking speed is 4 ft/sec. A traffic study has shown a peak hour flow rate of 460 vph. The study determined the following occurrences of gaps in the traffic at the crosswalk during the peak 10 min flow period.

gap length (sec)	no. of occurrences
10	5
11	5
12	4
13	3
≥14	2

(a) What is most nearly the minimum acceptable gap necessary for one person to cross the major street?

(A) 10 sec

(B) 11 sec

(C) 12 sec

(D) 14 sec

(b) If no platooning occurs, what is most nearly the average number of pedestrians waiting to cross during each acceptable gap in the traffic?

(A) 3 ped/gap

(B) 5 ped/gap

(C) 7 ped/gap

(D) 9 ped/gap

(c) What is the level of service for the intersection?

(A) LOS B

(B) LOS C

(C) LOS D

(D) LOS F

(d) What is most nearly the longest acceptable delay per pedestrian?

(A) 20 sec/ped

(B) 30 sec/ped

(C) 45 sec/ped

(D) 60 sec/ped

2. An off-street, two-way, shared bicycle and walking trail is attracting many weekend users. Complaints have risen about congestion on the trail due to overcrowding. The trail authority is considering widening the trail from a two-lane, two-way trail to a three-lane, two-way trail. This would widen the trail from the current 7 ft to 10 ft. The third lane would be used in alternate directions by pedestrians and bicycles as platooning permits. A recent traffic count showed flow rates of 200 pedestrian users and 304 bicycle users during the peak 1 hr period. The directional split is 50:50, and the peak hour factor is assumed to be 1.0 based on uniform traffic flow during the peak hour.

(a) What is the current level of service?

(A) LOS C

(B) LOS D

(C) LOS E

(D) LOS F

(b) Using the current ratio of pedestrians to bicycle users, what is most nearly the current capacity of the trail?

(A) 150 ped + 195 bicycles

(B) 187 ped + 284 bicycles

(C) 200 ped + 304 bicycles

(D) 214 ped + 326 bicycles

(c) Using the current traffic flow and ratio of pedestrians to bicycle users, what would the level of service be if the trail were widened to 10 ft with three lanes of flow?

(A) LOS B

(B) LOS C

(C) LOS D

(D) LOS F

3. Service disruption caused by a blockage occurs on a light rail transit line running at full capacity. The blockage requires all passengers to be transferred to buses for a portion of their trip. The LRT line operates with two-car trains in each direction, and each car has a capacity of 255 passengers/car. The schedule calls for a maximum of 28 trains/hr arriving at the station where passengers are transferred to buses. Each bus has a maximum capacity of 65 passengers. Traffic signals along the major artery traveled by the buses are timed to allow up to 10 buses in the traffic stream per 90 sec cycle. There are no bus stops at signalized intersections, all buses travel straight through the intersections, and there is no other scheduled route service along the arterial. Bus boarding times at the transfer location are as follows.

1st passenger	2.0 sec
2nd through 30th passenger	0.7 sec
31st passenger to capacity	1.4 sec
door opening	1.5 sec
door closing	2.0 sec
average interval for the next bus to move into boarding position	30 sec

(a) What is most nearly the minimum required width of the at-grade concourse walkway connecting the LRT platform and the bus platform if all passengers arriving on a two-car train are to be moved to the bus platform within 2 min of arrival?

(A) 11 ft

(B) 15 ft

(C) 23 ft

(D) 28 ft

(b) Most nearly how many bus trips will be required to carry the maximum hourly directional flow rate of passengers if each bus has a capacity of 65 passengers?

(A) 110 bus trips/hr

(B) 112 bus trips/hr

(C) 168 bus trips/hr

(D) 220 bus trips/hr

(c) Assuming the number of bus trips required by (b), what is most nearly the average number of buses traveling through the signal per cycle?

(A) 3.7 buses/cycle

(B) 5.5 buses/cycle

(C) 10 buses/cycle

(D) 40 buses/cycle

(d) What is most nearly the minimum number of bus berths required to handle the expected passenger flow rate?

(A) 2 berths

(B) 5 berths

(C) 7 berths

(D) 9 berths

SOLUTIONS

1. (a) Use Eq. 3.23 to determine the critical gap. The end clearance time can be considered near zero for the minimum acceptable gap.

$$t_c = \frac{L}{S_p} + t_s = \frac{30 \text{ ft}}{4.0 \ \frac{\text{ft}}{\text{sec}}} + 3.2 \text{ sec} = 10.7 \text{ sec} \quad (11 \text{ sec})$$

A clearance time of 11 sec is acceptable as the critical gap.

The answer is (B).

(b) The total number of expected gaps of 11 sec or longer is

$$N_g = 5 + 4 + 3 + 2 = 14 \text{ gaps}$$

Since arrivals are random, the average number of pedestrians waiting for each acceptable gap can be found from Eq. 3.25.

$$\begin{aligned} N_p = \frac{v_p}{N_g} &= \frac{\left(250 \ \frac{\text{ped}}{\text{hr}}\right)\left(\frac{10 \text{ min}}{60 \ \frac{\text{min}}{\text{hr}}}\right)}{14 \text{ gaps}} \\ &= 2.98 \text{ ped/gap} \quad (3 \text{ ped/gap}) \end{aligned}$$

The answer is (A).

(c) The vehicle flow rate per second is

$$v_v = \frac{460 \ \frac{\text{veh}}{\text{hr}}}{3600 \ \frac{\text{sec}}{\text{hr}}} = 0.128 \text{ veh/sec}$$

Using Eq. 3.28 and the critical gap found in Sol. 1(a) as the group critical gap, t_G, the average delay is

$$\begin{aligned} d_p &= \frac{1}{v_v}\left(e^{v_v t_G} - v_v t_G - 1\right) \\ &= \left(\frac{1}{0.128 \ \frac{\text{veh}}{\text{sec}}}\right)\left(\begin{array}{l} e^{(0.128 \text{ veh/sec})(11 \text{ sec})} \\ \quad - \left(0.128 \ \frac{\text{veh}}{\text{sec}}\right)(11 \text{ sec}) - 1 \end{array}\right) \\ &= 13.1 \text{ sec} \end{aligned}$$

From Table 3.9, the LOS is C.

The answer is (B).

(d) From Table 3.9, the longest delay acceptable (LOS E) is 45 sec/ped. This includes a high likelihood of pedestrian risk-taking behavior.

The answer is (C).

Pedestrian/ Transit Analysis

2. (a) The LOS is determined by the number of passing events that occur during the peak 15 min period. When the PHF is 1.0, the 15 min flow rates are in direct proportion to the full peak hour flow rate.

$$\left(200\ \frac{\text{ped}}{\text{hr}}\right)\left(\frac{15\ \text{min}}{60\ \text{min}}\right) = 50\ \text{ped/hr}$$

$$\left(304\ \frac{\text{bicycles}}{\text{hr}}\right)\left(\frac{15\ \text{min}}{60\ \text{min}}\right) = 76\ \text{bicycles/hr}$$

The directional split is 50:50, meaning 50% of the peak flow moves in each direction. Determine the number of events per hour that occur during the peak period using Eq. 3.34 and multiplying each flow rate by a factor of 0.5.

$$\begin{aligned} F_s &= 2.5v_{p,o} + v_{b,o} + 3v_{p,s} + 0.18v_{b,s} \\ &= (2.5)(0.5)\left(50\ \frac{\text{ped}}{\text{hr}}\right) + (0.5)\left(76\ \frac{\text{bicycles}}{\text{hr}}\right) \\ &\quad + (3)(0.5)\left(50\ \frac{\text{ped}}{\text{hr}}\right) \\ &\quad + (0.18)(0.5)\left(76\ \frac{\text{bicycles}}{\text{hr}}\right) \\ &= 182\ \text{events/hr} \end{aligned}$$

Referring to Table 3.12, the LOS is E for a two-lane, two-way shared trail with an event frequency of 182 events/hr.

The answer is (C).

(b) The capacity of the trail is the maximum flow rate at LOS E. From Table 3.12, the maximum flow rate for a two-way, two-lane trail is a rate of 195 events/hr. The proportionally increased flow is

$$\left(\frac{195\ \frac{\text{events}}{\text{hr}}}{182\ \frac{\text{events}}{\text{hr}}}\right)(200\ \text{ped}) + \left(\frac{195\ \frac{\text{events}}{\text{hr}}}{182\ \frac{\text{events}}{\text{hr}}}\right)(304\ \text{bicycles}) = 214\ \text{ped} + 326\ \text{bicycles}$$

The answer is (D).

(c) From Table 3.12, the LOS would be C for a two-way, three-lane trail.

The answer is (B).

3. (a) Use Eq. 3.55 to determine the headway of the trains.

$$h = \frac{60}{N_t} = \frac{60\ \frac{\text{min}}{\text{hr}}}{28\ \frac{\text{trains}}{\text{hr}}} = 2.14\ \text{min/train}$$

Determine the maximum number of passengers traveling through the concourse in each 2 min interval, $N_{p,2\,\text{min}}$.

$$\begin{aligned} N_{p,2\,\text{min}} &= \left(2\ \frac{\text{cars}}{\text{train}}\right)\left(255\ \frac{\text{passengers}}{\text{car}}\right) \\ &= 510\ \text{passengers/train} \end{aligned}$$

The pedestrian flow rate at transit terminals is found using the platoon-adjusted flow rate shown in Table 3.6. The maximum capacity of a transit concourse is 18 pedestrians per minute per foot of width for LOS E. The minimum walkway width, W_{min}, is

$$\begin{aligned} W_{\text{min}} &= \frac{510\ \text{ped}}{\left(18\ \frac{\text{ped}}{\text{min-ft}}\right)(2\ \text{min})} \\ &= 14.2\ \text{ft}\quad (15\ \text{ft}) \end{aligned}$$

The answer is (B).

(b) The maximum hourly directional passenger flow rate, $v_{p,\text{max}}$, is found from Eq. 3.49.

$$\begin{aligned} v_{p,\text{max}} &= N_t c_{\text{car}} N_{\text{cars}} \\ &= \left(28\ \frac{\text{trains}}{\text{hr}}\right)\left(255\ \frac{\text{passengers}}{\text{car}}\right)\left(2\ \frac{\text{cars}}{\text{train}}\right) \\ &= 14{,}280\ \text{passengers/hr} \end{aligned}$$

The hourly number of bus trips required to accommodate the maximum hourly directional flow rate is found from Eq. 3.50.

$$\begin{aligned} N_{\text{trips/hr}} &= \frac{v_{p,\text{max}}}{c_{\text{max}}} = \frac{14{,}280\ \frac{\text{passengers}}{\text{hr}}}{65\ \frac{\text{passengers}}{\text{bus}}} \\ &= 220\ \text{buses/hr} \end{aligned}$$

The answer is (D).

(c) The average bus flow rate per signal cycle is

$$\left(220\ \frac{\text{buses}}{\text{hr}}\right)\left(\frac{90\ \frac{\text{sec}}{\text{cycle}}}{3600\ \frac{\text{sec}}{\text{hr}}}\right) = 5.5\ \text{buses/cycle}$$

The answer is (B).

(d) Determine the dwell time for one bus.

$$\begin{aligned} t_{\text{dw}} &= \text{first passenger time} \\ &\quad + \text{2nd through 30th passenger time} \\ &\quad + \text{31st passenger to capacity time} \\ &\quad + \text{door opening time} \\ &\quad + \text{door closing time} \\ &= (1)(2.0\ \text{sec}) + (29)(0.7\ \text{sec}) + (35)(1.4\ \text{sec}) \\ &\quad + 1.5\ \text{sec} + 2.0\ \text{sec} \\ &= 74.8\ \text{sec/bus} \end{aligned}$$

Use Eq. 3.53 to determine the minimum service headway.

$$h_{\text{min}} = t_{\text{dw}} + t_{\text{ave}} = 74.8\ \frac{\text{sec}}{\text{bus}} + 30\ \frac{\text{sec}}{\text{bus}} = 104.8\ \text{sec/bus}$$

Use Eq. 3.54 to determine the passenger service volume, v_{ps}, at one berth.

$$\begin{aligned} v_{\text{ps}} &= \left(\frac{3600}{h_{\text{min}}}\right) c_{\text{max}} \\ &= \left(\frac{3600\ \frac{\text{sec}}{\text{hr}}}{104.8\ \frac{\text{sec}}{\text{bus}}}\right)\left(65\ \frac{\text{passengers-berth}}{\text{bus}}\right) \\ &= 2233\ \text{passengers-berth/hr} \end{aligned}$$

Given the known maximum flow rate of the system, the number of berths required is found from Eq. 3.52.

$$\begin{aligned} N_{\text{berths}} &= \frac{v_{p,\text{max}}}{v_{\text{ps}}} = \frac{14{,}280\ \frac{\text{passengers}}{\text{hr}}}{2233\ \frac{\text{passengers-berth}}{\text{hr}}} \\ &= 6.39\ \text{berths} \quad (7\ \text{berths}) \end{aligned}$$

The answer is (C).

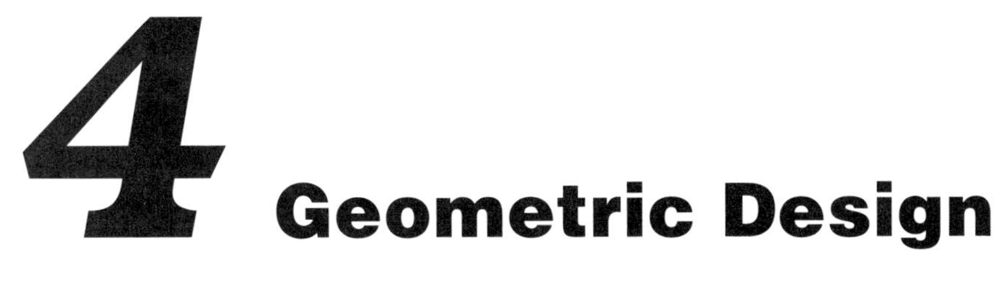

4 Geometric Design

Nomenclature

a	acceleration	ft/sec^2	m/s^2
a	given model parameter	–	–
A	absolute value of the algebraic difference in grades	%	%
A	azimuth	deg	deg
b_w	adjustment factor for number of lanes rotated	–	–
B	bearing angle	deg	deg
BC	beginning of curve	sta	sta
C	chord length	ft	m
C	maximum rate of change in lateral acceleration	ft/sec^3	m/s^3
C	vertical clearance	ft	m
d	distance	ft	m
D	absolute value of the algebraic difference in grades	decimal	decimal
D	degree of curve	deg	deg
D	distance	ft	m
e	superelevation rate	ft/ft	m/m
E	external distance	ft	m
E	relative elevation change	ft	m
EC	end of curve	sta	sta
f	coefficient of friction	–	–
F	force	lbf	N
g	gravitational acceleration, 32.2 (9.81)	ft/sec^2	m/s^2
g_c	gravitational constant, 32.2	lbm-ft/lbf-sec^2	n.a.
G	grade	%	%
h_1	driver's eye height	ft	mm
h_2	object height	ft	mm
HSO	horizontal sightline offset	ft	m
I	intersection angle	deg	deg
k	initial tangent distance	ft	m
K	length of vertical curve per percent grade difference	ft/%	m/%
L	length of curve	ft	m
m	difference in speed between passing vehicle and vehicle being passed	mph	kph
M	middle ordinate	ft	m
n	number of deflection angles	–	–
N	number	–	–
p	lateral offset between the tangent and circular curve	ft	m
PC	point of curvature	sta	sta
PI	point of intersection, vertex	sta	sta
PT	point of tangency	sta	sta
PVC	point of beginning of vertical curve	sta	sta
PVI	point of intersection of vertical curve tangents, vertex	sta	sta
PVT	point of end of vertical curve	sta	sta
R	curve radius	ft	m
R	rate of change in grade	%/sta, ft/sta^2	%/sta, m/sta^2
S	sight distance	ft	m
S	stopping distance	ft	m
T	tangent distance	ft	m
TC	tangent to curve	sta	sta
t	time	sec	s
v	speed	mph	kph
v	velocity	ft/sec	m/s
V	vertex	sta	sta
w	weight	lbf	N
W	width	ft	m
x	horizontal distance	ft	m
x	tangent distance	ft	m
y	tangent offset	ft	m

Symbols

α	angle	deg	deg
β	angle	deg	deg
Δ	maximum relative gradient	ft/ft	m/m
θ	central angle	deg	deg
θ	subtended angle	deg	deg
ϕ	angle of cross slope	deg	deg
ϕ	deflection angle	deg	deg

Subscripts

a	known
ave	average
B	backsight
c	central curve or centrifugal
D	design
F	foresight
l	central curve
L	lanes
max	maximum
min	minimum
N	north
NC	normal cross slope or normal crown
p	perception-reaction or pre-maneuver
r	roadway
R	runoff
s	side friction or spiral
S	south
t	total or tangent runout
x	desired
z	passing vehicle in opposing lane

1. INTRODUCTION

The geometric design of transportation facilities (e.g., highways, railways, or other linear facilities such as bikeways and airport taxiways) includes the design of horizontal and vertical alignments, cross sections and intersections, as well as other design details specific to the type of facility being constructed. Its goal is to maximize user safety and comfort, as well as to minimize environmental impact. As such, transportation facilities are designed in accordance with state and national standards, such as *A Policy on Geometric Design of Highways and Streets* (also known as the *Green Book*) and the *Roadside Design Guide*, both published by the American Association of State Highway and Transportation Officials (AASHTO).

This chapter covers basic elements of highway geometric design, including route and corridor alignment; horizontal, spiral, and vertical curves; superelevation; intersections and interchanges; as well as the geometric design of railways, bikeways, and mass transit systems. For more information about specific elements of the geometric design of a transportation facility, state and national standards should be consulted.

2. ROUTE AND CORRIDOR ALIGNMENT

A *route* is a series of straight lines (called *tangents*) and curves that are laid out in a continuous path. These geometric lines comprise the horizontal and vertical alignments of a transportation facility. The details of a route's geometric layout depend on the type of facility being designed. However, in general, the alignments connect the terminals of two other routes or extend or revise an existing route.

A route's horizontal and vertical alignments are designed within a *corridor*, which is a collective system of transportation facilities designed for travel between two or more points. A *corridor alignment study* assesses the feasibility of placing a particular route within a corridor. There are three distinct phases in a corridor alignment study: reconnaissance, the preliminary location survey, and the final location survey.

The first phase of a corridor alignment is *reconnaissance*, which is the collection and review of basic information that will set the approximate location of a route. This phase includes establishing the terminal ends of the alignment and the intervening connecting points. It also includes identifying major *topographic features*, such as rivers, existing railroads and highways, buildings, mountains, historical sites, architectural features, protected lands, and anything else that will have a physical effect on the route location.

Reconnaissance also includes a review of existing mapping or other location documentation, general traffic demands, funding parameters, and general government or institutional information that can be used to determine the suitability of a facility's alignment design. For example, in highway alignment design, the number and size of trucks in the traffic flow can determine the maximum grade and the location of access points.

In the past, reconnaissance required walking the proposed corridor with a design team to establish a general location of line and grade based on visual appearances. Measurements and notes were taken to record features affecting the alignment that may not be evident on the available mapping. Stakes and flagging were placed in the approximate location of the centerline or other horizontal reference line for future reference when returning to the site.

However, reconnaissance that formerly took days, weeks, or months can now be performed in a few hours using commercially available satellite and mapping imagery. Mapping websites, such as Google™ maps, also have extensive files that can be referenced and reproduced on working media for a relatively small cost. Ground level imagery of more populous areas is also available, so a preliminary record of the nature of topographic and human-made surface improvements can be prepared before ever venturing into the corridor itself.

The second phase of a corridor alignment study is a *preliminary location survey*. After the reconnaissance identifies a possible location, large-scale aerial mapping suitable for laying out the preliminary location is prepared. Maps are prepared for a broad area so that several alternative alignments can be studied by the surveyors. Maps created during this phase are typically at 1:2400 (1 in = 200 ft) scale or 1:1200 (1 in = 100 ft) scale for rural locations and 1:500 (1 in = 40 ft) scale for urban locations. Aerial mapping contours are the primary means of determining ground elevations during alignment studies.

During this phase, the surveyors set control points throughout the corridor to accurately measure the stakeout during later phases. The project control traverse is surveyed and mathematically connected to the aerial map traverse located objects on the aerial map in the field. In some cases, it may be possible to combine the aerial map traverse with the project control traverse to reduce the probability of a measurement error between mapping images and project control location.

Alternative routes are evaluated for the best fit of line and grade to the project criteria. For each alternative, earthwork quantities are projected, bridge structures are located, soil conditions are studied, preliminary right-of-way requirements are drawn, and economic studies are performed to determine the projected cost. During this phase, community hearings are held and a preliminary environmental impact analysis is prepared.

The third phase of a corridor alignment study is the *final location survey*, which begins the *final design*. For this phase, detailed aerial mapping is prepared with narrow coverage along the selected final route at a small scale. The scale of these maps is typically 1:250 (1 in = 20 ft) for congested locations, so that engineers can design connections to existing land features and utilities. During the final location survey, the line and grade are carefully staked out in the field using the preliminary alignment as a guide. Sufficient field information are prepared to verify mapping locations, and final grades are adjusted to determine cut and fill quantities. Bridge pier locations are carefully surveyed, and subsurface investigations are performed to determine final foundation design requirements.

During the final location survey, adjustments to the alignment are made to balance earthwork, match connections to adjoining work, and allow for needed clearances for objects intersecting or adjacent to the roadway (such as overpasses or utility poles). Homes and businesses in the final right-of-way are relocated. Community meetings are usually held prior to the final location survey to present to the community the impact of the project. These meetings sometimes lead to minor changes in the route geometry.

Periodically during the alignment process, the roadway template is applied to the line and grade in a three-dimensional model, including allowances for cut and fill slopes, drainage ditches, and cross pipes. Allowances are also made for geometric features of intersections and interchanges and, for railroads, the geometry of connecting spurs or yard trackage, all of which can have an effect on the location of the main roadway or trackway.

3. SPECIFYING LINE DIRECTION

The direction of any line can be specified by an angle between it and a reference line known as a *meridian*. If the meridian is arbitrarily chosen, it is called an *assumed meridian*. If the meridian is a true north-to-south line passing through the true north pole, it is called a *true meridian*. If the meridian is parallel to the earth's magnetic field, it is known as a *magnetic meridian*. If a rectangular grid is drawn over a map with any arbitrary orientation, the vertical lines are referred to as *grid meridians*.

The difference between a true meridian and a magnetic meridian is called the *declination* (*magnetic declination* or *variation*). An *isogonic line* connects locations in a geographical region that have the same magnetic declination. If the north end of a compass points to the west of true meridian, the declination is referred to as a *west declination* or *minus declination*. If the north end of a compass points to the east of true meridian, it is referred to as an *east declination* or *plus declination*.

The *azimuth* of a line is given as a clockwise angle from the reference direction, either "from the north" or "from the south" (for example, "NAz 320"). Azimuths never exceed 360°. Positive declinations are added to the magnetic compass azimuth, and negative declinations are subtracted from the magnetic compass azimuth. The result is called the true azimuth.

A *direction* (the variation of a line from its meridian) may be specified in several ways. (See Fig. 4.1.) Directions are specified as either bearings or azimuths referenced to either north or south. In the United States, most control work is performed using directions stated as azimuths. Most construction projects and property surveys specify directions as bearings.

Geometric Design

Figure 4.1 *Equivalent Bearing, Angle, and Azimuth Measurements*

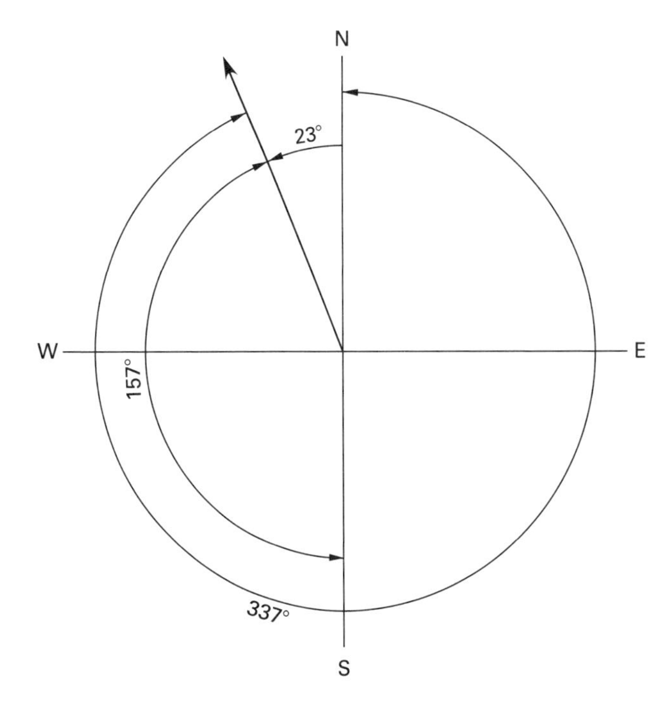

The angle between a line and the prolongation of a preceding line is called a *deflection angle*. Such measurements are labeled as Rt for clockwise angles and L for counterclockwise angles. A clockwise deflection angle measured from the preceding line (i.e., the back line) to the following line is called the *angle to the right* or the *azimuth from the back line*. The back line may be referenced from the north, which is most common, but the choice of the back line varies between jurisdictions. In Fig. 4.1, the angle to the right is measured from the south, or back line, to the azimuth.

The *bearing* of a line is referenced to the quadrant in which the line falls and the angle that the line makes with the meridian in that quadrant. It is necessary to specify the two cardinal directions (i.e., north or south and east or west) that define the quadrant in which the line is found. The north and south directions are specified first (for example, "N 45° E"). The angle of a bearing never exceeds 90°.

Bearings and azimuths are calculated from trigonometric and geometric relationships. From Fig. 4.2, the bearing angle, B, is given by Eq. 4.1.

$$\tan B = \frac{x_2 - x_1}{y_2 - y_1} \qquad 4.1$$

Figure 4.2 Calculation of Bearing Angle

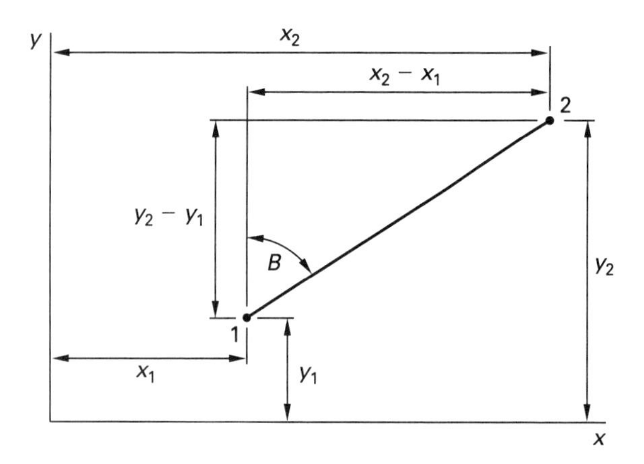

The distance, D, between two points is given by Eq. 4.2.

$$D = \sqrt{(y_2 - y_1)^2 + (x_2 - x_1)^2} \qquad 4.2$$

For the denominator in Eq. 4.1, the line from point 1 to point 2 bears north if positive and south if negative. Similarly, for the numerator, the line bears east if positive and west if negative. Determining the bearing and length of a line from the coordinates of two points is called *inversing the line*.

The azimuth of a line from point 1 to 2 measured from the north meridian, A_N, is given by Eq. 4.3.

$$\tan A_N = \frac{x_2 - x_1}{y_2 - y_1} \qquad 4.3$$

The azimuth of the same line from point 1 to 2 measured from the south meridian, A_S, is given by Eq. 4.4.

$$\tan A_S = \frac{x_1 - x_2}{y_1 - y_2} \qquad 4.4$$

Converting from bearings to azimuths is accomplished by selecting the bearing quadrant, then adding or subtracting 0°, 180°, or 360°, depending on the azimuth meridian and the bearing quadrant. Use the following equations to find a *north meridian azimuth* from a bearing.

NE bearing: azimuth = bearing angle

SE bearing: azimuth = 180° − bearing angle

SW bearing: azimuth = 180° + bearing angle

NW bearing: azimuth = 360° − bearing angle

Use the following equations to find a *south meridian azimuth* from a bearing.

NE bearing: azimuth = 180° + bearing angle

SE bearing: azimuth = 360° − bearing angle

SW bearing: azimuth = bearing angle

NW bearing: azimuth = 180° − bearing angle

Example 4.1

Convert (a) N 35° E, (b) S 45° 15′ E, (c) S 18° 30′ W, and (d) N 62° 45′ W to north meridian azimuths.

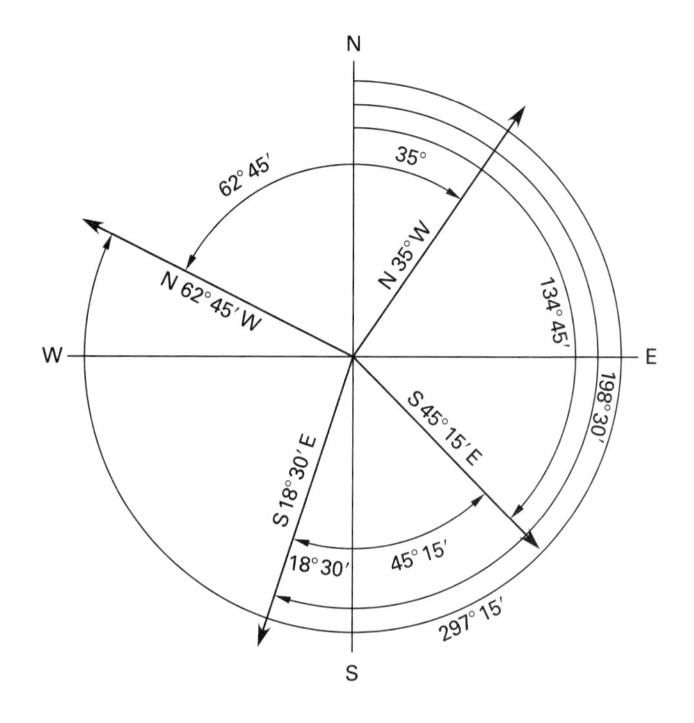

Solution

(a) N 35° E = 35° azimuth

(b) S 45° 15′ E = 180° − 45° 15′ = 134° 45′ azimuth

(c) S 18° 30′ W = 180° + 18° 30′ = 198° 30′ azimuth

(d) N 62° 45′ W = 360° − 62° 45′ = 297° 15′ azimuth

4. HORIZONTAL CURVES

Horizontal curves are used to change from one heading, or direction, to another. For most horizontal work, the *circular curve*, sometimes called a *simple curve*, is used as a smooth transition to the new direction. The main feature of a circular curve is that every point on its arc is precisely the same distance from the center point. That is, the radius is constant throughout the arc length. Figure 4.3 shows the elements of a horizontal circular curve.

Figure 4.3 *Elements of a Horizontal Circular Curve*

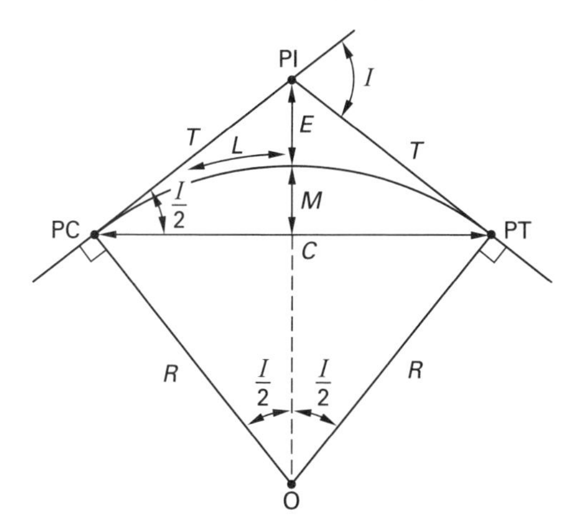

The *point of intersection* (PI, sometimes referred to as the *vertex*, V) is the intersection of the two tangents to which the curve is fitted. The *intersection angle*, I, also called the central angle or interior angle, is the angle between the curve tangents that deflect the direction from the previous bearing to a new bearing, or heading. The *point of curvature*, PC, is the point of the back tangent at which the circular curve begins. This point is also called TC, for *tangent to curve*, or BC, for *beginning of curve*. The *point of tangency*, PT, is the point at which the curve stops and the new or ahead tangent continues. Some texts refer to the PT as the TC, for tangent to curve, or EC, for *end of curve*. Lines extended at right angles to the tangents from the PC and PT intersect at the *center of the circular arc*, O. The intersection angle also appears at the center as the central angle subtended by the circular arc of the curve. The intersection angle is called *right intersection* or positive for a clockwise deflection, and *left intersection* or negative for a counterclockwise deflection. Some texts show the intersection angle as Δ and refer to the intersection angle as the *delta* of the curve. The *tangent distance*, T, is the distance from PI to PC, or the distance from PI to PT, and is calculated from Eq. 4.5. R is the radius of the curve.

$$T = R\tan\frac{I}{2} \qquad 4.5$$

The *external distance*, E, is the distance from the PI to the midpoint of the curve and is calculated using Eq. 4.6.

$$E = R\left(\sec\frac{I}{2} - 1\right) = R\,\text{exsec}\,\frac{I}{2} = R\tan\frac{I}{2}\tan\frac{I}{4} \qquad 4.6$$

$$\text{exsec}\,I = \sec I - 1 = \tan\frac{I}{2}\tan\frac{I}{4} \qquad 4.7$$

$$\sec I = \frac{1}{\cos I} \qquad 4.8$$

The *length of curve*, L, is the distance from PC to PT along the arc. It can be found using several relationships, as shown in Eq. 4.9.

$$\begin{aligned} L_{\text{ft}} &= \frac{2\pi R_{\text{ft}} I_{\text{deg}}}{360^\circ} = R_{\text{ft}} I_{\text{rad}} = \frac{(100\ \text{ft}) I_{\text{deg}}}{D_{\text{deg}}} \\ &= \frac{200\ \text{ft}}{D_{\text{deg}}} \arcsin\frac{C_{\text{ft}}}{2R_{\text{ft}}} \end{aligned} \qquad 4.9$$

The *long chord*, C_{long}, sometimes represented by LC to reduce confusion with other subchords, is the distance from the PC to the PT. It is the chord length for the intersection angle, I, and is calculated from Eq. 4.10.

$$C_{\text{long}} = 2R\sin\frac{I}{2} = 2T\cos\frac{I}{2} \qquad 4.10$$

The length of any chord (any straight line from one point on the curve to another), C, can be defined from the subtended intersection angle, θ, by substituting θ for I, as shown in Eq. 4.11.

$$C = 2R\sin\frac{\theta}{2} \qquad 4.11$$

The *middle ordinate*, M, is the length of the ordinate from the midpoint of the long chord, C_{long}, to the midpoint of the curve arc. The middle ordinate is found from Eq. 4.12 and Eq. 4.13.

$$M = R\left(1 - \cos\frac{I}{2}\right) = R\left(\text{vers}\,\frac{I}{2}\right) \qquad 4.12$$

$$M = \frac{C^2}{8R} = \left(\frac{C}{2}\right)\tan\frac{I}{4} \qquad 4.13$$

$$\text{vers}\,\frac{I}{2} = 1 - \cos\frac{I}{2} \qquad 4.14$$

$$\text{vers}\,I = 1 - \cos I \qquad 4.15$$

5. DEGREE OF CURVE

Curves are specified either by the radius, R, or *degree of curve*, D. Degree of curve is most commonly used when working in U.S. units, but it can be used to determine curvature when working in SI units as well. The degree of curve is defined either by the arc or the chord of a curve, as shown in Fig. 4.4. The *arc definition* is commonly used with highways and streets, where D is the central angle subtended by a 100 ft arc. This is the same as the deflection of a 100 ft long arc, as shown in Fig. 4.4(a).

Figure 4.4 *Degree of Curve Definition*

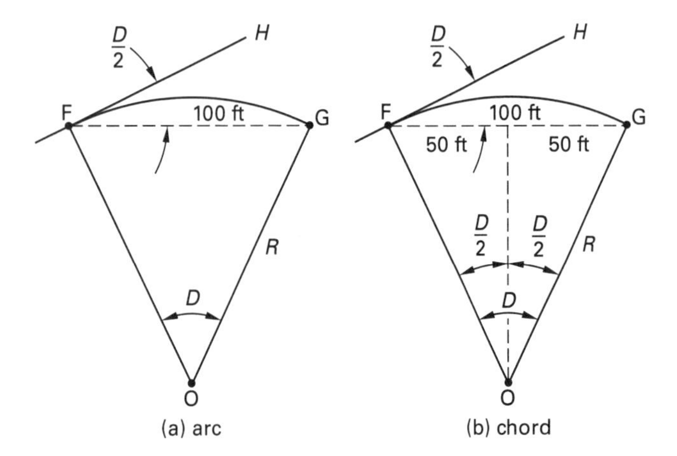

The *chord definition* (see Fig. 4.4(b)) is used with railroads and was used on pre-interstate highways. For chord definition curves, the central angle is subtended by a 100 ft chord. To differentiate the chord definition from the arc definition, degree of curve for the chord definition is shown as D_c on some plans.

In practice, when layout is performed using degree of curve instead of radius, approximate degrees such as 1°, 1° 30′, and so on are used. When layout is performed using the radius, approximate whole-numbered radii curves are used, such as 300 ft, 1000 ft, 10,000 ft, and so on. Using approximations expedites calculations and reduces the chance for error when defining curve segments. To find the radius when the degree of curve is given, use Eq. 4.16 for arc definition curves and Eq. 4.17 for chord definition curves.

$$R_{\text{ft}} = \frac{(100\text{ ft})\left(\frac{180°}{\pi}\right)}{D_{\text{deg}}} = \frac{5729.578\text{ ft-deg}}{D_{\text{deg}}}$$

$$\left[\begin{array}{c}\text{arc basis when } D=1°,\\ R=5729.58\text{ ft}\end{array}\right] \quad \text{[U.S. only]} \qquad 4.16$$

$$R_{\text{ft}} = \frac{50\text{ ft}}{\sin\frac{D_{\text{deg}}}{2}}$$

$$\left[\begin{array}{c}\text{chord basis when } D=1°,\\ R=5729.65\text{ ft}\end{array}\right] \quad \text{[U.S. only]} \qquad 4.17$$

Rearranging Eq. 4.16 and Eq. 4.17, use Eq. 4.18 and Eq. 4.19 to find D when R is known.

$$D_{\text{deg}} = \frac{(360°)(100\text{ ft})}{2\pi R_{\text{ft}}} = \frac{(180°)(100\text{ ft})}{\pi R_{\text{ft}}}$$

$$\text{[arc definition] [U.S. only]} \qquad 4.18$$

$$\sin\frac{D_{\text{deg}}}{2} = \frac{50\text{ ft}}{R_{\text{ft}}} \quad \text{[chord definition] [U.S. only]} \qquad 4.19$$

In jurisdictions using the metric system, distances are usually measured with a 20 m tape, though 30 m and 50 m tapes are sometimes used. Using the degree of curve, with the deflection angle for a 20 m chord, the degree of curve equations are Eq. 4.20, Eq. 4.21, and Eq. 4.22.

$$D_{\text{deg}} = \frac{(180°)(20\text{ m})}{\pi R_{\text{m}}} \quad \text{[arc definition] [SI only]} \qquad 4.20$$

$$\sin\frac{D_{\text{deg}}}{2} = \frac{10\text{ m}}{R_{\text{m}}} \quad \text{[chord definition] [SI only]} \qquad 4.21$$

$$L_{\text{m}} = (20\text{ m})\frac{I}{D} \quad \text{[SI only]} \qquad 4.22$$

$$R_{\text{m}} = \frac{(20\text{ m})\left(\frac{180°}{\pi}\right)}{D} \quad \text{[SI only]} \qquad 4.23$$

The length of the arc for arc or chord definition curves is slightly greater than the 100 ft chord length. The difference is greater for sharper curves and lesser for flatter curves. Arc lengths for chords on various degrees of curvature can also be found in most surveying texts.

Attempting to parallel a chord definition curve with an arc definition curve, such as fitting a highway curve adjacent to a railroad right-of-way, can be confusing. It is usually easier to convert the railroad degree of curve to a radius relationship first, then proceed with the highway curve and continue to work in radius relationships.

6. STATIONING ON A HORIZONTAL CURVE

When a route is initially laid out between PIs, the curve is undefined. The route distance is measured from PI to PI, but the route distance changes when the curve is laid out.

In route surveying, stationing is carried ahead continuously from a starting point or hub designated as station 0+00 and called *station zero plus zero zero*. The term *station* is applied to each subsequent 100 ft (or 100 m) length. Also, the term *station* is applied to any point whose position is given by its total distance from the beginning position. Thus, station 8+33.2 is a unique point 833.2 ft (or 833.2 m) from the starting point, measuring along the survey line. The partial length beyond the full station 8 is 33.2 ft (33.2 m) and is termed a *plus*. Moving or looking toward increasing stations is called *ahead stationing*. Moving or looking toward decreasing stations is called *back stationing*. Offsets from the centerline are either left or right looking ahead on stationing.

Normal curve layout follows the convention of showing ahead stationing from left to right on the plan sheet, or from the bottom to the top of the sheet. In ahead stationing, the first point of the curve is the point of curvature (PC), alternatively called the point of beginning of curve (BC). Stationing is carried ahead along the arc of the curve to the end point of the curve, called the point of tangency (PT). This end point is alternatively called the point of end of curve (EC). The PI is stationed ahead from the PC. Therefore, the difference in stationing along the back curve tangent is the *curve tangent length*. To avoid the confusion of creating two stations for the PC, the ahead tangent is rarely stationed.

From Eq. 4.24, the PT station is equal to the PC station plus the curve length, and from Eq. 4.25, the PC station is equal to the PI station minus the tangent distance.

$$\text{sta PT} = \text{sta PC} + L \qquad 4.24$$

$$\text{sta PC} = \text{sta PI} - T \qquad 4.25$$

Collectively, the stationing of the PI, PC, and PT, the intersection angle, the curve radius, the degree of curve, the tangent distance, the external distance, the chord length, and the length of the long chord are referred to as the complete curve specifications.

Example 4.2

Two horizontal tangents meet at sta 149+22.00 with a deflection angle of 65° 12′ Rt. The horizontal curve between the two tangents has a radius of 2500 ft. Using arc definition, what are the complete curve specifications of the horizontal curve?

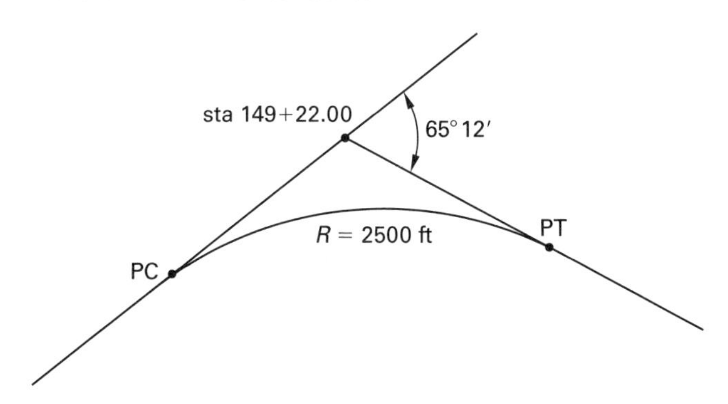

Solution

To station the PC and PT, the tangent distance and length of curve must be known. Find the nonstation characteristics first.

From Eq. 4.5, the tangent distance is

$$T = R\tan\frac{I}{2} = (2500 \text{ ft})\tan\frac{65^\circ\,12'}{2} = 1598.82 \text{ ft}$$

From Eq. 4.10, the length of the long chord is

$$C_{\text{long}} = 2R\sin\frac{I}{2} = (2)(2500 \text{ ft})\sin\frac{65^\circ\,12'}{2} = 2693.85 \text{ ft}$$

From Eq. 4.6, the external distance is

$$E = R\left(\sec\frac{I}{2} - 1\right) = (2500 \text{ ft})\left(\sec\frac{65^\circ\,12'}{2} - 1\right) = 467.53 \text{ ft}$$

Calculate the degree of curve using Eq. 4.18.

$$D_{\text{deg}} = \frac{(360^\circ)(100 \text{ ft})}{2\pi R_{\text{ft}}} = \frac{(360^\circ)(100 \text{ ft})}{(2\pi)(2500 \text{ ft})} = 2.292^\circ$$

From Eq. 4.9, the length of curve is

$$L_{\text{ft}} = \frac{200 \text{ ft}}{D_{\text{deg}}}\arcsin\frac{C_{\text{long,ft}}}{2R_{\text{ft}}} = \left(\frac{200 \text{ ft}}{2.292^\circ}\right)\left(\arcsin\frac{2693.85 \text{ ft}}{(2)(2500 \text{ ft})}\right) = 2844.67 \text{ ft}$$

Use Eq. 4.24 and Eq. 4.25 to station the PC and PT.

$$\begin{aligned} \text{sta PC} &= \text{sta PI} - T \\ &= \text{sta } 149+22.00 \\ &\quad\; -15+98.82 \\ &\quad\; \text{sta } 133+23.18 \\ \text{sta PT} &= \text{sta PC} + L \\ &= \text{sta } 133+23.18 \\ &\quad\; +28+44.67 \\ &\quad\; \text{sta } 161+67.85 \end{aligned}$$

The complete curve specifications are

$$PI = \text{sta } 149{+}22.00$$
$$PC = \text{sta } 133{+}23.18$$
$$PT = \text{sta } 161{+}67.85$$
$$I = 65^\circ\ 12'\ \text{Rt}$$
$$R = 2500\ \text{ft}$$
$$D = 2^\circ\ 17'\ 31'' = 2.292^\circ$$
$$T = 1598.82\ \text{ft}$$
$$E = 467.53\ \text{ft}$$
$$L = 2844.67\ \text{ft}$$
$$C_{\text{long}} = 2693.85\ \text{ft}$$

Example 4.3

In laying out the centerline of a roadway, two tangents intersect, with an intersection angle of 28° 18′. A curve with an external distance of approximately 40 ft must be laid out in order to clear an obstruction. Calculate the following using arc definition curves: (a) the degree of curvature to the nearest one-half degree, (b) the curve radius, (c) the actual external distance required to fit the selected curve, (d) the length of the curve, and (e) the tangential distance.

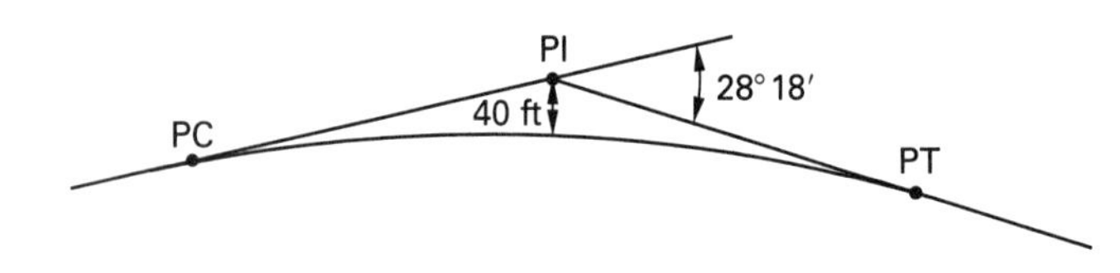

Solution

(a) The external distance is given as 40 ft.

Using Eq. 4.6, substitute for the secant and rearrange to find the radius.

$$R = \frac{E}{\frac{1}{\cos\frac{I}{2}} - 1} = \frac{40\ \text{ft}}{\frac{1}{\cos\frac{28^\circ\ 18'}{2}} - 1} = 1278.35\ \text{ft}$$

Rearranging Eq. 4.16, the degree of curve is

$$D = \frac{5729.58\ \text{ft-deg}}{R_{\text{ft}}} = \frac{5729.58\ \text{ft-deg}}{1278.35\ \text{ft}} = 4.48^\circ$$

The nearest half-degree is 4.5°.

(b) The radius of a 4.5° curve is found using Eq. 4.16.

$$R_{\text{ft}} = \frac{(100\ \text{ft})\left(\frac{180^\circ}{\pi}\right)}{D_{\text{deg}}} = \frac{(100\ \text{ft})\left(\frac{180^\circ}{\pi}\right)}{4.5^\circ} = 1273.24\ \text{ft}$$

(c) Using the radius found in part (b), the new external distance can be found from Eq. 4.6.

$$E = R\left(\frac{1}{\cos\frac{I}{2}} - 1\right) = (1273.24\ \text{ft})\left(\frac{1}{\cos\frac{28^\circ\ 18'}{2}} - 1\right) = 39.84\ \text{ft}$$

(d) The length of the curve is found using Eq. 4.9.

$$L_{\text{ft}} = \frac{(100\ \text{ft})I_{\text{deg}}}{D_{\text{deg}}} = \frac{(100\ \text{ft})(28^\circ\ 18')}{4.5^\circ} = 628.89\ \text{ft}$$

(e) The tangent length is found using Eq. 4.5.

$$T = R\tan\frac{I}{2} = (1273.24\ \text{ft})\tan\frac{28^\circ\ 18'}{2} = 321.00\ \text{ft}$$

7. LAYOUT OF HORIZONTAL CURVES

Horizontal curves for highways are designed based on the topography, design speed, desired sight distance around objects, and available right-of-way. All highway curves must conform with modern highway safety standards, which demand uniformity of elements throughout any given corridor.

The two most common methods of field layout are the *deflection angle method* and the *tangent offset method.* Although Global Positioning System (GPS) and Real Time Kinematic (RTK) satellite navigation systems have reduced the intensity of field work, the deflection angle method and the tangent offset method continue to be used in the field.

Topography and Fit of Alignment

Topography can be defined as level, rolling, or mountainous, as described in the *Highway Capacity Manual* (*HCM*). *Level terrain* is terrain with short grades no steeper than 1–2%. Although level terrain may be considered ideal for roadway design, large, flat regions usually require additional drainage facilities to carry rainwater away from the traveled surface.

Rolling terrain is terrain where heavy vehicles operate at reduced speeds compared to those of passenger cars but not for extended intervals. Generally, the roadway can be placed at grades of 2–4%. However, these grades do not extend more than 4000 ft (1220 m) in any one instance.

Mountainous terrain is terrain where grades occur such that heavy vehicles operate at substantially slower speeds than automobiles for extended distances. Grades steeper than 4% over a length of 4000 ft (1220 m) or more constitute mountainous terrain. The *HCM* and the *Green Book* have more information on mountainous terrain, along with recommendations for design provisions to handle traffic and safety concerns.

Design speed is a function of the terrain and the type and class of roadway to be built. All other design considerations are made in respect to the selected design speed. Interstate highways and freeways often use a design speed of 70 mph (113 kph) for level, relatively straight roadways in rural areas and 50 mph (80 kph) for urban locations. Multilane highways, two-lane highways, and urban streets have cross section and design speed parameters set forth by the responsible operating and funding agency. These speeds may range from freeway-like higher speed conditions to local feeder and access road conditions with design speeds as low as 15 mph (25 kph).

Design speeds can be influenced by available sight distance around buildings, rock croppings, or other obstructions. Design speed may have to factor in available right-of-way to preserve clearances for larger vehicles.

Once a roadway cross section is selected and the design speed and sight distance parameters are set, the centerline geometry can be laid out in preliminary form on topographical maps. This centerline is tested by taking copies of the preliminary map plans to the field for the surveyor and field engineer to locate and mark the centerline. Field staking begins at a known starting point or a connection to an existing facility. The tangents and curves are fitted together one by one until the opposite end is reached, usually connecting to another facility already in place. Field measurements are taken at critical control points, and the alignment is adjusted as needed to fit the required cross section to the control point field locations. After the field alignment is staked out and recorded, the field adjustments are taken back to the design office to be applied to the plan drawings.

The preliminary alignment plan, adjusted to meet field conditions, is presented for centerline alignment approval with walls, bridges, drainage, and other structures shown in their approximate locations on the plans. A preliminary design report is prepared, and the project is organized for cost estimation and funding agency approval.

After the preliminary alignment is approved, final field surveys are performed for structures and drainage facilities. Core borings for earth samples at foundation locations and for earthwork classification require a considerable investment and are usually not taken until after the grade and alignment approval to avoid unnecessary waste of funds should the alignment not be approved. Final geometry is prepared from the preliminary alignment, incorporating minor adjustments as necessary. These changes are kept to a minimum to avoid the costs and delays of reconfiguration.

The contractor is responsible for laying out the final alignment in the field using the final geometry. The designer is no longer involved except in extreme circumstances in which the layout on the plans is in error or a field discovery requires an adjustment. In either case, a change order is required and must be carefully documented so that any additional work relying on the centerline geometry uses the correct information.

Horizontal Alignment Layout and Location Control

Alignment layout begins with horizontal geometry, which is a series of tangents connected by curved lines, called circular arcs. The geometry conforms to a series of requirements that are based on a minimum design speed, such as minimum curve radius. Sometimes a *spiral curve*, also known as an *easement curve*, is introduced at the end of each circular curve in order to reduce driver discomfort or the off-tracking (e.g., onto the shoulder or into an adjacent lane) of vehicle wheels that often occurs at higher speeds when transitioning between a tangent and a curve.

Tangents and curves are tied to survey control points so that the alignment can be reproduced in the field for construction and ongoing maintenance during the life of the roadway.

Horizontal geometry is calculated as projected to a horizontal plane with no correction for the true length of slope distances. Adjustment for actual material quantities on slopes is made during construction estimation, but pay quantities are usually based on horizontal measurement only. These slope adjustments primarily affect pavement and pipe quantities.

Horizontal Curve Elements

Figure 4.5 shows the locations of additional horizontal curve elements and introduces terminology often referred to in alignment calculations. The tangents of a curve, T, are subtangents of the complete alignment, which extends beyond the limits of the curve (i.e., beyond the PC and PT). For simplicity, these subtangents are referred to as *curve tangents*, and a specific point on either of these tangents is called a *point on subtangent*, POST. This term should not be confused with a fence post or other type of post that may be referenced in deed descriptions. A specific point on either of the tangents that is normally beyond the curve limits is called a *point on tangent*, or POT.

Circular curves are usually laid out on plans with the stationing proceeding from left to right, as shown in Fig. 4.5, or from bottom to top. The curve tangent from the PI to the PC is called the *back tangent*, and the curve tangent from the PI to the PT is called the *ahead tangent*. A curve is stationed continuously from the tangent entering the curve at the PC, continuing ahead along the curve to the PT, and continuing ahead to the tangent ahead of the curve. Stationing continuously along the centerline of a highway or railroad ensures an exact measurement of the distance between any two points along the corridor. The back tangent is stationed from the PC to the PI in order to locate the exact distance to the PI, but the ahead tangent from the PI to the PT is rarely stationed so as to avoid having two stations for the PC.

Figure 4.5 *Additional Curve Elements*

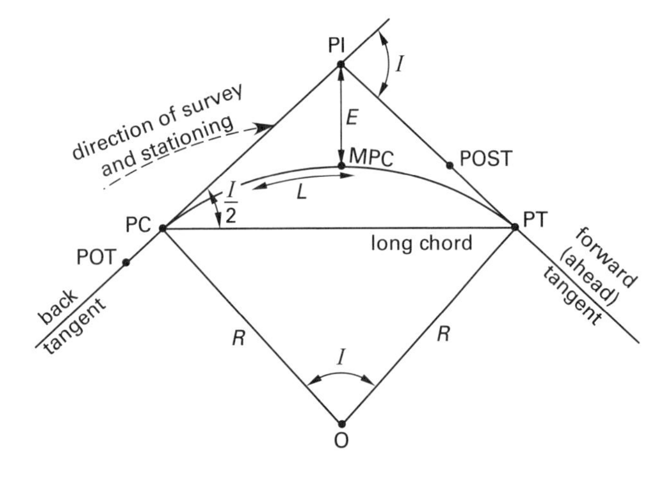

For horizontal curves laid out around an obstruction, the middle ordinate is sometimes referred to as the *horizontal sightline offset*, HSO. The point where the middle ordinate intersects the arc of the curve is the *midpoint of the curve* and can be labeled MPC.

Figure 4.6 shows the relationship of the various angles found in a circular curve layout. The deflection angle is positive if to the right (i.e., clockwise) and negative if to the left (i.e., counterclockwise).

Curve Layout by Deflection

One of the most widely used methods of field layout, or stakeout, is the deflection angle method from the PC or PT. The *deflection angle* is the angle between the tangent and the chord. In Fig. 4.7, the subtended angle, θ, of arc AB is twice the angle deflecting from the tangent $\overline{\text{AV}}$ to the chord $\overline{\text{AB}}$. The deflection method is popular because it involves the fewest instrument setups.

Starting from either the PC or the PT, stations are staked along the chord length. For full stations (i.e., $L = 100$ ft), $\theta = D$, as shown in Eq. 4.26. However, the

Figure 4.6 *Angles Used in Defining a Horizontal Circular Curve*

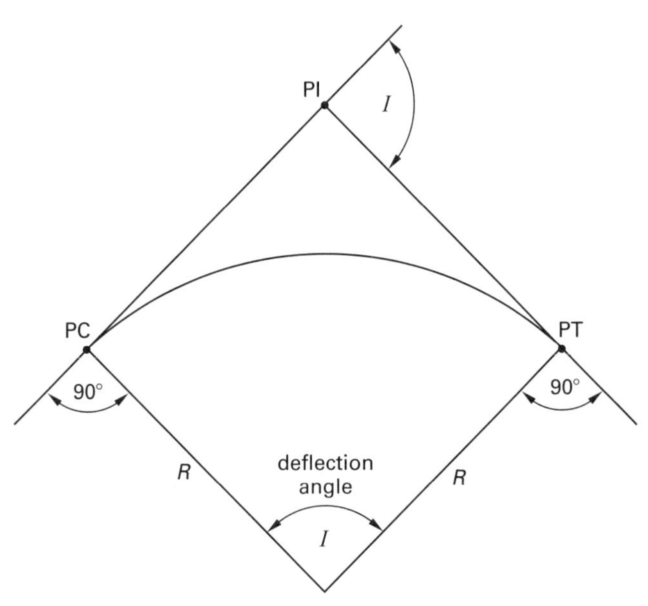

Figure 4.7 *Deflection Relationship in a Circular Curve*

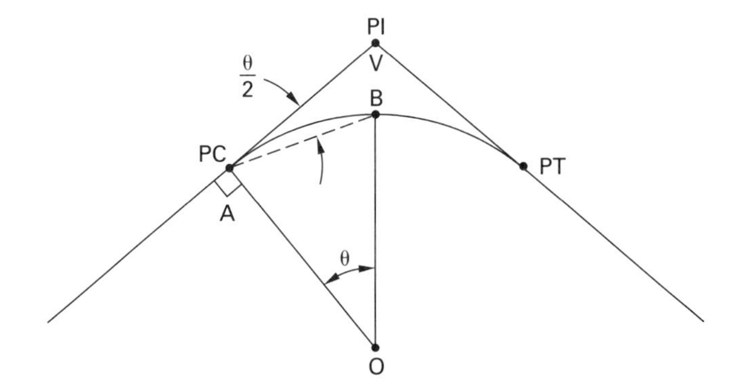

PC or PT do not necessarily fall on an even station, meaning the arc length will be less than 100 ft. An arc with a chord length less than 100 ft is known as a *subarc*. The deflection angle of the subarc is determined either by multiplying the deflection angle for a full station by the proportion of the chord length to a full station (i.e., $L/100$ ft) or by using Eq. 4.27. Subsequent stations are found by adding intervals of $D/2$ to the first deflection until the last full station before the PC or PT. The final $\theta/2$ increment should make the sum of the deflections equal to half of the full deflection of the curve, $I/2$, as shown in Eq. 4.28, which also serves as a check on calculations.

$$\theta = \frac{L_{\text{ft}} D_{\text{deg}}}{100 \text{ ft}} \quad \text{[U.S. only]} \qquad 4.26$$

$$\frac{\theta}{2} = \frac{L_{\text{ft}} D_{\text{deg}}}{200 \text{ ft}} \quad \text{[U.S. only]} \qquad 4.27$$

$$\frac{I}{2} = \sum_{1}^{n} \frac{\theta_1}{2} + \frac{\theta_2}{2} + \cdots + \frac{\theta_n}{2} \qquad 4.28$$

When staking out a curve, a surveying instrument setup on each station is usually not necessary, as the scope on the field instrument usually provides accurate readings up to at least 300 ft (91 m), and reducing the number of setups increases productivity as well as accuracy. For a typical curve layout, once the PC is staked, the next three stations can be staked before moving the instrument—that is, the setup is moved ahead three stations, the stakeout continues for another three stations, and so on, until the PT is reached. (See Fig. 4.8.) For a setup on an intermediate station, the angle relationships are illustrated in Fig. 4.9.

Figure 4.8 *Relationship of Chord Deflections to Total Curve Deflection*

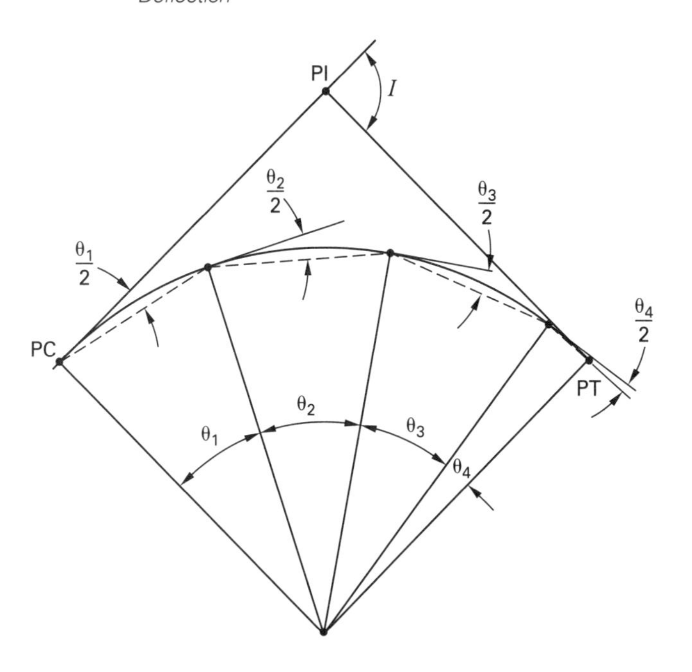

Figure 4.9 *Deflection Angles from an Intermediate Point on a Circular Curve*

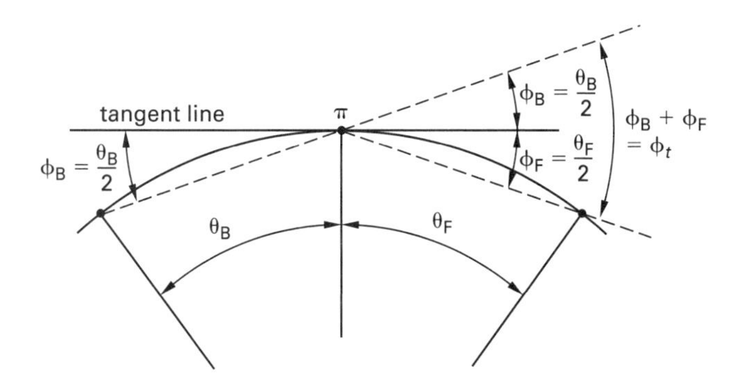

Once the PT is staked, the field instrument is moved over the PT stake. The $I/2$ angle is turned from the PC to the PI, and the scope is flopped (rotated vertically 180), allowing the forward tangent to be sighted and staked.

When the curve is long enough that an intermediate setup is required and the backsight is to the previously set point on the curve, the deflection of the back chord must be added to the deflection of the ahead chord to find the total deflection.

The points of measurement (i.e., the instrument locations) used in the field must be carefully recorded when using the deflection angle method. The instrument location is the point from which measurements are taken or about which the angle is turned. These points are referred to as stations, even if the instrument location is not recorded in units of stations (e.g., station A). Stations are frequently represented in field notes using the symbol $\overline{\wedge}$. Field notes are commonly recorded in tabular form, with the leftmost column heading labeled $\overline{\wedge}$ and each station listed underneath.

Example 4.4

A curve has a radius of 800 ft and a deflection angle of 32°. The PC station is 133+23.20, and the maximum sight line is 300 ft. Use sta 136+00 as the secondary instrument setup location. Using the deflection angle method, complete the tabulation of deflection angles to lay out the curve.

Solution

Calculate the necessary curve information, recording all data in tabular form as shown in *Illustration and Table for Example 4.4*. The leftmost column gives the station, $\overline{\wedge}$, of each instrument setup.

Determine the curve length using Eq. 4.9.

$$\begin{aligned} L_{\text{ft}} &= \frac{2\pi R_{\text{ft}} I_{\text{deg}}}{360°} = \frac{(2\pi)(800 \text{ ft})(32°)}{360°} \\ &= 446.80 \text{ ft} \end{aligned}$$

From Eq. 4.18, the degree of curve for arc definition is

$$\begin{aligned} D_{\text{deg}} &= \frac{(360°)(100 \text{ ft})}{2\pi R_{\text{ft}}} = \frac{(360°)(100 \text{ ft})}{(2\pi)(800 \text{ ft})} \\ &= 7.162° \end{aligned}$$

Use Eq. 4.24 to find the station of the PT.

$$\begin{aligned} \text{sta PT} &= \text{sta PC} + L = \text{sta } 133{+}23.20 + 446.80 \text{ ft} \\ &= \text{sta } 137{+}70.00 \end{aligned}$$

Deflections will first be made with the instrument set on the PC. To set the first full station, find the subarc length measured from the PC to sta 134+00.

$$\begin{aligned} L_{\text{subarc}} &= \text{sta } 134{+}00 - \text{sta } 133{+}23.20 \\ &= 76.80 \text{ ft} \end{aligned}$$

Calculate the deflection using Eq. 4.27.

$$\phi_{\text{sta }134+00} = \frac{\theta}{2} = \frac{L_{\text{subarc,ft}}D_{\text{deg}}}{200\text{ ft}} = \frac{(76.80\text{ ft})(7.162°)}{200\text{ ft}} = 2.750°$$

From Eq. 4.11, the chord length for a 76.80 ft arc is

$$C_{\text{sta }134+00} = 2R\sin\frac{\theta}{2} = (2)(800\text{ ft})\sin 2.750° = 76.77\text{ ft}$$

For sta 135+00, the deflection of a 176.80 ft arc (i.e., the subarc length plus one full station) is

$$\phi_{\text{sta }135+00} = \frac{\theta}{2} = \frac{L_{\text{ft}}D_{\text{deg}}}{200\text{ ft}} = \frac{(176.80\text{ ft})(7.162°)}{200\text{ ft}} = 6.331°$$

The chord length for a 176.80 ft arc is

$$C_{\text{sta }135+00} = 2R\sin\frac{\theta}{2} = (2)(800\text{ ft})\sin 6.331° = 176.44\text{ ft}$$

For sta 136+00, the deflection of a 276.80 ft arc from the tangent is

$$\phi_{\text{sta }136+00} = \frac{\theta}{2} = \frac{L_{\text{ft}}D_{\text{deg}}}{200\text{ ft}} = \frac{(276.80\text{ ft})(7.162°)}{200\text{ ft}} = 9.912°$$

The chord length for a 276.80 ft arc is

$$C_{\text{sta }136+00} = 2R\sin\frac{\theta}{2} = (2)(800\text{ ft})\sin 9.912° = 275.42\text{ ft}$$

Since the curve is longer than the maximum sight line of 300 ft, a second setup on the curve is necessary. The problem statement specified the second setup location as sta 136+00, but any station less than 300 ft from the PC could have been chosen. To locate sta 137+00, find the deflection angle of a 100 ft arc from a tangent located at the second setup point (i.e., sta 136+00).

$$\phi_{\text{sta }137+00} = \frac{\theta}{2} = \frac{L_{\text{ft}}D_{\text{deg}}}{200\text{ ft}} = \frac{(100\text{ ft})(7.162°)}{200\text{ ft}} = 3.581°$$

The chord length for a 100 ft arc is

$$C_{\text{sta }137+00} = 2R\sin\frac{\theta}{2} = (2)(800\text{ ft})\sin 3.581° = 99.94\text{ ft}$$

The back deflection from the tangent at sta 136+00 to the PC was found when locating sta 136+00. Add the back deflection to the ahead deflection (i.e., the deflection from the tangent at sta 136+00 to the PT) for the total deflection from the PC.

$$\phi_t = \phi_{\text{sta }136+00} + \phi_{\text{sta }137+00} = 9.912° + 3.581° = 13.493°$$

Working from sta 136+00 to the PT,

$$\text{arc length} = \text{sta }137+70.00 - \text{sta }136+00.00 = 170.00\text{ ft}$$

Calculate the deflection of a 170.00 ft arc from the tangent located at the setup point to the PT.

$$\phi_{\text{sta }137+70.00} = \frac{L_{\text{ft}}D_{\text{deg}}}{200\text{ ft}} = \frac{(170.00\text{ ft})(7.162°)}{200\text{ ft}} = 6.088°$$

The chord length for a 170.00 ft arc is

$$C_{\text{PT}} = 2R\sin\frac{\theta}{2} = (2)(800\text{ ft})\sin 6.088° = 169.69\text{ ft}$$

The table and curve are completed as shown in the *Illustration and Table for Example 4.4*.

The last line is a check of the total deflection, found by adding the angles shown in bold. The error shown is 1:114,000 deviations from 32.00000° due to the rounding in the calculations. This amounts to 1 sec of angle.

Curve Layout by Tangent Offset

The *tangent offset method* (also known as the *station offset method*) can also be used to lay out horizontal curves. This method is used to locate any point on the curve when working strictly from the tangent and PC and is typically used on short curves. The tangent offset method is well-suited to locating a single point with a high degree of accuracy or for approximating measurements using cloth, plastic, or steel tapes only. A curve's *tangent offset*, y, is the perpendicular distance from an extended tangent line to the curve. The *tangent distance*, x, is the distance along the tangent to a perpendicular point. (See Fig. 4.10).

Figure 4.10 can be mathematically represented as Eq. 4.29, where y is the tangent offset and x is the tangent distance.

$$y = R - \sqrt{R^2 - x^2} = R(1 - \cos 2\alpha) \qquad 4.29$$

Inaccessible PI

In order to lay out a complete curve, the PI must be located and used as the instrument setup location if

Illustration and Table for Example 4.4

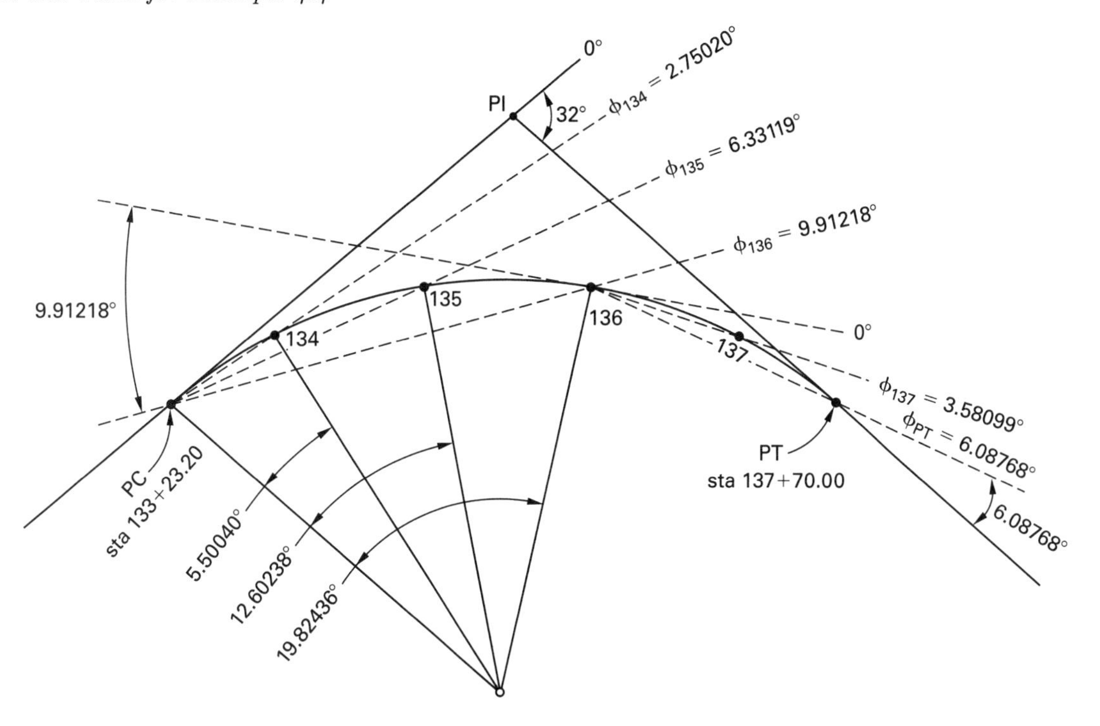

location, $\overline{\wedge}$ (sta)	backsight (sta)	ϕ_B	foresight (sta)	chord length (ft)	ϕ_F	ϕ_t
133+23.20 PC	tan	0°	134+00	76.77	2.750°	2.750°
133+23.20 PC	tan	0°	135+00	176.44	6.331°	6.331°
133+23.20 PC	tan	0°	136+00	275.42	9.912°	**9.912°**
136+00	133+23.20 PC	9.912°	137+00	99.94	3.581°	13.493°
136+00	133+23.20 PC	9.912°	137+70 PT	169.69	**6.088°**	**16.000°**
136+70.00 PT	136+00	6.088°	ahead tangent	–	**6.088°**	12.175°
check total deflection						32.000°

possible. The PI is used because it provides the most accurate calculation of the intersection angle and provides the longest sight tangent. If the PI can be occupied during curve layout, the PI is called accessible. However, the PI is at times inaccessible due to some physical obstruction or unreachable location, such as being located in a body of water, over a cliff, or in a building. The following procedure is used to compensate for an inaccessible PI.

A point on subtangent, POST, is located at any point A on the back subtangent that is visible from a point B on the ahead subtangent. The relationship between point A and point B is shown in Fig. 4.11. The sum of angles α and β must equal the intersection angle. From triangle PI-A-B, α and β can be calculated using the law of sines. The PI is found by adding the distance from A to the PI, or $\overline{\text{A PI}}$, to point A's station. After calculating the tangent distance, T, the PC and PT are found by measuring from points A and B to the PI. Therefore, the entire curve can be laid out accurately even though the PI cannot be occupied.

Figure 4.10 *Point on Curve Located by Tangent Offset*

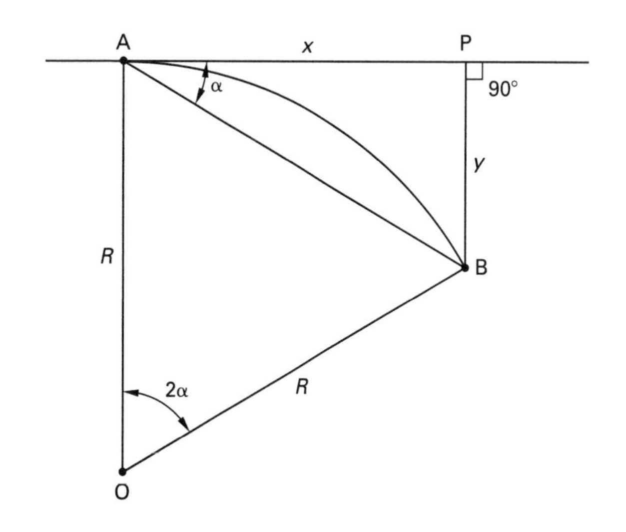

Figure 4.11 Inaccessible PI

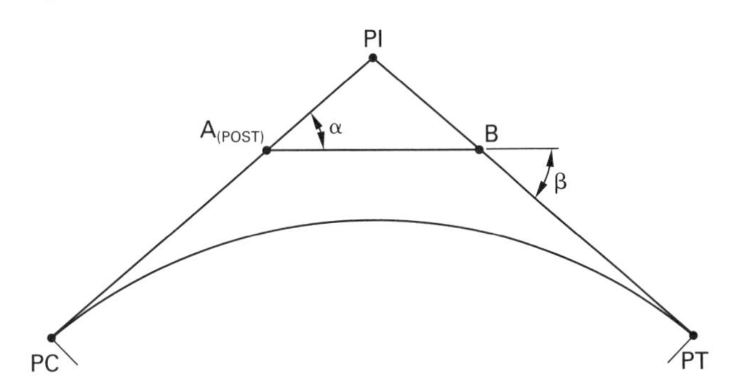

It is important to examine the location carefully before choosing the location of points A and B. Often, the subtangent $\overline{AB}$ is placed close to the curve so that it appears to be touching, or tangent to, the curve. Making this assumption without basis will yield an inaccurate result. Also, the distances $\overline{A\ PI}$ and $\overline{B\ PI}$ do not have to be equal and rarely are. Points A and B are simply set at convenient locations along the tangents.

8. SPIRAL CURVES

Spiral curves are introduced at the ends of a circular curve (i.e., at the PC and PT) in order to provide a transition, or *easement*, between the straight tangents and the circular parts of the curve. Without an easement, the sudden change from no lateral acceleration along the tangent to the lateral acceleration necessary to travel around a curve causes an increase in lateral force on the vehicle. Spiral curves were first used on railroads to reduce the jerk caused at the ends of curves. The principle of easement is a gradual increase in lateral acceleration by decreasing the curve radius until the central curve radius is reached. *Superelevation*, which is the difference between the inner and outer elevation of a track or roadway, is also transitioned along the easement, providing a smooth and comfortable ride for passengers by changing the cross slope in proportion to the change in curve radius. (See Sec. 4.12 for more information on superelevation.)

The spiral curve adopted by AASHTO for highways is modified from the standard railroad spiral curve. The degree of curve increases linearly along the spiral transition from zero at the TS to the degree of curve of the central curve at the SC. The TS is the point of change from the tangent to the spiral, and the SC is the point of change from the spiral curve to the circular curve. The degree of the central curve at the CS or SC, D_l, can be calculated from Eq. 4.30. D_c is the degree of curve for the entire curve, L_s is the length of the spiral (i.e., the entire curve length), and L is the length of the curve from the TS or ST to the CS or SC.

$$D_l = D_c\left(\frac{L}{L_s}\right) \qquad 4.30$$

The minimum length of a spiral is based on driver comfort and lateral shifts in the position of a vehicle within the curve. Spirals should be long enough that the increase in lateral acceleration as a vehicle enters the curve is comfortable to the driver. Spirals should also be long enough that the shift they cause in a vehicle's lateral position is similar to the shift produced by a vehicle's natural path. Therefore, the minimum length of a spiral should be the larger value found from the following equations.

$$\begin{aligned} L_{s,\text{min}} &= \sqrt{24p_{\text{min}}R} && [\textit{Green Book}\text{ Eq. 3-28}] \\ &= 0.0214\frac{\text{v}^3}{RC} && [\textit{Green Book}\text{ Eq. 3-29}] \end{aligned} \qquad \text{[SI]} \quad 4.31(a)$$

$$\begin{aligned} L_{s,\text{min}} &= \sqrt{24p_{\text{min}}R} && [\textit{Green Book}\text{ Eq. 3-28}] \\ &= 3.15\frac{\text{v}^3}{RC} && [\textit{Green Book}\text{ Eq. 3-29}] \end{aligned} \qquad \text{[U.S.]} \quad 4.31(b)$$

p_{min} is the minimum lateral offset between the tangent and the circular curve, v is the design speed in mph (kph), and C is the maximum rate of change in lateral acceleration. AASHTO suggests a value of 0.66 ft (0.20 m) for p_{min}, as this is representative of the minimum lateral shift that occurs from the natural steering behavior of most drivers. AASHTO recommends a value of 4.0 ft/sec^3 (1.2 m/s^3) for the maximum rate of change in lateral acceleration, C.

Spiral curves must also not be too long in relation to the length of the circular curve. A spiral curve that is too long can mislead drivers about the sharpness of the upcoming curve, causing drivers to approach the curve at unsafe speeds. The maximum length of a spiral curve can be calculated from Eq. 4.32.

$$L_{s,\text{max}} = \sqrt{24p_{\text{max}}R} \qquad [\textit{Green Book}\text{ Eq. 3-30}] \qquad 4.32$$

The maximum lateral offset between the tangent and circular curve, p_{max}, is suggested by AASHTO to be 3.3 ft (1.0 m), as this value is representative of the maximum lateral shift occurring as a result of the natural steering behavior of the majority of drivers.

Table 4.1 gives roadway spiral lengths recommended in the *Green Book*. The lengths given correspond to 2.0 sec of travel time at the design speed listed, which is representative of the natural spiral path used by most drivers.

Calculations involving components of a spiral curve require an additional significant digit to the right of the decimal. A general rule is to carry a value out to three decimals for customary U.S. units and four decimals for SI units. Carrying an additional significant digit is generally sufficient to ensure accuracy of the final measurements for layout purposes.

Table 4.1 Desirable Length of Spiral Curve Transition

SI units		customary U.S. units	
design speed (kph)	spiral length (m)	design speed (mph)	spiral length (ft)
20	11	15	44
30	17	20	59
40	22	25	74
50	28	30	88
60	33	35	103
70	39	40	117
80	44	45	132
90	50	50	147
100	56	55	161
110	61	60	176
120	67	65	191
130	72	70	205
		75	220
		80	235

From *A Policy on Geometric Design of Highways and Streets*, 2004, by the American Association of State Highways and Transportation Officials, Washington, D.C. Used by permission.

The geometric relationships given for horizontal curves, Eq. 4.7, Eq. 4.8, Eq. 4.14, and Eq. 4.15, also apply for spiral curves. The equations for degree of curve and curve radius for horizontal curves also apply to spiral curves. Spiral curve variables shared with other curve types are distinguished using a subscript s, except for the degree of curve, which uses D_c. Figure 4.12 illustrates a fully spiraled circular curve, and Fig. 4.13 demonstrates the details of a spiral curve. Spiral curve abbreviations and variables are given in Table 4.7.

Figure 4.12 Fully Spiraled Circular Curve Layout

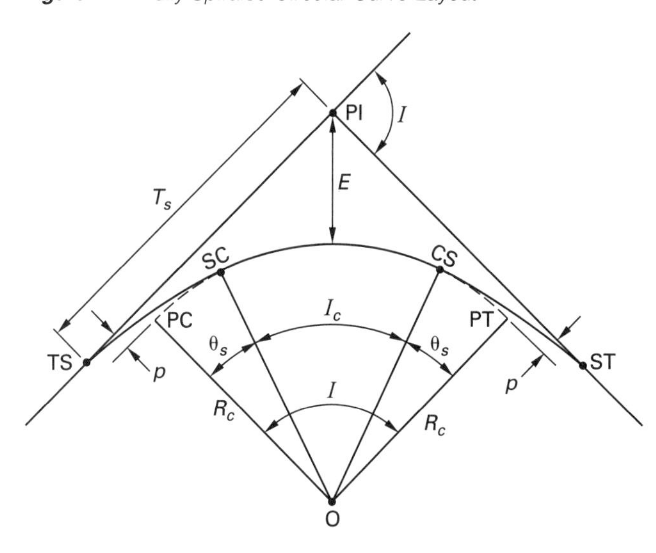

Unique to spiral curves are the points of change between spiral curves or between a spiral curve and a circular curve. SC is the point of change from a spiral curve to a circular curve, and TS is the point of change

Figure 4.13 Detail Elements of a Transition Spiral

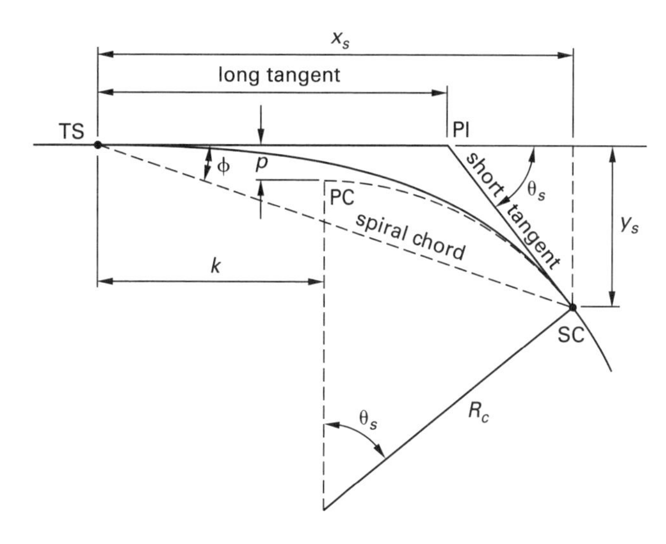

Table 4.2 Spiral Curves: Abbreviations and Terms

- CS point of change from circular curve to spiral
- D_c degree of curve of spiral at SC or CS, and of the central curve
- E_s external distance from the PI to the center of the circular arc (see Eq. 4.36)
- I intersection angle of the tangents of the entire curve
- I_c intersection angle of the included circular curve from the SC to the CS
- k distance along initial tangent extended from TS to point opposite PC of shifted curve (see Eq. 4.35)
- K rate of change of degree of curvature per foot (meter) of spiral
- L spiral arc from the TS or ST and any point on spiral
- L_s total spiral length from TS to SC, or CS to ST
- LT long tangent, distance along initial tangent extended from TS to spiral PI
- p offset from initial tangent to PC of shifted circular curve (see Eq. 4.34)
- R_c radius of central curve
- SC point of change from spiral curve to circular
- SCS point of change from one spiral to another, or an instant curve point between spirals
- SS point of change from one spiral to another, or an instant curve point between spirals
- ST short tangent, distance from spiral PI to SC
- ST point of change from spiral to tangent
- T_s tangent of a spiral curve (see Eq. 4.33)
- TS point of change from tangent to spiral
- x_s coordinate of SC from TS along tangent (see Eq. 4.41)
- y_s coordinate of SC from TS, offset distance (see Eq. 4.42)
- θ central angle of spiral arc of a given length, L (see Eq. 4.37)
- θ_s central angle of total spiral arc (see Eq. 4.38)
- ϕ spiral deflection angle at TS or ST from initial tangent to any point on spiral (see Eq. 4.39)
- ϕ_s spiral deflection angle to SC or CS (see Eq. 4.40)

from a tangent to a spiral curve. CS is the point of change from a circular curve to a spiral curve, and ST is the point of change from a spiral curve to a tangent. SS, or SCS, is the point of change from one spiral curve to another or an instant curve point between two spiral curves.

The tangent of a spiral curve, T_s, can be calculated using Eq. 4.33. R_c is the radius of the central curve, and p is the offset from the initial tangent to the PC of the shifted circular curve calculated from Eq. 4.34. Equation 4.35 is used to calculate k, which is the distance along the initial tangent from TS to the PC of the shifted curve. I is the intersection angle of the tangent and the entire curve, and I_c is the intersection angle of the circular curve measured from the SC to the CS.

$$T_s = (R_c + p)\tan\frac{I}{2} + k \quad 4.33$$

$$p = y_s - R_c \operatorname{vers}\theta_s \quad 4.34$$

$$k = x_s - R_c \sin\theta_s \quad 4.35$$

The external distance, E_s, measured from the PI to the center of the circular arc is found from Eq. 4.36.

$$E_s = (R_c + p)\operatorname{exsec}\frac{I}{2} + p \quad 4.36$$

The central angle of a spiral arc is calculated using Eq. 4.37 and Eq. 4.38. L is the spiral arc length from either the TS or ST to any point on the spiral. L_s is the total spiral length from TS to SC or from ST to CS. θ is the central angle of a spiral arc with a given length, L, and θ_s is the central angle of the total spiral arc length.

$$\theta = \left(\frac{L}{L_s}\right)^2 \theta_s \quad 4.37$$

$$\theta_s = \left(\frac{L_{s,\text{ft}}}{200\text{ ft}}\right) D_{c,\text{deg}} \quad \text{[U.S. only]} \quad 4.38$$

Equation 4.39 is used to calculate the spiral deflection angle, ϕ, at TS or ST from the initial tangent to any point on the spiral. The spiral deflection angle to SC or CS, ϕ_s, is found from Eq. 4.40.

$$\phi = \tfrac{1}{3}\left(\frac{L}{L_s}\right)^2 \theta_s \quad 4.39$$

$$\phi_s = \tfrac{1}{3}\theta_s \quad 4.40$$

The length between SC and TS measured along the tangent, x_s, is found from Eq. 4.41. The tangent offset, y_s, is the vertical distance between SC and TS and is calculated from Eq. 4.42.

$$x_{s,\text{ft}} = \frac{L_{s,\text{ft}}}{100\text{ ft}}\left(100\text{ ft} - \frac{(100\text{ ft})\left(\frac{\pi}{180^\circ}\right)^2 D_{c,\text{deg}}^2}{(5)(2!)} + \frac{(100\text{ ft})\left(\frac{\pi}{180^\circ}\right)^4 D_{c,\text{deg}}^4}{(9)(4!)} + \frac{(100\text{ ft})\left(\frac{\pi}{180^\circ}\right)^6 D_{c,\text{deg}}^6}{(13)(6!)} \cdots\right) \quad \text{[U.S. only]} \quad 4.41$$

$$y_{s,\text{ft}} = \frac{L_{\text{ft}}^3}{6R_{c,\text{ft}}L_{s,\text{ft}}} \quad \text{[U.S. only]} \quad 4.42$$

9. VERTICAL CURVES

Vertical curves are used to connect two vertical tangents in order to change the grade of a highway. Most vertical curve layouts use *parabolic curves*, which can be symmetrical or unsymmetrical. Symmetrical parabolic curves are most often used because they have a constant rate of grade change, which simplifies calculation. Calculations may also be simplified by basing control points on offsets from the vertical tangents. All calculations are based on a horizontal plane, projecting horizontal measurements up or down to the road surface, regardless of the slope of the grade. Slope distances are generally only used for quantity estimation purposes. Vertical alignment is shown as a vertical surface cut along the *horizontal control line*, which is normally the centerline of the roadway. Vertical curves are usually identified by the station of the point of vertical intersection, PVI, and the PVI elevation, such as "vertical curve 67+84 at elevation 100." The location and elevation of the PVI are the most important geometric points of a vertical curve, as the PVI location controls the grade lines.

A *sag vertical curve* is concave upward, and a *crest vertical curve* is concave downward. Grades are shown as positive (+) if the profile rises with advancing stations and negative (−) if the profile falls with advancing stations. Figure 4.14 illustrates three general cases of sag vertical curves, while Fig. 4.15 shows three general cases of crest vertical curves. The curves of Fig. 4.14(a) and Fig. 4.15(a) have *turning points*, or an instant level section, while the other curve conditions do not.

Figure 4.14 *Three Conditions of Sag Vertical Curves*

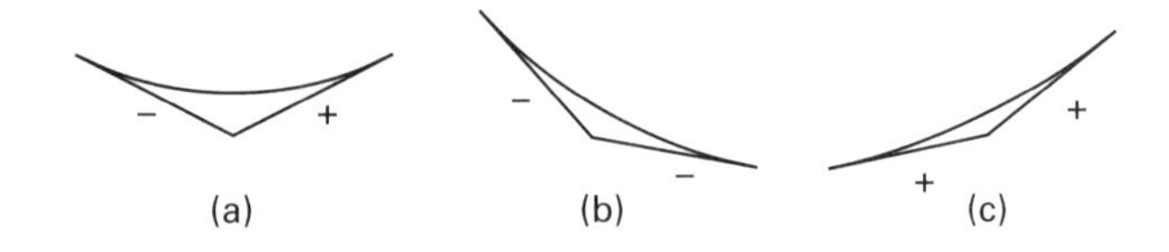

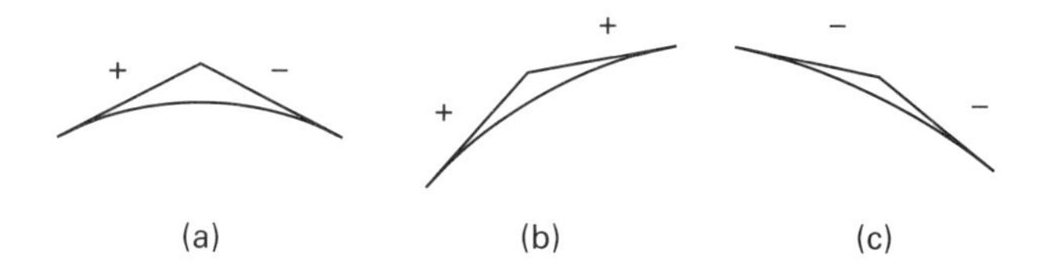

Figure 4.15 Three Conditions of Crest Vertical Curves

Parabolic Curves

The principle of a parabolic curve is applied directly to curve calculations. A parabolic curve consists of three basic points: the point of vertical curve, PVC, the point of vertical intersection, PVI, and the point of vertical tangency, PVT. The PVC is the beginning of the curve (also known as BVC), and the PVT is the end of the curve (also called EVC). The PVI is the intersection of the two tangents and is also referred to as the vertex, V. Offsets to a tangent, y, of a parabola are proportional to the squares of the distances from the point of tangency (i.e., the point where the curve and tangent meet), as shown in Fig. 4.16, and are calculated from Eq. 4.43. If the distance from the point of tangency is doubled, the offset from the tangent to the parabola becomes 2^2, or four times as large.

$$y = ax^2 \qquad 4.43$$

Figure 4.16 Tangent Offset Relationship to Distance in a Parabolic Curve

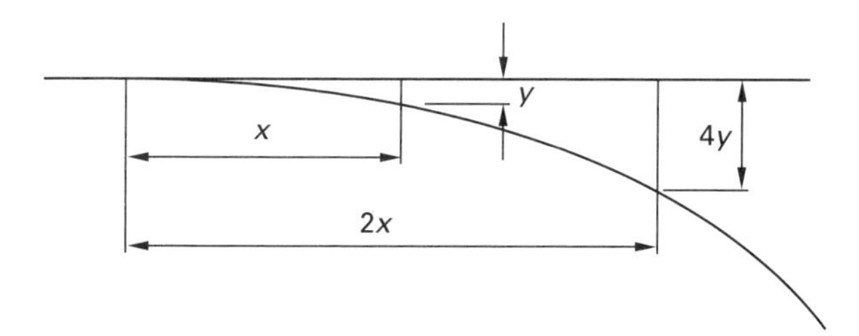

The middle ordinate and the external distance are equal for a parabolic curve, as shown in Fig. 4.17. However, the external distance is not frequently used in vertical curves. The middle ordinate, M, is found from Eq. 4.44, where A is the absolute value of the algebraic difference in grades.

$$E_{\text{ft}} = M_{\text{ft}} = \frac{|G_{2,\%} - G_{1,\%}|L_{\text{sta}}}{8\ \frac{\text{sta}}{\text{ft}}} = \frac{A_{\%}L_{\text{sta}}}{8\ \frac{\text{sta}}{\text{ft}}} \qquad \text{[U.S. only]} \qquad 4.44$$

The *rate of grade change per station*, R, can be determined using Eq. 4.45. In these equations, G_2 is the grade out of the curve, and G_1 is the grade into the curve. R can be expressed in units of %/sta or ft/sta^2, since 1 sta is equivalent to 100 ft. The rate, R, will be negative for crest vertical curves and positive for sag vertical curves.

Figure 4.17 Location of Middle Ordinate and External Distance on a Vertical Parabolic Curve

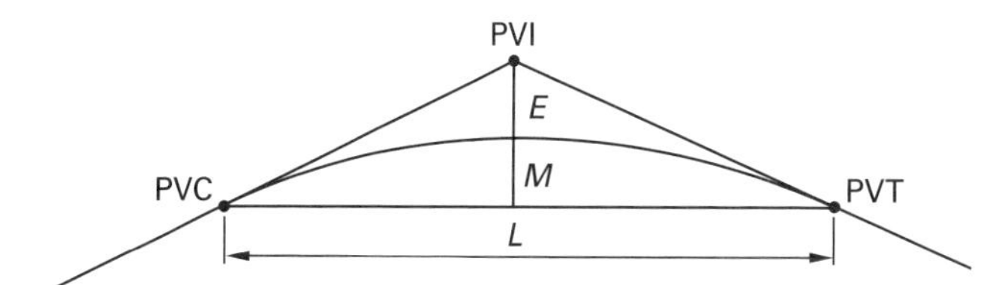

$$R = \frac{G_{2,\%} - G_{1,\%}}{L_{\text{sta}}} \qquad \text{[may be negative]} \qquad 4.45$$

An unknown elevation at point x, elev$_x$, on a vertical curve can be determined from a known elevation at point a, elev$_a$, and the rate of change in grade, R, as shown in Eq. 4.46.

$$\text{elev}_x = \frac{Rx^2}{2} + G_1x + \text{elev}_a \qquad 4.46$$

The x distance is the horizontal distance from the PVC, or any known elevation point on the curve, to the point of grade to be calculated. Equation 4.46 can be used to calculate the vertical curve elevations directly from a known elevation at one end. Equation 4.46 can be rewritten as Eq. 4.47, which is convenient for directly calculating the series of elevations needed to construct a curve.

$$\begin{aligned}\text{elev}_x &= \frac{Rx^2}{2} + G_{1,\text{ft/ft}}x + \text{elev}_a \\ &= \left(G_{1,\text{ft/ft}} + \frac{Rx}{2}\right)x + \text{elev}_a\end{aligned} \qquad 4.47$$

When the instant slope along the curve is equal to zero, the maximum or minimum elevations will occur. This high or low point, called the *turning point*, is determined by Eq. 4.48, where x is the horizontal distance from the PVC. The elevation of the turning point can be found using either the rate of grade change or the proportional offset method detailed in the following section.

$$x_{\text{turning point,sta}} = \frac{-G_{1,\%}}{R_{\%/\text{sta}}} \qquad 4.48$$

Design and Layout of Vertical Curves and Elevations

Vertical curves are considered symmetrical unless otherwise noted. The length of a symmetrical vertical curve is distributed equally on both sides of the PVI, so that the distance PVC–PVI is the same as the distance PVI–PVT. For example, a 500 ft long vertical curve would be comprised of two 250 ft half-curves located on each side of the PVI. The length of each half-curve is expressed as half the total curve length, $L/2$. The elevation of the PVC and PVT can be calculated if the PVI elevation is known by adding the product of the half-curve length, $L/2$, and the grade, G, along the curve length to the PVI elevation. Similarly, the product of the tangent length, T, and the grade can be added to or subtracted from the PVI elevation to determine the PVT and PVC elevations.

Unsymmetrical vertical curves are occasionally used to connect to existing roadways or to allow for special vertical clearance conditions, such as at an underpass (see Sec. 4.13). Unsymmetrical vertical curves are shown by designating the length of the back tangent and ahead tangent. If this information is not given, the curve is assumed to be a symmetrical vertical curve.

Both the tangent offset and the modified rate of change methods are effective ways to calculate elevations on vertical curves. However, the rate of change method is most often used when dealing with stations not at even pluses. Example 4.5 provides solutions using both methods.

Example 4.5

A symmetrical vertical curve is staked out between two grades, $\overline{AC}$ and $\overline{CB}$. The $\overline{AC}$ grade is −6.2%, and the $\overline{CB}$ grade is +3.8%. The PVI is located at sta 80+73.00 and has an elevation of 92.25 ft. The total length of the vertical curve is 2.50 sta. Write the notes for the grade stakes placed at 25 ft intervals along the curve, with the stakes not set at even pluses.

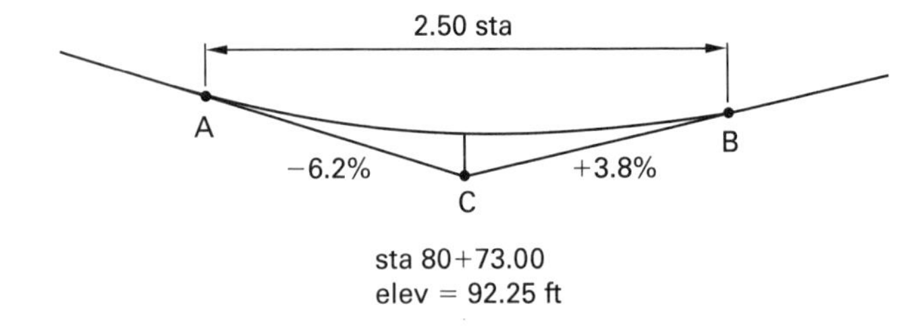

Solution

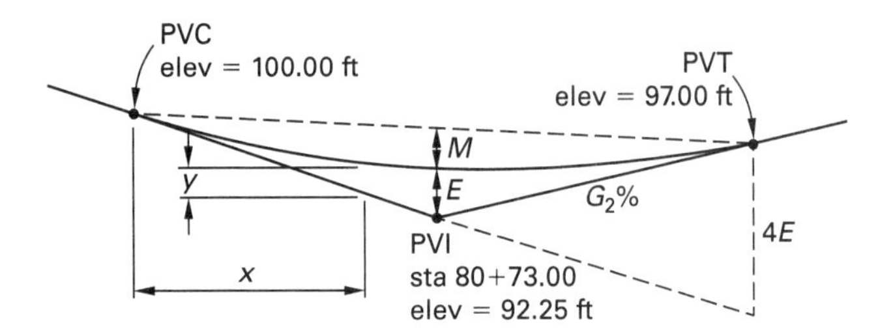

To use the method of proportional squared offsets, the solution can be worked from the PVC to the PVI and from the PVT to the PVI, or from either the PVC or PVT to the other end of the curve. This solution is worked from each end to the PVI. Tabulate all data for each station, spaced 25 ft apart, as shown in *Table for Example 4.5 Solution.*

First, the stations and elevations of the PVC and PVT are found. Since the curve is symmetrical, the PVI is at the center of the curve. Therefore, the length of the curve from the PVC to the PVI and from the PVI to the PVT is 2.50 sta/2 = 1.25 sta. The stations of the PVC and PVT are

$$\begin{aligned} \text{sta PVC} &= \text{sta PVI} - \frac{L}{2} \\ &= \text{sta } 80{+}73.00 - \left(\frac{2.50 \text{ sta}}{2}\right)\left(100 \ \frac{\text{ft}}{\text{sta}}\right) \\ &= \text{sta } 79{+}48.00 \end{aligned}$$

$$\begin{aligned} \text{sta PVT} &= \text{sta PVI} + \frac{L}{2} \\ &= \text{sta } 80{+}73.00 + \left(\frac{2.50 \text{ sta}}{2}\right)\left(100 \ \frac{\text{ft}}{\text{sta}}\right) \\ &= \text{sta } 81{+}98.00 \end{aligned}$$

Table for Example 4.5 Solution

	station	x (sta)	elevation on tangent	$C=(2x/L)^2$	offset = CM	elevation on curve
PVC	79+48.00	0	100.00	0	0	100.000
	+73.00	0.25	98.45	0.04	0.125	98.575
	+98.00	0.50	96.90	0.16	0.500	97.400
	80+23.00	0.75	95.35	0.36	1.125	96.475
	+48.00	1.00	93.80	0.64	2.000	95.800
PVI	+73.00	1.25	92.25	1.00	3.125	95.375
	+98.00	1.00	93.20	0.64	2.000	95.200
	81+23.00	0.75	94.15	0.36	1.125	95.275
	+48.00	0.50	95.10	0.16	0.500	95.600
	+73.00	0.25	96.05	0.04	0.125	96.175
PVT	+98.00	0	97.00	0	0	97.000

Noting that the grades when working from the PVI to the PVC and PVT are positive, the elevations of the PVC and PVT are

$$\text{elev}_{\text{PVC}} = \text{elev}_{\text{PVI}} + G_1\left(\frac{L}{2}\right)$$
$$= 92.25 \text{ ft} + (6.2\%)\left(\frac{2.50 \text{ sta}}{2}\right)\left(1 \frac{\text{ft}}{\%\text{-sta}}\right)$$
$$= 100.00 \text{ ft}$$

$$\text{elev}_{\text{PVT}} = \text{elev}_{\text{PVI}} + G_2\left(\frac{L}{2}\right)$$
$$= 92.25 \text{ ft} + (3.8\%)\left(\frac{2.50 \text{ sta}}{2}\right)\left(1 \frac{\text{ft}}{\%\text{-sta}}\right)$$
$$= 97.00 \text{ ft}$$

The middle ordinate is

$$M_{\text{ft}} = \frac{|G_{2,\%} - G_{1,\%}|L_{\text{sta}}}{8 \frac{\text{sta}}{\text{ft}}}$$
$$= \frac{|3.8\% - (-6.2\%)|(2.50 \text{ sta})}{8 \frac{\text{sta}}{\text{ft}}}$$
$$= 3.125 \text{ ft}$$

For a parabolic curve, $M = E$. Find the elevations on tangent, working from the PVC to the PVT. The grades are negative because they descend from the PVC and PVT. The grade point calculation for sta 80+23 is shown as an example of the calculations for each desired grade point.

$$\text{elev}_{x,\text{sta } 80+23} = G_1 x + \text{elev}_{\text{PVC}}$$
$$= (-6.2\%)(0.75 \text{ sta})\left(1 \frac{\text{ft}}{\%\text{-sta}}\right) + 100.00 \text{ ft}$$
$$= 95.35 \text{ ft}$$

The elevation on the curve is determined by adding the vertical offset correction, CM, to the elevation on vertical tangent. The vertical correction is calculated from the proportional square of the distance from the PVC to the PVI, and the PVT to the PVI multiplied by the middle ordinate. At sta 80+23,

$$C = \left(\frac{2x}{L}\right)^2 = \left(\frac{(2)(0.75 \text{ sta})}{2.50 \text{ sta}}\right)^2 = 0.36$$

Using the tangent elevation, elev_x, at sta 80+23,

$$\text{curve elevation}_{\text{sta } 80+23} = \text{elev}_x + CM$$
$$= 95.35 \text{ ft} + (0.36)(3.125 \text{ ft})$$
$$= 96.475 \text{ ft}$$

Other grade points at 25 ft intervals are calculated and tabulated similarly.

To find the low point, first calculate R using Eq. 4.45.

$$R = \frac{G_{2,\%} - G_{1,\%}}{L_{\text{sta}}} = \frac{3.8\% - (-6.2\%)}{2.50 \text{ sta}} = 4 \text{ \%/sta}$$

Using Eq. 4.48, the low point station is

$$\text{low point station} = \text{sta}_{\text{PVC}} + \frac{-G_1}{R}$$
$$= \text{sta } 79{+}48.00 + \frac{-(-6.2\%)}{4 \frac{\%}{\text{sta}}}$$
$$= \text{sta } 81{+}03.00$$

The curve is redrawn with the essential information included, and the elevation table is completed as shown in *Table for Example 4.5 Solution.*

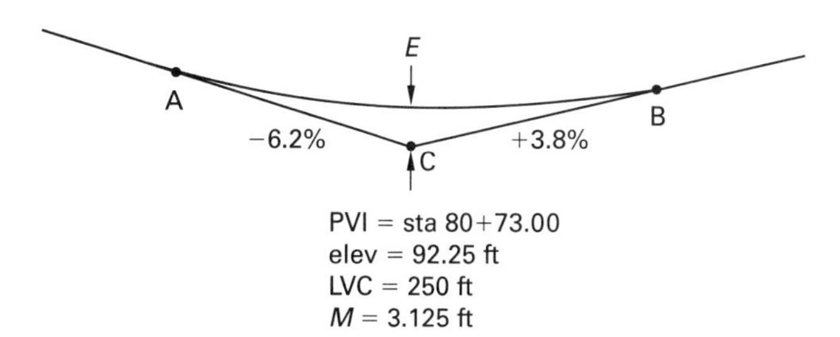

Alternate Solution

Tabulate all data for each station, spaced 25 ft apart, as shown in *Table for Example 4.5 Alternate Solution.*

Find the elevations directly from the back tangent using Eq. 4.45.

$$R = \frac{G_{2,\%} - G_{1,\%}}{L_{\text{sta}}} = \frac{3.8\% - (-6.2\%)}{2.50 \text{ sta}}$$
$$= 4 \text{ \%/sta}$$

Use Eq. 4.47 to find the elevation at the tangent distance. For example, to find the elevation at sta 80+23,

$$\text{elev}_x = \left(G_{1,\text{ft/ft}} + \frac{Rx}{2}\right)x + \text{elev}_a$$
$$= \left(-6.2\% + \frac{\left(4 \frac{\%}{\text{sta}}\right)(0.75 \text{ sta})}{2}\right) \times (0.75 \text{ sta})\left(1 \frac{\text{ft}}{\text{sta-\%}}\right) + 100.00 \text{ ft}$$
$$= 96.48 \text{ ft}$$

Table for Example 4.5 Alternate Solution

	station	x (sta)	$G_1 + Rx/2$ $= -6.2\%$ $+ 2x$	tangent offset $((x)(-6.2\% + 2x))$	elevation on curve
PVC	79+48.00	0	–6.2	0	100.00
	+73.00	0.25	–5.7	–1.425	98.575
	+98.00	0.50	–5.2	–2.600	97.400
	80+23.00	0.75	–4.7	–3.525	96.475
	+48.00	1.00	–4.2	–4.200	95.800
PVI	+73.00	1.25	–3.7	–4.625	95.375
	+98.00	1.50	–3.2	–4.800	95.200
	81+23.00	1.75	–2.7	–4.725	95.2875
	+48.00	2.00	–2.2	–4.400	95.600
	+73.00	2.25	–1.7	–3.825	96.175
PVT	+98.00	2.50	–1.2	–3.000	97.000
LP	81+03.00	1.55	–3.1	–4.805	95.195

Using Eq. 4.48, the low point is

$$\begin{aligned}\text{low point station} &= \text{sta}_{\text{PVC}} + \frac{-G_1}{R} \\ &= \text{sta } 79{+}48.00 + \frac{-(-6.2\%)}{4\ \frac{\%}{\text{sta}}} \\ &= \text{sta } 81{+}03.00\end{aligned}$$

From Eq. 4.47, the low point elevation is

$$\begin{aligned}\text{elev}_x &= \left(G_1 + \frac{Rx}{2}\right)x + \text{elev}_a \\ &= \left(-6.2\% + \frac{\left(4\ \frac{\%}{\text{sta}}\right)(1.55\ \text{sta})}{2}\right) \\ &\quad \times (1.55\ \text{sta})\left(1\ \frac{\text{ft}}{\text{sta-\%}}\right) + 100.00\ \text{ft} \\ &= 95.195\ \text{ft}\end{aligned}$$

The tabulation can be completed as shown in *Table for Example 4.5 Alternate Solution.*

10. SOURCES OF ERROR IN ROUTE ALIGNMENT

The tangents of a route alignment between two control points (i.e., points with known coordinates) become a traverse that can be checked for accuracy. A traverse with two known control points is known as a *closed traverse*. A traverse where one point has unknown coordinates is known as an *open traverse*. Open traverses cannot be checked for closure and are associated with a lower level of accuracy than a closed traverse. Confirming the accuracy of a traverse is known as *checking for closure* and involves verifying that the error between the calculated end point and the actual end point achieves the desired level of accuracy. The error is determined as a ratio of the difference between the calculated end point and the actual coordinates of the end point, divided by the total traverse length.

The acceptable error is determined by the desired level of accuracy, which is chosen before the survey begins. The *National Geodetic Survey*, NGS, established maximum error ratios for three levels (known as *orders*) of accuracy in horizontal route alignment. A first-order survey is the most accurate, and a third-order survey is the least accurate. Second- and third-order surveys are subdivided into class I and II, with class I representing a higher level of accuracy than class II. Each order and class has a maximum error ratio that cannot be exceeded in order to obtain the chosen level of accuracy. (For example, the maximum error ratio for a second-order, class I survey is 1:50,000.)

A route alignment's accuracy is influenced by the route layout method used. The most accurate method is using the PIs of each curve as the corner points. (This method uses the least number of traverse sides, which simplifies calculation.) Figure 4.18(a) illustrates the relationship of curves to a set of alignment tangents that have been adjusted using the PIs between point A and point F. Points B, C, D, and E are PIs of the route alignment.

Figure 4.18(b) is an exaggerated view of how the curve geometry is distorted if the traverse corners are set at the PCs and PTs of the curves with the long chord of each curve as a side of the control traverse. While the curve layout may be accurate between the PC and PT of an individual curve, the tangent headings between the curves are not the same as the subtangent headings of the adjoining curves, which introduces a small angular change at each PC and PT. This angular difference introduces a new PI at each of the curve points, which results in an inaccurate layout. The location errors caused by PCs and PTs not precisely placed on the

Figure 4.18 *Example of Route Alignment Layout Showing (a) Control by PI and (b) Control by PC and PT*

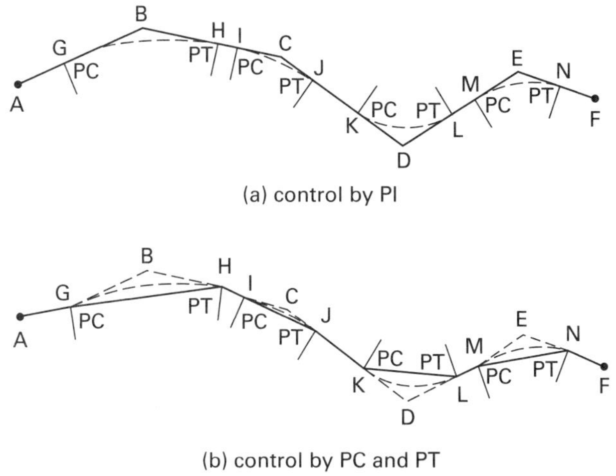

(a) control by PI

(b) control by PC and PT

overall alignment tangent are eliminated by controlling the location of the tangents.

Another source of route alignment error is caused by rounding tangent azimuths without recalculating the curve geometry. For example, a tangent is given a preliminary heading of 47° 52′ 25″. The curve geometry is calculated with this heading, and coordinates are set to the resulting curve points and listed on the plan. On the plan view, the heading is rounded to 47° 52′, but the coordinates are not recalculated with the rounded heading. The error in location of a 1000 ft tangent due to heading rounding is illustrated in Fig. 4.19.

Figure 4.19 *Location Error of 1000 ft Tangent Due to Heading Rounding*

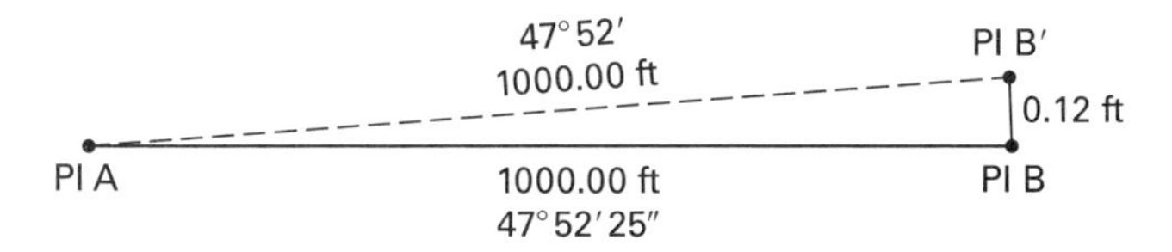

Rounding the heading 25″ shifts the location of point B by 0.12 ft. The PI intersects should be recalculated using the rounded heading as the new heading, which will accurately locate new PI coordinates. Once the curve is recalculated using the new PI coordinate, the geometry on the final plan can be recreated accurately as shown, which eliminates further layout errors and guesswork by the field engineers as to the intended final geometry.

11. SIGHT DISTANCE

Sight distance is the length of roadway a driver can see ahead of the vehicle. Within this sight distance, a driver must analyze upcoming road conditions and traffic situations, select appropriate actions or maneuvers, and then complete these actions or maneuvers. The design speed of horizontal curves and vertical curves and a driver's ability to see around obstructions are based on the maximum sight distance available to a driver.

AASHTO design criteria use three types of sight distance: stopping sight distance, decision sight distance, and passing sight distance.

Stopping Sight Distance

Safe *stopping sight distance* is the total distance required for a driver traveling at design speed to stop a vehicle before reaching an object in its path. Stopping sight distance is comprised of two distances: the perception-reaction distance and the braking distance. The *perception-reaction distance* (also called the *PIEV distance*) is the distance traveled at a constant approach speed during the perception-reaction time (known as the *PIEV time*), which is measured from the moment the driver sees an object requiring a stop to the time the brakes are applied. AASHTO refers to the PIEV distance as the *brake reaction distance* and uses an average of 2.5 sec as the *brake reaction time*, which is time it takes a driver to apply the brakes after seeing an object. Stopping sight distance calculations are determined using the driver's eye height set at 3.5 ft (1080 mm) and the object height at 2.0 ft (600 mm), which is equivalent to the height of a passenger car's taillight. Typical stopping sight distance values for various design speeds are given in Table 4.3.

The *braking distance* is the distance needed to stop the vehicle once the brakes have been applied. Braking distance varies according to the approach speed and the deceleration rate of the vehicle. Tables in Chap. 3 of the *Green Book* show braking distances for a deceleration rate of 11.2 ft/sec^2 (3.4 m/s^2), which is the maximum comfortable stopping rate. This rate takes into account wet pavement and tires with acceptable tread. Emergency stopping rates can be greater under ideal road and tire conditions but are not used for design purposes. While braking ability and stopping friction factors vary somewhat with speed, the variations are not easily quantifiable. There is sufficient variation in general traffic conditions that refinement would not warrant modifying design values. Therefore, most analysis uses a constant deceleration rate of 11.2 ft/sec^2 (3.4 m/s^2).

The braking distance required to bring a vehicle to a complete stop is determined from the design speed, v, in miles (kilometers) per hour and the deceleration rate, a, in feet (meters) per second squared, as shown in Eq. 4.49.

$$d_{\text{m}} = 0.039\left(\frac{\text{v}_{\text{kph}}^2}{a_{\text{m/s}^2}}\right) \quad \text{[\textit{Green Book} Eq. 3-1] [SI]} \qquad 4.49(a)$$

$$d_{\text{ft}} = 1.075\left(\frac{\text{v}_{\text{mph}}^2}{a_{\text{ft/sec}^2}}\right) \quad \text{[\textit{Green Book} Eq. 3-1] [U.S.]} \qquad 4.49(b)$$

Table 4.3 Stopping Sight Distance

SI units

design speed (kph)	braking reaction distance[a] (m)	braking distance on level (m)	stopping sight distance calculated[b] (m)	stopping sight distance design (m)
20	13.9	4.6	18.5	20
30	20.9	10.3	31.2	35
40	27.8	18.4	46.2	50
50	34.8	28.7	63.5	65
60	41.7	41.3	83.0	85
70	48.7	56.2	104.9	105
80	55.6	73.4	129.0	130
90	62.6	92.9	155.5	160
100	69.5	114.7	184.2	185
110	76.5	138.8	215.3	220
120	83.4	165.2	248.6	250
130	90.4	193.8	284.2	285

customary U.S. units

design speed (mph)	braking reaction distance[a] (ft)	braking distance on level (ft)	stopping sight distance calculated[b] (ft)	stopping sight distance design (ft)
15	55.1	21.6	76.7	80
20	73.5	38.4	111.9	115
25	91.9	60.0	151.9	155
30	110.3	86.4	196.7	200
35	128.6	117.6	246.2	250
40	147.0	153.6	300.6	305
45	165.4	194.4	359.8	360
50	183.8	240.0	423.8	425
55	202.1	290.3	492.4	495
60	220.5	345.5	566.0	570
65	238.9	405.5	644.4	645
70	257.3	470.3	727.6	730
75	275.6	539.9	815.5	820
80	294.0	614.3	908.3	910

[a]brake reaction distance predicated on a time of 2.5 sec
[b]deceleration rate of 11.2 ft/sec^2 (3.4 m/s^2) used to determine calculated stopping sight distance

From *A Policy on Geometric Design of Highways and Streets*, 2004, by the American Association of State Highway and Transportation Officials, Washington, D.C. Used by permission.

The braking distance for roadways on a grade is found from Eq. 4.50. The friction factor, f, is equal to the deceleration rate divided by the gravitational acceleration and can be substituted for a, as shown in Eq. 4.50.

$$d_{\text{m}} = \frac{\text{v}_{\text{kph}}^2}{254\left(\left(\frac{a_{\text{m/s}^2}}{g}\right) \pm G_{\%/100\,\text{m}}\right)} = \frac{\text{v}_{\text{kph}}^2}{254(f \pm G_{\%/100\,\text{m}})}$$

[*Green Book* Eq. 3-3] [SI] *4.50(a)*

$$d_{\text{ft}} = \frac{\text{v}_{\text{mph}}^2}{30\left(\left(\frac{a_{\text{ft/sec}^2}}{g}\right) \pm G_{\%/100\,\text{ft}}\right)} = \frac{\text{v}_{\text{mph}}^2}{30(f \pm G_{\%/100\,\text{ft}})}$$

[*Green Book* Eq. 3-3] [U.S.] *4.50(b)*

The total stopping sight distance, S, is calculated from Eq. 4.51 as the sum of the distance traveled during the perception-reaction time and the braking distance calculated from Eq. 4.49.

$$S_{\text{m}} = \left(0.278\ \frac{\frac{\text{m}}{\text{s}}}{\frac{\text{km}}{\text{h}}}\right)\text{v}_{\text{kph}}t_{p,\text{s}} + 0.039\left(\frac{\text{v}_{\text{kph}}^2}{a_{\text{m/s}^2}}\right)$$

[*Green Book* Eq. 3-2] [SI] *4.51(a)*

$$S_{\text{ft}} = \left(1.47\ \frac{\frac{\text{ft}}{\text{sec}}}{\frac{\text{mi}}{\text{hr}}}\right)\text{v}_{\text{mph}}t_{p,\text{sec}} + 1.075\left(\frac{\text{v}_{\text{mph}}^2}{a_{\text{ft/sec}^2}}\right)$$

[*Green Book* Eq. 3-2] [U.S.] *4.51(b)*

Decision Sight Distance

Decision sight distance is appropriate where hazards exist that require drivers to make decisions to perform maneuvers other than a stop, such as lane changes or exit ramp selections, in order for traffic to proceed in an orderly and smooth fashion. These decisions often include multiple actions to be taken simultaneously and may involve selection from several choices of action to be performed. Examples include approaches to complex intersections, multiple interchange ramps, toll booth plazas, restrictive sight distance locations, and instances in which the driver may need to be prepared for further alternative maneuvers in quick succession. Only providing sufficient sight distance for a hurried stop may increase the danger to other motorists, while not providing enough time to make an appropriate selection of an alternate path or course of action for an evasive maneuver. More decision time may also be needed where visual clutter exists, such as advertising signs and busy commercial activity found along a commercial corridor. Since decision sight distances include a margin of error in addition to the time necessary to make evasive maneuvers, decision sight distance values are typically much larger than stopping sight distances.

The response time is affected by information encountered by a driver, as illustrated in Fig. 4.20 and Fig. 4.21. The reaction time increases with the *information content*, or the amount and complexity of information, the driver must process before reacting. In order to quantify information, the unit bits is used. Expected and unexpected conditions also affect the reaction time required

Figure 4.20 Time to React to Expected and Unexpected Information, Median Driver

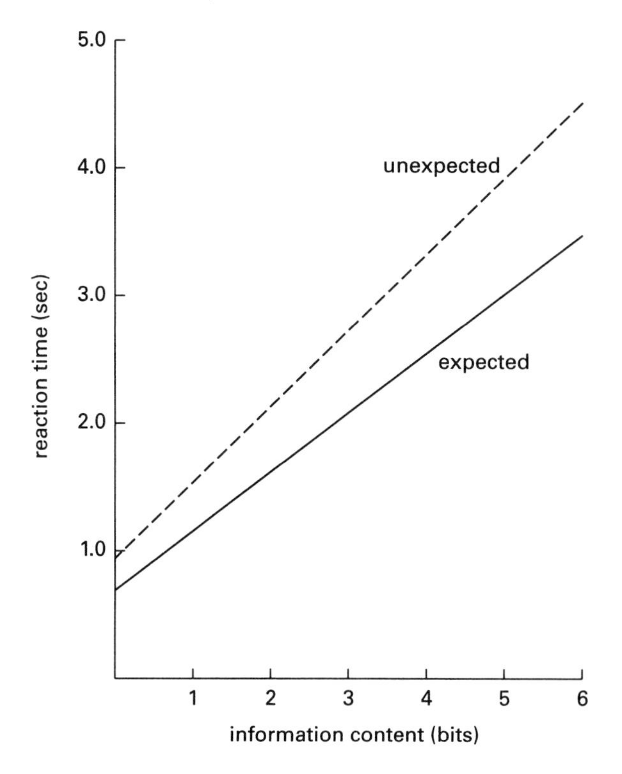

From *A Policy on Geometric Design of Highways and Streets*, 2004, by the American Association of State Highway and Transportation Officials, Washington, D.C. Used by permission.

Figure 4.21 Time to React to Expected and Unexpected Information, 85th Percentile Driver

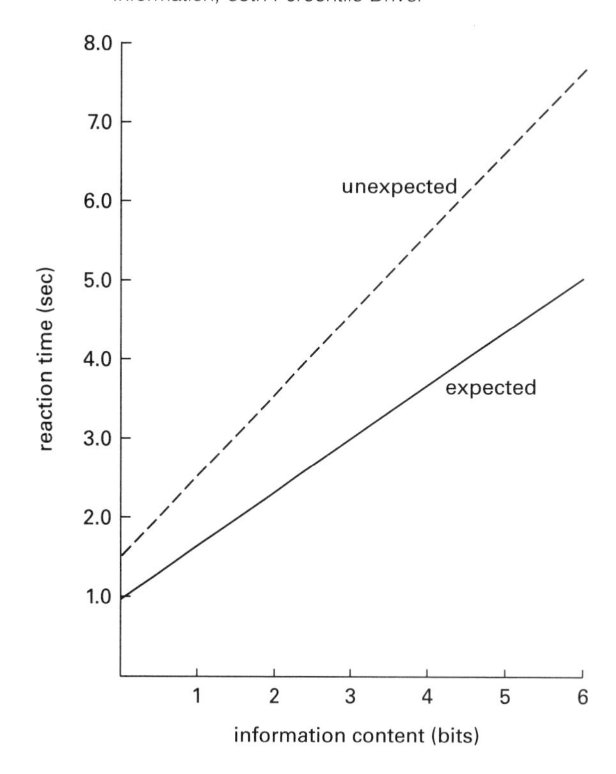

From *A Policy on Geometric Design of Highways and Streets*, 2004, by the American Association of State Highway and Transportation Officials, Washington, D.C. Used by permission.

Geometric Design

to process information. An expected condition, such as a signalized intersection familiar to the driver, takes less time to process and react to than an unexpected condition, such as a driver swerving into an adjacent lane.

Chapter 3 of the AASHTO *Green Book* gives approach maneuvers, which are used to calculate the decision sight distance. The decision sight distances classified by approach maneuver are shown in Table 4.4.

Avoidance maneuvers A and B are calculated using the pre-maneuver time, t_p,[1] which is the time needed for the driver to recognize the upcoming condition, identify alternative movements, and initiate a response. The pre-maneuver time, t_p, ranges from 3.0 sec to 9.1 sec, which is longer than the perception-reaction time used in stopping sight distance. Avoidance maneuvers C through E use t_t, which is the total of the pre-maneuver and maneuver times. Typical values of t_p and t_t are given in Table 4.4

Avoidance maneuver A is a stop condition on a rural road and uses a pre-maneuver time of 3.0 sec. *Avoidance maneuver B* is a stop condition on an urban road and uses a pre-maneuver time of 9.1 sec. The decision sight distances for maneuvers A and B are determined using Eq. 4.52, where t_p is the pre-maneuver time only, v is the design speed, and a is the driver deceleration. Most decision sight distances can be found from Table 4.4, but for cases where the given design speeds do not apply or for other special cases, Table 4.4 can be used in conjunction with Eq. 4.52.

$$d_{\text{m}} = \left(0.278\ \frac{\frac{\text{m}}{\text{s}}}{\frac{\text{km}}{\text{h}}}\right) \text{v}_{\text{kph}} t_{p,\text{s}} + 0.039\left(\frac{\text{v}_{\text{kph}}^2}{a_{\text{m/s}^2}}\right)$$

[*Green Book* Eq. 3-4] [SI] 4.52(*a*)

$$d_{\text{ft}} = \left(1.47\ \frac{\frac{\text{ft}}{\text{sec}}}{\frac{\text{mi}}{\text{hr}}}\right) \text{v}_{\text{mph}} t_{p,\text{sec}} + 1.075\left(\frac{\text{v}_{\text{mph}}^2}{a_{\text{ft/sec}^2}}\right)$$

[*Green Book* Eq. 3-4] [U.S.] 4.52(*b*)

Avoidance maneuver C is a speed, path, or direction change on a rural road, in which the total pre-maneuver and maneuver time, t_t, varies between 10.2 sec and 11.2 sec. *Avoidance maneuver D* is a speed, path, or direction change on a suburban road, in which t_t varies between 12.1 sec and 12.9 sec. *Avoidance maneuver E* is a speed, path, or direction change on an urban road where t_t varies between 14.0 sec and 14.5 sec. The decision sight

[1]Note that stopping sight distance uses the perception-reaction time, while decision sight distance uses the pre-maneuver time, which includes the perception-reaction time. Both use the variable t_p.

Table 4.4 Example Decision Sight Distances

SI units

design speed (kph)	decision sight distance (m) avoidance maneuver A	B	C	D	E
50	70	155	145	170	195
60	95	195	170	205	235
70	115	235	200	235	275
80	140	280	230	270	315
90	170	325	270	315	360
100	200	370	315	355	400
110	235	420	330	380	430
120	265	470	360	415	470
130	305	525	390	450	510

customary U.S. units

design speed (mph)	decision sight distance (ft) avoidance maneuver A	B	C	D	E
30	220	490	450	535	620
35	275	590	525	625	720
40	330	690	600	715	825
45	395	800	675	800	930
50	465	910	750	890	1030
55	535	1030	865	980	1135
60	610	1150	990	1125	1280
65	695	1275	1050	1220	1365
70	780	1410	1105	1275	1445
75	875	1545	1180	1365	1545
80	970	1685	1260	1455	1650

Avoidance Maneuver A: Stop on rural road. $t_p = 3.0$ sec
Avoidance Maneuver B: Stop on urban road. $t_p = 9.1$ sec
Avoidance Maneuver C: Speed/path/direction change on rural road. t_t varies between 10.2 sec and 11.2 sec
Avoidance Maneuver D: Speed/path/direction change on suburban road. t_t varies between 12.1 sec and 12.9 sec
Avoidance Maneuver E: Speed/path/direction change on urban road. t_t varies between 14.0 sec and 14.5 sec

From *A Policy on Geometric Design of Highways and Streets*, 2004, by the American Association of State Highway and Transportation Officials, Washington, D.C. Used with permission.

distances for maneuvers C, D, and E are calculated using Eq. 4.53.

$$d_{\text{m}} = 0.278\text{v}_{\text{m/s}}t_{t,\text{s}} \quad [\textit{Green Book}\text{ Eq. 3-5}]\ [\text{SI}] \qquad 4.53(a)$$

$$d_{\text{ft}} = 1.47\text{v}_{\text{ft/sec}}t_{t,\text{sec}} \quad [\textit{Green Book}\text{ Eq. 3-5}]\ [\text{U.S.}] \qquad 4.53(b)$$

The criteria of 1080 mm (3.5 ft) eye height and 600 mm (2.0 ft) object height used for stopping sight distance are the same for decision sight distance.

Passing Sight Distance

Passing sight distance is applicable on two-lane highways when there are sufficient gaps in opposing flows to allow passing maneuvers to occur and when there are few access points, with only occasional entering traffic and sufficient sight distance to observe traffic in both lanes and both directions. Occasional access includes residential and rural driveways, minor side roads, and driveways to adjoining land uses. Passing sight distance is not applicable to multilane highways. Passing sight distances for two-lane highways are shown in Table 4.5.

Table 4.5 Passing Sight Distance for Design of Two-Lane Highways

SI units

design speed (kph)	assumed speeds (kph) passed vehicle	passing vehicle	passing sight distance (m) from *Green Book* Exh. 3-6	rounded for design
30	29	44	200	200
40	36	51	266	270
50	44	59	341	345
60	51	66	407	410
70	59	74	482	485
80	65	80	538	540
90	73	88	613	615
100	79	94	670	670
110	85	100	727	730
120	90	105	774	775
130	94	109	812	815

customary U.S. units

design speed (mph)	assumed speeds (mph) passed vehicle	passing vehicle	passing sight distance (ft) from *Green Book* Exh. 3-6	rounded for design
20	18	28	706	710
25	22	32	897	900
30	26	36	1088	1090
35	30	40	1279	1280
40	34	44	1470	1470
45	37	47	1625	1625
50	41	51	1832	1835
55	44	54	1984	1985
60	47	57	2133	2135
65	50	60	2281	2285
70	54	64	2479	2480
75	56	66	2578	2580
80	58	68	2677	2680

From *A Policy on Geometric Design of Highways and Streets*, 2004, by the American Association of State Highway and Transportation Officials, Washington, D.C. Used with permission.

A *passing maneuver* is a maneuver in which a faster-moving vehicle overtakes a slower-moving vehicle. The driver of the faster-moving vehicle must have sufficient sight distance to see oncoming traffic, be able to complete the passing maneuver, and be able to return to the right-hand lane without forcing the opposing vehicle to slow down or cutting off the overtaken vehicle by returning to the right-hand lane before fully passing. The passing maneuver is described as the total of four distances, or stages, as shown in Fig. 4.22. The stages are identified by the travel distances involved in each stage, d_1, d_2, d_3, and d_4.

Figure 4.22 *Two-Lane Highway Passing Maneuver Stages*

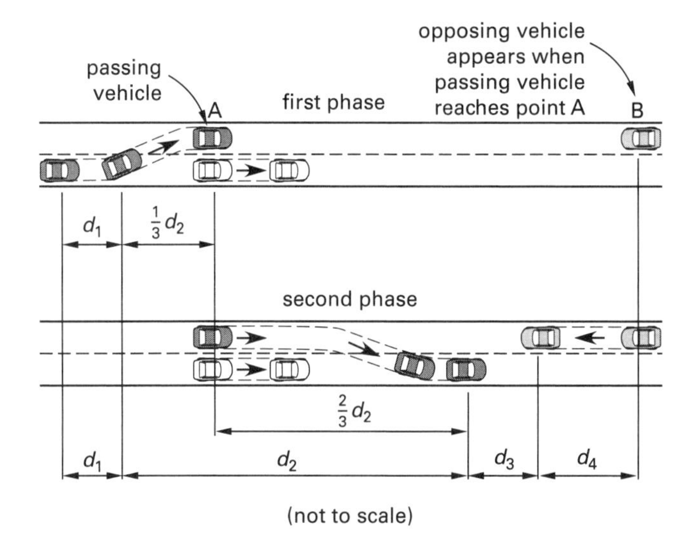

d_1 is the *initial stage distance.* During this stage, the overtaking driver observes a sufficient length of clear roadway ahead. The driver moves the vehicle toward the oncoming lane and begins accelerating. As the distance closes with the slower-moving vehicle, the driver makes a final commitment to pass or to back off and return to the right lane if the available distance is insufficient to complete the passing maneuver safely. The initial stage continues until the passing vehicle begins to encroach on the opposing lane.

The initial stage distance, d_1, is determined by Eq. 4.54. v is the velocity of the car in miles (kilometers) per hour, a is the average acceleration in miles (kilometers) per hour per second, and t_1 is the initial maneuver time in seconds. m is the difference between the speeds of the passing vehicle and the overtaken vehicle in miles (kilometers) per hour. Average acceleration values are given in Table 4.6.

$$d_{1,\mathrm{m}} = 0.278 t_{1,\mathrm{s}} \left(\mathrm{v_{kph}} - m_{\mathrm{kph}} + \frac{a_{\mathrm{kph/s}} t_{1,\mathrm{s}}}{2} \right)$$

[*Green Book* Eq. 3-6] [SI] *4.54(a)*

$$d_{1,\mathrm{ft}} = 1.47 t_{1,\mathrm{sec}} \left(\mathrm{v_{mph}} - m_{\mathrm{mph}} + \frac{a_{\mathrm{mph/sec}} t_{1,\mathrm{sec}}}{2} \right)$$

[*Green Book* Eq. 3-6] [U.S.] *4.54(b)*

d_2 is the *passing stage distance.* Once committed to passing, the passing vehicle accelerates and moves fully into the opposing lane, passes the slower vehicle, and then returns to the right traveled lane. Moving into the left lane and accelerating to a point parallel to the slower vehicle takes about one-third of the passing stage distance, d_2. During this first third, the passing vehicle can return to its position behind the slower vehicle should an opposing vehicle appear before the actual pass takes place. Moving ahead of the slower vehicle and returning to the right lane takes the remaining two-thirds of d_2.

The passing stage distance is based on the time the passing vehicle occupies the opposing lane, t_2, and is found using Eq. 4.55. The opposing vehicle and the passing vehicle are assumed to be traveling at the same speed.

$$d_{2,\mathrm{m}} = 0.287 \mathrm{v_{m/s^2}} t_{2,\mathrm{s}} \quad \text{[\textit{Green Book} Eq. 3-7] [SI]} \qquad 4.55(a)$$

$$d_{2,\mathrm{ft}} = 1.47 \mathrm{v_{ft/sec^2}} t_{2,\mathrm{sec}} \quad \text{[\textit{Green Book} Eq. 3-7] [U.S.]} \qquad 4.55(b)$$

d_3 is the *clearance stage distance.* This is the distance between the passing vehicle and the opposing vehicle when the passing vehicle returns to the right lane. This clearance distance is mostly a matter of the safety and comfort of both passing and opposing drivers, which can vary considerably with the density of traffic and the aggressiveness of the drivers. Design values show increased clearance time with increased speed and vary from 100 ft to 250 ft (30 m to 75 m). Values of d_3 can be determined from Table 4.6.

d_4 is the *opposing vehicle travel distance* during the passing maneuver. This distance is based on the assumption that the passing and opposing vehicle are traveling at the same speed and is determined from the time the opposing lane is actually blocked, or two-thirds of d_2, as shown in Eq. 4.56.

$$d_4 = \tfrac{2}{3} d_2 \qquad 4.56$$

The total passing sight distance, d, is $d_1 + d_2 + d_3 + d_4$.

The recommended design values, given in Table 4.6, are based on the condition of a single vehicle passing another single vehicle, and the following assumptions.

1. The overtaken vehicle maintains a constant speed.
2. The passing vehicle reduces speed and trails the overtaken vehicle as it enters a passing section.
3. When the passing section is reached, the passing driver needs a short period of time to perceive the clear passing section and to react to start the passing maneuver.
4. Passing is accomplished under a delayed start, due to the perception-reaction time, and a hurried return, due to opposing traffic. While in the left lane, the passing vehicle travels at an average speed of 10 mph (15 kph) or more than the speed of the overtaken vehicle and accelerates during thc maneuver.
5. After the passing vehicle returns to the right lane, there is suitable clearance distance between it and the opposing vehicle in the other lane.

Geometric Design

Table 4.6 Elements of Safe Passing Sight Distance for Design of Two-Lane Highways

	SI units				customary U.S. units			
	speed range (kph)				speed range (mph)			
	50–65	66–80	81–95	96–110	30–40	40–50	50–60	60–70
	average passing speed (kph)				average passing speed (mph)			
component of passing maneuver	56.2	70.0	84.5	99.8	34.9	43.8	52.6	62.0
initial maneuver								
a = average acceleration*	2.25	2.30	2.37	2.41	1.40	1.43	1.47	1.50
t_1 = time (sec)*	3.6	4.0	4.3	4.5	3.6	4.0	4.3	4.5
d_1 = distance traveled	45	66	89	113	145	216	289	366
occupation of left lane								
t_2 = time (sec)*	9.3	10.0	10.7	11.3	9.3	10.0	10.7	11.3
d_2 = distance traveled	145	195	251	314	477	643	827	1030
clearance length								
d_3 = distance traveled*	30	55	75	90	100	180	250	300
opposing vehicle								
d_4 = distance traveled	97	130	168	209	318	429	552	687
total distance ($d_1 + d_2 + d_3 + d_4$)	317	446	583	726	1040	1468	1918	2383

*For consistent speed relation, observed values adjusted slightly.

Note: In the SI portion of the table, speed values are in kilometers per hour, acceleration rates are in kilometers per hour per second, and distances are in meters. In the customary U.S. portion of the table, speed values are in miles per hour, acceleration rates are in miles per hour per second, and distances are in feet.

From *A Policy on Geometric Design of Highways and Streets*, 2004, by the American Association of State Highway and Transportation Officials, Washington, D.C. Used by permission.

Design values of d_1, d_2, d_3, and d_4 are plotted, along with the total of the four values, in *Green Book* Exh. 3-6. The values in Table 4.6 are derived from these graphical plots and can be used as design values for a wide range of speeds.

Determining Vertical Curve Length

Stopping sight distance is used to determine the lengths of crest and sag vertical curves. (The passing sight distance can also be used, but the minimum curve length would be much greater than the length based on only the stopping sight distance and would be more costly.) The stopping sight distance can be determined by the length of a straight line between the driver's eye and an object on the roadway ahead. AASHTO typically sets the driver's eye height at 3.5 ft (1080 mm), and the object ahead is set at 2.0 ft (600 mm) above the roadway. Design values for stopping sight distance are given in Table 4.3.

A simplified method to determine the stopping sight distance for a crest or sag vertical curve is to use the *K-value method* presented in the *Green Book*. The length of vertical curve per percent of grade difference, K, is the inverse of the rate of change, R, and is the ratio of the curve length, L, to the absolute value of the algebraic grade difference, A, as shown in Eq. 4.57.

$$K = \frac{L}{A} = \frac{L}{|G_2 - G_1|} \quad \text{[always positive]} \qquad 4.57$$

A typical design situation will give two of the variables, with the third variable being the unknown. In some cases, one or more of the known conditions may have a range of available conditions, and careful selection of a value in the range will be required. This type of problem requires some logical thinking and deductive reasoning, such as determining whether the maximum or the minimum grade change is applicable to the condition presented.

The *Green Book* presents graphs of K-values versus design speeds based on comfort criteria and stopping sight criteria. These graphs are general guidelines for transportation engineers, and data from these graphs are presented in Table 4.7.

The minimum recommended K-values are based on sight distance. Applying K-values uniformly throughout an undulating roadway segment can improve both ride comfort and geometric appearance. While tolerance for vertical acceleration normally controls the rate of vertical change, the appearance of direction changes can be disconcerting for drivers and passengers. A graceful and smooth appearance minimizes distress, and fewer accidents are caused by startled or confused drivers.

Visually, when a curve is too short, the appearance is of no curve transition at all. It is important to also take into account the length of the approach tangent, especially on sag vertical curves. Longer approach tangents tend to draw the eye farther ahead, while a sudden increase in grade shortens the perceived sight distance. In the most extreme cases, the driver envisions the approaching grade increase with no vertical curve at all and will tend to slow considerably on the approach. This problem can be reduced by providing a vertical

Table 4.7 Design Controls for Crest and Sag Vertical Curves

SI units

design speed (kph)	crest vertical curve design K-value	crest vertical curve minimum length (m)	sag vertical curve design K-value	sag vertical curve minimum length (m)
20	1	10	3	12
30	2	20	6	18
40	4	25	9	24
50	7	30	13	30
60	11	46	18	36
70	17	42	23	42
80	26	48	30	48
90	39	54	38	54
100	52	60	45	60
110	74	66	55	66
120	95	74	63	72
130	124	80	73	80

customary U.S. units

design speed (mph)	crest vertical curve design K-value	crest vertical curve minimum length (ft)	sag vertical curve design K-value	sag vertical curve minimum length (ft)
15	3	40	10	40
20	7	60	17	60
25	12	80	26	80
30	19	90	37	95
35	29	110	49	105
40	44	120	64	120
45	61	135	79	135
50	84	150	96	150
55	114	170	115	165
60	151	180	136	180
65	193	195	157	195
70	247	205	181	210
75	312	215	206	220
80	384	230	231	230

From *A Policy on Geometric Design of Highways and Streets*, 2004, by the American Association of State Highway and Transportation Officials, Washington, D.C. Used by permission.

curve that is considerably longer than recommended. For smaller differences in grades, the length can be as much as four or five times longer than the normal recommendations. This usually does not raise the cost of construction, but may reduce costs by reducing the amount of excavation necessary to construct a sag vertical curve. Sag vertical curves with longer lengths are shallower and therefore require less excavation than a curve with a shorter length.

The upper range of K-values is controlled by the ability to drain the roadway surface. AASHTO recommends a drainage maximum of 167 ft/% (51 m/%) for crest and sag curves. Should higher K-values be necessary, other measures, such as increasing the cross slope, are necessary to adequately drain the pavement.

For lower-speed roadways where appearance is less often a factor, there are two conditions of sight distance criteria that control the minimum length of a curve: when the sight distance is shorter than the length of the vertical curve ($S < L$), and when the sight distance is longer than the length of the vertical curve ($S > L$). The equations shown in Table 4.8 are used to determine the required curve length, L, and must be initially calculated for both the $S < L$ and the $S > L$ cases. Constants are based on the driver's eye height and the object height, both in feet (millimeters).

Equation 4.58 and Eq. 4.59 may also be used to determine the curve length for sight distance on crest vertical curves.

$$L = \frac{A_{\%}S^2}{200(\sqrt{h_1}+\sqrt{h_2})^2} \quad [S < L] \text{ [\textit{Green Book} Eq. 3-41]} \qquad 4.58$$

$$L = 2S - \frac{200(\sqrt{h_1}+\sqrt{h_2})^2}{A_{\%}} \quad [S > L] \text{ [\textit{Green Book} Eq. 3-41]} \qquad 4.59$$

Stopping sight distance on sag vertical curves can be analyzed using headlight sight distances. This method assumes the maximum upward projection of the headlight high-beam will illuminate a distance of roadway

Table 4.8 AASHTO Criteria for Minimum Vertical Curve Lengths Based on Sight Distance[a]

	stopping sight distance[b] (crest curves)	passing sight distance[c] (crest curves)	stopping sight distance (sag curves)
	SI units		
$S<L$	$L=\frac{AS^2}{658}$	$L=\frac{AS^2}{864}$	$L=\frac{AS^2}{120+3.5S}$
$S>L$	$L=2S-\frac{658}{A}$	$L=2S-\frac{864}{A}$	$L=2S-\frac{120+3.5S}{A}$
	customary U.S. units		
$S<L$	$L=\frac{AS^2}{2158}$	$L=\frac{AS^2}{2800}$	$L=\frac{AS^2}{400+3.5S}$
$S>L$	$L=2S-\frac{2158}{A}$	$L=2S-\frac{2800}{A}$	$L=2S-\frac{400+3.5S}{A}$

[a]$A=|G_2-G_1|$, absolute value of the algebraic difference in grades, in percent.
[b]The drivers's eye is 3.5 ft (1080 mm) above the road surface, viewing an object 2.0 ft (600 mm) high.
[c]The drivers's eye is 3.5 ft (1080 mm) above the road surface, viewing an object 3.5 ft (1080 mm) high.

Compiled from *A Policy on Geometric Design of Highways and Streets*, Chap. 3, copyright © 2004 by the American Association of State Highway and Transportation Officials, Washington, D.C.

ahead equal to the stopping sight distance, and oncoming traffic or roadway lighting will illuminate the roadway ahead when low-beams are used. As in crest vertical curves, the conditions of $S<L$ and $S>L$ need to be considered for headlight sight distance.

With sag curves, both gravitational and centrifugal forces act on the driver and passengers, making comfort the controlling factor in the design. Equation 4.60 can be used to calculate the length of the curve so that the added acceleration is kept below 1 ft/sec^2 (0.3 m/s^2).

$$L_{\text{m}}=\frac{A_{\%}\text{v}_{\text{kph}}^2}{395}\quad [\textit{Green Book}\ \text{Eq. 3-51}]\quad [\text{SI}]\qquad 4.60(a)$$

$$L_{\text{ft}}=\frac{A_{\%}\text{v}_{\text{mph}}^2}{46.5}\quad [\textit{Green Book}\ \text{Eq. 3-51}]\quad [\text{U.S.}]\qquad 4.60(b)$$

Example 4.6

A downgrade of 3% intersects an upgrade of 5% at an elevation of 100.00 ft at sta 67+84.00, where a 200 ft curve will be fitted. Show complete curve specifications, including the formula used to determine each item. Include the station and elevation of the PVC and the PVT, curve length, K-value, middle ordinate, and low point station and elevation.

Solution

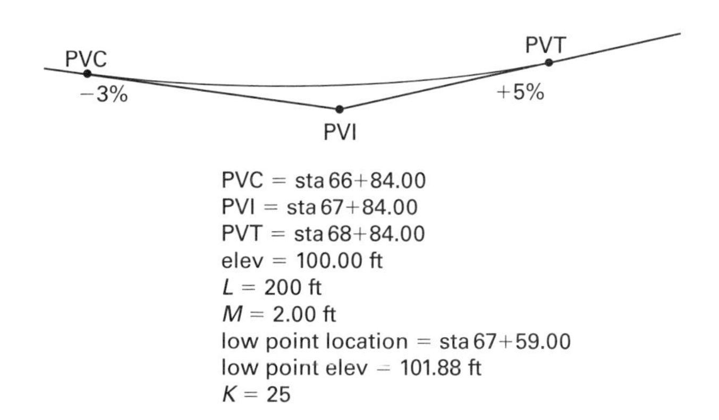

The previous tabulation shows how the vertical curve information would appear on highway plans in many jurisdictions. The calculations to obtain the information are shown as follows.

From Eq. 4.45,

$$R_{\%/\text{sta}}=\frac{G_{2,\%}-G_{1,\%}}{L_{\text{sta}}}=\frac{5\%-(-3\%)}{\frac{200\ \text{ft}}{100\ \frac{\text{ft}}{\text{sta}}}}=4\ \%/\text{sta}\quad [\text{same as } 4\ \text{ft/sta}^2]$$

As shown in Eq. 4.45 and Eq. 4.57, the K-value is the inverse of R. (Note that the units are dropped when K is shown on highway plans.)

$$K=\frac{1}{R_{\%/\text{sta}}}=\frac{(1)\left(100\ \frac{\text{ft}}{\text{sta}}\right)}{4\ \frac{\%}{\text{sta}}}=25\ \text{ft}/\%$$

From Eq. 4.44,

$$M_{\text{ft}}=\frac{|G_{2,\%}-G_{1,\%}|L_{\text{sta}}}{8\ \frac{\text{sta}}{\text{ft}}}=\frac{|5\%-(-3\%)|(200\ \text{ft})}{\left(8\ \frac{\text{sta}}{\text{ft}}\right)\left(100\ \frac{\text{ft}}{\text{sta}}\right)}=2.0\ \text{ft}$$

The station of the PVC is

$$\text{sta PVC} = \text{sta PVI} - \frac{L}{2} = \text{sta 67+84.00} - \frac{200\text{ ft}}{2} = \text{sta 66+84.00}$$

The elevation of the PVC is

$$\text{elev}_{\text{PVC}} = \text{elev}_{\text{PVI}} + G_1\left(\frac{L}{2}\right) = 100.00\text{ ft} + (0.03)\left(\frac{200\text{ ft}}{2}\right) = 103.00\text{ ft}$$

The station of the PVT is

$$\text{sta PVT} = \text{sta PVI} + \frac{L}{2} = \text{sta 67+84.00} + \frac{200\text{ ft}}{2} = \text{sta 68+84.00}$$

Using Eq. 4.48, find the station of the low point of the curve.

$$\begin{aligned}\text{low point station} &= \text{sta PVC} + x_{\text{turning point,sta}} \\ &= \text{sta PVC} + \frac{-G_{1,\%}}{R_{\%/\text{sta}}} \\ &= \text{sta 66+84.00} + \frac{-3\%}{4\ \frac{\%}{\text{sta}}} \\ &= \text{sta 66+83.25}\end{aligned}$$

Using Eq. 4.47 and the known PVC elevation for elev_a, the elevation of the low point of the curve is

$$\begin{aligned}\text{elev}_{\text{low point}} &= \frac{Rx^2}{2} + G_{1,\text{ft/ft}}x + \text{elev}_{\text{PVC}} \\ &= \frac{\left(4\ \frac{\text{ft}}{\text{sta}^2}\right)(0.75\text{ sta})^2}{2} + (-0.03)(0.75\text{ sta}) \\ &\quad \times \left(100\ \frac{\text{ft}}{\text{sta}}\right) + 103.00\text{ ft} \\ &= 101.88\text{ ft}\end{aligned}$$

12. SUPERELEVATION

The difference between a curve's inside and outside elevations is known as the *superelevation*, *e*. A roadway is superelevated to resist the centrifugal force acting on a vehicle as it rounds a curve, which allows the driver to comfortably maneuever the curve at a higher speed. The *angle of the slope*, ϕ, also called *curve banking* or *cross slope*, is related to the design speed, v, and the curve radius, R, as shown by Eq. 4.61.

$$e = \tan\phi = \frac{\text{v}^2}{gR} \qquad 4.61$$

When used in route design, the slope angle is usually disregarded, and the slope is described using the tangent values of the angle showing a rise to run relationship, such as 0.02 ft/ft, 1/4 in/ft, 20 mm/m, and so on.

Lateral Forces on a Moving Vehicle

A vehicle on a sloped roadway is illustrated with a free-body diagram, as shown in Fig. 4.23. Vehicles moving along a curved path experience several lateral forces. Tire friction, gravity, and the centrifugal tendency of a vehicle work in concert to provide a balance of forces within the safe and comfortable operating range of speed for each curve.

A few relationships apply to the elements shown in Fig. 4.23. Equation 4.62 illustrates the relationship between the mass and the weight of an object.

$$m = \frac{w}{g} \qquad \text{[SI]} \qquad 4.62(a)$$

$$m = \frac{wg_c}{g} \qquad \text{[U.S.]} \qquad 4.62(b)$$

Figure 4.23 Forces Acting on a Vehicle While Rounding a Curve

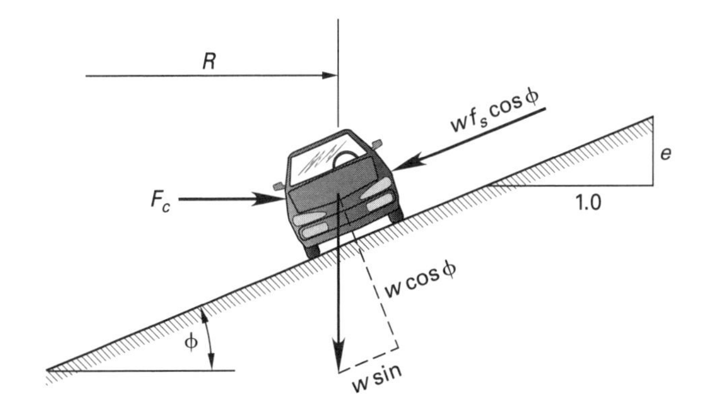

w = weight of vehicle
f_s = coefficient of side friction
F_c = centrifugal force
R = radius of curve
ϕ = angle of cross slope
e = rate of superelevation, $\tan\phi$

The force acting on the vehicle is calculated from the acceleration and the object's mass, as shown in Eq. 4.63.

$$F = ma \qquad \text{[SI]} \qquad 4.63(a)$$

$$F = \frac{ma}{g_c} \qquad \text{[U.S.]} \qquad 4.63(b)$$

When the vehicle is steered along a curved path, the act of steering exerts a *lateral acceleration* inward, also called *radial* or *centripetal acceleration*. Lateral acceleration incorporates the friction force exerted by the vehicle's tires against the pavement and the vertical gravity acting on the mass of the vehicle. The lateral acceleration is counteracted by the *centrifugal force* exerted by the vehicle's mass.

Centripetal acceleration and centrifugal force must be properly balanced in order for the vehicle to stay in the curve and continue through it safely. In cases where steering alone is unlikely to keep the vehicle on the curve, the roadway can be sloped toward the inside of the curve to increase the vertical gravity acting on the vehicle. However, a certain amount of centrifugal force is necessary to help drivers control their vehicles while rounding the curve and to help drivers and passengers maintain a comfortable position in their seats. For this reason, curved roadways should never be sloped such that centrifugal force is completely negated.

The centrifugal force, F_c, is calculated from Eq. 4.64.

$$F_c = \frac{m\mathrm{v}_t^2}{R} \qquad \text{[SI]} \qquad 4.64(a)$$

$$F_c = \frac{m\mathrm{v}_t^2}{g_c R} = \frac{w\mathrm{v}_t^2}{gR} \qquad \text{[U.S.]} \qquad 4.64(b)$$

The proper balance of forces acting on a vehicle while rounding a curve can be found from Eq. 4.65. R is the curve radius; v is the vehicle speed; f_s is the side friction; e is the rate of superelevation, $\tan\phi$; and g is gravitational acceleration, 32.2 ft/sec^2 (9.81 m/s^2).

$$R \approx \frac{\mathrm{v}_t^2}{g(e+f_s)} \qquad 4.65$$

Substituting $g = 32.2$ ft/sec^2 (9.81 m/s^2) into Eq. 4.65 and using units of miles per hour (kilometers per hour) for v gives the minimum curve radius, Eq. 4.66.

$$R_{\mathrm{m}} = \frac{\mathrm{v}_{\mathrm{kph}}^2}{127(e+f_s)}$$

[*Green Book* Eq. 3-10] [SI] 4.66(a)

$$R_{\mathrm{ft}} = \frac{\mathrm{v}_{\mathrm{mph}}^2}{15(e+f_s)}$$

[*Green Book* Eq. 3-10] [U.S.] 4.66(b)

Side Friction Factor

The *side friction factor*, f_s, is dependent on the lateral traction available where the tires meet the roadway. The side friction factor is sometimes called the *comfort factor*, as this is the unbalanced superelevation force a passenger feels as a vehicle travels around a horizontal curve. This friction factor varies with tire tread design, tire wear, roadway surface conditions, and travel speed. The general design conditions given in the AASHTO *Green Book*, Chap. 3, for the side friction factor include wet pavement (either asphalt or concrete) and average tire tread wear.

Table 4.9 extrapolates recommended design side friction factors for different types of facilities and conditions from the graphs in *Green Book* Exh. 3-11. For rural and high speed urban highways, design friction factors range from 0.17 for a design speed of 10 mph (0.18 for a design speed of 20 kph) to 0.08 for a design speed of 80 mph (130 kph). For intersections, friction factors range from 0.38 at 10 mph (0.35 at 20 kph) to 0.15 at 45 mph (70 kph). Friction factors range from 0.30 at 20 mph (0.32 at 30 kph) to 0.16 at 45 mph (70 kph) for low-speed urban streets.

Table 4.9 *Assumed Side Friction Factors, f_s, for Design Speeds*

SI units

speed (kph)	rural and high-speed urban highways*	intersections*	low-speed urban streets*	new tires on wet concrete pavements
20	0.18	0.35		
30	0.17	0.28	0.32	0.50
40	0.16	0.23	0.25	0.48
50	0.16	0.19	0.22	0.46
60	0.15	0.17	0.19	0.43
70	0.14	0.15	0.16	0.41
80	0.14			0.39
90	0.13			0.37
100	0.11			0.35
110	0.10			
120	0.09			
130	0.08			

customary U.S. units

speed (mph)	rural and high-speed urban highways*	intersections*	low-speed urban streets*	new tires on wet concrete pavements
10	0.17	0.38		
15	0.17	0.32		
20	0.16	0.26	0.30	0.50
25	0.16	0.23	0.25	0.48
30	0.16	0.20	0.22	0.46
35	0.15	0.18	0.20	0.44
40	0.15	0.16	0.18	0.42
45	0.14	0.15	0.16	0.40
50	0.14			0.39
60	0.12			0.35
70	0.10			
80	0.08			

*assumed for design

From *A Policy on Geometric Design of Highways and Streets*, 2004, by the American Association of State Highway and Transportation Officials, Washington, D.C. Used by permission.

The maximum side friction factor for any design speed is based on the *point of impending skid*, the point at which a vehicle's tires would begin to skid. Although AASHTO gives studies with friction factors in the range of 0.34–0.36 for smooth tires on wet pavement at 45 mph (80 kph), recommended design values are far below this range for reasons of safety and the variability of field conditions, as well as comfort and safety. Skidding should be avoided, and a margin for error should be provided for safety reasons, so design friction factors are far below factors that approach the point of impending skid.

The friction factor reduces as speed increases. Friction factors are generally independent of vehicle weight and are instead more dependent on the roughness or smoothness of the pavement surface, the presence of surface contaminants such as oil or a film of dirt and water, tread design, and the amount of tread wear. Tire pressure above or below the tire manufacturer's recommendations change the footprint of tire contact and also affect the friction factor.

When high friction is required to keep a vehicle on a curve, swerving becomes perceptible, the drift angle increases, and drivers experience a sensation of increased need for intense concentration that is considered undesirable. For these reasons, the AASHTO design friction factor values are conservative.

Design Speed versus Average Running Speed

Not all drivers in a traffic stream operate their vehicles at the same speed for a given design condition. Drivers seek their own comfortable speed according to many factors, such as driver capability, vehicle size and weight, familiarity with the road, and so on. All roadways have both an average running speed and a design speed. The *average running speed* is found by adding the distances traveled by all vehicles on a roadway during a period of time and dividing this number by the sum of the vehicles' running times (see Chap. 2). *Design speed* is chosen during the design of a new roadway and is the controlling factor in determining a roadway's geometric features. The average running speed of all vehicles is usually less than the design speed, with the difference between design speed and average running speed increasing as the design speed increases. The average running speed is less than the 85th percentile speed. Table 4.10 compares the distribution of average running speeds and design speeds up to 80 mph (130 kph), as described in the *Green Book*.

Table 4.10 *Design Speeds and Average Running Speeds*

SI units		customary U.S. units	
design speed (kph)	average running speed (kph)	design speed (mph)	average running speed (mph)
20	20	15	15
30	30	20	20
40	40	25	24
50	47	30	28
60	55	35	32
70	63	40	36
80	70	45	40
90	77	50	44
100	85	55	48
110	91	60	52
120	98	65	55
130	102	70	58
		75	61
		80	64

From *A Policy on Geometric Design of Highways and Streets*, 2004, by the American Association of State Highway and Transportation Officials, Washington, D.C. Used by permission.

Distribution of Superelevation and Side Friction for Curve Design

The superelevation rate, e, and side friction factor, f_s, combine to keep a vehicle on a curved roadway. For many curves with radii greater than the recommended minimum for a given design speed, superelevation at the maximum slope is not necessary, nor is it always desirable. There are five methods of distributing e and f_s for curve design recognized by AASHTO.

Method 1: For any radius, superelevation and side friction are equal to each other. The values of f_s and e vary in proportion to the inverse of the radius.

Method 2: As the radius decreases for a given speed, superelevation is not introduced until f_s reaches maximum, then f_s remains at maximum as e increases to maximum for that speed. This method is used with urban street settings at lower speeds where drivers expect greater side friction forces, and there are more constraints involved with providing full superelevation.

Method 3: Side friction remains at zero as the radius decreases for the design speed, until e reaches its maximum. Then, f_s increases to its maximum at the minimum radius for that speed.

Method 4: This method is the same as method 3 except it is based on average running speed instead of design speed.

Method 5: The values of f_s and e vary in inverse proportion to the radius, but in a curvilinear fashion. The distribution curve between f_s and e increases the superelevation slightly over method 1, but not nearly as much as method 3, due to the parabolic shape of the distribution curve. This method is commonly used on roadways with higher speeds, including rural highways, urban freeways, and high-speed urban streets, and roadways with radii greater than the minimum for a given design speed.

Method 1 is often used because it is simplest to apply. Methods 2, 3, and 4 are less commonly used because of the tendency to choose superelevation rates that are either too high or too low for the conditions. A low superelevation rate can be taken too fast, leading to erratic driver control. A high superelevation rate can lead to negative side friction, which decreases driver and passenger comfort. Method 5 is generally favored overall for higher speed roadway conditions. Figure 4.24 shows graphical plots of the five methods. Advantages and disadvantages of all five methods are discussed in further detail in *Green Book* Chap. 3.

Figure 4.24 *Superelevation and Side Friction Distribution Methods*

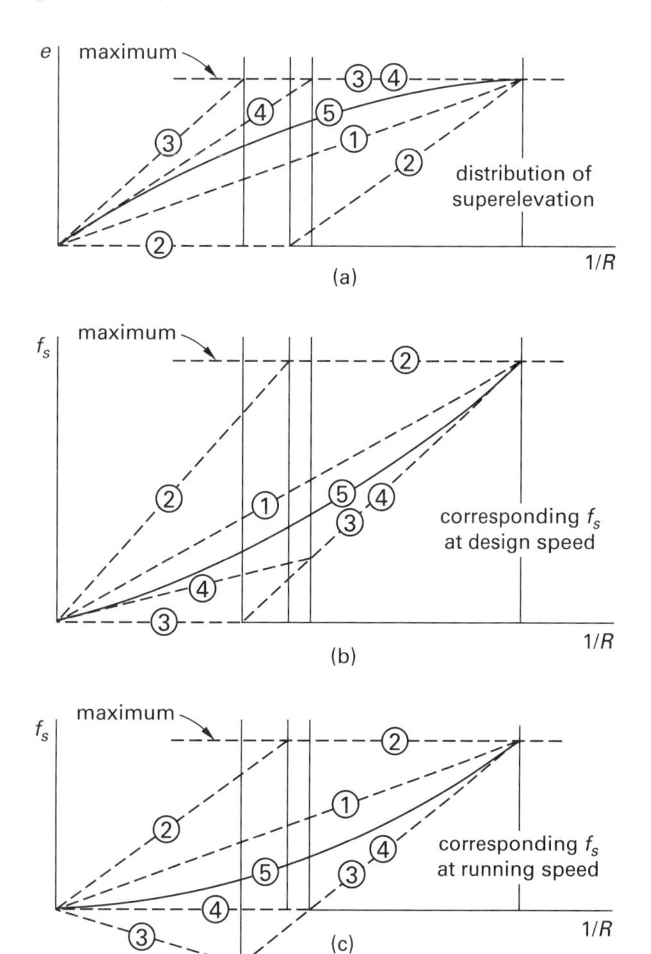

From *A Policy on Geometric Design of Highways and Streets*, 2004, by the American Association of State Highway and Transportation Officials, Washington, D.C. Used by permission.

Superelevation Procedure Development for Method 5

The distribution of e and f_s for method 5 is a curvilinear plot. The AASHTO method used to define this curve can be studied in the *Green Book* should a more precise e and f_s distribution for various curve radii be desired.

To create a distribution curve of e and f_s for a given curve, the running speeds on the curve are first plotted in relation to design speeds using the values shown in Table 4.10. In a "balanced" condition, the superelevation rate (in ft/ft or m/m) and the side friction factor for a given curve radius, the maximum superelevation rate, and the design speed combine to exactly balance the centrifugal force exerted on a vehicle traveling over the curve. For curves with a radius greater than the minimum for the design speed and maximum superelevation rate, as shown in Table 4.11, the appropriate side friction factor from Table 4.9 is subtracted from the total superelevation. (It is assumed when plotting these distribution curves that the side friction factor for all curves is 0 until a factor from Table 4.9 is applied.) The difference between the centripetal force produced by side friction and the centripetal force needed to create a balanced condition is established by the superelevation rate, and the force generated by the superelevation is shown as the finalized distribution curve of superelevation values in ft/ft (m/m).

The shape of the f_s distribution curve is parabolic and can be analyzed similarly to vertical curve geometry. Three points can be established to develop the parabolic distribution curve of e and f_s. See the *Green Book* for more information on method 5.

Maximum Superelevation Rates

In general, maximum superelevation rates for highways depend on climate and terrain conditions, whether the highway is in a rural or urban area, and the frequency of slower-moving vehicles (these vehicles' operation may be impaired by high superelevation rates). Using the curve shown in Fig. 4.24 and applying superelevation method 5, AASHTO has developed a series of superelevation distribution curves and tables to assist in selecting a superelevation rate for a particular design curve. These plots and tables are reproduced in App. 4.A and App. 4.B.

The first step in selecting a superelevation rate is to select the maximum superelevation rate for the road conditions at hand. The highest superelevation rate commonly used for highways is 10%. Rates as high as 12% have been used in some cases. However, some truck and trailer combinations have a tendency toward overturning at these steep cross slopes. 8% is considered to be a reasonable maximum rate, particularly in locations with snow and ice.

Where traffic congestion occurs regularly or where vehicles are expected to travel at very low speeds or remain stopped for long periods of time, the maximum rate is reduced to 4% or 6% in order to reduce discomfort of passengers at steeper cross slopes. This maximum range can be applied in urban areas and along roadway segments with many driveway connections. Intersections where there are a large number of vehicles turning and crossing also have maximum superelevation rates between 4% and 6%, because the high rate of traffic movements tends to result in lower running speeds.

Superelevation is limited by how easily pavement can be drained, and areas where it is difficult to achieve proper pavement drainage cannot be superelevated. No superelevation may be applied where it is difficult to warp

Table 4.11 Minimum Radius Using Maximum Values of e and f_s

SI units						customary U.S. units					
design speed (kph)	maximum e^* (%)	maximum f_s	total $(e+f_s)$	calculated radius (m)	rounded radius (m)	design speed (mph)	maximum e^*(%)	maximum f_s	total $(e+f_s)$	calculated radius (ft)	rounded radius (ft)
15	4.0	0.40	0.44	4.0	4	10	4.0	0.38	0.42	15.9	16
20	4.0	0.35	0.39	8.1	8	15	4.0	0.32	0.36	41.7	42
30	4.0	0.28	0.32	22.1	22	20	4.0	0.27	0.31	86.0	86
40	4.0	0.23	0.27	46.7	47	25	4.0	0.23	0.27	154.3	154
50	4.0	0.19	0.23	85.6	86	30	4.0	0.20	0.24	250.0	250
60	4.0	0.17	0.21	135.0	135	35	4.0	0.18	0.22	371.2	371
70	4.0	0.15	0.19	203.1	203	40	4.0	0.16	0.20	533.3	533
80	4.0	0.14	0.18	280.0	280	45	4.0	0.15	0.19	710.5	711
90	4.0	0.13	0.17	375.2	375	50	4.0	0.14	0.18	925.9	926
100	4.0	0.12	0.16	492.1	492	55	4.0	0.13	0.17	1186.3	1190
15	6.0	0.40	0.46	3.9	4	10	6.0	0.38	0.44	15.2	15
20	6.0	0.35	0.41	7.7	8	15	6.0	0.32	0.38	39.5	39
30	6.0	0.28	0.34	20.8	21	20	6.0	0.27	0.33	80.8	81
40	6.0	0.23	0.29	43.4	43	25	6.0	0.23	0.29	143.7	144
50	6.0	0.19	0.25	78.7	79	30	6.0	0.20	0.26	230.8	231
60	6.0	0.17	0.23	123.2	123	35	6.0	0.18	0.24	340.3	340
70	6.0	0.15	0.21	183.7	184	40	6.0	0.16	0.22	484.8	485
80	6.0	0.14	0.20	252.0	252	45	6.0	0.15	0.21	642.9	643
90	6.0	0.13	0.19	335.7	336	50	6.0	0.14	0.20	833.3	833
100	6.0	0.12	0.18	437.4	437	55	6.0	0.13	0.19	1061.4	1060
110	6.0	0.11	0.17	560.4	550	60	6.0	0.12	0.18	1333.3	1330
120	6.0	0.09	0.15	755.9	756	65	6.0	0.11	0.17	1656.9	1660
130	6.0	0.08	0.14	950.5	951	70	6.0	0.10	0.16	2041.7	2040
						75	6.0	0.09	0.15	2500.0	2500
						80	6.0	0.08	0.14	3047.6	3050
15	8.0	0.40	0.48	3.7	4	10	8.0	0.38	0.46	14.5	14
20	8.0	0.35	0.43	7.3	7	15	8.0	0.32	0.40	37.5	38
30	8.0	0.28	0.36	19.7	20	20	8.0	0.27	0.35	76.2	76
40	8.0	0.23	0.31	40.6	41	25	8.0	0.23	0.31	134.4	134
50	8.0	0.19	0.27	72.9	73	30	8.0	0.20	0.28	214.3	214
60	8.0	0.17	0.25	113.4	113	35	8.0	0.18	0.26	314.1	314
70	8.0	0.15	0.23	167.8	168	40	8.0	0.16	0.24	444.4	444
80	8.0	0.14	0.22	229.1	229	45	8.0	0.15	0.23	587.0	587
90	8.0	0.13	0.21	303.7	304	50	8.0	0.14	0.22	757.6	758
100	8.0	0.12	0.20	393.7	394	55	8.0	0.13	0.21	960.3	960
110	8.0	0.11	0.19	501.5	501	60	8.0	0.12	0.20	1200.0	1200
120	8.0	0.09	0.17	667.0	667	65	8.0	0.11	0.19	1482.5	1480
130	8.0	0.08	0 .16	831.7	832	70	8.0	0.10	0.18	1814.8	1810
						75	8.0	0.09	0.17	2205.9	2210
						80	8.0	0.08	0.16	2666.7	2670
15	10.0	0.40	0.50	3.5	4	10	10.0	0.38	0.48	13.9	14
20	10.0	0.35	0.45	7.0	7	15	10.0	0.32	0.42	35.7	36
30	10.0	0.28	0.38	18.6	19	20	10.0	0.27	0.37	72.1	72
40	10.0	0.23	0.33	38.2	38	25	10.0	0.23	0.33	126.3	126
50	10.0	0.19	0.29	67.9	68	30	10.0	0.20	0.30	200.0	200
60	10.0	0.17	0.27	105.0	105	35	10.0	0.18	0.28	291.7	292
70	10.0	0.15	0.25	154.3	154	40	10.0	0.16	0.26	410.3	410
80	10.0	0.14	0.24	210.0	210	45	10.0	0.15	0.25	540.0	540
90	10.0	0.13	0.23	277.3	277	50	10.0	0.14	0.24	694.4	694
100	10.0	0.12	0.22	357.9	358	55	10.0	0.13	0.23	876.8	877
110	10.0	0.11	0.21	453.7	454	60	10.0	0.12	0.22	1090.9	1090
120	10.0	0.09	0.19	596.8	597	65	10.0	0.11	0.21	1341.3	1340
130	10.0	0.08	0.18	739.3	739	70	10.0	0.10	0.20	1633.3	1630
						75	10.0	0.09	0.19	1973.7	1970
						80	10.0	0.08	0.18	2370.4	2370
15	12.0	0.40	0.52	3.4	3	10	12.0	0.38	0.50	13.3	13
20	12.0	0.35	0.47	6.7	7	15	12.0	0.32	0.44	34.1	34
30	12.0	0.28	0.40	17.7	18	20	12.0	0.27	0.39	68.4	68
40	12.0	0.23	0.35	36.0	36	25	12.0	0.23	0.35	119.0	119
50	12.0	0.19	0.31	63.5	64	30	12.0	0.20	0.32	187.5	188
60	12.0	0.17	0.29	97.7	98	35	12.0	0.18	0.30	272.2	272
70	12.0	0.15	0.27	142.9	143	40	12.0	0.16	0.28	381.0	381
80	12.0	0.14	0.26	193.8	194	45	12.0	0.15	0.27	500.0	500
90	12.0	0.13	0.25	255.1	255	50	12.0	0.14	0.26	641.0	641
100	12.0	0.12	0.24	328.1	328	55	12.0	0.13	0.25	806.7	807
110	12.0	0.11	0.23	414.2	414	60	12.0	0.12	0.24	1000.0	1000
120	12.0	0.09	0.21	539.9	540	65	12.0	0.11	0.23	1224.6	1220
130	12.0	0.08	0.20	665.4	665	70	12.0	0.10	0.22	1484.8	1480
						75	12.0	0.09	0.21	1785.7	1790
						80	12.0	0.08	0.20	2133.3	2130

*In recognition of safety considerations, use of $e_{max} = 4.0\%$ should be limited to urban conditions.

From *A Policy on Geometric Design of Highways and Streets*, 2004, by the American Association of State Highway and Transportation Officials, Washington, D.C. Used by permission.

pavement for drainage. In some cases, the application of the normal pavement cross section results in negative superelevation, such as in a housing plan or industrial development area.

Table 4.11 shows examples of the minimum curve radius for a superelevation rates of 4%, 6%, 8%, 10%, and 12% using speed-adjusted maximum friction factors and can be used to select the minimum curve radius for a given design speed.

Minimum Radius Without Superelevation

Applying Eq. 4.66 using the maximum value for f_s and the normal superelevation rate yields the minimum radius without superelevation for a given design speed. For the outside lanes on a normal sloped roadway (i.e., the lanes are sloped down from the centerline), a negative superelevation value is used. Appendix 4.C and App. 4.D give the minimum radii for low-speed roadways and urban streets. Negative superelevation rates of 1.5%–2.5% occur on the outside lanes of a normally crowned street without superelevation. For minimum radii of higher speed roadways, the reduced friction factors shown in App. 4.D need to be considered.

Superelevation Axis of Rotation

Before superelevation of a highway begins, the axis of rotation must be chosen. The *axis of rotation* is the point on the cross section about which the roadway will be rotated to achieve the full superelevation rate. For two-lane highways, three axes of rotation are typically used: the centerline; the lower (inside) pavement edge; or the upper (outside) pavement edge. (See Fig. 4.25.) For multilane highways separated by a median where each directional roadway is on a straight cross slope, the edges of the median are typically chosen as the axes of rotation. Therefore, each directional roadway is rotated about the median edge while the median remains level.

Each axis of rotation is suited to specific situations, and no one choice is applicable to all conditions. Roadway-edge appearance, sight lines, drainage, and other vertical geometry conditions usually establish which axis of rotation is selected. Using the lower edge as the axis is most common for two-lane highways because it is easier to control drainage and usually requires less excavation.

Superelevation Transitions

Superelevation is usually carried uniformly throughout the curve and is transitioned smoothly at each end from the *normal crown*, or normal cross section. The transition from the normal crown to full superelevation happens in several steps. (See Fig. 4.26.) The outer lane is first raised to a level section. This is called removing *adverse crown*, and the segment over which the removal occurs is the *tangent runout*. Next, the outer lane continues to rotate to match the slope of the inner lane. The entire roadway cross section is rotated to achieve the full superelevation. The distance required to rotate from the point of no adverse crown to full superelevation is called the *superelevation transition length*, or the *length of runoff*. This length is determined by the design speed of the roadway and the change in slope required to achieve the full superelevation rate.

Figure 4.25 *Typical Axes of Rotation for Highways*

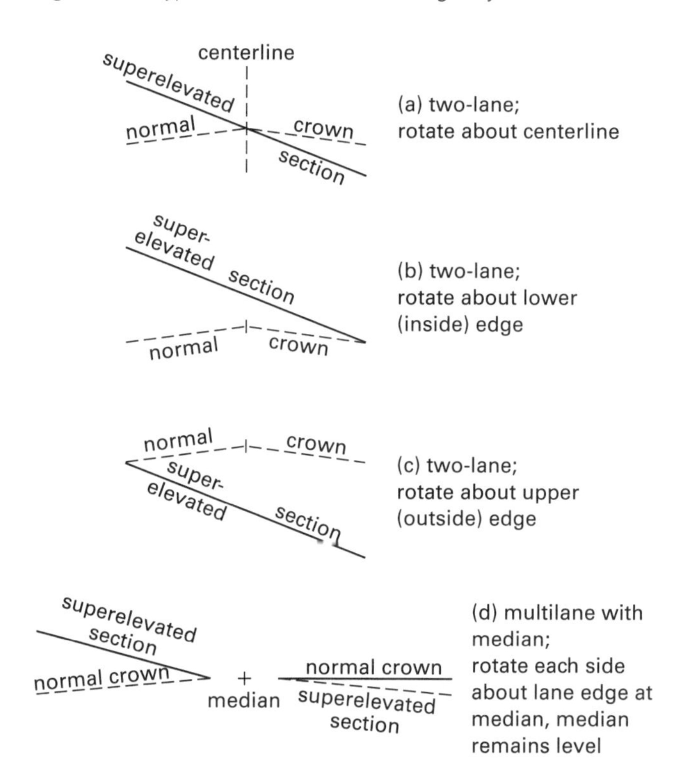

When a vehicle transitions from a normal cross section to a superelevated section, the driver has to make adjustments to steering, and passengers experience a noticeable roll of the vehicle about its axis. The adjustments found to be acceptable by drivers and passengers are those carried out over a period of approximately 2 sec. One method of establishing the minimum transition length is to use the distance a vehicle travels in 2 sec at the design speed. However, this method may worsen existing drainage problems along the transition length. For this reason, AASHTO recommends using the relative difference in profile grade method.

The relative difference in profile grade between the pavement edge and the axis of rotation is used to determine the transition length to ensure the profile edges appear smooth. The *Green Book* recommends establishing a *maximum relative gradient* for the outer and inner edge, or outer edge and centerline, to avoid creating a rough transition with sudden, uncomfortable up and down movements of the vehicle. Table 4.12 shows the current AASHTO recommendations for maximum relative gradients according to design speed. Transition lengths are directly proportional to the total superelevation, which is a product of the lane width and the superelevation rate (see Eq. 4.69). The relative gradients shown are for two-lane roadways.

Figure 4.26 Roadway Profile Around Circular Curve

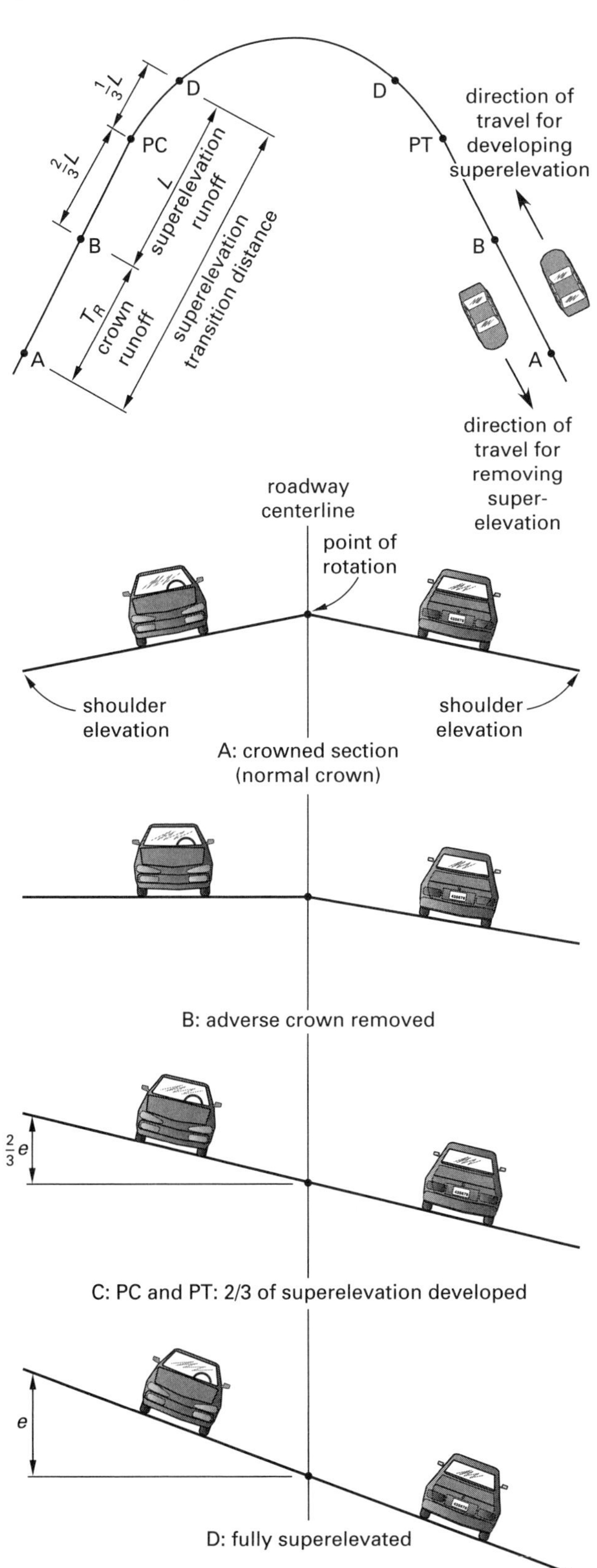

From *Civil Engineering Reference Manual*, by Michael R. Lindeburg, copyright © 2011, Professional Publications, Inc. Reproduced with permission.

Table 4.12 Maximum Relative Gradients of Pavement Edges, $\Delta_{\%}$, for Two-Lane Highways

SI units

design speed (kph)	maximum relative gradient (%)	equivalent maximum relative slope
20	80	1:125
30	75	1:133
40	70	1:143
50	65	1:154
60	60	1:167
70	55	1:182
80	50	1:200
90	47	1:213
100	44	1:227
110	41	1:244
120	38	1:263
130	35	1:286

customary U.S. units

design speed (mph)	maximum relative gradient (%)	equivalent maximum relative slope
15	78	1:128
20	74	1:135
25	70	1:143
30	66	1:152
35	62	1:161
40	58	1:172
45	54	1:185
50	50	1:200
55	47	1:213
60	45	1:222
65	43	1:233
70	40	1:250
75	38	1:263
80	35	1:286

From *A Policy on Geometric Design of Highways and Streets*, 2004, by the American Association of State Highway and Transportation Officials, Washington, D.C. Used by permission.

Another method, often used on multilane highways and other wide roads, is to run a profile along the extreme edge of the pavement and design a comfortable vertical curve into the edge transition. The transition should be checked for drainage slope, as it is possible to create level sections in the transition zone that do not drain well.

For any method used when transitioning from a negative slope to a positive slope, or vice versa, there will be a short section of pavement that is level or near level. This level section will trap water on the pavement surface. A shallow cross slope combined with a roadway profile grade of less than 0.5% can cause ponding of water, which can in turn cause drivers to lose control of their vehicles. Ponding can be minimized by keeping the transition length as short as possible, by moving the transition location ahead or back on stationing, or by changing the vertical profile of the edges of the level section.

Proportioning transition lengths according to lane widths (e.g., doubling the length for four lanes) may result in transitions far too long for practical application. To compensate for lane width, Eq. 4.67 can be used to determine the minimum runoff length, L_R. W is the width of one lane, N_L is the number of lanes rotated, and b_w is the adjustment factor for number of lanes rotated (see Fig. 4.27). $\Delta_{\%}$ is the maximum relative gradient (i.e., the relative difference in longitudinal grade between the axis of rotation and the pavement edge), expressed as a percentage, and e is the design superelevation rate.

$$L_R = \left(\frac{WN_Le}{\Delta_{\%}}\right)b_w \quad \text{[Green Book Eq. 3-25]} \qquad 4.67$$

Equation 4.68 shows a simplified method of finding the transition length using the elevation change of the outer pavement edge determined from Eq. 4.69 and the maximum relative slope. The *maximum relative slope*, Δ, is the maximum relative gradient expressed in foot per foot (meter per meter). L_R is the minimum runoff length, and E is the elevation change for a fully superelevated lane.

$$L_{R,\text{m}} = E_\text{m}\Delta_{\text{m/m}} \qquad \text{[SI]} \qquad 4.68(a)$$

$$L_{R,\text{ft}} = E_\text{ft}\Delta_{\text{ft/ft}} \qquad \text{[U.S.]} \qquad 4.68(b)$$

$$E = We \qquad 4.69$$

Example 4.7

A freeway has three lanes, 12 ft (3.7 m) wide, in each direction. A curve with a design speed of 70 mph (110 kph) is to be superelevated to a 3.0% cross slope. Using the maximum relative gradient, determine the runoff length.

SI Solution

From Table 4.12, the maximum relative gradient for a design speed of 110 kph is 0.41%. The adjustment fac-

Figure 4.27 *Superelevation Transition Adjustment Factor for Number of Lanes Rotated*

number of lanes rotated (N_L)	adjustment factor* (b_w)	length increase relative to one-lane rotated (N_Lb_w)
1	1.00	1.0
1.5	0.83	1.25
2	0.75	1.5
2.5	0.70	1.75
3	0.67	2.0
3.5	0.64	2.25

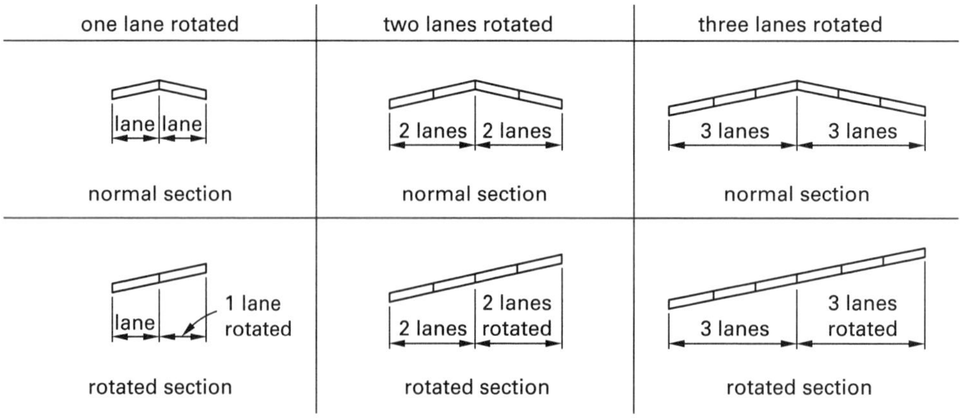

*$b_w = (1 + 0.5(N_L - 1))/N_L$

From *A Policy on Geometric Design of Highways and Streets*, 2004, by the American Association of State Highway and Transportation Officials, Washington, D.C. Used by permission.

tor, b_w, for three rotated lanes is 0.67. Using Eq. 4.67, the runoff length is

$$L_R = \left(\frac{WN_Le}{\Delta_\%}\right)b_w = \left(\frac{(3.7\text{ m})(3)(0.03)}{0.41\%}\right)(0.67)$$
$$= 54.4\text{ m}$$

Customary U.S. Solution

From Table 4.12, the maximum relative gradient for a design speed of 70 mph is 0.40%. The adjustment factor, b_w, for three rotated lanes is 0.67. Using Eq. 4.67, the runoff length is

$$L_R = \left(\frac{WN_Le}{\Delta_\%}\right)b_w$$
$$= \left(\frac{(12\text{ ft})(3)(0.03)}{0.40\%}\right)(0.67)$$
$$= 180.9\text{ ft}$$

Example 4.8

A two-lane roadway is to be superelevated at a rate of 0.04 ft/ft (0.04 m/m). The design speed is 60 mph (100 kph), and the transition is to be designed using the equivalent maximum relative slope. The lanes are each 12 ft (3.7 m) wide. Determine the runoff length.

SI Solution

Determine the relative elevation change of the outer pavement edge for the fully elevated lane using Eq. 4.69.

$$E = We = (3.7\text{ m})\left(0.04\ \frac{\text{m}}{\text{m}}\right)$$
$$= 0.148\text{ m}$$

From Table 4.12, the maximum relative slope for a 100 kph design speed is 1:227 or 227 m/m. Determine the runoff length using Eq. 4.68.

$$L_{R,\text{m}} = E_\text{m}\Delta_{\text{m/m}} = (0.148\text{ m})\left(227\ \frac{\text{m}}{\text{m}}\right)$$
$$= 33.6\text{ m}$$

Customary U.S. Solution

Determine the relative elevation change of the outer pavement edge for a fully elevated lane using Eq. 4.69.

$$E = We = (12\text{ ft})\left(0.04\ \frac{\text{ft}}{\text{ft}}\right)$$
$$= 0.48\text{ ft}$$

From Table 4.12, the maximum relative slope for a 60 mph design speed is 1:222 or 222 ft/ft. Determine the transition length using Eq. 4.68.

$$L_{R,\text{ft}} = E_\text{ft}\Delta_{\text{ft/ft}} = (0.48\text{ ft})\left(222\ \frac{\text{ft}}{\text{ft}}\right)$$
$$= 106.6\text{ ft}$$

Minimum Tangent Runout Length

Tangent runout is the section of a roadway over which the adverse crown is removed. In order to achieve a smooth transition, the rate at which the adverse crown is removed should equal the maximum relative gradient used to determine the minimum runoff length, L_R. Therefore, the minimum length of tangent runout, L_t, can be determined by Eq. 4.70. e is the superelevation rate, and e_{NC} is the normal cross slope rate, both in decimals.

$$L_t = \left(\frac{e_{\text{NC}}}{e}\right)L_R \quad [\textit{Green Book}\text{ Eq. 3-26}] \qquad 4.70$$

Appendix 4.E gives a tabulation of minimum runoff lengths and tangent runout lengths for various superelevation rates and design speeds.

Example 4.9

A freeway with three 12 ft (3.7 m) wide lanes in each direction has a curve with a superelevation of 3.0%. The design speed is 70 mph (110 kph), and the minimum runoff length is 181 ft (55.1 m). If the normal section cross slope is 1.5%, what is the minimum tangent runout length?

SI Solution

Using Eq. 4.70, the minimum tangent runout length is

$$L_t = \left(\frac{e_{\text{NC}}}{e}\right)L_R = \left(\frac{0.015}{0.03}\right)(55.1\text{ m})$$
$$= 27.6\text{ m}$$

Customary U.S. Solution

Using Eq. 4.70, the minimum tangent runout length is

$$L_t = \left(\frac{e_{\text{NC}}}{e}\right)L_R = \left(\frac{0.015}{0.03}\right)(181\text{ ft})$$
$$= 90.5\text{ ft}$$

Location of Transition With Respect to the End of Curve

Transitions Without Spirals

In tangent-to-curve design, roadway designers must decide where to place the superelevation transition length in respect to the PC. Neither placing the transition entirely on the approach tangent nor placing it entirely on the circular curve is desirable. For general overall conditions, roadway engineers tend to place two-thirds (67%) of the superelevation transition length on the tangent and the remaining transition length on the curve. (See Fig. 4.26.) This practice provides a reasonable compromise between lateral acceleration and vehicle lateral motion as the driver corrects for the curve and highway cross slope for a single lane-width transition. However, this convention may not provide sufficient cross slope for wider pavements and lower design speeds. AASHTO recommends increasing the length of superelevation runoff on the tangent in some situations, as shown in Table 4.13.

Table 4.13 *Portion of Runoff in Tangent to Minimize Vehicle Lateral Motion*

SI units

design speed (kph)	portion of runoff located prior to the curve — no. of lanes rotated 1.0	1.5	2.0–2.5	3.0–3.5
20–70	0.80	0.85	0.90	0.90
80–130	0.70	0.75	0.80	0.85

customary U.S. units

design speed (mph)	portion of runoff located prior to the curve — no. of lanes rotated 1.0	1.5	2.0–2.5	3.0–3.5
15–45	0.80	0.85	0.90	0.90
50–80	0.70	0.75	0.80	0.85

From *A Policy on Geometric Design of Highways and Streets*, 2004, by the American Association of State Highway and Transportation Officials, Washington, D.C. Used by permission.

Transitions With Spirals

The most effective spiral curve design for a given situation is one that closely approximates the natural spiral path drivers tend to adopt in that situation. A curve design that feels natural to drivers will result in increased driver comfort and vehicle control, both of which increase roadway safety.

The most natural spiral curves are those in which the length of the spiral and the length of the superelevation transition are equal. If a tangent runout is needed, it is most commonly placed on a length of roadway beyond the end of the spiral. Proper selection of spiral length is covered in Sec. 4.8.

13. VERTICAL AND HORIZONTAL CLEARANCES

Clearances on Vertical Curves

General procedures for determining vertical clearance above or below vertical curve grade lines are covered in many codebooks and design manuals, such as the *Green Book*. One complicated design condition worthy of note is the issue of sight distances for sag vertical curves at undercrossings. Figure 4.28 shows a sag vertical curve at an undercrossing. As shown in Fig. 4.28, the fascia of the undercrossing may block the line of sight at the point where the *critical clearance distance*, C, intersects the fascia, reducing available sight distance. This limited sight distance is particularly a problem when it occurs at two-lane undercrossings where passing sight distance is needed.

Figure 4.28 *Sight Distance on Sag Vertical Curves at Undercrossings*

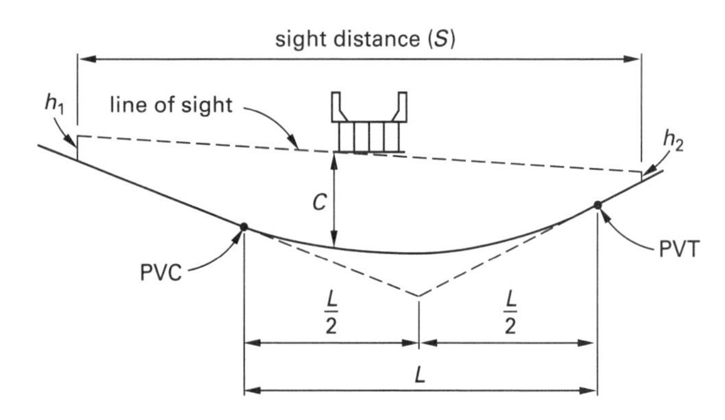

The available sight distance can be found graphically or by calculation. An approximation of the available sight distance can be found using a simplified geometric construction, as shown in Fig. 4.29, with the assumption that the fascia on the approach side is the location of the critical clearance distance, C, and the sight distance, S, is greater than the length of the curve, L. A sight line can be drawn between the point where the driver's eye height, h_1, intersects the back grade point and the height of the structure, h_2, intersects the ahead grade point. The length of the sight line is the available sight distance, S. S is proportional to L, and the sum of the external distance, E, and the distance O_s is proportional to double the external distance, $2E$, as shown in Fig. 4.29 and Eq. 4.71.

Figure 4.29 *Line of Sight Below Underpass When $S > L$*

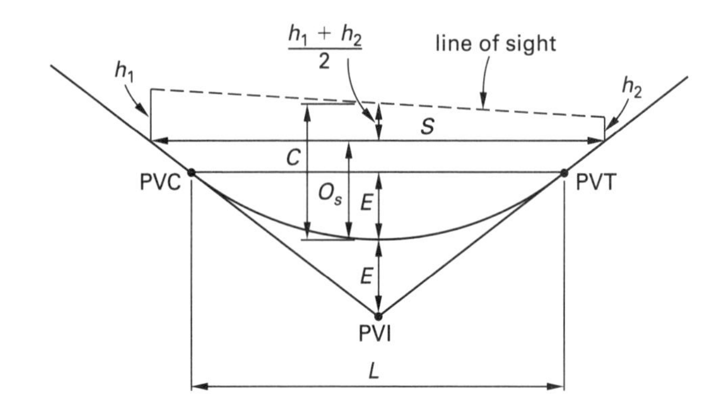

$$\frac{S}{L} = \frac{E + O_s}{2E} = \frac{1}{2} + \frac{O_s}{2E} \qquad 4.71$$

The external distance of the curve is found using Eq. 4.72. A is the absolute value of the algebraic difference in grades in percent, and L is the length of the curve in stations.

$$E_{\text{ft}} = \frac{A_{\%} L_{\text{sta}}}{8\ \frac{\text{sta}}{\text{ft}}} \qquad \text{[U.S. only]} \qquad 4.72$$

As shown in Fig. 4.29, the O_s distance is

$$O_s = C - \frac{h_1 + h_2}{2} \qquad 4.73$$

The O_s distance can be further simplified by substituting the default values of h_1 (3.5 ft or 1080 mm) and h_2 (2.0 ft or 600 mm) into $h_1 + h_2/2$, which gives an averaged value of 2.75 ft (840 mm). Replacing O_s in Eq. 4.71 with $C-$ 2.75 ft (840 mm) yields Eq. 4.74, which can be rearranged into Eq. 4.75 and Eq. 4.76 to find the curve length and sight distance, respectively.

$$\frac{S_{\text{m}}}{L_{\text{m}}} = \frac{1}{2} + \frac{C_{\text{m}} - 840 \text{ mm}}{2E_{\text{m}}} \qquad \text{[SI]} \qquad 4.74(a)$$

$$\frac{S_{\text{ft}}}{L_{\text{ft}}} = \frac{1}{2} + \frac{C_{\text{ft}} - 2.75 \text{ ft}}{2E_{\text{ft}}} \qquad \text{[U.S.]} \qquad 4.74(b)$$

$$L_{\text{m}} = \frac{2S_{\text{m}}}{1 + \frac{C_{\text{m}} - 840 \text{ mm}}{E_{\text{m}}}} \qquad \text{[SI]} \qquad 4.75(a)$$

$$L_{\text{ft}} = \frac{2S_{\text{ft}}}{1 + \frac{C_{\text{ft}} - 2.75 \text{ ft}}{E_{\text{ft}}}} \qquad \text{[U.S.]} \qquad 4.75(b)$$

$$S_{\text{m}} = \frac{L_{\text{m}}}{2}\left(1 + \frac{C_{\text{m}} - 840 \text{ mm}}{E_{\text{m}}}\right) \qquad \text{[SI]} \qquad 4.76(a)$$

$$S_{\text{ft}} = \frac{L_{\text{ft}}}{2}\left(1 + \frac{C_{\text{ft}} - 2.75 \text{ ft}}{E_{\text{ft}}}\right) \qquad \text{[U.S.]} \qquad 4.76(b)$$

When the sight distance below an underpass is less than the length of the curve below the underpass, the parabolic curve between PVC and PVT resembles a half circle with radius R. (See Fig. 4.30.)

Using h_1 = 3.5 ft (1080 mm) and h_2 = 2.75 ft (840 mm), the equations for L and S become Eq. 4.77 and Eq. 4.78, respectively.

$$L_{\text{m}} = \frac{S_{\text{m}}^2 A_{\%}}{8(C_{\text{m}} - 840 \text{ mm})} \qquad \text{[SI]} \qquad 4.77(a)$$

$$L_{\text{ft}} = \frac{S_{\text{ft}}^2 A_{\%}}{8(C_{\text{ft}} - 2.75 \text{ ft})} \qquad \text{[U.S.]} \qquad 4.77(b)$$

Figure 4.30 *Line of Sight Below Underpass When S < L*

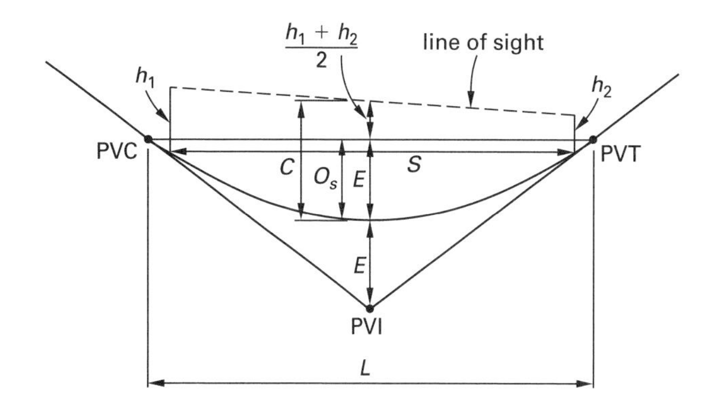

$$S_{\text{m}} = \sqrt{\frac{8L_{\text{m}}(C_{\text{m}} - 840 \text{ mm})}{A_{\%}}} \qquad \text{[SI]} \qquad 4.78(a)$$

$$S_{\text{ft}} = \sqrt{\frac{8L_{\text{ft}}(C_{\text{ft}} - 2.75 \text{ ft})}{A_{\%}}} \qquad \text{[U.S.]} \qquad 4.78(b)$$

When the obstruction created by the overhead structure (known as the critical clearance distance) is not located directly above the PVI, Eq. 4.74 through Eq. 4.78 remain accurate as long as the critical clearance distance is within 200 ft (60 m) of the PVI. When the critical clearance distance is located more than 200 ft (60 m) from the PVI, calculations using these equations are more approximate than definite.

Vertical Clearances from Overhead Obstructions

Codes governing the interstate highway system require newly constructed and reconstructed mainline routes to have 16.5 ft (5.0 m) vertical clearance. These clearances must be maintained at all times, even after a roadway has been repaved. The clearance must be across the entire roadway width, including the entire width of the shoulders. Most states allow load heights of 13.5 ft (4.1 m) without permits and 14.5 ft (4.4 m) load heights with permits. It is common practice to provide at least 1 ft (0.3 m) additional clearance above the maximum load height permitted. For locations where the clearance is less than 14.5 ft (4.4 m), advance warning signs are required, and the low-clearance point must be marked clearly with zebra or chevron safety stripes.

Local jurisdictions over local and nonmajor traffic routes often set the vertical clearance minimum as low as 12.5 ft (3.8 m). Occasionally, for locations where traffic is restricted to passenger cars only, and in some parts of older cities where reconstruction costs are very high, access for taller vehicles must be provided over alternate routes or, in extreme cases, prohibited entirely.

Where a bridge or overhead obstruction is installed after an approach of many miles, or where conditions are such that drivers might perceive an overhead obstruction as too low to accommodate tall loads, there is an increased risk that truckers or others carrying tall loads will think there is insufficient clearance for their vehicle and brake suddenly, which slows traffic and increases the risk of collisions. To mitigate this problem, the vertical clearance beneath these obstructions should be increased by a minimum of 2 ft (0.6 m).

Overhead signs, utility wire crossings, and traffic signals must be designed to these requirements as well. Usually, anything that is suspended on wires and can move around in the wind is installed with a higher clearance. Many power, telephone, and cable TV companies design for at least 17 ft (5.2 m) of clearance over major arterials when installing a new crossing. Over time, cables have a tendency to sag, reducing the clearance. It is the responsibility of the utility owner to maintain the minimum clearance required by local regulations. Overhead signs rigidly mounted on frames may need to be installed with additional clearance to accommodate light fixtures and maintenance catwalks. These devices must be installed below the sign face itself to avoid blocking the view of oncoming drivers.

Clearances on Horizontal Curves

When designing horizontal curves, attention must be paid to any sight obstructions that may exist. Roadway engineers should check the conditions of each curve and make adjustments as necessary to provide adequate sight distance for the curve's design speed. The sight distance typically refers to the driver's line of sight, as shown in Fig. 4.31. Sight distances on the inside of a curve can be equated to the chord of a circular arc centered at the closest point of obstruction to the edge of the traveled lane. (See Fig. 4.31.) The radial offset distance at the point of obstruction to the centerline of the traveled lane can be considered the middle ordinate of the circular arc. The *horizontal sightline offset*, HSO, is the same as the middle ordinate, *M*, but is typically referred to as the horizontal sightline offset when calculating sight distance.

$$\text{HSO} = \frac{C_{\text{long}}^2}{8R} \qquad 4.79$$

Usually a radius, *R*, is given, and it is necessary to find either the length of the long chord, C_{long}, or the horizontal sightline offset, HSO. The length of the chord is often used as the line of sight. The stopping sight distance required around an obstruction is determined as the arc length of the curve and is therefore slightly longer than the chord length, providing a margin of safety.

For analysis, the radius of the centerline of the extreme right lane must be determined. The rightmost lane is sometimes called the "inside lane," although the designation of "inside" versus "outside" varies depending on actual conditions and the jurisdiction. (For instance, on a divided highway, the leftmost lane adjacent to the median barrier is called the

Figure 4.31 *Horizontal Sight Distance Components on the Inside of a Curve*

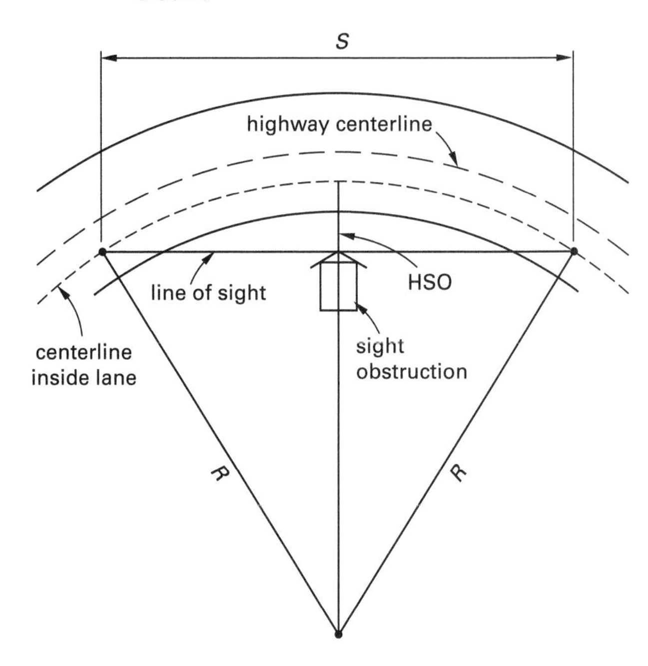

inside lane in some jurisdictions.) Whatever the terminology, it is important that the lane closest to any obstruction (e.g., buildings, walls, slopes, or any other sight-blocking object) have the shortest sight distance. Where there is a cut slope, the horizontal sightline offset is taken at a point 2.75 ft (840 mm) above the roadway, which is the average of 3.5 ft (1080 mm) eye height and 2.0 ft (600 mm) object height. Stopping sight distance, *S*, and the horizontal sightline offset, HSO, are related using Eq. 4.80.

$$\text{HSO} = R\left(1 - \cos\frac{28.65S}{R}\right) \quad \textit{[Green Book Eq. 3-38]} \qquad 4.80$$

Horizontal Clearance from Obstructions

All streets and highways are required to have a certain amount of lateral clearance. In general, all streets should have a minimum clearance of 1.5 ft (0.5 m) between the curb face and any obstructions. On streets with a continuous parking lane, a clearance of 1.5 ft (0.5 m) between the parking lane and any obstructions is desirable but not required. Rural collector streets with a design speed of 45 mph (70 kph) or less, as well as urban collector streets with shoulders but no curbs, should have *clear zones* (recovery areas free of fixed objects) of at least 10 ft (3 m). For more information on clear zones, see Chap. 6.

14. ACCELERATION AND DECELERATION

Average traffic accelerates at a rate of approximately 4.4 ft/sec^2, or 30 mph/sec^2 (1.3 m/s^2 or 4.8 kph/s^2), at speeds up to 30 mph (50 kph). The acceleration rate decreases to 2.8 ft/sec^2 (1.9 mph/sec^2) as speeds approach 70 mph (110 kph). Highway design must consider acceleration and deceleration rates and the distances necessary to safely achieve these rates. The distance required to reach a certain speed will increase for upgrade acceleration and decrease for downgrade acceleration due to driver behavior and the effects of gravity. Drivers tends to accelerate at a lower rate on upgrades and at a higher rate on downgrades. AASHTO design curves illustrating acceleration and deceleration distances are given in App. 4.F and App. 4.G. The acceleration and deceleration rates are based on low horsepower-to-weight ratio vehicles and represent the lower limit of design conditions. Both acceleration and deceleration rates will vary depending on local conditions, such as approach grade and driver familiarity, and with changing vehicle technology.

The mix of vehicles in the traffic flow can have a significant effect on acceleration rates. Modern trucks and intercity buses accelerate at nearly the same rates as automobiles, but heavily loaded large trucks and recreational vehicles tend to have much lower acceleration rates than standard automobiles. Traffic flow that contains a large number of recreational vehicles or heavily loaded large trucks will typically have lower acceleration rates than traffic flow containing only automobiles, trucks, and buses. However, a very small proportion of slower vehicles will have little effect on acceleration rates.

Deceleration rate is usually dictated by the comfort of the driver and passengers, not the braking ability of the vehicle. A comfortable deceleration rate for most drivers (approximately 90%) is 11.2 ft/sec^2 (3.4 m/s^2), which is the default deceleration value recommended by AASHTO. Rapid deceleration is discomforting to most people and tends to be used only when necessary or by select types of drivers. Deceleration rates greater than 14.8 ft/sec^2 (4.5 m/s^2) are seen only in cases where drivers are making panic stops.

When designing for deceleration, all vehicles have nearly the same capability, regardless of size or weight. For heavy vehicles such as commercial semitrucks, the braking distance is nearly the same as for smaller vehicles. However, the total stopping distance is slightly greater because of the delay in applying air brake systems and minor variations between braking rates of multiple axle systems. Large commercial vehicles are generally driven by specially trained drivers, who know to compensate for the increased brake application time of their vehicles. Furthermore, the elevated cabs of these commercial vehicles afford the drivers an increased line of sight relative to drivers of standard automobiles. These factors are assumed to compensate for the increased brake application time, so the same stopping sight distances are used for all vehicles.

15. INTERSECTIONS AND INTERCHANGES

Intersection Geometry

An intersection of two or more roadways can create traffic flow conflicts and delays, as well as collisions. Fundamental horizontal alignment principles are explained in Chap. 1 and Chap. 9 of the *Green Book*. The primary principle governing intersection geometry is the elimination of acute angles at the intersection, which increases sight distance in one or more directions for approaching drivers and provides sufficient width for turning of the design vehicle.

Horizontal geometry involves setting the inner edge of the pavement radius to match the design speed and widening the pavement to allow for the swept path of longer vehicles. The design guides are illustrated in the Appendices as follows.

- AASHTO design vehicle dimensions are shown in App. 4.H, with the minimum turning radii shown in App. 4.I.
- Selected turning path templates are shown in App. 4.J through App. 4.P.
- Turning templates and edges-of-traveled-way for various turn angles and design vehicles are given in App. 4.Q and App. 4.R.
- Curve radii for simple curves, curves with tapers, and three-centered compound curves are found in App. 4.S.
- The effect of curb radii on right-turning paths is illustrated in App. 4.T.
- Cross street widths occupied by turning vehicles are tabulated in App. 4.U.
- The effect of crosswalk lengths and corner setbacks are illustrated in App. 4.V and App. 4.W.

Example 4.10

A cul-de-sac is being installed at the end of a suburban street to allow a conventional school bus (S-BUS36 or S-BUS11) to turn around without a backing maneuver. What is the minimum outside radius required to accommodate the school bus without allowing overhang?

Solution

From App. 4.P, select the minimum turning radius (i.e., the path of the left front wheel), which shows a radius of 38.9 ft (11.86 m).

Example 4.11

A 120° intersection turn will have transit buses turning from the lane closest to the curb onto a cross street with 12 ft (3.6 m) wide lanes. Buses are expected to turn from the proper position in the lane and swing wide on the cross street. The curb radius is 20 ft (6 m). How wide must the cross street be for a bus to complete its turning movement?

Solution

From App. 4.U, with a transit bus turning from its proper position in the lane, case A applies. For a 120° turn and a 20 ft (6 m) curb radius, the width occupied on the cross street is 40 ft (12.2 m).

Example 4.12

A crosswalk at a 90° intersection is 3 m wide, the corner curb radius is 3 m, and the right-of-way is 3 m from the face of the curb at the traveled way. If the curb radius is increased to 8.0 m, how much length will be added to the crosswalk at each corner?

Solution

From App. 4.V, with a radius of 3 m and a crosswalk width of 3 m, the crosswalk distance added is 0.8 m for the curb return. For curb radius of 9 m, the distance added for the curb return is 8.0 m. The additional crosswalk distance necessary to increase the curb radius to 8.0 m is

$$8.0\ \text{m} - 0.8\ \text{m} = 7.2\ \text{m}$$

Interchange Geometry

Grade separation structures for crossing movements are used in interchange design to alleviate the conflict delays and safety demands caused by high volumes of traffic traveling through at-grade intersections. There are a variety of ramp configurations that can be used to connect a roadway to a crossing structure. The ramp configuration is often determined by topography or property constraints, but desired traffic operation is the most important consideration. Some configurations are proven to be more effective than others, while some configurations are actively discouraged as design standards evolve. Interchange configurations are illustrated in *Green Book* Exh. 10-1.

Left-hand ramp exits require slower-moving vehicles, such as trucks and buses, to move into the left lanes, which are generally used by higher-speed vehicles. Therefore, left-hand exits are typically avoided for high-speed, high-volume roadways. *Inside loop ramps* have tighter radius curves, which slow traffic and reduce capacity. Furthermore, large trucks tend to slow more on tight inside loop ramps that require longer deceleration and acceleration tapers with the main roadways. *Outside loop ramps* have larger radii, higher design speeds, and higher capacities. Inside loop ramps should be used for lower volume movements and outside loop ramps for higher volume movements.

Partial or full cloverleaf designs are generally discouraged because of the short weaving distance between the entering and exiting loop ramps, additional travel distance for left-turning traffic, and the relatively large amount of land required. However, cloverleaf designs remain popular as they are easier for drivers to understand than more complex ramp configurations. For this reason, cloverleaf designs can, and do, work well in rural or suburban locations with low to moderate traffic volumes and nearly equal traffic movements in all turning directions.

A fully directional interchange can be designed with all ramps at a sufficiently broad radius so that traffic can move from one freeway to another with little reduction in speed. This type of interchange provides the least delay in heavily trafficked segments of a freeway system and is used extensively where real estate and topography allow.

Advantages and disadvantages of the major interchange types are discussed in further detail in *Green Book* Chap. 10.

Intersection and Grade Separation Warrants

Interchanges and grade separations can be useful solutions to reduce traffic congestion and improve safety. However, they are also costly. To determine whether or not an interchange is justified, the *Green Book* presents six warrants to consider. Warrants for grade separations are more general than signal warrants.

1. *Design designation:* This is the universal standard for the interstate highway system. It allows no at-grade intersection to occur at any point along a designated interstate highway. Although there are many factors that influence safety, such as access control, provision of medians, and elimination of parking and pedestrian traffic, grade separation yields the largest increment of safety. Therefore, any intersection that is part of a freeway should consider grade separation.
2. *Bottleneck* or *spot congestion relief:* Bottlenecks and spot congestion occur when there is insufficient capacity at intersections along heavily traveled roads. Interchanges and partial interchanges are often justified to eliminate delays caused by necessary signalization. However, it is a poor use of construction and maintenance resources to use grade separation when lesser means of access are adequate. Therefore, this warrant is usually satisfied by thorough traffic capacity and economic analysis.
3. *Safety improvement:* This warrant is often related to other conditions, such as poor sight distance or poor topographic conditions, that result in crash-prone intersection designs. A grade separation or interchange may be warranted if other more expensive methods of eliminating crashes are impractical or

not likely to be effective. Safety is an especially significant factor when considering crossing grade separation of railroads and highways.

4. *Site topography:* There are locations, particularly in hilly or mountainous areas, where an at-grade intersection cannot be made with a reasonably acceptable alignment to accommodate the expected traffic. In these conditions, and when alternate access is not available, a grade separation may be warranted.
5. *Road-user benefits:* The user cost of additional travel distance caused by interchange geometry can be outweighed by the cost of delay caused by signalization at a busy intersection. Using benefit-cost analysis, the road-user benefits are compared to the cost of construction. The greater the ratio of benefits to costs, the greater the justification for construction of the interchange.
6. *Traffic volume:* Current traffic volume or projected increased volume may be sufficient to warrant an intersection or grade separation. At-grade intersections that have volumes in excess of capacity and cross streets with high volumes of traffic benefit the most from improvements in the movement of traffic.

Underpass and Overpass Roadways

When a highway grade separation is proposed, the engineer must decide whether an overpass roadway or an underpass roadway should be built. Topography must be considered when making this decision. When an engineer looks at the topography of a site, one of three things will generally happen: (1) the influence of the topography will be great, making it necessary to closely fit the design to it; (2) the topography will favor neither an overpass nor an underpass; or (3) the alignment and gradeline of one of the highways will predominate, making it necessary to design with respect to that highway's alignment instead of the topography of the site.

When topography is not the governing factor, there are other factors that may be considered. Underpass and overpass roadways each have advantages and disadvantages. While the discussion of relative merit can be rather extensive, the basic advantages and disadvantages are as follows.

Overpass roadways

- provide greater visibility to traffic conditions ahead of the structure
- provide greater vertical clearance for large loads
- reduce the advance visibility of the upcoming interchange to approaching drivers, reducing advanced notice of upcoming road conditions
- reduce the visibility of approaching and departing ramps to and from the roadway below
- increase the propensity for icing of the bridge deck (i.e., road surface) compared to the adjoining at-grade roadway
- require sturdy bridge railings to guide the movement of errant vehicles and prevent them from driving off the structure
- require careful control of drainage to avoid discharge onto the roadway below and to avoid undermining of slopes and footings from concentrated discharge points
- require more ownership and maintenance responsibility than approach roadways

Underpass roadways

- reduce the visibility of the roadway ahead of the bridge structure
- reduce the vertical clearance for large loads
- increase the advance visibility of the upcoming interchange to approaching drivers, alerting drivers to possible traffic condition changes
- increase the visibility of the departure and arrival ramps, which assist the anticipation of upcoming traffic events
- decrease the initial tendency of the roadway surface under the bridge to form ice, but also reduce the melting effect of heat from the sun
- require adequate abutment and pier setbacks, along with roadside safety protection devices for immovable piers and wall faces
- may require supplemental underbridge lighting during the daytime
- tend to collect trash and debris more than open roadways
- may require additional drainage structures, especially if the underpassing roadway dips down for additional clearance under the bridge
- may require less ownership and maintenance responsibility than an overhead structure at the same location

16. RAILROAD ENGINEERING

Railroad Design

Since railroads predate highways, railroads and highways share the same fundamentals of curves, spirals, grades, and vehicle dynamics, with the exception of the following two principles.

- Railroad tracks have a fixed *gauge*, or distance, between rails. Most measurements along the track take into account the fixed gauge and use one or the other rail as the baseline for geometry.

- Railroad (and street railway) equipment can tolerate less abrupt changes in geometry than vehicles on a roadway. To accommodate these limitations, grades are not as steep, curves are broader, and transitions are longer.

An overriding factor with all railroad design is the limited friction generated by a steel wheel on a steel rail compared to a vehicle's rubber tire on hard pavement. This factor, along with much larger vehicle masses and greater pounds per motive horsepower of trains, directly affects considerations given to grades, stopping distances, acceleration rates, and the effects of vehicle momentum. Certain rules of thumb prevail with railroad designs, such as chord definition curve layouts rather than arc definition curve layouts and superelevation based on inches (millimeters) of outer rail elevation above the inner rail. Because axle loadings have increased greatly since the steam locomotive era, accurate rail alignment and grade control are increasingly important on modern railroads.

Typical railroad corridor design starts with chord definition curves. Grades and profiles may be evaluated using economic analysis over a several mile (kilometer) long segment to determine the most economical construction and operating cost. If the rail line is to haul passengers, an analysis of station spacing, line haul speeds, and capacity may be required.

Horizontal Curves

Chord definition curves are widely used in railroad design. The chord definition uses deflection of a 100 ft (30 m) chord instead of an arc. (See Sec. 4.5 for chord definition equations). Railroad surveyors have easy access to tables used for layout by chord deflections, precluding the need for error-prone calculations while in the field. Modern technology is also playing a role in decreasing errors, as the newest innovations in field layout employ GPS, RTK, and other remote sensing systems, which greatly improve productivity and decrease track downtime. Remote sensing is particularly applicable to railroad conditions because of the large distances involved along narrow corridors that often limit traditional line-of-sight methods.

Transitions into and out of superelevation are accomplished more gently than on highways. Additionally, when placing reverse curves (i.e., a track curving one way, then instantly curving the opposite direction) back-to-back, it is necessary to introduce a section of tangent track in order to accomplish tangent runout for each curve and also to allow the car couplers to align themselves for the next curve without causing a derailment. The minimum tangent between reverse curves is usually as long as the longest car, or the sum of the two superelevation runouts, whichever is longer. A rule of thumb is to use 100 ft (30 m) of tangent between curves, including spiral transitions.

Railroad designers must be aware of potential derailment points, such as at a *point of continuing curve*, PCC, or a PT with a sharp curve adjacent to a broad curve or tangent. When a long car (e.g., an 80 ft (24.5 m) automobile carrier) is coupled to a short car, the drawbar develops a sharp angle with the car centerline, which can pull the long car off the track at the end of the curve if no spiral easement is provided. Designs where this problem may occur must either add transitions at these points of derailment or lengthen existing transitions to reduce the risk of derailment. These design changes can be expensive, so engineers should do their best to make sure they are included in the initial design.

Grades

Grades have a greater effect on railroads than on roadways because of the higher horsepower requirements necessary to move heavy loads and the limits of friction on steel rail. With heavy freight and high-speed passenger lines, grades of 1% are considered maximum, if not excessive, for the longest and heaviest trains. Branch lines and feeder lines using short trains and with industrial access may have grades of 4% or more for short stretches. Rapid transit lines can operate on somewhat steeper grades than locomotive-hauled commuter trains, although grades are usually limited to a maximum of 4%. Grades on street railway operations are much steeper and can reach 15% for short distances. As a practical limit, street railways try to keep grades below 8%. When operation on steep or very steep grades is necessary, the vehicles must have large, powerful motors, and every axle must be powered.

When designing railroad grades, the rise in feet per mile (meters per kilometer) is often cited, such as a 528 ft rise per mi (10 m rise per km) for a 1% grade. When a 1 mi (1.6 km) long train is climbing a 1% grade, the locomotive is more than 500 ft (152 m) above the last car, meaning the locomotive horsepower must be sufficient to lift the entire train 500 ft (152 m) for each mile traveled. The grade and total train weight determine the horsepower needed to pull the load at a given speed or efficiency, and grade determinations on railroads often involve a thorough analysis to balance a variety of requirements, such as the cost of building a bridge or tunnel to change a grade versus the horsepower and operating cost requirements of moving the load.

Train operations in hilly country involve a series of carefully planned brake applications and full throttle operations so that train momentum can carry the train over short, steep segments. The locomotive may put out full power while going downhill to haul the rear of the train up the last hill and may apply brakes while going uphill in preparation for the next speed restriction several miles ahead. Improper braking technique or throttle application can derail a train through a sharp curve or break the couplings (i.e., connections) between cars.

Should a train be required to stop on a steep uphill grade, it may be necessary to uncouple cars from the rear to allow the locomotive to pull the attached cars uphill. The locomotive must then uncouple from the cars and return to pull the remainder of the train uphill. This is a time-consuming operation and is highly undesirable. Therefore, the signal location and spacing must be carefully planned in order to avoid stopping a train on an uphill climb. In contrast, for rapid transit or light rail operations, the maximum grade at any point on the line should not exceed the capacity of the vehicle or train to start from a standing stop. Backing a stalled passenger train down a hill against the prevailing traffic is an invitation for disaster that no transit operator wants to face.

17. BIKEWAY GEOMETRIC DESIGN

Bikeway Widths and Clearances

The minimum operating space for a bicycle rider should be 3.5 ft (1.1 m) wide and 8.0 ft (2.4 m) high, as suggested by the FHWA's guidelines for LOS C conditions. The minimum operating width for a single-lane, one-way bikeway should be 5.0 ft (1.5 m). For a path accommodating a bicycle lane and a pedestrian lane, the minimum width should be 6.5 ft (2.0 m), which allows a 3.0 ft (0.9 m) lane width for pedestrians.

For two or more lanes of bicycle traffic, multiples of the 3.5 ft (1.1 m) operating space width should be used. While reduction in lane width is allowed for pedestrian facilities, reductions are not recommended for multilane bikeways. Bicyclists need more space than pedestrians, as bicycling requires a certain amount of lateral movement for bicyclists to regain and maintain balance. Therefore, lateral obstructions (e.g., walls, light poles, railings, and vegetation) should be kept at least 2 ft (0.6 m) away from the edge of the pathway.

Opposing bicycle traffic designs should provide more lateral space between the opposing lanes than between same-direction, multilane paths. The minimum recommended width for a two-way, two-lane bikeway is 8.0 ft (2.4 m), increasing to 10 ft (3.0 m) in locations where there are higher volumes, higher speeds, or sharper curves. The minimum width for a three-lane, one-way bikeway should be 10 ft (3.0 m). Minimum vertical clearance of 8.0 ft (2.4 m) is necessary for seated riders. Preferred vertical clearance is 10 ft (3.0 m), which allows riders to pass under an obstruction while standing.

Design Speeds

Actual travel speeds vary greatly for bicycle riders. Recreational riders and family groups may travel at a leisurely 8–10 mph (13–16 kph) on level paths, while more serious riders may travel at speeds ranging from 14 mph to 25 mph (23 to 40 kph). Serious riding enthusiasts who are used to traveling long distances and/or performing physical workouts may regularly travel at speeds of 30 mph (48 kph) or greater.

A design speed of 20 mph (32 kph) will accommodate the majority of riders on level or nearly level ($<1.5\%$) grades with smooth surfaces. For unpaved or rough surfaces along level paths, the design speed can be reduced to 15 mph (24 kph) because riding speeds tend to be lower along these surfaces. Usually, experienced, higher-speed bicyclists will compensate quickly for varying travel path conditions or sudden restrictions in geometry.

For grades greater than 4%, the minimum design speed should be increased to 30 mph (48 kph). Uphill riders may operate at lower speeds, while downhill riders may take advantage of the grade and operate at higher speeds.

Stopping Distance

Bicycles require the same stopping distances as automobiles, and the formulas used to determine bicycle stopping distances are the same as AASHTO's formulas for automobile stopping distances, with the exception that the coefficient of friction, f, is limited to 0.25 to account for the poor-weather braking characteristics of many bicycles.

A friction factor of 0.25 is equivalent to a deceleration rate of 0.25 g (8 ft/sec^2; 2.4 m/s^2), which is the recommended deceleration rate for bicycles. Using a more rapid deceleration rate, such as the 0.35 g (11.2 ft/sec^2; 3.4 m/s^2) deceleration rate used to determine stopping sight distances for highways (see *Green Book* Exh. 3-1), is not recommended for bicycles. Stopping at greater rates requires more skill to modulate the brakes in order to avoid skidding, and less experienced bicyclists may have difficulty maintaining balance should a skid maneuver be required. Furthermore, the small footprint of bicycle tires provides a less reliable braking surface than automobile tires.

Equation 4.81 gives the stopping distance for bicycles. The decimal grade, G, will be positive for an ascending grade and negative for a descending grade, which will decrease or increase the braking distance, respectively. The second term of Eq. 4.81 determines the distance traveled during 2.5 sec of perception-reaction time before braking begins. Stopping distances from selected speeds, v, are shown in Table 4.14 using the default deceleration rate of 0.25 g.

$$S_{\text{m}} = \frac{\text{v}_{\text{kph}}^2}{254(f \pm G_{\text{dec}})} + \frac{\text{v}_{\text{kph}}}{1.4} \quad \text{[SI]} \quad 4.81(a)$$

$$S_{\text{ft}} = \frac{\text{v}_{\text{mph}}^2}{30(f \pm G_{\text{dec}})} + 3.67\text{v}_{\text{mph}} \quad \text{[U.S.]} \quad 4.81(b)$$

Example 4.13

Using a friction factor of 0.25, what is the design stopping distance for a bicycle traveling at a speed of 20 mph on a 4% downgrade?

Geometric Design

Table 4.14 Bicycle Stopping Distance for Selected Design Speeds

	grade			
speed (mph)	−10%	−5%	0%	4%
10	60 ft	54 ft	50 ft	48 ft
15	105 ft	93 ft	85 ft	81 ft
20	165 ft	140 ft	127 ft	120 ft
25	230 ft	196 ft	175 ft	164 ft
30	310 ft	260 ft	230 ft	214 ft

(Multiply ft by 0.305 to obtain m.)
(Multiply mph by 1.61 to obtain kph.)

Solution

From Eq. 4.81(b),

$$\begin{aligned} S_{\text{ft}} &= \frac{\text{v}_{\text{mph}}^2}{30(f \pm G_{\text{dec}})} + 3.67\text{v}_{\text{mph}} \\ &= \frac{\left(20\ \frac{\text{mi}}{\text{hr}}\right)^2}{(30)(0.25 - 0.04)} + (3.67)\left(20\ \frac{\text{mi}}{\text{hr}}\right) \\ &= 137\ \text{ft} \end{aligned}$$

Horizontal Alignment

The minimum radius of horizontal curves depends on the design speed, v, and superelevation, e, of the bikeway. The equation relating the radius of the curve to design speed and superelevation is the same equation as is used with highways, Eq. 4.66.

Superelevation rates should be limited to a maximum of 5% (0.05 ft/ft or 0.05 m/m) for general purpose bikeways. Bicycles are harder to control on slippery surfaces than four-wheeled vehicles, so the minimum superelevation should be 2% (0.02 ft/ft or 0.02 m/m) to provide adequate drainage to eliminate standing water on the riding surface.

Side friction factors, f_s, should be no greater than 0.22–0.30 for hard paved surfaces and 0.11–0.15 for unpaved surfaces.

Sight distance around an obstruction on a horizontal curve can be determined using the sight distance equations for horizontal curves, Eq. 4.58 and Eq. 4.59. The actual stopping distance along the arc of the curve will be slightly greater than the straight-line sight distance past the obstruction.

Vertical Alignment

Vertical curves on bicycle lanes that follow street profiles will have adequate, if not generous, sight distance since streets are usually designed for speeds greater than those achieved by bicyclists. On the other hand, bikeways that don't parallel other facilities may be designed with much more abrupt vertical and horizontal alignments. Therefore, a bikeway should always be checked to ensure adequate sight distance.

Sight distance formulas for crest vertical curves (such as Eq. 4.82 and Eq. 4.83) are used with the eye height of the bicyclist set at 4.5 ft (1400 mm) and the height of the object set at 0. A is the absolute value of the algebraic difference in grade.

When the sight distance, S, is greater than the curve length, L, use Eq. 4.82 to calculate the minimum curve length.

$$L_{\text{m}} = 2S_{\text{m}} - \frac{900}{A_{\%}} \quad \text{[SI]} \qquad 4.82(a)$$

$$L_{\text{ft}} = 2S_{\text{ft}} - \frac{280}{A_{\%}} \quad \text{[U.S.]} \qquad 4.82(b)$$

When $S < L$, use Eq. 4.83.

$$L_{\text{m}} = \frac{A_{\%} S_{\text{m}}^2}{900} \quad \text{[SI]} \qquad 4.83(a)$$

$$L_{\text{ft}} = \frac{A_{\%} S_{\text{ft}}^2}{280} \quad \text{[U.S.]} \qquad 4.83(b)$$

Grades should not exceed 5% for long, unbroken distances. In hilly terrain, grades should be broken into 400–700 ft (122–213 m) segments of alternating steeper and flatter grades. Table 4.15 gives guidelines for determining maximum grades for extended segments of bikeways.

Table 4.15 Maximum Grade for Extended Bikeway Segments

length of segment (100 ft)	desirable maximum grade (%)	acceptable maximum grade (%)
0	7	10.8
1	6	9.5
2	5	8.3
3	4.3	7.4
4	3.5	6.6
5	3.2	6.2
6	2.9	5.7
7	2.8	5.4
8	2.7	5.2
9	2.6	5.0
10	2.5	4.8
12	2.4	4.6
14	2.2	4.3
16	2.0	4.0
18	1.8	3.8
20	1.6	3.5

(Multiply ft by 0.305 to obtain m.)

Example 4.14

A bicycle path has an upgrade of 2% meeting a downgrade of 3%. The design speed is 15 mph, and the friction factor is 0.25. What is the minimum vertical curve length at this location?

Solution

This is a crest vertical curve. The total grade change is

$$\begin{aligned} A &= |G_2 - G_1| = |-3\% - 2\%| \\ &= 5\% \end{aligned}$$

Use the total decimal grade change value, −0.05, for G. From Eq. 4.81(b), the stopping distance is

$$\begin{aligned} S_{\text{ft}} &= \frac{\text{v}_{\text{mph}}^2}{30(f \pm G_{\text{dec}})} + 3.67\text{v}_{\text{mph}} \\ &= \frac{\left(15\ \frac{\text{mi}}{\text{hr}}\right)^2}{(30)(0.25 - 0.05)} + (3.67)\left(15\ \frac{\text{mi}}{\text{hr}}\right) \\ &= 92.6\ \text{ft} \end{aligned}$$

Assume the stopping distance is greater than the length of the vertical curve. From Eq. 4.82(b),

$$\begin{aligned} L_{\text{ft}} &= 2S_{\text{ft}} - \frac{280}{A_{\%}} = (2)(92.6\ \text{ft}) - \frac{280}{5\%} \\ &= 129\ \text{ft} \end{aligned}$$

Check for $S < L$ using Eq. 4.83(b).

$$\begin{aligned} L_{\text{ft}} &= \frac{A_{\%} S_{\text{ft}}^2}{280} = \frac{(5\%)(92.6\ \text{ft})^2}{280} \\ &= 153\ \text{ft} \end{aligned}$$

Since 92.6 ft is less than 153 ft (i.e., $S < L$), the required length of the vertical curve is 153 ft.

18. MASS TRANSIT GEOMETRIC DESIGN

Because transit involves moving people, parameters of human comfort are the design criteria for acceleration, deceleration, and lateral forces on vehicles. Urban public transit usually has more restrictive design limits than automobiles to accommodate the comfort and safety of standing passengers. As a general rule, average accelerations should be limited to ±0.1 g in any direction, but design values should be selected carefully based on the type of service being provided.

Acceleration when starting and deceleration when stopping are also limited by passenger comfort. Starting acceleration can range from 0.05 g to 0.1 g for rapid transit, light rail, and buses, with the higher range of acceleration limited by horsepower and traction rates. Passengers have a somewhat higher tolerance for deceleration during braking situations. Common deceleration rates range from 0.1 g to 0.3 g. The higher range is limited by braking power, traction (on rail vehicles), and passenger comfort. Overhead electrically powered vehicles usually have greater acceleration capability than diesel-only powered vehicles, while rubber-tired vehicles have greater deceleration capabilities than rail-mounted vehicles.

Buses operating on city streets and highways follow the existing road geometry, with operators trained to adjust for the needs of passenger comfort. Busways and BRT roadways can be designed more like rail lines for improved comfort. Horizontal curves follow usual highway or railroad practice for superelevation and curvature. For highway curves, the design limits of comfort are covered in the AASHTO *Green Book*, and Eq. 4.66 can be used. For BRT, the side friction factor can be reduced from the values used on highways to reduce sidesway and improve the comfort of higher level seating found on buses.

The transition from an upgrade to a downgrade, and vice versa, is accomplished through a vertical curve. For train passengers on vertical curves, the vertical acceleration tolerated is less than that tolerated for highway design. The American Railway Engineering and Maintenance-of-Way Association (AREMA) has suggested Eq. 4.84 be used, where L is the length of the vertical curve, D is the absolute value of the algebraic difference in grades, expressed in decimals, v is the speed in miles per hour, and a is the vertical acceleration in feet per second squared. The recommended vertical acceleration, a, is 0.6 ft/sec^2 (0.18 m/sec^2) for passenger lines, including transit.

$$L_{\text{ft}} = D\text{v}_{\text{mph}}^2 \left(\frac{2.15}{a_{\text{ft/sec}^2}}\right) \quad \text{[U.S. only]} \qquad 4.84$$

Longer vertical curve lengths are often rounded to the nearest 100 ft (30 m) for ease of construction when there is negligible effect on the vertical acceleration.

Example 4.15

A bus travels at 40 mph on a curve that is superelevated 3%. The side friction factor is 0.05. What is the minimum radius for this curve?

Solution

From Eq. 4.66(b), the minimum curve radius is

$$\begin{aligned} R_{\text{ft}} &= \frac{\text{v}_{\text{mph}}^2}{15(e + f_s)} = \frac{\left(40\ \frac{\text{mi}}{\text{hr}}\right)^2}{(15)(0.03 + 0.05)} \\ &= 1333\ \text{ft} \end{aligned}$$

Example 4.16

A +2.0% grade meets a −1.5% grade on a transit line with a design speed of 70 mph. Using the recommended vertical acceleration of 0.6 ft/sec^2, what is the length of vertical curve required?

Solution

Using Eq. 4.84, the vertical curve length required is

$$L_{\text{ft}} = D\text{v}^2_{\text{mph}}\left(\frac{2.15}{a_{\text{ft/sec}^2}}\right)$$

$$= |-0.015 - 0.02|\left(70\ \frac{\text{mi}}{\text{hr}}\right)^2\left(\frac{2.15}{0.6\ \frac{\text{ft}}{\text{sec}^2}}\right)$$

$$= 615\ \text{ft}$$

The length of curve needed is 615 ft. However, the length can be rounded to 600 ft for ease of construction with negligible effect on the vertical acceleration.

PRACTICE PROBLEMS

1. An existing 1.5° curve will be shortened by intersecting a new back tangent at a bearing 15° north from the existing back tangent, which is at a bearing of N 80° E. The PC will relocate to PC′, and the PI will relocate to PI′. The PT will remain in its existing location at N 10,000 ft and E 50,000 ft, and the curve from the new PC location to the PT will also remain in place. The PT is at sta 38+27.00, and the ahead tangent is at bearing N 55° E.

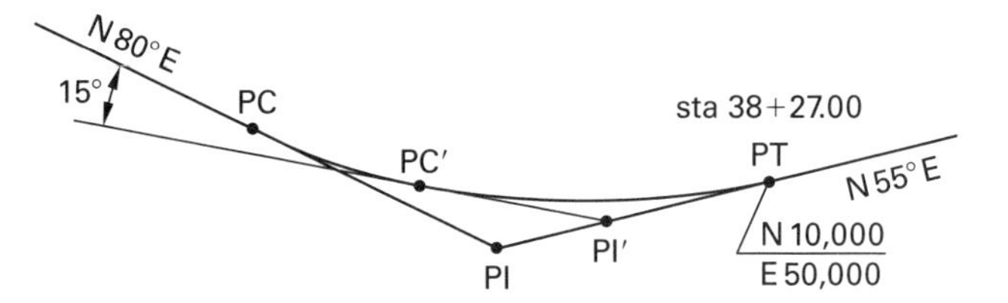

(a) The new curve length is most nearly

(A) 15 ft

(B) 670 ft

(C) 1700 ft

(D) 3800 ft

(b) The radius of the curve is most nearly

(A) 330 ft

(B) 670 ft

(C) 3800 ft

(D) 8600 ft

(c) The north coordinate of the shifted PI′ is most nearly

(A) 330 ft

(B) 9400 ft

(C) 9700 ft

(D) 9800 ft

(d) The station of the PI′ is most nearly

(A) sta 28+26

(B) sta 31+60

(C) sta 34+93

(D) sta 34+95

2. A 2100 ft radius curve is being fitted to a PI located at sta 234+34.10, with the back tangent bearing at N 80° E and the ahead tangent at S 65° E. Offset to the right of, and parallel to, the back tangent is a historic structure that has a 150 ft clearance buffer. The PI is located 100 ft back from the offset point of the corner of the structure. The proposed road construction consists of a six-lane roadway with 12 ft lanes, 10 ft shoulders, and a 16 ft wide median. The edge of the paved shoulder cannot encroach on the buffer zone.

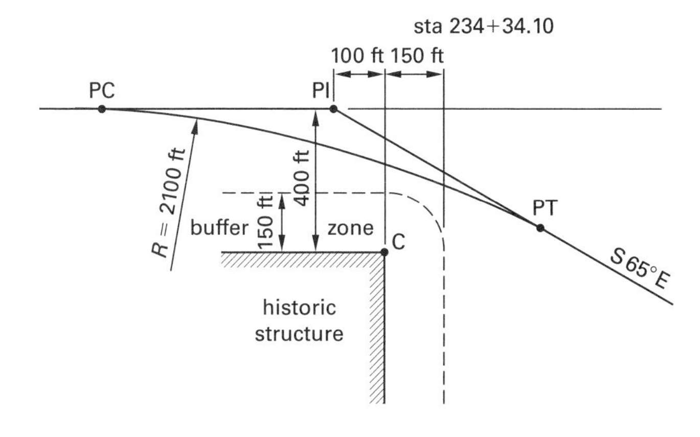

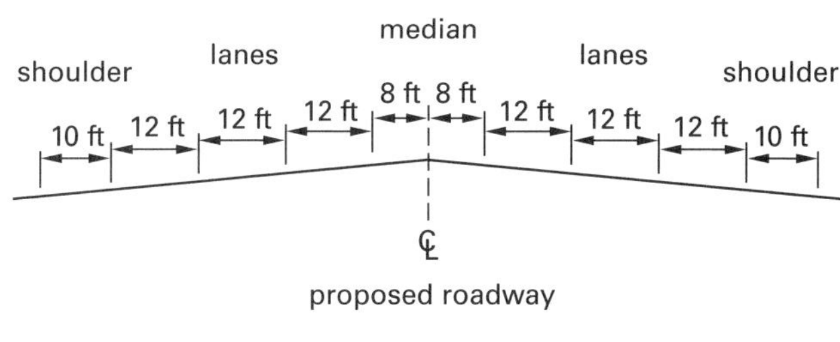

(not to scale)

(a) The actual clearance to the corner of the historical building is most nearly

(A) 110 ft

(B) 180 ft

(C) 200 ft

(D) 240 ft

(b) The station of the closest point to the building corner is most nearly

(A) sta 233+05

(B) sta 236+44

(C) sta 236+57

(D) sta 239+95

(c) The offset distance from the back tangent to the closest point to the building on the curve centerline is most nearly

(A) 140 ft

(B) 160 ft

(C) 180 ft

(D) 260 ft

3. An existing 250 ft long crest vertical curve will be connected to new roadway construction, replacing the existing part of the curve from the PVI to the PVT. The new ahead curve tangent will be 350 ft long on a −1.5% grade, replacing the existing 125 ft tangent on a −3.0% grade. The existing pavement will remain from the PVC to the PVI location, which is at elevation 105.25 ft. An existing utility pipe crosses the centerline at 70 ft ahead from the PVI location. The pipe has a cover of 2.0 ft to the existing grade line.

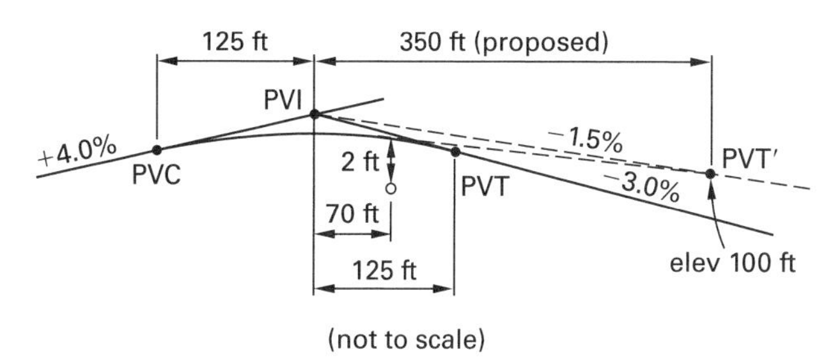

(not to scale)

(a) The middle ordinate of the existing curve is most nearly

(A) 2.2 ft

(B) 2.5 ft

(C) 2.9 ft

(D) 3.7 ft

(b) The elevation of the top of the pipe for the existing curve is most nearly

(A) 99.9 ft

(B) 100.7 ft

(C) 102.7 ft

(D) 103.2 ft

(c) The new grade line will provide a new pipe cover of most nearly

(A) 2.1 ft

(B) 5.1 ft

(C) 7.1 ft

(D) 13 ft

(d) The design speed for the new curve based on sight distance is most nearly

(A) 40 mph

(B) 50 mph

(C) 55 mph

(D) 60 mph

4. A horizontal curve with full spiral transitions has a center curve of 3° and is fitted with 150 ft spirals. The curve deflection is 31° to the right. Superelevation is to be transitioned uniformly along the spiral, with a maximum superelevation rate of 8%. The normal straight cross slope for this roadway is 2%.

(a) The curve tangent of this curve is most nearly

(A) 150 ft

(B) 530 ft

(C) 610 ft

(D) 760 ft

(b) The instant radius at the midpoint of the spiral is most nearly

(A) 960 ft

(B) 1900 ft

(C) 2500 ft

(D) 3800 ft

(c) The superelevation at a point 100 ft from the TS or ST is most nearly

(A) 0.03 ft/ft

(B) 0.06 ft/ft

(C) 0.07 ft/ft

(D) 0.08 ft/ft

5. An exit ramp from an urban freeway terminates on a 6% downgrade. The ramp has a 300 ft radius with a 75° curve to the right and a superelevation of 8.0%. There is a stop-controlled cross street intersection 200 ft downstream of the PT of the ramp curve. The intersection is not visible from the ramp until a vehicle is about two-thirds around the curve. The tangent is 250 ft long between the gore nose of the ramp (i.e., the intersection of the freeway and ramp pavements) and the PC of the curve. The ramp is also on a 200 ft vertical curve, approximately aligned with the horizontal curve. The freeway and ramp grade is +0.25% leading into the vertical curve. The ramp is elevated on a structure with 42 in high parapets on each side and a width of 18 ft between parapets. The ramp exit lane is 1500 ft long and is an extension of an upstream entrance ramp. Design speed and prevailing traffic speed on the adjoining freeway through lanes is 50 mph. Several serious collisions have occurred, including vehicles skidding into the parapet on the curve due to maintaining a speed of 50 mph and vehicles on the exit ramp broadsiding cross traffic at the stop-controlled intersection. The exit ramp has a posted speed limit of 30 mph using *MUTCD* standard exit sign positions. The ramp is being investigated for possible design deficiencies.

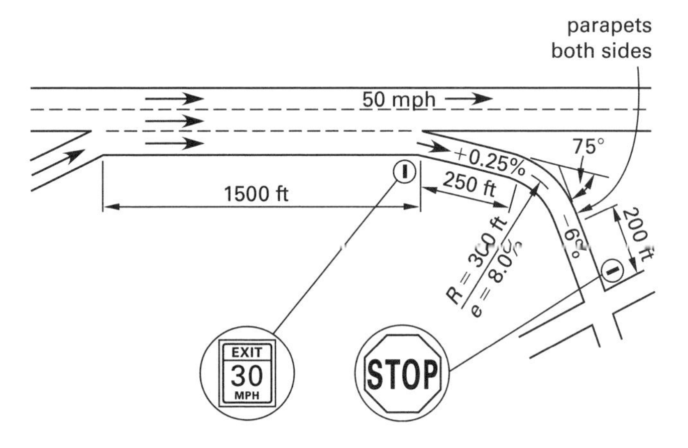

(a) The exit ramp design speed is most nearly

(A) 25 mph

(B) 30 mph

(C) 35 mph

(D) 40 mph

(b) If drivers cannot see over the ramp parapets, what is most nearly the stopping sight distance to the intersection?

(A) 70 ft

(B) 130 ft

(C) 200 ft

(D) 330 ft

(c) Using actual vehicle speed on the ramp, what is most nearly the decision sight distance for the ramp curve?

(A) 620 ft

(B) 720 ft

(C) 890 ft

(D) 1000 ft

(d) For vehicles on the exit ramp, what is most nearly the decision sight distance to the stop-controlled intersection?

(A) 220 ft

(B) 490 ft

(C) 620 ft

(D) 910 ft

(e) Based on the decision sight distance, what speed should be posted on the exit ramp?

(A) 15 mph

(B) 25 mph

(C) 30 mph

(D) 35 mph

6. An urban freeway with two lanes in each direction has two closely spaced exit ramps to local streets. The freeway also splits into separate single-lane roadways that connect to other highways. The exit ramps and roadway splits occur within a 1200 ft segment. The exit ramps are level and terminate in stop-controlled intersections, while the lane connections to other highways are free-flowing except for traffic backups. Traffic on the exit ramps backs up onto the main roadway at various times throughout the day. The backups do not necessarily occur simultaneously on both ramps, forcing approaching drivers to choose between ramp options. The approach speed on the freeway is 50 mph, the connections to other highways operate at 40 mph, and the ramps have a posted speed limit of 25 mph.

(a) What is most nearly the minimum stopping sight distance for the exit ramps to local streets?

(A) 120 ft

(B) 160 ft

(C) 200 ft

(D) 250 ft

(b) What is most nearly the decision sight distance required for the exit ramps?

(A) 430 ft

(B) 490 ft

(C) 930 ft

(D) 1000 ft

(c) What is most nearly the stopping sight distance for the ramps connecting onto other highways?

(A) 310 ft

(B) 430 ft

(C) 590 ft

(D) 690 ft

(d) What is most nearly the decision sight distance required for the lane connections to other highways?

(A) 310 ft

(B) 620 ft

(C) 830 ft

(D) 1000 ft

(e) What is most nearly the decision sight distance for the main freeway approach to the ramps and lane splits?

(A) 200 ft

(B) 620 ft

(C) 830 ft

(D) 1000 ft

SOLUTIONS

1. (a) Find the deflection angle of the curve, which is the same as the intersection angle of the curve tangents, I. First, the bearing of the new back tangent must be found. The existing back tangent lies in the NE quadrant. Therefore, shifting the back tangent 15° north results in a counterclockwise shift.

$$\text{N}\,80^\circ\,\text{E} - 15^\circ = \text{N}\,65^\circ\,\text{E}$$

The deflection angle of the new curve is found by subtracting the back tangent from the ahead tangent.

$$\text{N}\,55^\circ\,\text{E} - \text{N}\,65^\circ\,\text{E} = -10^\circ$$

A negative difference in the NE quadrant indicates that the deflection is to the left, or counterclockwise.

The length of the curve arc is determined using Eq. 4.9.

$$\begin{aligned} L_{\text{ft}} &= \frac{(100\text{ ft})I}{D_{\text{deg}}} = \frac{(100\text{ ft})(10^\circ)}{1.5^\circ} \\ &= 666.67\text{ ft} \quad (670\text{ ft}) \end{aligned}$$

The answer is (B).

(b) The radius of the curve can be found using Eq. 4.16.

$$\begin{aligned} R_{\text{ft}} &= \frac{(100\text{ ft})\left(\frac{180^\circ}{\pi}\right)}{D_{\text{deg}}} \\ &= \frac{(100\text{ ft})\left(\frac{180^\circ}{\pi}\right)}{1.5^\circ} \\ &= 3819.72\text{ ft} \quad (3800\text{ ft}) \end{aligned}$$

The answer is (C).

(c) The length of the curve tangent is found using Eq. 4.5.

$$\begin{aligned} T &= R\tan\frac{I}{2} = (3819.72\text{ ft})\tan\frac{10^\circ}{2} \\ &= 334.18\text{ ft} \end{aligned}$$

The difference in the north coordinate is found by multiplying the tangent distance by the cosine of the bearing of the heading. The bearing of the heading from the PT to the shifted PI is S 55° W.

$$\begin{aligned} \text{PI}'\text{ N} &= \text{PT N} - T\cos(\text{heading}) \\ &= 10{,}000.00\text{ ft} - (334.18\text{ ft})\cos 55^\circ \\ &= 9808.32\text{ ft} \quad (9800\text{ ft}) \end{aligned}$$

The answer is (D).

(d) The station of the PI′ is found by subtracting the length of the curve from the PT station to get the PC′ station, then adding the tangent length to the PI′ station.

$$\begin{aligned} \text{sta PT} - L + T &= \text{sta } 38{+}27.00 - 666.7\text{ ft} + 334.18\text{ ft} \\ &= \text{sta } 34{+}94.51 \quad (\text{sta } 34{+}95) \end{aligned}$$

The answer is (D).

2. (a) It is necessary to determine the tangent of the curve, which is found from the intersection angle and the radius. The intersection angle is found by subtracting the bearing angles.

$$I = (180^\circ - 80^\circ) - 65^\circ = 35^\circ$$

The curve tangent length can be determined using Eq. 4.5.

$$T = R\tan\frac{I}{2} = (2100\text{ ft})\tan\frac{35^\circ}{2} = 662.13\text{ ft}$$

A triangle is constructed using known distances. Side $\overline{\text{HC}}$ is along the historical structure face and is parallel to the back tangent of the curve.

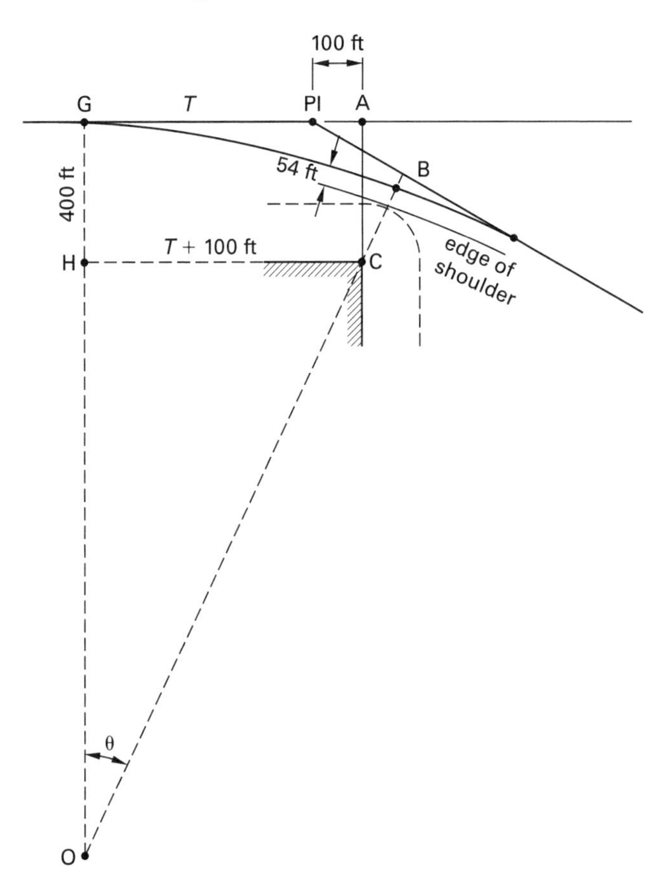

$$\overline{GA} = T + \overline{PI\ A} = 662.13\text{ ft} + 100\text{ ft} = 762.13\text{ ft}$$

$$\overline{OH} = R - \overline{GH} = 2100\text{ ft} - 400\text{ ft} = 1700\text{ ft}$$

$$\theta = \arctan\frac{\overline{GA}}{\overline{OH}} = \arctan\frac{762.13\text{ ft}}{1700\text{ ft}} = 24.15°$$

$$\overline{OC} = \sqrt{\overline{GA}^2 + \overline{OH}^2} = \sqrt{(762.13\text{ ft})^2 + (1700\text{ ft})^2} = 1863.02\text{ ft}$$

$$\overline{CB} = R - \overline{OC} = 2100\text{ ft} - 1863.02\text{ ft} = 236.98\text{ ft}$$

The clearance length, L, is $\overline{CB}$, 236.98 ft, minus the width of the roadway, W_r, measured from the centerline to the edge of the paved shoulder. As shown in the roadway profile illustration, the median width is the distance from the centerline to the edge of the median, 8 ft.

$$W_r = 8\text{ ft} + (3)(12\text{ ft}) + 10\text{ ft} = 54\text{ ft}$$

$$L = \overline{CB} - W_r = 236.98\text{ ft} - 54\text{ ft} = 182.98\text{ ft}\quad(180\text{ ft})$$

The answer is (B).

(b) The arc distance from the PC to the radial clearance point, B, is

$$\overline{PC\ B} = \frac{\theta R}{\frac{180°}{\pi}} = \frac{(24.15°)(2100\text{ ft})}{\frac{180°}{\pi}} = 885.14\text{ ft}$$

The station of the closest radial point is

$$\begin{aligned}\text{sta PI} - T + \text{PC B} &= \text{sta } 234{+}34.10 - 662.13\text{ ft} + 885.14\text{ ft}\\ &= \text{sta } 236{+}57.11\quad(\text{sta } 236{+}57)\end{aligned}$$

The answer is (C).

(c) The offset from the tangent to the radial point on the curve is shown as $\overline{EB}$ in the following illustration. The distance $\overline{AE}$ is the distance beyond the offset point to the corner of the historical structure.

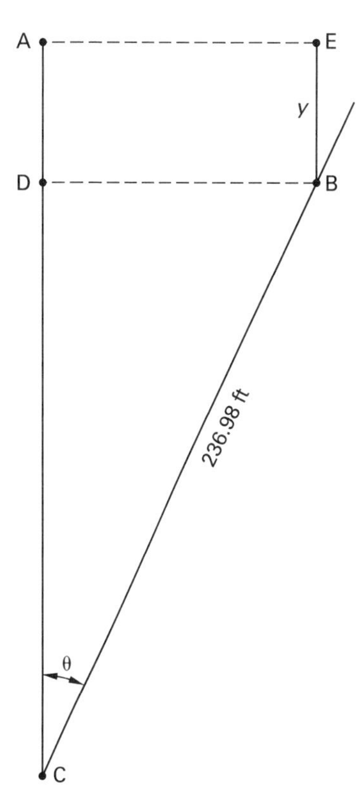

$$\overline{AE} = \overline{CB}\sin\theta = (236.98\text{ ft})\sin 24.15° = 96.95\text{ ft}$$

$$\overline{PI\ A} + \overline{AE} = 100\text{ ft} + 96.95\text{ ft} = 196.95\text{ ft}$$

The distance from the PC to point E is

$$x = T + (\overline{PI\ A} + \overline{AE}) = 662.13\text{ ft} + 196.95\text{ ft} = 859.08\text{ ft}$$

The offset, y, from a tangent can be found using Eq. 4.29.

$$\begin{aligned}y &= R - \sqrt{R^2 - x^2}\\ &= 2100\text{ ft} - \sqrt{(2100\text{ ft})^2 - (859.08\text{ ft})^2}\\ &= 183.76\text{ ft}\quad(180\text{ ft})\end{aligned}$$

The answer is (C).

3. (a) The middle ordinate is determined using Eq. 4.44.

$$M_{\text{ft}} = \frac{|G_{2,\%} - G_{1,\%}|L_{\text{sta}}}{8\ \frac{\text{sta}}{\text{ft}}} = \frac{|-3.0\% - 4.0\%|(250\text{ ft})}{\left(8\ \frac{\text{sta}}{\text{ft}}\right)\left(100\ \frac{\text{ft}}{\text{sta}}\right)} = 2.188\text{ ft}\quad(2.2\text{ ft})$$

The answer is (A).

Geometric Design

(b) The elevation of the top of the pipe can be found after determining the elevation of the PVC.

$$\begin{aligned} \text{elev}_{\text{PVC}} &= \text{elev}_{\text{PVI}} - G_{1,\text{ft/ft}} T_{1,\text{ft}} \\ &= 105.25 \text{ ft} - \left(0.04 \ \frac{\text{ft}}{\text{ft}}\right)(125 \text{ ft}) \\ &= 100.25 \text{ ft} \end{aligned}$$

Use Eq. 4.45 to find the rate of change in grade.

$$\begin{aligned} R &= \frac{G_{2,\%} - G_{1,\%}}{L} = \frac{-3.0\% - 4.0\%}{250 \text{ ft}} \\ &= -0.028 \ \%/\text{ft} \end{aligned}$$

Use Eq. 4.47 to find the elevation of the curve above the pipe. The pipe is located 125 ft + 170 ft = 195 ft from the PVC.

$$\begin{aligned} \text{elev}_x &= \frac{R_{\%/100 \text{ ft}} x_{\text{ft}}^2}{2} + G_{1,\text{ft/ft}} x_{\text{ft}} + \text{elev}_{\text{PVC}} \\ &= \frac{\left(\frac{-0.028\%}{100 \text{ ft}}\right)(195 \text{ ft})^2}{2} + \left(0.04 \ \frac{\text{ft}}{\text{ft}}\right)(195 \text{ ft}) \\ &\quad + 100.25 \text{ ft} \\ &= 102.73 \text{ ft} \end{aligned}$$

The top of the pipe elevation is

$$102.73 \text{ ft} - 2.0 \text{ ft} = 100.73 \text{ ft} \quad (100.7 \text{ ft})$$

The answer is (B).

(c) The elevation of the PVI and the grade line at the PVI are remaining the same. Therefore, the existing middle ordinate must be used to determine the grade points on the new half of the curve. Determine the elevation of the new tangent above the pipe location.

$$\begin{aligned} \text{elev}_{\text{PVT}'} &= \text{elev}_{\text{PVI}} + G_2(70 \text{ ft}) \\ &= 105.25 \text{ ft} + (-0.015)(70 \text{ ft}) \\ &= 104.20 \text{ ft} \end{aligned}$$

To determine the elevation on the new grade line, subtract the proportional square of the distance ratio multiplied by the middle ordinate from the elevation on the tangent. Subtract the elevation of the top of the pipe from the elevation on the new grade line to determine the new pipe cover.

$$\begin{aligned} &\text{elev}_{x',\text{new grade}} - \text{elev}_{\text{top of pipe}} \\ &\quad = \text{elev}_{\text{PVT}'} - (\text{distance ratio})^2 M - \text{elev}_{\text{top of pipe}} \\ &\quad = 104.20 \text{ ft} - \left(\frac{350 \text{ ft} - 70 \text{ ft}}{350 \text{ ft}}\right)^2 (2.188 \text{ ft}) \\ &\qquad - 100.73 \text{ ft} \\ &\quad = 2.07 \text{ ft} \quad (2.1 \text{ ft}) \end{aligned}$$

The answer is (A).

(d) To find the value of K, use Eq. 4.57.

$$\begin{aligned} K &= \frac{L}{|G_2 - G_1|} = \frac{125 \text{ ft} + 350 \text{ ft}}{|-1.5\% - 4.0\%|} \\ &= 86.36 \end{aligned}$$

Referring to Table 4.7, a K-value between 84 and 114 for a crest vertical curve indicates a 50 mph design speed.

The answer is (B).

4. (a) The components of the curve tangent equation are found first. As a matter of practice, calculations involving components of a spiral transition must carry an additional significant digit to the right of the decimal. Usually, carrying out to three decimals in the customary U.S. system and four decimals in the SI system is sufficient to control location tolerance of the final measurement for layout purposes.

Use Eq. 4.41 to find the coordinate along the tangent.

$$\begin{aligned} x_{s,\text{ft}} &= \frac{L_{s,\text{ft}}}{100 \text{ ft}} \left(\begin{array}{l} 100 \text{ ft} - \dfrac{(100 \text{ ft})\left(\frac{\pi}{180^\circ}\right)^2 D_{c,\text{deg}}^2}{(5)(2!)} \\ \quad + \dfrac{(100 \text{ ft})\left(\frac{\pi}{180^\circ}\right)^4 D_{c,\text{deg}}^4}{(9)(4!)} \\ \quad - \dfrac{(100 \text{ ft})\left(\frac{\pi}{180^\circ}\right)^6 D_{c,\text{deg}}^6}{(13)(6!)} \end{array} \right) \\ &= \frac{150 \text{ ft}}{100 \text{ ft}} \left(\begin{array}{l} 100 \text{ ft} - \dfrac{(100 \text{ ft})\left(\frac{\pi}{180^\circ}\right)^2 (3^\circ)^2}{(5)(2!)} \\ \quad + \dfrac{(100 \text{ ft})\left(\frac{\pi}{180^\circ}\right)^4 (3^\circ)^4}{(9)(4!)} \\ \quad - \dfrac{(100 \text{ ft})\left(\frac{\pi}{180^\circ}\right)^6 (3^\circ)^6}{(13)(6!)} \end{array} \right) \\ &= 149.95 \text{ ft} \end{aligned}$$

Geometric Design

Calculate the central angle of the total spiral arc using Eq. 4.38.

$$\theta_s = \left(\frac{L_{s,\text{ft}}}{200\ \text{ft}}\right) D_{c,\text{deg}} = \left(\frac{150\ \text{ft}}{200\ \text{ft}}\right)(3^\circ)$$
$$= 2.25^\circ$$

Calculate the radius of the central curve using Eq. 4.16.

$$R_{c,\text{ft}} = \frac{(100\ \text{ft})\left(\frac{180^\circ}{\pi}\right)}{D_{c,\text{deg}}} = \frac{(100\ \text{ft})\left(\frac{180^\circ}{\pi}\right)}{3^\circ}$$
$$= 1909.86\ \text{ft}$$

Find the extended distance along the tangent of the shifted curve using Eq. 4.35.

$$k = x_s - R_c \sin\theta_s = 149.96\ \text{ft} - (1909.86\ \text{ft})(\sin 2.25^\circ)$$
$$= 74.97\ \text{ft}$$

Calculate the offset distance of the full spiral length using Eq. 4.42.

$$y_{s,\text{ft}} = \frac{L_{\text{ft}}^3}{6R_{c,\text{ft}}L_{s,\text{ft}}} = \frac{(150\ \text{ft})^3}{(6)(1909.86\ \text{ft})(150\ \text{ft})}$$
$$= 1.963\ \text{ft}$$

To find the distance of the shifted center curve, use Eq. 4.34.

$$p = y_s - R_c \operatorname{vers}\theta_s = 1.963\ \text{ft} - (1909.86\ \text{ft})(\operatorname{vers} 2.25^\circ)$$
$$= 0.491\ \text{ft}$$

The spiral curve tangent can now be calculated using Eq. 4.33.

$$T_s = (R_c + p)\tan\frac{I}{2} + k$$
$$= (1909.86\ \text{ft} + 0.491\ \text{ft})\tan\frac{31^\circ}{2} + 74.97\ \text{ft}$$
$$= 604.76\ \text{ft} \quad (610\ \text{ft})$$

The answer is (C).

(b) The degree of the central curve can be calculated from Eq. 4.30. The length of the spiral, L_s, is 150 ft, so the length of the curve, L, from the TS or ST to the midpoint of the curve is 150 ft/2 = 75 ft.

$$D_l = D_c\left(\frac{L}{L_s}\right) = (3^\circ)\left(\frac{75\ \text{ft}}{150\ \text{ft}}\right) = 1.5^\circ$$

Use Eq. 4.16 to calculate the curve radius.

$$R_c = \frac{(100\ \text{ft})\left(\frac{180^\circ}{\pi}\right)}{D_{l,\text{deg}}} = \frac{(100\ \text{ft})\left(\frac{180^\circ}{\pi}\right)}{1.5^\circ}$$
$$= 3819.72\ \text{ft} \quad (3800\ \text{ft})$$

The answer is (D).

(c) Superelevation is transitioned uniformly along the spiral from the normal cross slope of 2% to the maximum 8%. At a point 100 ft from the TS or ST, the transition will be

$$\Delta e = \frac{L_p}{L_s}(e_{\text{max}} - e_{\text{NC}}) + e_{\text{NC}}$$
$$= \left(\frac{100\ \text{ft}}{150\ \text{ft}}\right)\left(0.08\ \frac{\text{ft}}{\text{ft}} - 0.02\ \frac{\text{ft}}{\text{ft}}\right) + 0.02\ \frac{\text{ft}}{\text{ft}}$$
$$= 0.06\ \text{ft/ft}$$

The answer is (B).

5. (a) The design speed is limited by the ramp geometry. The most restrictive element of the geometry is the curve radius. From Table 4.11, the minimum curve radius for a design speed of 30 mph and a superelevation of 8.0% is 214 ft, which is less than the curve radius of 300 ft. Therefore, 30 mph is not the maximum design speed. Referring again to Table 4.11, the next possible design speed is 35 mph, which has a minimum curve radius of 314 ft, so the design speed is between 30 mph and 35 mph.

However, the maximum speed for the curve is controlled by the superelevation and the side friction factor. From Table 4.9, the side friction factor is 0.20 for a design speed of 30 mph and 0.18 for a design speed of 35 mph. Determine the design speed using Eq. 4.66(b) for a friction factor of 0.19, the average of the two possible side friction factors.

$$\text{v}_{\text{mph}} = \sqrt{R_{\text{ft}}(15)(e + f_s)}$$
$$= \sqrt{(300\ \text{ft})(15)(0.08 + 0.19)}$$
$$= 34.9\ \text{mph} \quad (35\ \text{mph})$$

The design speed is most nearly 35 mph.

The answer is (C).

Geometric Design

(b) Find the curve length using Eq. 4.9.

$$L_{\text{ft}} = \frac{2\pi R_{\text{ft}} I_{\text{deg}}}{360°} = \frac{(2\pi)(300\ \text{ft})(75°)}{360°} = 392.7\ \text{ft}$$

The stopping sight distance is limited by the high parapets to a point two-thirds around the 300 ft radius curve. Therefore, the stopping sight distance is found by adding one-third of the curve length to the intersection approach length, 200 ft.

$$S = \left(\tfrac{1}{3}\right)(392.7\ \text{ft}) + 200\ \text{ft} = 330.9\ \text{ft} \quad (330\ \text{ft})$$

The answer is (D).

(c) From the footnote to Table 4.4, the avoidance maneuver for a direction change on an urban freeway is avoidance maneuver E. Using Table 4.4, for avoidance maneuver E and a 50 mph approach speed, the decision sight distance is 1030 ft (1000 ft).

The answer is (D).

(d) From Table 4.4, the stop-controlled intersection is avoidance maneuver B. The decision sight distance for the stop-controlled intersection uses the current posted speed limit of 30 mph. For avoidance maneuver B and a 30 mph design speed, the decision sight distance is 490 ft.

The answer is (B).

(e) The decision sight distance provides additional time for complex situations where drivers must recognize an obstacle, choose a course of action, and then maneuver their vehicles accordingly. However, for a ramp leading to a stop-controlled intersection, drivers only need to recognize that a stop is required. Therefore, the stopping sight distance can be used as the decision sight distance, 330.9 ft.

The footnote in Table 4.4 gives the pre-maneuver time for avoidance maneuver B as 9.1 sec. The maximum speed can be found by dividing the stopping sight distance by the pre-maneuver time.

$$\begin{aligned} v &= \frac{S}{t_p} = \frac{(330.9\ \text{ft})\left(3600\ \frac{\text{sec}}{\text{hr}}\right)}{(9.1\ \text{sec})\left(5280\ \frac{\text{ft}}{\text{mi}}\right)} \\ &= 24.8\ \text{mph} \quad (25\ \text{mph}) \end{aligned}$$

A speed of 25 mph should be posted on the ramp.

The answer is (B).

6. (a) The ramp is posted for 25 mph, which can be presumed to be the design speed. Based on Table 4.3, the minimum stopping sight distance for a design speed of 25 mph is 155 ft (160 ft).

The answer is (B).

(b) Based on Table 4.4, the decision sight distance would fall under avoidance maneuver B. A stop on an urban road uses a pre-maneuver time of 9.1 sec. Values for a design speed of 25 mph are not given. However, approaching traffic would be reducing speed from 50 mph when approaching the ramp and may not reach 25 mph when the decision point is reached. Therefore, using the minimum decision sight distance for a design speed of 30 mph, 490 ft, is acceptable.

The answer is (B).

(c) From Table 4.3, the design stopping sight distance for a design speed of 40 mph is 305 ft (310 ft).

The answer is (A).

(d) The decision sight distance for lane changes on ramps is avoidance maneuver E, speed or path changes on urban roads. From Table 4.4, avoidance maneuver E requires 825 ft (830 ft) of decision sight distance for a 40 mph design speed.

The answer is (C).

(e) From Table 4.4, the decision sight distance for vehicles approaching a main freeway is avoidance maneuver E (speed or path changes on urban roads), which requires 1030 ft (1000 ft) at a design speed of 50 mph.

The answer is (D).

5 Construction

Nomenclature

AC	asphalt content	%	%
C	cohesiometer value	–	–
C	cost	\$	\$
D	deflection	in	mm
D	duration	various	various
DP	dust-to-binder ratio	–	–
EF	earliest finish	–	–
ES	earliest start	–	–
G	specific gravity	–	–
H	height	in	n.a.
L	mass	lbm	g
L_n	haul length	sta	sta
LF	latest finish	–	–
LS	latest start	–	–
m	mass	lbm	kg
N	number	–	–
p	pressure	lbf/in^2	MPa
P	percentage	%	%
Q	quantity	–	–
s	change in soil volume (shrinkage or swell)	–	–
s	temperature standard deviation	°F	°C
S	stability	–	–
T	pavement design temperature	°F	°C
T_{air}	seven-day average air temperature	°F	°C
T_{min}	minimum pavement design temperature	°F	°C
V	volume	ft^3	m^3
V	volumetric ratio	–	–
VFA	voids filled with asphalt	%	%
VMA	voids in mineral aggregate	%	%
VTM	volume of air voids	%	%
w/c	water to cement ratio	–	–
W	weight	lbf	N
W	width	in	n.a.

Symbols

γ	specific weight	lbf/ft^3	n.a.
ρ	density	lbm/ft^3	kg/m^3

Subscripts

a	air voids
ave	average
b	asphalt, base, borrow, or bulk
ba	absorbed asphalt
be	effective asphalt
c	crew, cut, or concrete
ca	coarse aggregate
cap	capacity
e	effective, embankment, or excavation
f	fill
fa	fine aggregate
h	haul or horizontal
max	maximum
mb	bulk
mm	maximum (zero air voids)
p	in place
s	aggregate or supervision
sa	apparent
sb	bulk
se	effective
t	total
v	vertical
w	waste

1. INTRODUCTION

Construction design of transportation facilities is primarily concerned with providing facilities that allow vehicles and/or pedestrians to navigate safely and easily, with a minimum amount of delay. The aspects of transportation of most concern to a transportation engineer are earthwork, pavement design, and project management.

2. EXCAVATION AND EMBANKMENT

The alignment of a highway or railroad is established to provide connections between two or more points. When laying out highways or railroad alignments, designers attempt to follow existing topography as much as possible to minimize the construction costs and produce visually pleasing finished projects. Grade and design of horizontal and vertical curves are based primarily on the design speed, but also include adjustments for connecting roadways, drainage, right-of-way availability, and considerations for appearance, ride quality, and environmental impact.

In order to accommodate design features, the existing topography must almost always be modified. Limits of available right-of-way and clearances from topographic features, such as rivers, cliffs, and scenic views, may move the alignment to a less than desirable location, and because the natural surface of the ground is not sufficiently uniform to simply be paved, the ground surface must be altered to fit the location and shape of the new roadway. In these situations, ground that is too high must be cut down, and ground that is too low must be filled in. This shaping of the ground surface is known as *earthwork*. Excavation of existing material is known as a *cut*, and filling in a low point with an embankment is known as a *fill*.

Earthwork is often the largest and most time-consuming work element of a transportation project, and carefully fine-tuning the alignment during design to balance earthwork quantities can have a large impact on the overall cost of a job. Hauling and mass diagrams can be used to determine the most economical method of performing necessary cut and fill.

The first step in developing a cross section is drawing the *roadway template*, a visual representation of the shape planned for the roadway. While a roadway template and a cross section are related, the two concepts are not synonymous. The cross section shows the original grade lines and other existing landscape features, while the roadway template illustrates the major features of the proposed roadway only, such as the number of lanes. The roadway template is then superimposed over the cross section, which is used to determine the necessary earthwork quantities.

Cross Sections

Cross sections are plotted at right angles to the centerline, baseline, or grade line, at intervals of 50 ft or 100 ft (10 m or 30 m). Extending from the edges of the roadway template are cut and fill lines that intersect the original ground. The template is set according to the roadway requirements, showing the number of lanes, the widths of shoulders, the widths of medians (for divided highways), drainage swales, or other features affecting the basic cross section. Figure 5.1 shows a typical highway cross section with the grade line near the ground line.

Cut and fill slopes are determined by testing soil samples from core borings drilled along the alignment. For instance, cuts in hard rock may allow a steeper slope than cuts in soft rock or soil. Fills made from granular material with little clay can be made steeper than fills made from soil that has a large amount of clay or silt. The soil is tested in moist conditions, such as would occur in the field at the project site. The steepest slope that will be stable for the life of the project is established, and this slope is used on the cross sections to determine the minimum amount of cut or fill that must be performed.

Figure 5.1 *Typical Two-Lane Roadway Cross Section, Side Hill Cut*

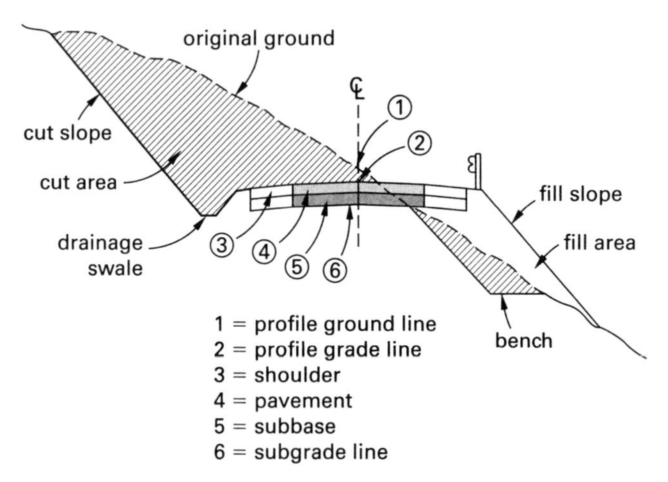

Figure 5.2(a) shows a cross section that is entirely in cut. The cut slopes are set at 1.5:1 horizontal:vertical for the cut slope. Figure 5.2(b) is a fill section with side slopes of 2:1 horizontal:vertical. The point where the side slope intersects the original ground is the *limit of earthwork*.

Figure 5.2 *Cut and Fill Sections*

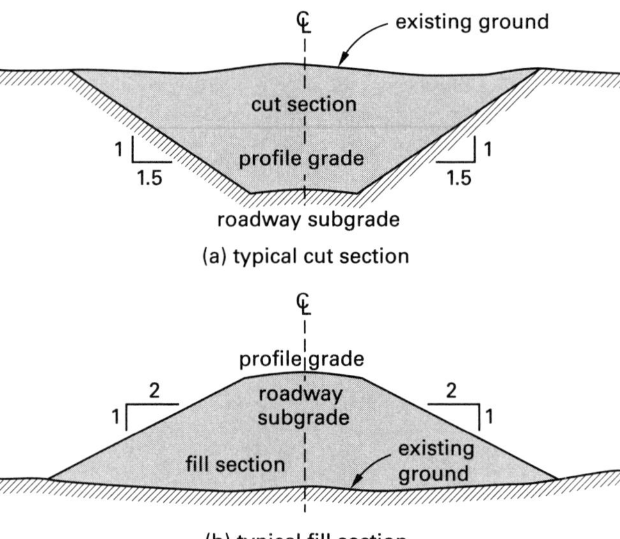

The area of cut and fill at each cross section is measured using closed polygons that are bounded by the cut and fill lines and the original ground line. The cut and fill quantities are determined individually for each cross section by the *average end area method*, or *prismoidal method*, and, after adjusting for shrinkage, plotted on a mass-haul diagram to show cumulative cut and fill relationships to distance along a pavement.

Roadway Cross Sections

There are five major elements that make up a *roadway cross section*, as shown in Fig. 5.3. These elements include the subbase, subgrade, surface pavement, shoulders and curbs, and drainage systems.

Figure 5.3 *Elements of a Typical Roadway Cross Section*

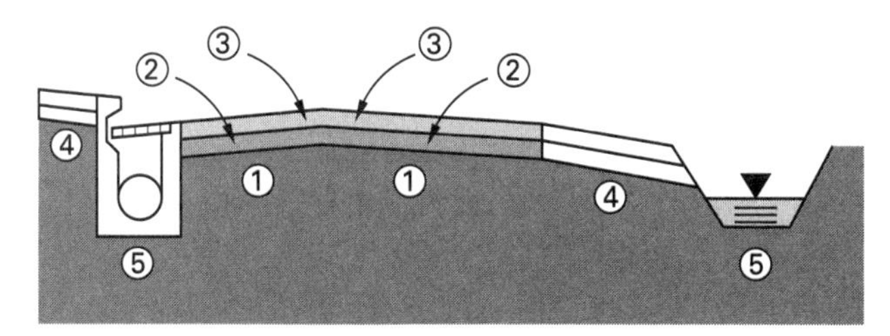

1 = subgrade
2 = subbase
3 = surface pavement
4 = shoulder and curb
5 = drainage system

The *subbase*, or aggregate base, is composed of a free-draining, frost-free material, and is used to transmit and spread the pavement load to the subgrade. It also drains the water away from the underside of the pavement.

The *subgrade* is the soil layer that is directly below the granular subbase. It is carefully graded and compacted, and of a drainable soil type that provides the foundation for the road. Subgrade design involves soil classification and/or the modification of soil to provide minimum bearing capabilities, stability requirements, and resistance control.

The *surface pavement* forms the roadway surface and also transmits wheel loads to the aggregate base. In addition to load-carrying capability, the surface pavement requires durability to withstand the fatigue of frequent axle-load passages. Surface pavement design begins with subgrade bearing capacity and applies standard loading conditions to a base and pavement combination. The loading conditions include both frequency and weight of axle passages.

Shoulders and *curbs* provide lateral support for the surface pavement edges. Shoulders carry water away from the roadway, provide a space for snow to be moved off of the roadway, and serve as a pull-off area for vehicles. Curbs control drainage flow along the pavement edge, help support sidewalks, and provide guidance for traffic flow.

Drainage systems, such as side ditches, curb gutters, inlets, and pipes, carry and keep excess water away from the roadway surface, as well as under the pavement.

Railroad Track Cross Sections

Railroad cross sections have elements performing similar functions as roadway cross sections, as shown in Fig. 5.4.

The *sub-ballast*, often made from the same material as the top ballast, is used to transfer and spread the top ballast load to the subgrade, as well as drain water away from the ballast.

As with roadways, the railroad *subgrade* is a carefully graded, compacted, and drainable type of soil that provides the foundation for the track structure. The subgrade can be sloped at 48:1 to both sides of the trackway, or entirely to one side, as the track slope does not depend on the slope of the subgrade. The subgrade can be 4 ft (1.2 m) or more below the top of the rail.

The *top ballast*, along with the track rail and tie system, forms the riding surface for rail vehicles. The rail, tie, and top ballast transmit the wheel loads to the sub-ballast, and the top ballast serves to hold the track in alignment. The track structure and top ballast are designed to withstand expected wheel loadings. Generally, the top ballast is at least 12 in (30 cm) deep at the minimum point, from the top of the sub-ballast to the bottom of the lowest tie. This depth provides both structural support integrity and clearance for mechanical tamping equipment that can adjust the ballast under the tie when track lining is performed.

For a *railroad drainage system*, the ballast acts as a porous drain, carrying surface water down to the subgrade. Water from the subgrade must be carried away from the track structure alongside ditches and cross

Figure 5.4 *Elements of a Typical Railroad Track Cross Section*

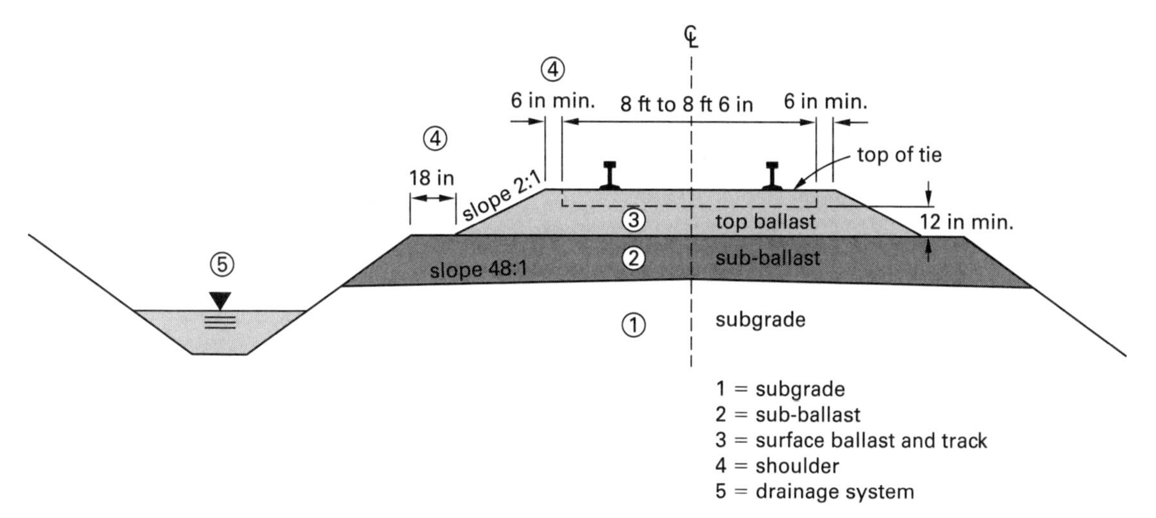

1 = subgrade
2 = sub-ballast
3 = surface ballast and track
4 = shoulder
5 = drainage system

pipes. It is important to keep the track structure free of water as much as possible so that freeze-thaw cycles and the pumping action of passing rail-car axles do not shift the track alignment.

Railroad *ballast shoulders* support the ballast edges and help spread the load of succeeding layers of ballast to the subgrade. The shoulders also help restrain the ties from lateral movement under wheel loads.

Determining Cut and Fill Quantities

Generally, in determining the earthwork areas on cross sections, a horizontal line is drawn at the subgrade elevation. The subgrade line is considered the excavation or embankment limit. In a cut condition, all material above the subbase line is removed. In a fill condition, fill material is placed on the existing ground up to the subbase line. Additional excavation is needed for drainage swales and slope benches. Additional excavation may also be needed to remove unsuitable material, rock quantities that interfere with subgrade construction, detention sumps, or other proposed underground work. This unsuitable material is replaced with suitable fill.

Rock fill may be included on steep slopes, or crushed rock layers may be included for underground water drainage. Buffer zones between the roadway shoulder and pedestrian or housing areas often use landscaped berm fill sections. Berms and dikes may also be constructed to control drainage and can be a useful way to dispose of excess cut material. Unsuitable material is disposed of off-site and is not included in engineered fill sections.

The total amounts of material that must be excavated as cut or placed as fill are referred to as the *earthwork mass-haul quantities* for an earthwork project. Subbase, base, and pavement surface materials are not included in the earthwork mass-haul quantities, but are instead included in the pavement calculations. Pipe excavation is not included in the earthwork quantities, as it is normally paid for with the pipe quantity bid item.

When material is removed from its natural position, its density decreases because it occupies more volume, and it is said to *swell*. When material is compacted into a fill, its density increases. Compaction reduces the volume of material from the hauled volume, and it is said to *shrink*. Generally, soil will shrink and rock will swell when compacted in a fill. Figure 5.5 illustrates these volume relationships. The in-place (*in situ* or *bank*) volume is shown as a solid line, and the dashed lines represent the change in volume from the in-place volume. Figure 5.6 further illustrates volume and density changes during earthwork activities. The percentages of change between the three material conditions shown in Fig. 5.6 are intended only as examples and are much larger than the percentages of change encountered during actual earthwork. When the cut and fill volumes are tabulated for each cross section, they need to be adjusted for swell and shrinkage before being plotted on a mass diagram (see Sec. 5.3.). Cut in

Figure 5.5 *Excavation Volume Relationships*

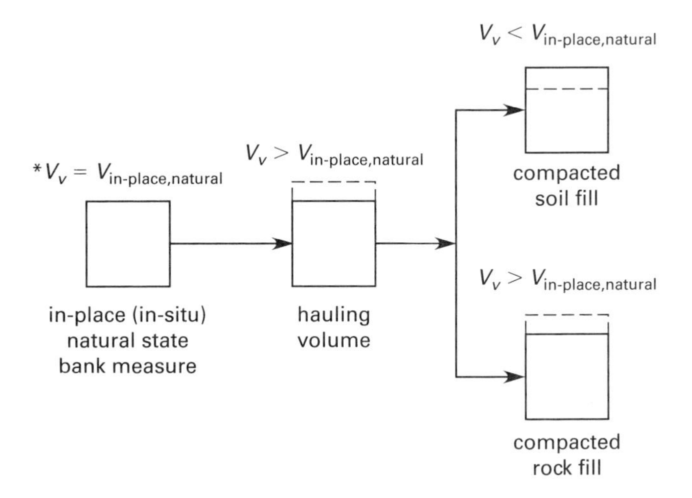

Figure 5.6 *Typical Material Volume Change During Earthmoving*

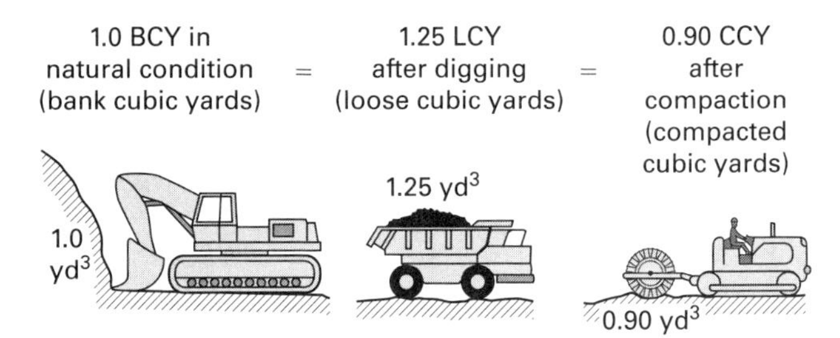

excess of what is needed as fill, or that is unsuitable to use for fill, is called *waste*. *Borrow* refers to any additional material brought from off-site to complete a fill.

The shrinkage and swell factors for a given material are determined by soil tests on the material, and the following formulas are used to find the volume of cut or fill. s is the change in soil volume, either as shrinkage, which is negative, or swell, which is positive. The volume of fill, V_f, can be found from Eq. 5.1.

$$V_f = (1+s)V_c \qquad 5.1$$

Equation 5.1 can be rearranged into Eq. 5.2 to find the volume of cut, V_c.

$$V_c = \frac{V_f}{1+s} \qquad 5.2$$

Equation 5.1 and Eq. 5.2 determine the amount of cut or fill available at a site and represent the theoretical cut and fill volumes adjusted for shrinkage or swell. In order to determine the actual amount of material available for fill, the amount of cut produced is reduced by the volume of unsuitable excavation, which is usually wasted off-site. *Unsuitable excavation* is any material considered unsuitable for fill, such as peat or sod. The amount of fill is then adjusted for shrinkage and swell and compared to the available volume of cut to determine whether waste will be produced or borrow will be needed. If the adjusted amount of fill required is greater than the available volume of cut, the job is a borrow job (i.e., borrow is required). If the adjusted amount of fill is less than the available amount of cut, then the job is a waste job. In situations where all of the cut is waste (e.g., a site's soil is unsuitable), the volume of cut is zero, and the volume of borrow is equal to the volume of fill adjusted for shrinkage and swell.

The volume of borrow, V_b, or waste, V_w, is calculated from Eq. 5.3.

$$V_b = -V_w = V_c - \frac{V_f}{1+s} \qquad 5.3$$

Example 5.1

Earthwork for a section of highway has 7330 m^3 of cut and 6960 m^3 of fill. The cut material shrinks 9% when compacted into fill. How much borrow or waste occurs?

Solution

Determine the amount of borrow or waste using Eq. 5.3.

$$\begin{aligned} V_b &= V_c - \frac{V_f}{1+s} \\ &= 7330 \text{ m}^3 - \frac{6960 \text{ m}^3}{1+(-0.09)} \\ &= -318 \text{ m}^3 \end{aligned}$$

Borrow is negative. Therefore, this is a borrow job. The amount of borrow required is 318 m^3.

Hauling and Material Handling

For earthwork, truck or jobsite equipment hauling involves four parts. The first part is excavating and loading. This can be accomplished by one piece of equipment, or one piece of equipment can dig up the material while another performs loading duties.

The second part is the actual hauling. The hauling capacity of trucks and equipment is specified as both a volume and a weight capacity. For soils of low density, including loam and organic material, the load volume capacity is often reached before the maximum hauling weight. When excavating rock or other dense material, the weight capacity most often limits the size of the load. Both capacities should be checked to determine the minimum number of load trips so that trucks and other equipment are not overloaded. The change in volume or density is compared to the original in situ density, regardless of the amount of shrinkage or swell that occurs during excavation and compaction.

Equation 5.4 can be used to find the hauling volume, V_h, and Eq. 5.5 is used to find the hauling mass, m_h. ρ_b is the in situ soil mass density. s is the shrinkage or swell of the material.

$$V_h = V_b(1+s) \qquad 5.4$$

$$m_h = \frac{\rho_b}{1+s} \qquad 5.5$$

The third part is dumping or unloading the trucks. Most on-road and off-road trucks dump the load by lifting and tilting the hauling end toward the rear. The load is allowed to fall out of the rear as the truck moves ahead. For large volumes, this saves time, as the trucks can keep moving through the dumping area. Some trucks unload by opening up doors in the bottom of the bed. This type of unloading is generally faster than end dumping and is safer because the truck body does not have to be elevated, which can interfere with overhead obstructions. Other off-road vehicles, such as earthmover scrapers, unload by pushing the load forward out of the scraper bowl and over the cutting edge as the scraper moves ahead. It is important to allow sufficient time for unloading when analyzing a hauling cycle to avoid congestion at the dump site.

The fourth part is the return of haulers to the loading point. Empty trucks can travel faster than loaded trucks, which means the return trip may not take as much time as the original haul trip.

Equipment costs are usually calculated on a per-hour basis for work within the jobsite. Costs of over-the-road travel, such as delivering materials and supplies from outside vendors, are often calculated on a per-vehicle-mile basis. Labor costs are usually on a per-hour basis and are added to the equipment cost. When comparing costs for material supplies, the delivery costs include total truck and driver time to make a round trip from the supplier's stockpile or warehouse to the jobsite.

Often, more than one material supplier is being considered or is necessary to fill the order. Typical comparisons done to choose between suppliers include the cost of material from each source and the load-haul-unload-return cycle for each truck. The most cost-effective supplier for the job is the one that can provide the lowest cost for materials delivered to the jobsite at the proper time.

Example 5.2

Earthwork requires 30,000 yd^3 of borrow fill. A borrow pit is 4 mi away and has soil with an in situ density of 135 lbm/ft^3. The soil in the borrow pit expands 25% when hauled and shrinks 11% when compacted in the fill.

Two contractors are being considered, each with a different type of truck to haul the material. Contractor A has a truck with a 55,000 lbm capacity, a 15 yd^3 load bed, and a cost of $3.50 per mile to operate. Contractor B has a truck with an 82,000 lbm capacity, a 25 yd^3 load bed, and a cost of $4.75 per mile to operate. What is the lowest cost of hauling the borrow fill?

Solution

Determine how much borrow is needed using Eq. 5.3.

$$V_b = V_c - \frac{V_f}{1+s} = 0 - \frac{30{,}000 \text{ yd}^3}{1+(-0.11)} = -33{,}700 \text{ yd}^3$$

Therefore, the amount of borrow required is 33,700 yd^3. Determine the hauling volume using Eq. 5.4.

$$\begin{aligned} V_h &= V_b(1+s) = (33{,}700 \text{ yd}^3)(1+0.25) \\ &= 42{,}125 \text{ yd}^3 \end{aligned}$$

Determine the mass of the hauled material per cubic yard using Eq. 5.5.

$$\begin{aligned} m_h &= \frac{\rho_b}{1+s} = \left(\frac{135 \frac{\text{lbm}}{\text{ft}^3}}{1+0.25}\right)\left(27 \frac{\text{ft}^3}{\text{yd}^3}\right) \\ &= 2916 \text{ lbm/yd}^3 \end{aligned}$$

Determine the cost per contractor.

Contractor A

Check the capacity of the truck to determine whether mass or volume governs.

$$\begin{aligned} V_{\text{cap}} m_h &= (15 \text{ yd}^3)\left(2916 \frac{\text{lbm}}{\text{yd}^3}\right) \\ &= 43{,}740 \text{ lbm} \quad [< 55{,}000 \text{ lbm}] \end{aligned}$$

43,740 lbm is less than 55,000 lbm. Therefore, volume governs.

The hauling cost will be the number of truck loads times the number of miles (round-trip) times the cost per mile.

$$\left(\frac{42{,}125 \text{ yd}^3}{15 \frac{\text{yd}^3}{\text{truckload}}}\right)(2)(4 \text{ mi})\left(3.50 \frac{\$}{\text{mi}}\right) = \$78{,}633$$

Contractor B

Check the capacity of the truck.

$$\begin{aligned} V_{\text{cap}} m_h &= (25 \text{ yd}^3)\left(2916 \frac{\text{lbm}}{\text{yd}^3}\right) \\ &= 72{,}900 \text{ lbm} \quad [< 82{,}000 \text{ lbm}] \end{aligned}$$

72,900 lbm is less than 82,000 lbm. Therefore, volume governs.

The hauling cost is

$$\left(\frac{42{,}125 \text{ yd}^3}{25 \frac{\text{yd}^3}{\text{truckload}}}\right)(2)(4 \text{ mi})\left(4.75 \frac{\$}{\text{mi}}\right) = \$64{,}030$$

The lowest cost is $64,030 using contractor B's trucks.

3. MASS DIAGRAMS

Designers use a mass diagram when setting the grade line to balance the cut and fill quantities. A *mass diagram*, or *mass-haul diagram*, is a plotted curve showing the cumulative volume of excavated materials in a cut and the volume necessary for fill.

Mass diagrams are useful tools to find the most economical hauling distance of cut and fill materials. Mass diagrams use units of station-yards (station-meters), indicating one cubic yard (cubic meter) moved one station (100 ft (100 m)). Distances are plotted in stations from left to right along a horizontal axis. The adjusted cross section quantities are plotted cumulatively on the vertical axis, using positive values for excavation and negative values for fill. A rising line indicates an excess of cut over fill, and a falling line indicates an excess of fill over cut.

A mass diagram is not the same as a profile drawing, though both are used in the planning of earthwork. A *profile drawing* depicts a vertical surface cut through the centerline of the baseline of a roadway on which the original ground surface and the vertical position of the roadway are plotted. Figure 5.7 shows a typical profile of the vertical differences between a roadway grade line and a ground line with a mass diagram. The grade line is relatively straight, but the ground line rises and falls above and below the grade line. In section B, the ground must be cut down to the grade line, and in sections A and C, the ground must be filled in to meet the grade line.

On either diagram, a *balance line* is drawn between two adjacent *balance points* in a crest or sag area. The volumes of excavation between the balance points are equal, so a contractor can use the earth volume from the excavation on one side of the grade point for the embankment on the other side of the grade point. The balance line in Fig. 5.7 intersects the mass diagram at two points, showing that the cut soil for section A can be used as fill soil for section C (assuming the soil is suitable for use as fill).

Figure 5.8 depicts a mass diagram showing the subbases for a roadway project, with distances provided in customary U.S. units. The balance point for this diagram falls between station 13 and station 14. The ordinate of station 13 is +1519, and the ordinate of station 14 is −254, so the balance line falls by 1519 yd^3 + 254 yd^3 = 1773 yd^3 every 100 ft or 17.73 yd^3/ft. The curve crosses the baseline at a distance of 1519 yd^3/(17.73 yd^3/ft) = 86 ft from station 13 (written as sta 13+86).

Figure 5.7 Balance Line Between Two Points

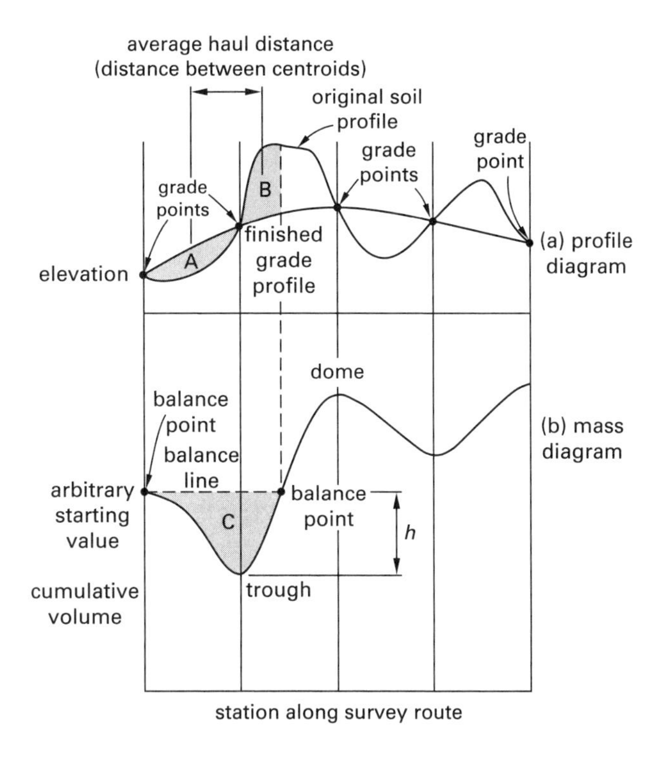

Figure 5.8 Mass Diagram Showing Subbases

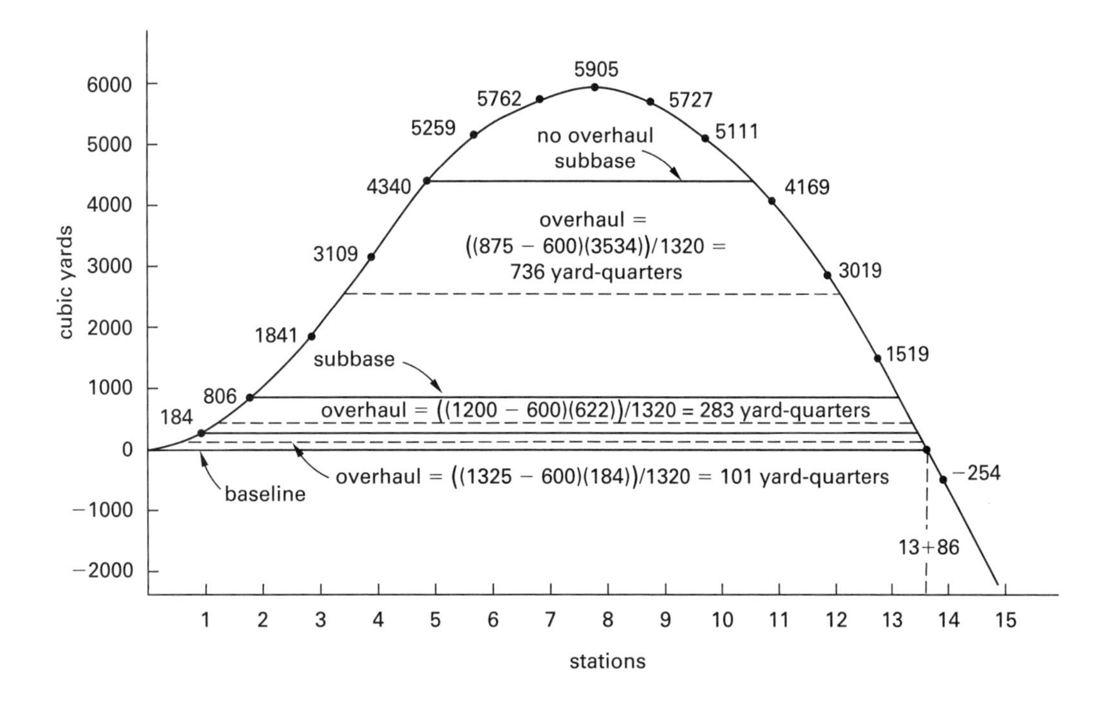

Once a job is bid, the contractor is given a *freehaul distance*, included in the bid price, that specifies the distance excavated materials will be moved by the contractor free of charge. Material moved farther than the freehaul distance is termed *paid overhaul*. Costs for overhaul will be provided in the contractor's bid. Waste and fill materials are usually balanced so that the amount of necessary haul materials is not excessive. Figure 5.9 shows a mass diagram and a profile diagram depicting the freehaul distance and overhaul volume.

Figure 5.9 *Freehaul and Overhaul*

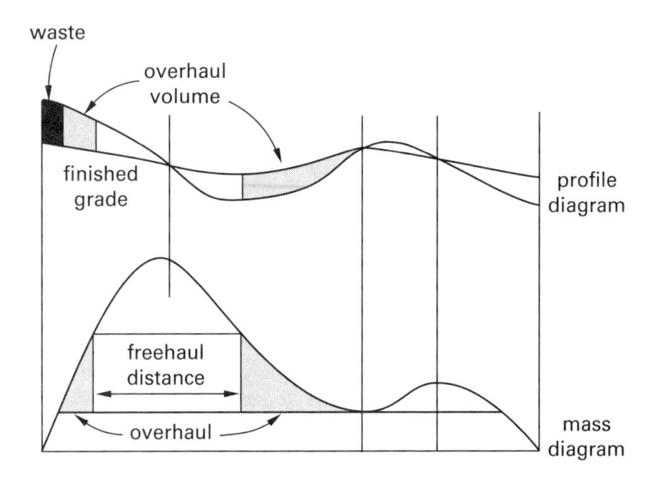

The *economic overhaul limit*, also called the *limit of profitable haul*, C_{limit}, is the limit at which it is more profitable to borrow and waste rather than haul from another part of the job. This can be determined without a mass diagram, using the following equations instead. C_e is the cost of one cubic yard (cubic meter) of excavation or embankment.

$$C_{\text{limit}} = 2C_e \qquad 5.6$$

When excavation is hauled into embankment, use Eq. 5.7. C_h is the cost to haul one cubic yard (cubic meter) one station, or 100 ft (100 m), and L_n is the length of haul in 100 ft (100 m) stations.

$$C = C_e + C_h L_n \qquad 5.7$$

To find the limit of profitable haul, compare the cost calculated from Eq. 5.7 with the value of $2C_e$. If they are equal, the limit of profitable haul has been reached.

Rearranging Eq. 5.7, Eq. 5.8 can be used to find the haul length.

$$L_n = \frac{C_e}{C_h} \qquad 5.8$$

Generally, it is more common in rural areas than in urban areas to find convenient locations for waste sites and borrow pits. Freehaul distances such as 300 ft, 500 ft, or 900 ft (91.4 m, 152 m, or 274 m) were often found in specifications for railroads and interstate highways during the early to mid-20th century. However, with the completion of the interstate highway system and with fewer new railroad mainlines being constructed, a large proportion of work now involves making improvements to shorter stretches of the alignment—frequently while maintaining traffic on the roadway. While it has become increasingly difficult to balance cuts and fills within the limits of a single job, equipment improvements have resulted in faster handling of earthwork within the jobsite, as well as making longer hauls more manageable for on-highway equipment.

Example 5.3

The excavation or embankment cost for an earthwork project is \$10 per yd^3, and the hauling cost is \$0.65 per yd^3-sta. What is the limit of profitable haul?

Solution

Using Eq. 5.8, the haul length is

$$L_n = \frac{C_e}{C_h} = \frac{10\ \frac{\$}{\text{yd}^3}}{0.65\ \frac{\$}{\text{yd}^3\text{-sta}}} = 15.4 \text{ sta}$$

Check that the limit of profitable haul has been reached by comparing the cost calculated from Eq. 5.6 and Eq. 5.7.

$$C_{\text{limit}} = 2C_e = (2)\left(10\ \frac{\$}{\text{yd}^3}\right) = 20.0$$

$$\begin{aligned} C &= C_e + C_h L_n \\ &= 10\ \frac{\$}{\text{yd}^3} + \left(0.65\ \frac{\$}{\text{yd}^3\text{-sta}}\right)(15.4 \text{ sta}) \\ &= 20.0 \end{aligned}$$

Therefore, the limit of profitable haul is 15.4 sta.

4. PAVEMENT DESIGN

Pavement design involves matching a project's strength and durability requirements to a given material. The base soil, or foundation, is analyzed, and the depths of various layers of pavement are designed based on the loads applied, the subgrade strength, and the frost penetration depth.

Pavement design's primary concern is vertical loads plus impacts. Designs are based on a combination of maximum expected wheel or axle loadings and the expected frequency of groups of axle loadings. *Horizontal loads*, such as those resulting from braking, acceleration, or turning movements, are not usually considered for concrete, but are included for Superpave asphalt. (See Sec. 5.8.) Pavement loadings in areas where vehicles will stand for extended periods of time, such as in parking lots, do not need to account for the frequency of wheel-load repetitions, but are instead governed by the maximum load applied for concrete and the distribution of load for asphalt.

Pavement Thickness Design

Concrete slab thickness is determined primarily by the maximum single load applied or by the repetition of loads over a long period of time, depending on which is greater. Generally, the repetition of loads over a period of time requires greater strength than a maximum single load because a highway slab with a depth of no more than 6 in (152 mm) can occasionally sustain axle loadings at low speeds considerably in excess of legal limits. A slab thickness of 5 in (127 mm) is the nominal minimum thickness for lightly traveled streets and driveways. Additionally, the relative strength of the subbase using R-values or CBR values influences slab thickness.

Highway pavement thickness is commonly specified in 1 in (25 mm) increments from 5 in to 10 in (130 mm to 250 mm) or greater. For heavily traveled interstates, pavement thicknesses of 12 in (300 mm) or more are used to help prevent costly traffic congestion caused by pavement rehabilitation. Where frost is a factor, highway subbase thickness should be at least 6 in and can range up to a foot or more depending on the subsoil and load applied.

In order to simplify the variety of vehicle combinations riding on highways, vehicle traffic for pavement thickness design is usually grouped according to axle load weights, spacing of axle combinations, and frequency of application. Because of the exponential relationship of axle load to pavement life, the heaviest and most frequent load combinations are used to determine the design strength. Traffic is commonly broken down into classifications of automobiles, light trucks, heavy single-unit trucks, and heavy multiple-unit trucks. Truck traffic can be further classified as commercial or industrial to indicate expected loading needs along with appropriate axle load weight classifications.

Pavement thickness design is a complex and involved process that is thoroughly covered in publications written by the American Association of State Highway and Transportation Officials (AASHTO), the American Concrete Pavement Association, the National Pavement Association, the Portland Cement Association (PCA), and the Asphalt Institute.

Soils and Subgrade

Because the soil layer immediately below a pavement base has a direct effect on the performance of the pavement, the characteristics of this soil must be known in order to design a pavement that is economical and able to carry the expected loads. The best soil types have good load strengths, drain well, are not affected by frost or other types of heaving, are free of organic material, and can be shaped and compacted into a stable mass. In the AASHTO classification system, soils in the A-l and A-2 classifications are considered proper subgrade materials. Soils in the A-1 and A-2 classifications have less than 35% *fines*, or material passing the no. 200 sieve. The AASHTO classification system is given in App. 5.A.

Soil classification information may be obtained from a number of different systems. Generally, both pavement bearing pressure and design charts are based on Hveem's resistance value (*R-value*) or the *California bearing ratio* (CBR) index. When a classification is given in one system, its name in another system may be approximated using charts such as the USDA triangle or graphs relating one system to another. PCA publishes *A Soil Primer*, which also contains conversions between classification systems. The USDA triangle is given in App. 5.B.

Usually, subgrade soil is specified to be carefully excavated, compacted, and graded as a separate operation, often with its own unit price for payment. If suitable soil is not available for subgrade, cement treating or other modifications to available soil may be necessary. Once a subgrade design strength is identified, the design of the granular drainage base, subbase, and surface pavement may proceed.

Subbase Materials

Aggregate materials used for subbase construction are often locally produced to minimize hauling cost. Aggregate must be free of fines, relatively uniform in gradation, and not too large in size. A common method is to specify AASHTO no. 65 or no. 67, or some other readily available standard designation of aggregate size classification.

In addition to size classification, subbase materials are tested for maximum organic material, acceptable chemical composition, soundness, minimum fractured faces (river gravel), frost susceptibility, shrinkage and swell, minimum crushing strength, absorption, color (architectural surfaces), alkali reaction, and other appropriate parameters.

Pavement Materials

Highway pavement materials are relatively few in number, yet their makeup and composition are directly responsible for the successful performance of a paved surface. All high-type pavements have coarse aggregate, fine aggregate, and a cementitious material to bond the aggregates.

The most common choice for high strength and durability is *concrete pavement*. Concrete is highly resistant to abrasion, chemical attack, and spilled fuel solvents. When properly placed, concrete has been known to last well over 100 years with little or no maintenance. However, concrete requires specialized knowledge to handle its unforgiving qualities (i.e., once placed, setting and curing occur within a few hours, and any mistakes made are permanently captured), and it is susceptible to the abuse of excessive deicing salts, which soften the surface and weaken the joint edges. It also can have a higher cost compared with bituminous paving mixtures with equal strength or extended life cycles. Figure 5.10 illustrates the typical concrete surfacing layers. Table 5.1 gives the commonly used maximum aggregate sizes.

Figure 5.10 Typical Concrete Surfacing Layers

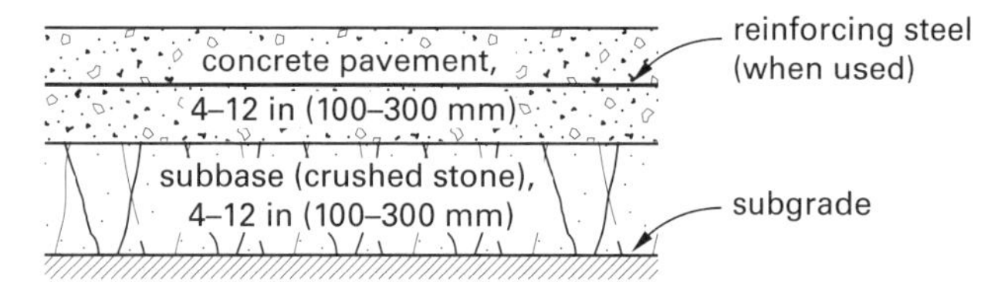

Table 5.1 Typical Aggregate Maximum Sizes for Concrete Pavements

aggregate type	size in	size mm
concrete pavement	1.5–2	37.5–50
subbase	2.5	64
subgrade	4	100

Bituminous Pavement

Bituminous pavement (commonly referred to as *asphalt*, *asphalt pavement*, and *flexible pavement*) is composed of upper layers of aggregate mixed with a bituminous binder, such as asphalt. It is often preferred to concrete because it can be rapidly placed and requires less skill. It may also be installed on weaker subgrade sections, which decreases the amount of overlaying needed and defers some capital cost. However, if asphalt is not placed well, its edges can take on a ragged appearance. It also raises environmental concerns due to its hydrocarbon solvent emissions and safety concerns for workers who can become injured due to the high temperatures and irritating solvents used during the installation process. Additionally, asphalt is highly susceptible to sun damage in low traffic applications and to solvent action resulting from fuel spills. Construction costs vary regionally, making it difficult to select either asphalt or concrete pavement based on empirical data and cost alone. Figure 5.11 illustrates the layers of asphalt paving surfaces, though not all layers are present in every design. Table 5.2 gives the commonly used maximum aggregate sizes.

Asphalt pavement is called flexible pavement because the surface material adjusts to the load condition imposed, then elastically rebounds. Asphalt (including bitumen) is used to bond aggregate together to avoid permanent displacement under loads. Asphalt also creates an impermeable layer, protecting the base and subbase from water. The most common causes of asphalt pavement failure include

- water (frost) penetration through surface cracks
- subgrade/subbase failure due to overload
- surface cracking due to aging caused by oxidation of the asphalt cement and the effects of the sun

Figure 5.11 Typical Asphalt Surfacing Layers

wearing surface,
1–3 in (25–75 mm) asphalt mat

binder course,
$1\frac{1}{2}$–4 in (38–100 mm) asphalt mat

asphalt treated base (black),
3–5 in (75–125 mm)

untreated base* (crushed stone),
3–8 in (75–200 mm)

subbase* (crushed stone or blended materials),
4–12 in (100–300 mm)

subgrade

*These layers may be combined according to local practice.

Table 5.2 Typical Aggregate Maximum Sizes for Asphalt Pavements

layer	size in	size mm
wearing course	0.5	12.5
binder course	0.75	19
base course	1–2	25–50
stone base	3	75
subgrade	4	100

Materials Procedures

Pavement design usually requires calculating weights for each ingredient based on a predicted range of outcomes that are selected from known data. Bituminous mixes have a relatively narrow range of strengths and applications. Therefore, design becomes more a matter of total in-place volume. Concrete, on the other hand, can be used in a variety of ways and can have a wide range of strengths based on how its ingredients are proportioned. A commonly used reference is the AASHTO *Guide for Design of Pavement Structures*. Additional reference material is published by trade organizations that concentrate on providing technical data for their members. For concrete, PCA's *Design and Control of Concrete Mixtures* is often used as a manual of concrete mix design and should be a part of any review of mix design procedures. For asphalt pavements, the Asphalt Institute's *Mix Design Methods for Asphalt Concrete and Other Hot-Mix Types* (*MS-2*) is often referenced.

The procedure for listing data and calculating material problems is best aided by using a tabular format of the four or five main ingredients and working across the page, relating weights, volumes, densities, and correction factors.

Concrete mixtures for pavement are usually specified as requiring both a minimum compressive strength of 2500 psi to 5000 psi (17 MPa to 34 MPa), depending on the expected loads on the pavement, and a minimum entrained air content of 1.0% to 7.5%, depending on the expected weather exposure. Asphalt pavement is usually specified as requiring a percent compaction based on a maximum laboratory density, but is increasingly being specified using strength criteria based on the expected load, weather, and subgrade conditions.

For design of concrete mixes, it is necessary to know the amount of water stored in the bulk aggregate and the amount of water necessary for saturation. The water absorbed in the aggregate up to the point of saturation is not available for hydration, and water may need to be added to the total mix water to compensate for absorption, whereas water contained in a bulk pile that exceeds the saturation requirements is available for hydration and will need to be subtracted from the amount of water needed before water is added to the mix.

The amount of free water available for hydration is called the *water-cement ratio*. The water-cement ratio is used to determine strength qualities, workability, and dissolvability of a mix. Higher water-cement ratios improve workability, but reduce strength. Lower water-cement ratios increase strength, but require more power for equipment to be used in the placing and finishing of concrete slabs. Water-cement ratios of 0.4 to 0.6 by weight are common in concrete pavement design.

Water must be removed from bituminous mixtures and so is not a factor in design. Bituminous pavement design is instead performed based on the quality of the bituminous material and the quality of the aggregate available for the mix.

5. HOT MIX ASPHALT

Hot mix asphalt (HMA), also known as bituminous concrete, hot plant mix, asphalt-concrete paving, asphalt concrete, flexible pavement, blacktop, macadam, or simply "asphalt," consists of a mixture of asphalt binder and mineral aggregate (stone and sand). The asphalt cement acts as a binding agent that glues the aggregate particles into a dense mass and waterproofs the mixture. The aggregate, bound together by asphalt, becomes a stone framework with toughness able to support loads and maintain its shape. The performance of the binder and aggregate system is affected by both the properties of the individual components and their combined reaction to the system.

Asphalt is called a *viscoelastic* material because it has both viscous and elastic properties. At intermediate temperatures of asphalt paving application, the mixture displays characteristics of a low-viscosity fluid and an elastic solid.

There are several methods of designing a hot mix asphalt concrete mix. The two traditional methods are the Marshall method and the Hveem method, which use viscosity penetration-graded asphalt specifications to design asphalt concrete mixes. However, in 1987, the *Strategic Highway Research Program* (SHRP) began a five-year, $150 million research program to improve the performance and durability of asphalt roads in the United States. A major focus of the program was developing performance-graded asphalt specifications that resulted in mixes with field performances closer to the performances measured under laboratory conditions.

The result of SHRP's research was *Superpave*, short for *superior performing asphalt pavement*. Superpave is covered in more detail in Sec. 5.8. This section focuses on general information and terminology relating to HMA mix design, regardless of method.

The material presented in this section is consistent with the methods presented in *The Asphalt Handbook* (*MS-4*). Information on Superpave mix design is derived mostly from the Asphalt Institute's *Superpave Mix Design, Superpave Series No. 2* (*SP-2*). Superpave binder specifications and tests can be found in the Asphalt Institute's *Performance Graded Asphalt Binder Specification and Testing, Superpave Series No. 1* (*SP-1*).

Asphalt Pavement Behavior

Wheel loads applied to asphalt pavement produce vertical compressive stress and shear stress within the asphalt layer, and horizontal tensile stress at the bottom of the asphalt layer. The primary asphalt pavement distress conditions are permanent deformation, fatigue cracking, and low temperature cracking. HMA design should analyze these distresses in preparing a design mix.

When asphalt cement pavement hardens, either due to lower temperature or to age, it cracks more readily. When asphalt cement softens, primarily due to increased temperature of the ground surface, it tends to flow laterally under load, which causes rutting or other surface irregularities. Hardening and softening can be controlled by varying the binder mix ingredients.

Permanent deformation (*rutting*) is characterized by rutting along the wheel paths and is primarily caused by two conditions. Rutting can come from a weak subgrade where the asphalt concrete surface remains essentially intact, but the subgrade becomes overloaded and fails. It can also come from a weak asphalt layer, where repeated wheel loads cause lateral displacement of the asphalt pavement. The asphalt surface will migrate and form a hump above the original grade line adjacent to the wheel paths, making the rutting more pronounced.

Weak asphalt rutting is more pronounced during summer under higher temperatures, while *weak subgrade rutting* can occur whenever the subgrade is weakened by water intrusion. The traditional approach to reducing rutting is to increase the thickness of the asphalt pavement layers. Superpave approaches this problem by increasing control over aggregate properties to provide more friction and strength and by increasing control over binder properties to provide a wider range of temperature performance.

Fatigue cracking (also known as *alligator cracking*) is a progressive process that results from overstressing of the asphalt materials. Fatigue cracking that occurs near the end of the pavement design cycle normally indicates that the pavement has served its useful life. If it occurs earlier in the design cycle, the cracking may indicate that design loads were significantly underestimated. An early sign of fatigue cracking is longitudinal cracks that form along the wheel paths in the pavement's weakest areas. Over time, individual cracks merge together and additional cracks develop parallel and adjacent to the initial cracks, developing into cross patterns. The result is a checkerboard pattern covering the entire asphalt surface. When chunks of pavement are dislodged, potholes develop. Through these cracks and potholes, water intrudes, further weakening the subbase.

There are several ways to reduce fatigue cracking.

- Accurately predict heavy design loads during the design period.
- Use thicker pavements.
- Reduce subgrade moisture.
- Specify pavement materials that have high wet-condition strengths.
- Increase the resilience of the HMA mixture to withstand deflections.
- Use softer asphalts, which behave more like elastic materials. Hard asphalts are stiffer and have less tensile strength to resist cracking.

Low-temperature cracking is caused by adverse environmental conditions. It is characterized by intermittent transverse cracks, usually across the entire pavement width and occurring at regular intervals. Low-temperature cracks are often called *shrinkage cracks* and occur during cold weather when the tensile strength of the asphalt is less than the tensile stress. Shrinkage cracks occur more often in hard asphalts and in aged pavements. Low-temperature cracking can be reduced by specifying softer binders and controlling the air-void content of the pavement, which reduces the oxidation aging rate of the pavement.

While fatigue cracking is greatly affected by pavement structures, subgrade conditions, and traffic, permanent deformation of rutting is directly a function of shear strength and aggregate properties. Low temperature cracking also correlates closely with binder properties.

Mix Design

The exact methods and tests used for design of hot mix asphalt for a given roadway vary depending on the method being employed for the project. All mix design methods are dependent upon the properties of the aggregates and asphalt binder used in the mix, but which properties and how they are implemented is different from method to method. The rest of this section discusses some of the general properties commonly discussed in mix design, and the following sections cover each of the three most common mix design methods and the material properties that govern the designs.

Aggregates

Mineral aggregate for asphalt pavements can come from a wide variety of sources. Two of the most common sources are natural aggregates and processed aggregates. *Natural aggregates* are mined from river or glacial deposits. They are washed and screened, and are used without further processing. *Processed aggregates* have been quarried, crushed, screened, washed, and otherwise processed to achieve specified performance criteria. *Synthetic aggregates* are made from materials not mined or quarried and are often an industrial by-product, such as slag from steel making. Other products are sometimes introduced to synthetic aggregates, such as expanded clay or shale, to increase skid resistance, or fiber reinforcement, to increase shear strength. *Reclaimed asphalt pavement* (RAP) is another source of aggregate for asphalt pavements. Using RAP as a source for aggregate is becoming common practice as costs and environmental practice make RAP more competitive.

The aggregate properties most important to mix design are shear strength and gradation density. *Shear strength*, the total load the aggregate can resist before it fails in shear, is necessary to resist repeated load applications and is a function of both the inherent strength of the aggregate and the nesting of the aggregate pieces (a "knitting action" primarily dependent on the internal friction of the compacted aggregate matrix). Rough-textured faces that are made when aggregate is crushed provide more resistance than the rounded, smooth-textured aggregate faces prevalent with river gravel. Crushed aggregate also has fewer voids than rounded aggregate, requiring less binder.

The amount of voids in aggregate is controlled by *gradation*, or the process of selecting a mixture of sizes ranging from the largest design size to the smallest to fit together in a dense, nearly solid mass. The amount of natural sand is often limited because natural sand tends to be rounded and have poor internal friction. Gradation uses a 0.45 power gradation technique to define cumulative particle size distribution. This method graphs percent passing against sieve size in millimeters raised to the 0.45 power, which results in a linear curve for most aggregates. For example, for a 19 mm maximum size, the percent passing curve would look like Fig. 5.12. The percent passing is calculated using Eq. 5.9.

$$\text{percent passing} = \left(\frac{\text{actual particle size}}{\text{maximum aggregate size}}\right)^{0.45} \qquad 5.9$$

Figure 5.12 *Maximum Density Gradation for 19 mm Maximum Size Using 0.45 Power for Sieve Opening Size*

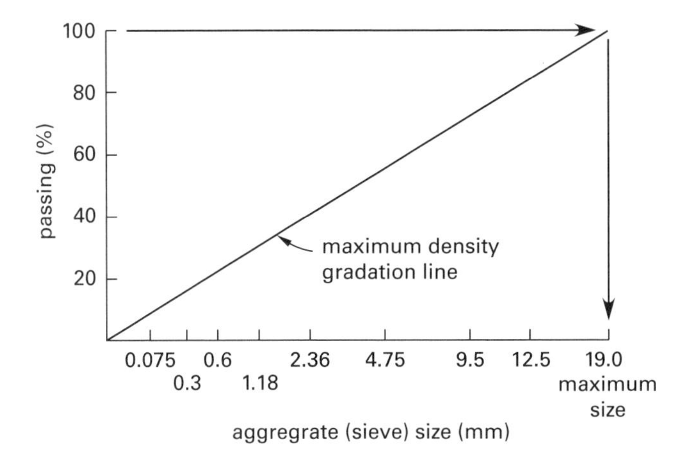

Reprinted with permission from the Asphalt Institute, *Superpave Mix Design* (*SP-2*), 2001, Fig. 3.9.

The *maximum density gradation line* (also known as the *0.45 power line*) shown in Fig. 5.12 shows a gradation where the aggregates fit together in the densest configuration. Superpave size definitions include

- *maximum size:* one size larger than the nominal maximum size
- *nominal maximum size:* one sieve size larger than the first sieve to retain more than 10%

Superpave aggregates are specified by setting control points on the curve, as shown in Fig. 5.13, at the nominal maximum size, an intermediate sieve size (2.36 mm), and at the smallest sieve size (0.075 mm). The control points determine master gradation ranges. A *restricted zone* is a band set between an intermediate sieve and the 0.03 mm sieve.

Avoiding the maximum density line in the restricted zone is recommended so the mixture does not possess an excess of fine sand in relation to the total sand. An excess of fine sand creates a "sanded" mixture that can have inadequate voids in the mineral aggregate to allow enough asphalt for adequate durability.

Figure 5.13 *Superpave Gradation Limits for 12.5 mm Superpave Mixture*

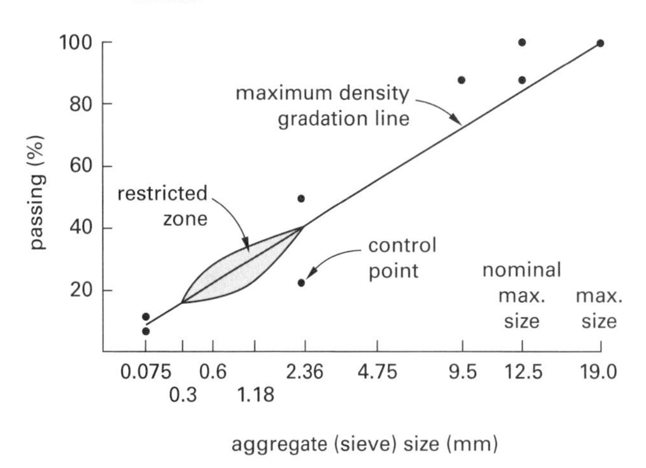

Reprinted with permission from the Asphalt Institute, *Superpave Mix Design* (*SP-2*), 2001, Fig. 3.10.

The original Superpave specifications included the restricted zone as a guideline for HMA mixes. However, the restricted zone is not a requirement, and the value of the concept has been disputed recently. NCHRP Report 464, *The Restricted Zone in the Superpave Aggregate Gradation Specification*, suggests that the restricted zone is unnecessary if the fine aggregate angularity and all other volumetric requirements are satisfied. Studies have shown that mixtures using aggregates that fall within the restricted zone that otherwise meet all Superpave requirements can perform as well as aggregates outside the restricted zone. The restricted zone was removed from AASHTO's Superpave specification (AASHTO M323) based on the recommendations given in NCHRP Report 464.

The design aggregate structure is a distribution of aggregate sizes that meets the Superpave gradation requirements. Table 5.3 shows the five commonly used Superpave mix designations. The recommended maximum aggregate size is 0.33–0.40 times the minimum lift thickness for Superpave design and 0.40–0.50 times the minimum lift thickness for Marshall or Hveem mix designs. This maximum size is necessary to achieve the workability needed for proper compaction. Superpave aggregates require more careful stockpiling, surge-bin operations, and handling to prevent segregation, especially the larger size aggregate mixtures, in order to meet the mix performance requirements.

Binder

Asphalt binder has three major properties: viscoelasticity, temperature susceptibility, and aging. Viscoelasticity of asphalt cement is discussed in Sec. 5.5. *Temperature susceptibility* means that the asphalt stiffness changes with variations in temperature. For this reason, almost every asphalt cement and mixture

Table 5.3 Superpave Aggregate Mixture Gradations

Superpave designation	nominal maximum size (mm)	maximum size (mm)
37.5 mm	37.5	50.0
25.0 mm	25.0	37.5
19.0 mm	19.0	25.0
12.5 mm	12.5	19.0
9.5 mm	9.5	12.5
4.75 mm	4.75	9.5

Reprinted with permission from the Asphalt Institute, *The Asphalt Handbook, Manual Series No. 4 (MS-4)*, Table 4.3, © 2007.

test is performed with a specified temperature component. Performance is also affected by the duration loading. That is, slow loading can be simulated by higher temperatures, and a fast loading rate can be simulated by lower temperatures.

The process of *oxidation* causes aging in asphalt binders, which are organic substances. Oxidation causes hardening, making the asphalt concrete pavement more brittle and less elastic. The aging process occurs more quickly at higher temperatures, making oxidation more of a concern in hot desert climates.

Asphalt Mixture Volumetric Properties

HMA mixture design, including Superpave, follows volumetric procedures for mix design. Volume quantities are used primarily because air voids cannot be weighed. Once the mix is designed, the volume quantities can be converted to mass to provide a job-mix formula. The definition of volumetric properties applies both to paving mixtures that have been compacted in the laboratory and to the undisturbed samples that have been cut from a pavement in the field.

Mineral aggregate used in HMA is porous and has the ability to absorb water and asphalt. The amount of absorption varies with the type of aggregate. Water can be absorbed into the surface pores, as well as internal pores. Asphalt, due to its lower viscosity, can be absorbed partially into the surface pores, but rarely permeates into the interior pores of denser aggregates. Asphalt also coats the aggregate surface. This coating seals the aggregate from moisture penetration and provides the cementitious action to bind the aggregate particles together. Figure 5.14 illustrates the coating action of aggregate in asphalt.

The following measurements and calculations are needed to determine the volumetric properties of a compacted paving mixture.

1. Measure the bulk specific gravity of the coarse aggregate and the fine aggregate.
2. Measure the specific gravity of the asphalt cement and the mineral filler.
3. Calculate the bulk specific gravity of the aggregate combination in the paving volume.

Figure 5.14 Asphalt Binder Coating and Aggregate Absorption

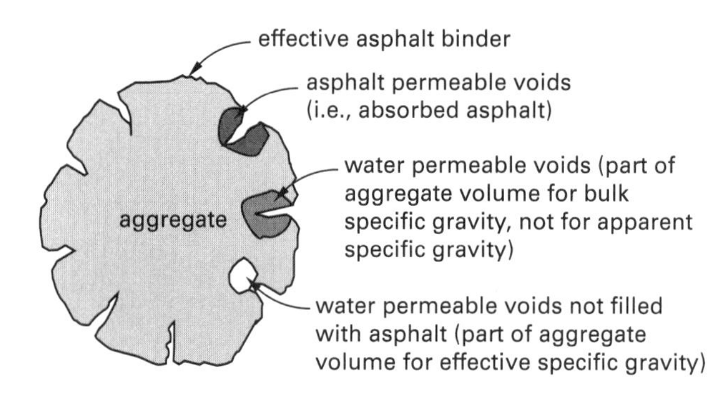

Reprinted with permission from the Asphalt Institute, *Superpave Mix Design (SP-2)*, 2001, Fig. 4.1.

4. Measure the maximum specific gravity of the loose paving mixture.
5. Measure the bulk specific gravity of the compacted paving mixture.
6. Calculate the effective specific gravity of the aggregate.
7. Calculate the maximum specific gravity at other asphalt contents.
8. Calculate the asphalt absorption of the aggregate.
9. Calculate the effective asphalt content of the paving mixture.
10. Calculate the percentage of voids in mineral aggregate in the compacted paving mixture.
11. Calculate the percentage of air voids in the compacted paving mixture.
12. Calculate the percentage of voids filled with asphalt in the compacted paving mixture.

Aggregate is described using three specific gravity terms: bulk specific gravity, apparent specific gravity, and effective specific gravity. For all three, the aggregate mass and the water mass are at a specified temperature, the aggregate and the water are in air of equal density, and the water is gas-free distilled.

Bulk specific gravity, G_{sb}, is the ratio of the mass of a unit volume of permeable material to the mass of an equal volume of water. Voids in the aggregate include both those that are permeable and those that are impermeable normal to the material. The bulk specific gravity of mineral filler is difficult to determine accurately. The error is usually negligible if the apparent specific gravity of the filler is substituted for the bulk specific gravity. To calculate the bulk specific gravity of aggregate, use Eq. 5.10.

$$G_{sb} = \frac{P_1 + P_2 + \cdots + P_n}{\frac{P_1}{G_1} + \frac{P_2}{G_2} + \cdots + \frac{P_n}{G_n}} \qquad 5.10$$

G_{sb} is the bulk specific gravity for the total aggregate; P_1, P_2, and P_n are individual percentages of mass of aggregate; and G_1, G_2, and G_n are individual bulk specific gravities of aggregate (e.g., coarse, fine).

Apparent specific gravity, G_{sa}, is the ratio of the mass of a unit volume of an impermeable material to the mass of an equal volume of water. The apparent specific gravity is found using Eq. 5.11, where m is the mass and W is weight.

$$G_{sa} = \frac{m}{V_{\text{aggregate}}\rho_{\text{water}}} \quad \text{[SI]} \quad 5.11(a)$$

$$G_{sa} = \frac{W}{V_{\text{aggregate}}\gamma_{\text{water}}} \quad \text{[U.S.]} \quad 5.11(b)$$

Effective specific gravity, G_{se}, is the ratio of the mass of a unit volume of a permeable material, excluding voids permeable to asphalt, to the mass and equal volume of water. Use Eq. 5.12 to determine the effective specific gravity of aggregate.

$$G_{se} = \frac{P_{mm} - P_b}{\frac{P_{mm}}{G_{mm}} - \frac{P_b}{G_b}} \quad 5.12$$

G_{se} is the effective specific gravity of aggregate. G_{mm} is the maximum specific gravity of paving mixture excluding air voids. P_{mm} is the percentage of mass of total loose mixture, which equals 100%. P_b is the asphalt content, the percentage of the total mass of the mixture. G_b is the specific gravity of asphalt.

The test for *maximum specific gravity*, G_{mm}, described in AASHTO T209 or ASTM D2041 should be performed on at least two, or even three, specimens that are at or very close to the estimated optimum asphalt content. Since asphalt absorption varies very little with changes in asphalt content, the effective specific gravity of the aggregate is considered constant. Averaging the G_{se} results, the maximum specific gravity for other than the optimum asphalt content can be obtained from Eq. 5.13.

$$G_{mm} = \frac{P_{mm}}{\frac{P_s}{G_{se}} - \frac{P_b}{G_b}} \quad 5.13$$

P_s is the percentage of aggregate in the total mix and is found by subtracting the percentage of asphalt, P_b, from 100%.

The volume of asphalt binder absorbed by an aggregate is almost always less than the volume of water absorbed. As a result, the value of the effective specific gravity for an aggregate should be between its bulk and apparent specific gravities. Should the effective specific gravity fall outside of these limits, its value must be assumed to be incorrect, and the composition of the mix in terms of aggregate and total asphalt content should be rechecked to find the source of the error.

The volumetric relationship of compacted asphalt is illustrated in Fig. 5.15. There are three types of voids: air voids, voids filled with asphalt, and voids filled with absorbed asphalt.

Figure 5.15 *Volume Components of Compacted Asphalt*

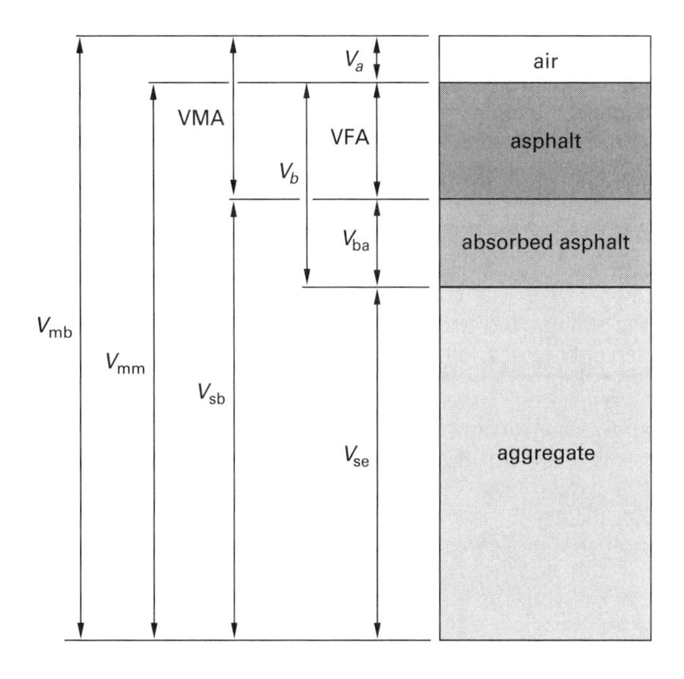

VMA = voids in mineral aggregate
V_{mb} = bulk volume of compacted mix
V_{mm} = voidless volume of paving mix
VFA = voids filled with asphalt
V_a = air voids
V_b = asphalt
V_{ba} = absorbed asphalt
V_{sb} = mineral aggregate (by bulk specific gravity)
V_{se} = mineral aggregate (by effective specific gravity)

Voids in the mineral aggregate, VMA, is the volume of the intergranular void space between the aggregate particles of a compacted paving mixture, including the air voids and the effective asphalt content. VMA is expressed as a percentage of the sample's total volume.

VMA is calculated based on the bulk specific gravity of the aggregate by subtracting the volume of the aggregate, determined by its bulk specific gravity, from the bulk volume of the compacted paving mixture. The mix composition is determined as a percentage of the total volume mass, using Eq. 5.14.

$$\text{VMA} = 100\% - \frac{G_{mb}P_s}{G_{sb}} \quad 5.14$$

Construction

When the mix composition is determined as a percentage of the aggregate mass, Eq. 5.15 is used.

$$\mathrm{VMA} = 100\% - \left(\frac{G_{\mathrm{mb}}}{G_{\mathrm{sb}}}\right)\left(\frac{100\%}{100\% + P_b}\right) \times 100\% \qquad 5.15$$

The *effective asphalt content*, P_{be}, of a paving mixture is the asphalt coating on the outside of the aggregate. It serves as the glue that holds the matrix of aggregates together and determines the performance of the mixture. The effective asphalt content is the outside coating minus the quantity lost by absorption and can be found from Eq. 5.16.

$$P_{\mathrm{be}} = P_b - \frac{P_{\mathrm{ba}} P_s}{100\%} \qquad 5.16$$

Air voids, V_a, is the total volume of the small pockets of air between the coated aggregate particles throughout the compacted paving mixture. V_a is expressed as a percentage of the bulk volume of the compacted paving mixture.

Asphalt absorption, P_{ba}, is expressed as a percentage of the mass of the aggregate rather than as a percentage of the total mass of the mixture. It can be found using Eq. 5.17.

$$P_{\mathrm{ba}} = \frac{G_b(G_{\mathrm{se}} - G_{\mathrm{sb}})}{G_{\mathrm{sb}} G_{\mathrm{se}}} \times 100\% \qquad 5.17$$

The *percentage of air voids,* P_a, or the *voids in the total mixture*, VTM, is calculated using Eq. 5.18.

$$P_a = \mathrm{VTM} = \frac{G_{\mathrm{mm}} - G_{\mathrm{mb}}}{G_{\mathrm{mm}}} \times 100\% \qquad 5.18$$

Percentage of voids filled with asphalt, VFA, is the percentage of the volume of intergranular void space between the aggregate particles that is occupied by the effective asphalt. VFA does not include the absorbed asphalt and is found using Eq. 5.19.

$$\mathrm{VFA} = \frac{\mathrm{VMA} - \mathrm{VTM}}{\mathrm{VMA}} \times 100\% \qquad 5.19$$

Superpave asphalt mixtures use several AASHTO and ASTM test standards for finding the specific gravity of the mix components, as shown in Table 5.4.

Dust is calculated as the percentage of the material's mass passing through the 0.075 mm sieve (using wet-sieve analysis) divided by the effective asphalt binder content. The *dust-to-binder ratio*, also known as the *dust proportion*, DP, is calculated as

$$\mathrm{DP} = \frac{P_{0.075\,\mathrm{mm}}}{P_{\mathrm{be}}} \qquad 5.20$$

Table 5.4 *Specific Gravity Test Standards for Asphalt Paving Mixture Components*

material	measured characteristic	symbol	test method AASHTO	test method ASTM
asphalt content	bulk specific gravity	P_b	T228	D70
mineral filler	bulk specific gravity	G_{sb}	T100	D854
coarse aggregate	bulk specific gravity	P_1	T85	C127
fine aggregate	bulk specific gravity	P_2	T84	C128
loose paving mixture	maximum specific gravity	G_{mm}	D2041	T209
compacted paving mixture	bulk specific gravity	G_{mb}	D1188 D2726	T166

$P_{0.075\,\mathrm{mm}}$ is the aggregate content passing through the 0.075 mm sieve.

Some calculations for design of a concrete mix require the specific weight of the materials be known. The *specific weight* of a material, γ, is a customary U.S. value measured in lbf/ft^3. Specific weight of a material can be found using Eq. 5.21. V is the *volumetric ratio* of the material, a dimensionless value representing the ratio of the volume of the material to the total volume of the concrete mix. G_{sb} is the bulk specific gravity of the material, and γ_{water} is the specific weight of water, 62.4 lbf/ft^3.

$$\gamma = V G_{\mathrm{sb}} \gamma_{\mathrm{water}} \qquad \text{[U.S. only]} \qquad 5.21$$

The weight of each material per sack of cement can be found from Eq. 5.22. γ is the specific weight of the material. 94 lbf/sack is the standard weight of concrete.

$$Q = \gamma \frac{94\ \frac{\mathrm{lbf}}{\mathrm{sack}}}{\gamma_c} \qquad \text{[U.S. only]} \qquad 5.22$$

In some cases, water absorption in a concrete mix is less than the amount of moisture included with the aggregates. In these cases, there will be excess water introduced into the mix along with the aggregate, and the excess water must be subtracted from the mix. Excess water is found by simply subtracting the water absorbed by the aggregates from the total water added to the mix.

Example 5.4

Using the following table of paving mixture data, calculate the bulk specific gravity, the effective specific gravity, the maximum specific gravity, the asphalt absorption, the effective asphalt content, the percentage of air voids, the VMA, and the VFA for the aggregate mixture.

component	specific gravity	percentage of total mix (%)	percentage of total aggregate (%)
asphalt content	$G_b = 1.030$	$P_b = 5.3$	$P_b = 5.6$
coarse aggregate	$G_1 = 2.716$	$P_1 = 47.4$	$P_1 = 50.0$
fine aggregate	$G_2 = 2.689$	$P_2 = 47.3$	$P_2 = 50.0$
loose paving mixture	$G_{\text{mm}} = 2.535$	–	–
compacted paving mixture	$G_{\text{mb}} = 2.442$	–	–

Solution

Use the given table to determine values for the aggregate mixture. Using Eq. 5.10, the bulk specific gravity is

$$\begin{aligned} G_{\text{sb}} &= \frac{P_1 + P_2 + \cdots + P_n}{\dfrac{P_1}{G_1} + \dfrac{P_2}{G_2} + \cdots + \dfrac{P_n}{G_n}} \\ &= \frac{50.0\% + 50.0\%}{\dfrac{50.0\%}{2.716} + \dfrac{50.0\%}{2.689}} \\ &= 2.703 \end{aligned}$$

Using Eq. 5.12, the effective specific gravity is

$$\begin{aligned} G_{\text{se}} &= \frac{P_{\text{mm}} - P_b}{\dfrac{P_{\text{mm}}}{G_{\text{mm}}} - \dfrac{P_b}{G_b}} \\ &= \frac{100\% - 5.3\%}{\dfrac{100\%}{2.535} - \dfrac{5.3\%}{1.030}} \\ &= 2.761 \end{aligned}$$

The percentage of aggregate in the mix is

$$\begin{aligned} P_s &= 100\% - P_b \\ &= 100\% - 5.3\% \\ &= 94.7\% \end{aligned}$$

Using Eq. 5.13, the maximum specific gravity is

$$\begin{aligned} G_{\text{mm}} &= \frac{P_{\text{mm}}}{\dfrac{P_s}{G_{\text{se}}} - \dfrac{P_b}{G_b}} \\ &= \frac{100\%}{\dfrac{94.7\%}{2.761} - \dfrac{5.3\%}{1.030}} \\ &= 3.430 \end{aligned}$$

Using Eq. 5.17, the asphalt absorption is

$$\begin{aligned} P_{\text{ba}} &= \frac{G_b(G_{\text{se}} - G_{\text{sb}})}{G_{\text{sb}} G_{\text{se}}} \times 100\% \\ &= \frac{(1.030)(2.761 - 2.703)}{(2.703)(2.761)} \times 100\% \\ &= 0.8\% \end{aligned}$$

Using Eq. 5.16, the effective asphalt content is

$$\begin{aligned} P_{\text{be}} &= P_b - \frac{P_{\text{ba}} P_s}{100\%} \\ &= 5.3\% - \frac{(0.8\%)(94.7\%)}{100\%} \\ &= 4.54\% \end{aligned}$$

Using Eq. 5.18, the percentage of air voids between the coated aggregate particles is

$$\begin{aligned} \text{VTM} &= \frac{G_{\text{mm}} - G_{\text{mb}}}{G_{\text{mm}}} \times 100\% \\ &= \frac{2.535 - 2.442}{2.535} \times 100\% \\ &= 3.68\% \end{aligned}$$

Using Eq. 5.14, the percentage of VMA is

$$\begin{aligned} \text{VMA} &= 100\% - \frac{G_{\text{mb}} P_s}{G_{\text{sb}}} \\ &= 100\% - \frac{(2.442)(94.7\%)}{2.703} \\ &= 14.4\% \end{aligned}$$

Using Eq. 5.19, the percentage of VFA is

$$\begin{aligned} \text{VFA} &= \frac{\text{VMA} - \text{VTM}}{\text{VMA}} \times 100\% \\ &= \frac{14.4\% - 3.68\%}{14.4\%} \times 100\% \\ &= 74.4\% \end{aligned}$$

6. MARSHALL MIX DESIGN

The Marshall method was developed in the late 1930s and was subsequently modified by the U.S. Army Corps of Engineers for higher airfield tire pressures and loads. It was subsequently adopted by the entire U.S. military. Although the method disregards shear strength, it considers strength, durability, and voids. It also has the advantages of being a fairly simple procedure, not requiring complex equipment, and being portable.

The primary goal of the Marshall method is to determine the optimum asphalt content. The method starts by determining routine properties of the chosen asphalt and aggregate. The specific gravities of the components are determined by standard methods, as are the mixing and blending temperatures from the asphalt viscosities. (See Fig. 5.16.) Trial blends with varying asphalt contents are formulated, and samples are heated and compacted. Density and voids are determined.

Figure 5.16 Determining Asphalt Mixing and Compaction Temperatures*

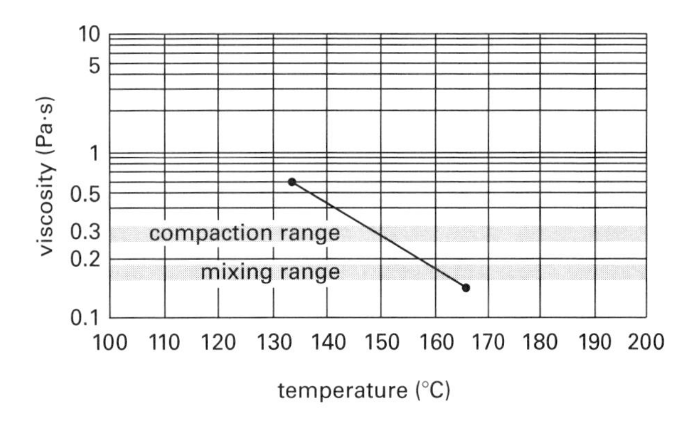

*Compaction and mixing temperature are found by connecting two known temperature-viscosity points with a straight line.

The stability portion of the Marshall procedure measures the maximum load supported by the test specimen at a loading rate of 2 in/min (50.8 mm/min). The load is increased until it reaches a maximum, then when the load just begins to decrease, the loading is stopped and the maximum load is recorded. During the loading, an attached dial gauge measures the specimen's plastic flow. The flow value is recorded in 0.01 in (0.25 mm) increments at the same time the maximum load is recorded.

Various parameters are used to determine the best mix. Three asphalt contents are averaged to determine the *target optimum asphalt content*: the 4% air voids content, the maximum stability content, and the maximum density (unit weight) content. The target optimum asphalt content is subsequently validated by checking against flow and VMA minimums.

The *Marshall test method* is a density voids analysis and a stability-flow test of the compacted test specimen. The size of a test specimen for a Marshall test is 2.5 in × 4 in (64 mm × 102 mm) diameter × length. Specimens are prepared using a specified procedure for heating, mixing, and compacting the asphalt aggregate mixtures. When preparing data for a Marshall mix design, the stability of the samples that are not 2.5 in (63.5 mm) high must be multiplied by the correlation ratios in Table 5.5. The Marshall test device and typical results are shown in Fig. 5.17.

Table 5.5 Stability Correlation Factors

specimen volume (cm^3 (in^3))	approximate specimen thickness (mm (in))	correlation ratio
406–420 (25.0–25.6)	50.8 (2.0)	1.47
421–431 (25.7–26.3)	52.4 (2.06)	1.39
432–443 (26.4–27.0)	54.0 (2.13)	1.32
444–456 (27.1–27.8)	55.6 (2.25)	1.25
457–470 (27.9–28.6)	57.2 (2.25)	1.19
471–482 (28.7–29.4)	58.7 (2.31)	1.14
483–495 (29.5–30.2)	60.3 (2.37)	1.09
496–508 (30.2–31.0)	61.9 (2.44)	1.04
509–522 (31.1–31.8)	63.5 (2.5)	1.00
523–535 (31.9–32.6)	64.0 (2.52)	0.96
536–546 (32.7–33.3)	65.1 (2.56)	0.93
547–559 (33.4–34.1)	66.7 (2.63)	0.89
560–573 (34.2–34.9)	68.3 (2.69)	0.86
574–585 (35.0–35.7)	71.4 (2.81)	0.83
586–598 (35.8–36.5)	73.0 (2.87)	0.81
599–610 (36.6–37.2)	74.6 (2.94)	0.78
611–625 (37.3–38.1)	76.2 (3.0)	0.76

(Multiply cm^3 by 0.061 to obtain in^3.)
(Multiply mm by 0.0394 to obtain in.)

The objective of a Marshall test is to find the optimum asphalt content for the blend or gradation of aggregates. To accomplish this, a series of test specimens is prepared for a range of different asphalt contents so that when plotted, the test data will show a well-defined optimum value. Tests are performed in 1/2% increments of asphalt content, with at least two asphalt content samples above the optimum and at least two below.

After each sample is prepared, it is placed in a mold and compacted with 35, 50, or 75 hammer blows, as specified by the design traffic category. The compaction hammer is dropped from a height of 18 in (457 mm), and after compaction, the sample is removed and subjected to the bulk specific gravity test, stability and flow test, and density and voids analysis. Results of these tests are plotted, a smooth curve giving the best fit is drawn for each set of data, and an optimum asphalt content is determined that meets the criteria of (a) maximum stability and unit weight and (b) median of limits for percent air voids from Table 5.6. (See Fig. 5.18 and Fig. 5.19.)

The *stability* of the test specimen is the maximum load resistance in pounds that the test specimen will develop at 140°F (60°C). The *flow value* is the total movement (strain) in units of 1/100 in occurring in the specimen between the points of no load and maximum load during the stability test.

The *optimum asphalt content* for the mix is the numerical average of the values for the asphalt content. This value represents the most economical asphalt content that will satisfactorily meet all of the established criteria.

Table 5.6 Marshall Mix Design Criteria

mix criteria	light traffic ESALs $< 10^4$ surface and base min–max	medium traffic $10^4 <$ ESALs $< 10^6$ surface and base min–max	heavy traffic ESALs $> 10^6$ surface and base min–max
compactive effort, no. of blows/face	35	50	75
stability, N (lbf)	3336 (750)–NA	5338 (1200)–NA	8006 (1800)–NA
flow, 0.25 mm (0.01 in)	8–18	8–16	8–14
air voids, %	3–5	3–5	3–5
VMA, %	Varies with aggregate size. (See Fig. 5.19.)		

(Multiply N by 0.225 to obtain lbf.)

Reprinted with permission of the Asphalt Institute from *The Asphalt Handbook, Manual Series No. 4 (MS-4)*, 7th ed., Table 4.6, © 2007.

Figure 5.17 Marshall Testing

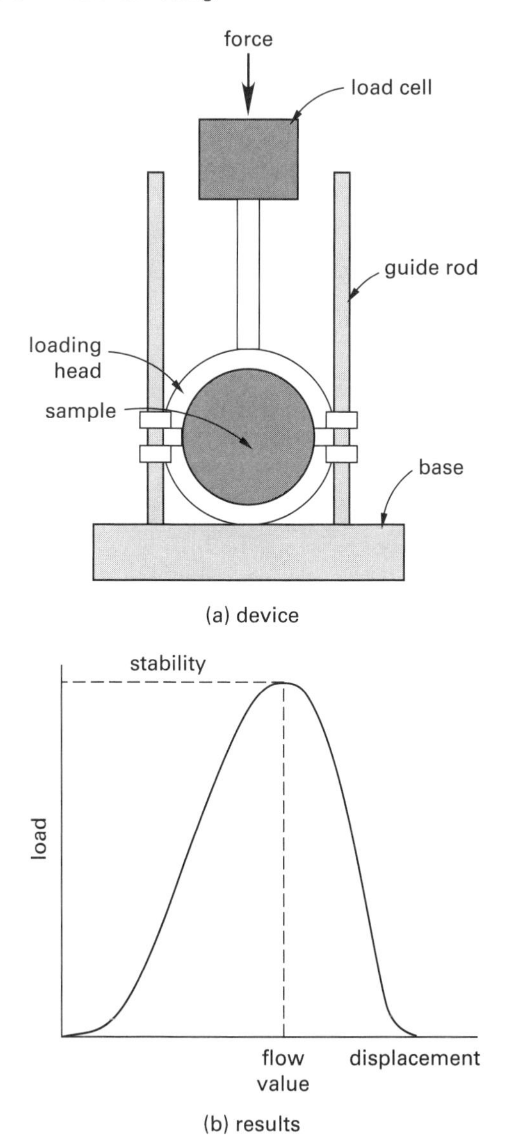

Figure 5.18 Typical Marshall Mix Design Test Results

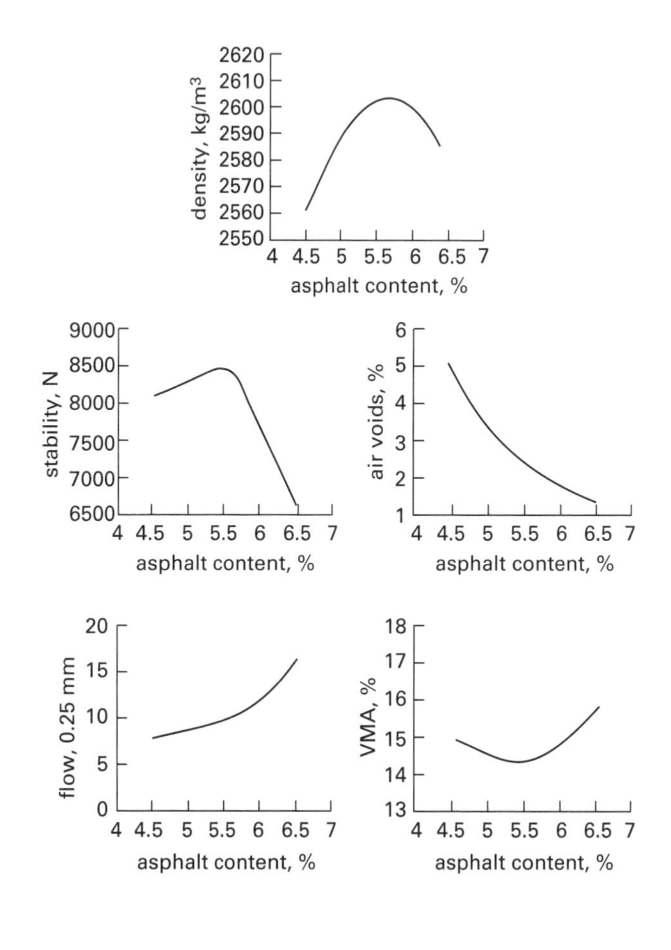

Construction

Figure 5.19 VMA Criterion for Mix Design

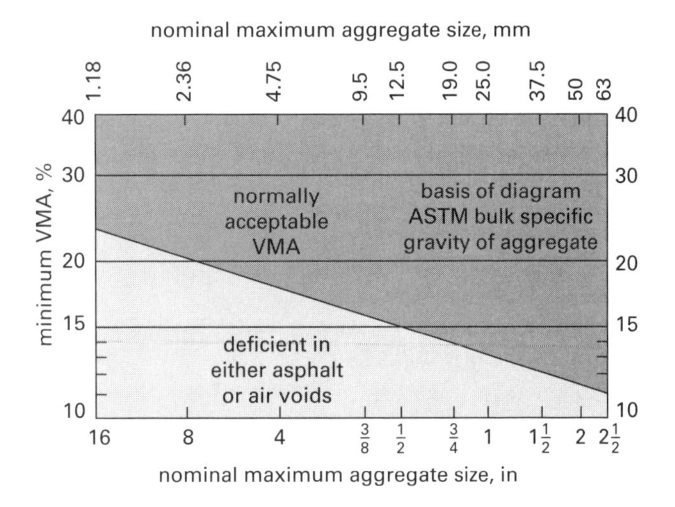

Use the following procedure to determine the optimum asphalt content for the mix. Examples of Marshall mix design graphs are given in Fig. 5.18.

step 1: Calculate the air voids in the mix, VTM, for each asphalt content using Eq. 5.18. Graph the VTM values, as percentages, against the asphalt content percentages.

step 2: Calculate the bulk specific gravity, G_{sb}, using Eq. 5.10, then determine the voids in mineral aggregate, VMA, for each asphalt content from Eq. 5.15. Graph the VMAs against the asphalt content percentages.

step 3: Graph the remaining data results, including the Marshall stability, theoretical density, and flow against the asphalt content percentages.

step 4: Using the graphs, determine the asphalt contents that coincide with the maximum stability, maximum unit weight, and the median air voids. The median air voids can be determined from Table 5.6, but a value of 4% is typically used. The asphalt content for VTM corresponds to the median air voids percentage.

step 5: Determine the optimum asphalt content by adding the asphalt contents determined in step 4 and dividing by three.

step 6: Verify the optimum asphalt content criteria for stability and flow against Table 5.6, then check the voids in mineral aggregate against Fig. 5.16. The optimum asphalt content is determined when all criteria are met.

Example 5.5

An asphalt mix uses an aggregate blend of 56% coarse aggregate (specific gravity of 2.72) and 44% fine aggregate (specific gravity of 2.60). The maximum aggregate size for the mixture is 3/4 in (19 mm). The resulting asphalt mixture is to have a specific gravity of 2.30. The asphalt content is to be selected on the basis of medium traffic and the following Marshall test data.

maximum content, % by weight of mix	Marshall stability (N)	flow (0.25 mm)	theoretical density $(kg/m)^3$	mixture specific gravity
4	4260	10.0	2300	2.42
5	4840	12.3	2330	2.44
6	5380	14.4	2340	2.44
7	5060	16.0	2314	2.40
8	3590	19.0	2250	2.30

Solution

step 1: Calculate air voids for each asphalt content and graph these values versus asphalt content. For example, for the 4% asphalt mixture, using Eq. 5.18,

$$\begin{aligned}\text{VTM}_{4\%} &= \frac{G_{\text{mm}} - G_{\text{mb}}}{G_{\text{mm}}} \times 100\% \\ &= \frac{2.42 - 2.30}{2.42} \times 100\% \\ &= 4.96\%\end{aligned}$$

step 2: Calculate the voids in mineral aggregate for each asphalt content. Use Eq. 5.10.

$$\begin{aligned}G_{\text{sb}} &= \frac{100\%}{\frac{P_{\text{ca}}}{G_{\text{ca}}} + \frac{P_{\text{fa}}}{G_{\text{fa}}}} = \frac{100\%}{\frac{56\%}{2.72} + \frac{44\%}{2.60}} \\ &= 2.67\end{aligned}$$

Use Eq. 5.14.

$$\begin{aligned}\text{VMA}_{4\%} &= 100\% - \frac{G_{\text{mb}} P_s}{G_{\text{sb}}} \\ &= 100\% - \frac{(2.30)(96\%)}{2.67} \\ &= 17.3\%\end{aligned}$$

step 3: Graph the results.

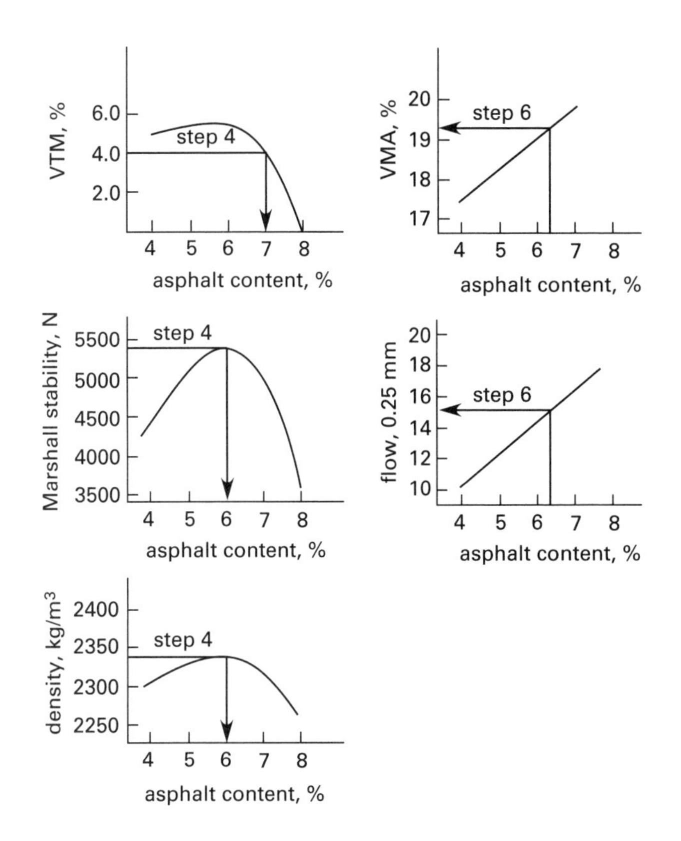

step 4: Obtain asphalt contents for maximum stability, maximum unit weight, and median air voids (4% from Table 5.6) from the graphs.

step 5: Determine the optimum AC.

$$\text{optimum AC} = \frac{\text{AC}_{\text{stability}} + \text{AC}_{\text{density}} + \text{AC}_{4\%\,\text{air voids}}}{3} = \frac{6.0\% + 6.0\% + 7.0\%}{3} = 6.33\%$$

step 6: Check stability against the criteria for light traffic. Stability ≈ 5280 > 3336 [OK]. Check VMA = 19.25 against Fig. 5.16 [OK]. Check flow = 14.8 against Table 5.6 [OK].

7. HVEEM MIX DESIGN

The Hveem mix design method was developed in the 1920s and has been extensively used in some western states. Like the Marshall method, the goal is to determine the optimum asphalt content. Unlike the Marshall method, it has the additional sophistications of measuring resistance to shear and considering asphalt absorption by aggregates. It has the disadvantage of requiring more specialized and nonportable equipment for mixing, compaction, and testing.

The method makes three assumptions: (1) The optimum asphalt binder content depends on the aggregate surface area and absorption, (2) Stability is a function of aggregate particle friction and mix cohesion, and (3) HMA durability increases with asphalt binder content.

After selecting the materials, the *centrifuge kerosene equivalent* (CKE) of the fine aggregate and the retained surface oil content of the coarse aggregate are determined. These measures of surface absorption are used to estimate the optimum asphalt content.[1] Trial blends are formulated over the asphalt range of [CKE–1%, CKE+2%] in 0.5% increments. Then, specimens of the trial blends are prepared by heating and densifying in a California kneading compactor.[2]

A Hveem stabilometer, a closed-system triaxial test device, is used to determine the stability (i.e., the horizontal deformation under axial load). The stabilometer applies an increasing load to the top of the sample at a predetermined rate. As the load increases, the lateral pressure is read at specified intervals. The stability (*stabilometer value*), S, is calculated as

$$S = \frac{22.2}{\frac{p_h D}{p_v - p_h} - 0.222} \qquad 5.23$$

In Eq. 5.23, p_v is the vertical pressure, which is typically 400 lbf/in^2 (2.8 MPa), and p_h is the corresponding horizontal pressure. D is the deflection in units of 0.01 in (0.25 mm). The stabilometer value ranges from 0 to 90. Zero represents a liquid, a condition where lateral pressure is equal to vertical pressure. 90 represents an incompressible solid, a condition where there is no lateral pressure regardless of the vertical pressure. Minimum stabilities vary according to a roadway's traffic volume, measured in *equivalent single-axle loads* (ESALs). Minimum stabilities of 30, 35, and 37 are typically specified for light (ESALs $< 10^4$), medium ($10^4 <$ ESALs $< 10^6$), and heavy (ESALs $> 10^6$) traffic, respectively.

[1]It is important that the trial blends include asphalt contents greater than and less than the optimum content. The purpose of determining the CKE and surface oil content is to estimate the optimum content prior to formulating trial blends. However, it is also practical to simply formulate trial blends within a range of 4–7% of asphalt, varying each blend by an increment such as 0.5%. Because of this alternative methodology of bracketing the optimum asphalt content, the CKE process has become essentially obsolete.

[2]The kneading compactor produces compactions that are more similar to pavements that have been roller-compacted with steel- and rubber-tired rollers.

HMA mixtures rarely fail from cohesion, but oil mixtures may require subsequent testing in a cohesiometer to determine cohesive strength. Basically, the sample used in the stability test is loaded as a cantilevered beam until it fails. The cohesiometer value, C, is determined from the mass of the load (shot), L, in grams, the width (diameter) of the sample, W, in inches, and the height of the specimen, H, in inches. A minimum cohesiometer value of 100 is typical for HMA, while acceptance values for oil mixtures may be 50.

$$C = \left(\frac{L}{W}\right)(0.20H + 0.044H^2) \quad \text{[U.S. only]} \qquad 5.24$$

Visual observation, volumetrics (air voids, voids filled with asphalt, and voids in the mineral aggregate), and stability are used to determine the optimum asphalt content. The design asphalt content is selected as the content that produces the highest durability without dropping below a minimum allowable stability. Essentially, as much asphalt binder as possible is used while still meeting minimum stability requirements. The *pyramid method* can be used to select the optimum asphalt content.

step 1: Place all sample asphalt percentages in the base of the pyramid.

step 2: Eliminate any samples with moderate to severe surface *flushing*.[3]

step 3: Place the three highest asphalt percentages in the next pyramid level.

step 4: Eliminate any samples that do not meet the stability requirements.

step 5: Place the two highest asphalt percentages in the next pyramid level.

step 6: Choose the mixture that has the highest asphalt content while still having 4% air voids. This is the optimum mixture.

Example 5.6

An HMA sample with a minimum Hveem stability of 37 is required. The following specimens were prepared. What is the optimum asphalt content?

sample	asphalt content	stability	air voids content	comments
1	4%	37	3.9%	
2	4.5%	38	4.0%	
3	5%	39	4.1%	
4	5.5%	35	3.9%	
5	6%	38	3.8%	moderate surface flushing noted

Solution

Use the pyramid method.

step 1: Place all sample asphalt percentages in the base of the pyramid.

step 2: Eliminate any samples with more than moderate surface flushing. Sample 5 had moderate surface flushing noted, so it should be excluded.

step 3: Of the asphalt contents remaining after step 2, add the three highest percentages in the next level. This eliminates 4% from the pyramid.

step 4: Check the stability of each sample remaining after step 3 and eliminate any sample with a stability less than the stability requirements. A minimum Hveem stability of 37 is required, which eliminates sample 4, 5.5%, which has a stability of 35.

step 5: Place the highest remaining asphalt percentages on the next level. Since only 4.5% and 5% remain, move both to the next level.

step 6: The optimum asphalt content is the highest asphalt content that still has 4% air voids. Therefore, the optimum mixture has an asphalt content of 5%.

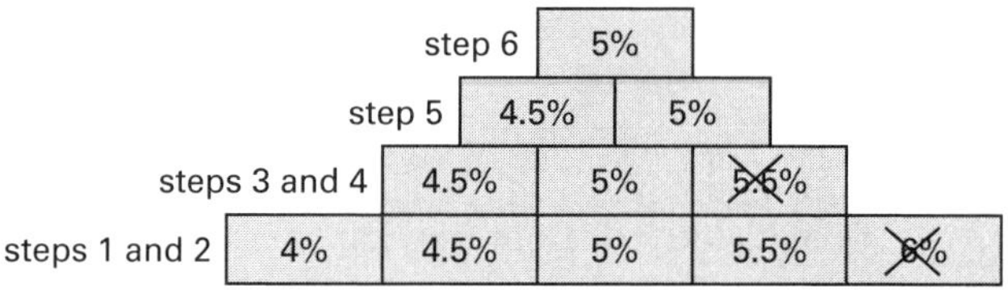

step 6			5%		
step 5		4.5%	5%		
steps 3 and 4		4.5%	5%	~~5.5%~~	
steps 1 and 2	4%	4.5%	5%	5.5%	~~6%~~

8. SUPERPAVE MIX DESIGN PROCEDURES

Superpave is a performance-based system for designing asphalt material with characteristics that eliminate environmental factors that decrease pavement life and increase maintenance costs, such as pavement rutting, cracking due to high or low temperatures, fatigue cracking, and heavy traffic. The Superpave process has three major components: asphalt binder specification, mixture design and analysis, and a computer software system.[4]

Superpave mix design builds upon many traditional asphalt procedures. It emphasizes a closer relationship between laboratory test procedures and in-place asphalt performance. Applicable AASHTO procedures include AASHTO T312, *Standard Method for Preparing and Determining the Density of Hot Mix Asphalt* (HMA) *Specimens by Means of the SHRP Gyratory Compactor*, and AASHTO R30, *Standard Practice for Mixture*

[3]Another name for flushing is *bleeding*. This is an indication of excessive asphalt. A specimen with light flushing will have sheen. With moderate flushing, paper will stick to the specimen. With heavy flushing, surface puddling or specimen distortion will be noted.

[4]A full review of the research can be found in the *National Cooperative Highway Research Program* (NCHRP) *Report 539*, published by the Transportation Research Board, www.TRB.org.

Conditioning of Hot Mix Asphalt (HMA). In general, the procedures described by the Asphalt Institute involve

- selecting asphalt and aggregate materials that meet respective criteria
- developing several aggregate trial blends to meet Superpave gradation requirements
- blending asphalt with trial blends and short-term oven aging the mixtures
- compacting specimens and analyzing the volumetrics of the trial blends
- selecting the best trial blend as the design aggregate structure, then compacting samples of the design aggregate structure at several asphalt contents to determine the design asphalt content

An outline of Superpave mix design procedures is shown in App. 5.D.

Aggregate Properties for Superpave

The Superpave study surveyed pavement experts to determine which aggregate properties and specifications were the most effective. These properties, called *consensus aggregate properties*, were selected based on their use and specified values. Consensus aggregate properties include coarse aggregate angularity, fine aggregate angularity, flat and elongated particles in coarse aggregate, and clay content. Table 5.7 gives the required minimum values as a function of traffic level and position within the pavement. Superpave measures a roadway's traffic volume using the design ESAL, which is the estimated volume for a 20 yr design period. ESALs are used regardless of what the roadway's actual design life is and are estimated as outlined in AASHTO's *Guide for Design of Pavement Structures*. Detailed explanations of the aggregate properties can be found in *Superpave Mix Design, Superpave Series No. 2 (SP-2)*.

Coarse aggregate angularity is the percentage by weight of aggregates that are less than 0.19 in (4.75 mm) and have at least one fractured face. It has a high degree of internal friction to help prevent permanent deformation and can be measured by test procedures in ASTM D5821, *Determining the Percentage of Fractured Particles in Coarse Aggregate.*

Uncompacted void content of fine aggregate is the percentage of air voids present in loosely compacted aggregates smaller than 0.09 in (2.36 mm). Higher percentages of air voids indicate more fractured faces and can be determined using AASHTO T304, *Uncompacted Void Content of Fine Aggregate.*

Flat and elongated particles, which are characteristic of coarse aggregate, are the percentage by mass of the coarse aggregates with a length to thickness ratio greater than five. Because they tend to break during construction or under loads, it is undesirable to have large amounts in a mixture. Test methods are performed on particles greater than 4.75 mm (0.19 in) and follow procedures described in ASTM D4791, *Flat and Elongated Particles in Coarse Aggregate.*

Clay content is the percentage of clay material by weight contained in the aggregate fraction that is finer than a 4.75 mm sieve. Tests are performed according to AASHTO T176, *Plastic Fines in Graded Aggregates and Soils by Use of the Sand Equivalent Test* (ASTM D2419). The terms sand equivalent and clay content are used interchangeably when referring to the sand equivalent test.

Properties are source-specific and are critical to Superpave performance. Source properties include toughness, soundness, and deleterious materials. These property values are normally set by local jurisdictions depending on the availability of aggregate material within economical reach.

Toughness is the percentage loss of material from an aggregate blend during the *Los Angeles abrasion test* (AASHTO T96, ASTM C131, or ASTM C535). For this test, coarse aggregate larger than 2.36 mm (0.09 in) is impacted and ground by steel spheres.

Table 5.7 *Superpave Aggregate Consensus Property Minimum Requirements*

design ESALs[a] (million)	minimum coarse aggregate angularity (%)		minimum uncompacted void content of fine aggregate (%)		minimum sand equivalent (%)	maximum flat and elongated[c] (%)
	≤100 mm	> 100 mm	≤100 mm	> 100 mm		
<0.3	55/–	–/–	–	–	40	–
0.3 to < 3	75/–	50/–	40	40	40	10
3 to < 10	85/80[b]	60/–	45	40	45	10
10 to < 30	95/90	80/75	45	40	45	10
≥30	100/100	100/100	45	45	50	10

(Multiply mm by 3.937×10^{-2} to obtain in.)

[a]Design ESALs (equivalent single-axle loads) are the anticipated project traffic level expected on the design lane over a 20 year period. Regardless of the actual design life of the roadway, determine the design ESALs for 20 years and choose the appropriate N_{design} level.

[b]85/80 denotes that 85% of the coarse aggregate has one fractured face and 80% has two or more fractured faces.

[c]Criterion based upon a 5:1 maximum-to-minimum ratio.

(If less than 25% of a layer is within 3 in (100 mm) of the surface, the layer may be considered to be below 3 in (100 mm) for mixture design purposes.)

Reprinted with permission from the Asphalt Institute, *The Asphalt Handbook, Manual Series No. 4 (MS-4)*, 7th ed., Table 4.2, © 2007.

The mass percentage of coarse material lost is then measured. An aggregate blend usually has a loss range of 35–40%.

Soundness is the percentage loss of material from weathering, simulated by immersing the test sample in saturated solutions of sodium sulfate or magnesium sulfate, followed by oven drying. The immersion and drying cycle is repeated, and the sample is measured over a required number of cycles. The solution salts penetrate into the void spaces of the aggregate, then rehydrate after drying and exert expansive forces similar to freezing water. Typical values are 10–20% loss over five cycles. The test is performed according to AASHTO T104 or ASTM C88.

Deleterious materials are contaminants such as clay lumps, wood, shale, mica, and coal. The analysis can be performed on both coarse and fine aggregate, and is covered by AASHTO T112 or ASTM C142. Using wet sieving through specified sieves, the mass percentage of material lost is reported as the percentage of clay lumps and friable particles. Allowable percentages vary from a low of 0.2% to a high of 10%, depending on the exact composition of the contaminant.

Binding Properties for Superpave

Performance testing of the binder is an important element of Superpave. AASHTO MP1 specifications (shown in App. 5.C) are selected based on the ESALs and climate conditions of the finished pavement. AASHTO asphalt binder ratings are based on the temperature requirements placed on the binder. Performance grade binders are selected in six-degree intervals (e.g., starting from a high temperature grade of 40°C and a low temperature grade of −10°C). For example, a binder classified as PG 58-22 means that the binder must meet physical property high temperature requirements greater than or equal to 58°C, and low temperature physical requirements less than or equal to −22°C. In addition to the seven common grades shown in App. 5.C, the high and low temperatures extend as far as necessary in standard six-degree increments.

There are three methods of selecting the binder grade.

- *geographic area:* An agency develops a map showing the binder grade that the designer will use based on weather and/or policy decisions.
- *pavement temperature:* The designer determines design pavement temperatures.
- *air temperature:* The designer determines design air temperatures, which are converted to design pavement temperatures.

Asphalt binder temperature grades are selected using the temperatures at 0.79 in (20 mm) below the pavement surface for the high temperature and at the pavement surface for the low temperature. For the high temperature, Eq. 5.25 is used to convert the seven-day high air temperature to the high pavement design temperature. Equation 5.25 is based on the following standard values: heat radiation to the atmosphere (0.70), radiation transmission through air (0.81), solar gain (0.90), and wind speed (14 ft/sec or 4.5 m/s).

$$T_{20\,\text{mm}} = \begin{pmatrix} T_{\text{air}} - (0.00618)(\text{lat})^2 \\ + (0.2289)(\text{lat}) + 42.2^\circ\text{C} \end{pmatrix}(0.9545) - 17.78^\circ\text{C} \quad 5.25$$

$T_{20\,\text{mm}}$ is the high pavement design temperature at a depth of 0.79 in (20 mm), T_{air} is the seven-day, average high air temperature in degrees Celsius, and lat is the geographical latitude of the project in degrees.

For the low design pavement temperature, the preferred method is to use the LTPPBind software developed by the Long-Term Pavement Performance (LTPP) program of the FHWA Turner-Fairbank Highway Research Center. The software deals strictly with the PG binder selection and is a Windows-based program that helps highway agencies choose the most effective and cost-effective Superpave asphalt binder performance grade for a particular site.

Temperature information is available in LTPPBind for more than 6000 weather stations in the United States and Canada and includes more than 20 years of data. The software allows input variations so that the user can quickly compare "what if" scenarios and can customize the inputs at a specific location for a specific project.

For initial estimating purposes without using the software, the pavement surface can be assumed to be equal to the low air temperature. This is a conservative approach because the pavement surface temperature is almost always warmer than the air temperature in cold weather. Another method is to use Eq. 5.26, which is used in the Asphalt Institute manuals.

$$T_{\text{min}} = 0.859\,T_{\text{air}} + 1.7^\circ\text{C} \quad 5.26$$

Reliability refers to the percent probability in a single year that the actual temperature (a one-day low or a seven-day high) will not exceed the design temperature calculated from Eq. 5.25 and Eq. 5.26. The design pavement temperatures represent a 50% reliability level. Standard deviations, in degrees Celsius, can be used to achieve other reliability levels. A 95% reliability level is obtained from the design pavement temperatures plus or minus two standard deviations, $2s$, and a 98% reliability from plus or minus three standard deviations, $3s$. Agencies may select a reliability level in order to maximize specific goals. For example, funding restrictions may be such that 95% reliability is used in order to stretch available tax dollars and gain additional miles of roadway.

As described in Sec. 5.5, asphalt pavements react in a viscoelastic manner according to the duration of applied loads and the frequency of loadings. Binder grades can be adjusted to accommodate expected high traffic volumes and speeds. This practice is known as *grade bumping*, which adjusts the binder selection based exclusively on climate to ensure adequate performance. For example, the binder for slow-moving loads on a roadway with a design traffic of 3,000,000 ESALs should be one high-temperature grade higher than the binder grade selected for climate, as shown in Table 5.8. If the original binder grade were PG 52, PG 58 would be used instead. These adjustments can reduce, for example, the washboard effect caused by a large volume of trucks waiting at a signalized intersection or buses waiting with their engines running to allow passengers at a heavily patronized bus stop to board. Table 5.8 summarizes AASHTO's grade-bumping policy as presented in the Asphalt Institute's *SP-2*.

Table 5.8 *Binder High Temperature Grade Selection Increase Based on Traffic Speed and Traffic Load*[a]

design ESALs[b] (millions)	adjustment to binder PG grade[f] traffic load rate		
	standing[c]	slow[d]	standard[e]
<0.3	–[g]		
0.3 to < 3	2	1	
3 to < 10	2	1	
10 to < 30	2	1	–[g]
≥30	2	1	1

[a]Practically, performance graded binders stiffer than a PG 82-XX should be avoided. In cases where the required adjustment to the high temperature binder grade would result in a grade higher than a PG 82, consideration should be given to specifying a PG 82-XX and increasing the design ESALs by one level (e.g., 10 to <30 million increased to ≥30 million).
[b]Design ESALs are the anticipated project traffic level expected on the design lane over a 20 year period. Regardless of the actual design life of the roadway, determine the design ESALs for 20 years and choose the appropriate N_{design} level.
[c]Standing traffic—where the average traffic speed is less than 11 mph (20 kph).
[d]Slow traffic—where the average traffic speed ranges from 11 mph to 38 mph (20 kph to 70 kph).
[e]Standard traffic—where the average traffic speed is greater than 38 mph (70 kph).
[f]Increase the high temperature grade by the number of grade equivalents indicated (one grade equivalent to 42°F or 6°C). Do not adjust the low temperature grade.
[g]Consideration should be given to increasing the high temperature grade by one grade equivalent.

Reprinted with permission from the Asphalt Institute, *Superpave Mix Design* (*SP-2*), 2001, Table 3.1.

Other factors may influence pavement performance, which the engineer should take into consideration when selecting a binder. Unlike fatigue cracking, which is typically affected by pavement structure, subgrade conditions, and traffic loadings, permanent deformation (i.e., rutting) is a function of shear strength and aggregate properties. Low temperature cracking also correlates closely with binder properties. Careful selection of binder properties can reduce pavement damage and extend the life of the asphalt.

Software such as LTPPBind can be used to help determine the appropriate grade, as can the AASHTOWare, which assists in the entire Superpave mix design process, including quality control and quality assurance guidelines.

Example 5.7

A midwestern location with a latitude of 41.42° has a mean seven-day maximum air temperature of 31°C and a standard deviation of 2°C. In an average year, there is a 50% chance that the seven-day maximum air temperature will exceed 31°C. The low design temperature in the same location has a one-day minimum temperature of −20°C and a standard deviation of 4°C. What are the design temperatures for 50% and 98% reliabilities?

Solution

For 50% reliability, calculate the design high temperature using Eq. 5.25.

$$\begin{aligned} T_{20\,\text{mm}} &= \begin{pmatrix} T_{\text{air}} - (0.00618)(\text{lat})^2 \\ +\,(0.2289)(\text{lat}) + 42.2^\circ\text{C} \end{pmatrix}(0.9545) \\ &\quad - 17.78^\circ\text{C} \\ &= \begin{pmatrix} 31^\circ\text{C} - (0.00618)(41.42^\circ)^2 \\ +\,(0.2289)(41.42^\circ) + 42.2^\circ\text{C} \end{pmatrix}(0.9545) \\ &\quad - 17.78^\circ\text{C} \\ &= 51^\circ\text{C} \end{aligned}$$

For the design low temperature, use Eq. 5.26.

$$\begin{aligned} T_{\text{min}} &= 0.859\,T_{\text{air}} + 1.7^\circ\text{C} \\ &= (0.859)(-20^\circ\text{C}) + 1.7^\circ\text{C} \\ &= -15^\circ\text{C} \end{aligned}$$

For the 50% reliability level, the design pavement temperatures are 51°C and −15°C. A 98% reliability level is obtained from the mean plus or minus three standard deviations.

$$\begin{aligned} 51^\circ\text{C} + 3s &= 51^\circ\text{C} + (3)(2^\circ\text{C}) \\ &= 57^\circ\text{C} \\ -15^\circ\text{C} - 3s &= -15^\circ\text{C} - (3)(4^\circ\text{C}) \\ &= -27^\circ\text{C} \end{aligned}$$

Therefore, for the 98% reliability level, the design pavement temperatures are 57°C and −27°C.

Superpave Gyratory Compactor

Most of the test equipment required for the preparation of a Superpave asphalt-mixture design is carried over from traditional asphalt test equipment, such as ovens, mixers, scales, and so forth. One of the major elements introduced for the Superpave mix design process is laboratory testing with a *Superpave gyratory compactor* (SGC). The SGC was developed to emulate field conditions of wheel pass repetitions. It involves placing a 4.6 in (150 mm) diameter asphalt mix sample in a gyrating cylinder and applying an 87 psi (600 kPa) load during rotational gyrations. The sample is angled at 1.25° from the axis of the compactor and is rotated at 30 rpm. The test specimen's height is measured throughout the test, and a compaction characteristic is developed for the subject sample.

The SGC is designed to impart a certain degree of lateral loading on the test sample to emulate rutting caused by tire movement, in addition to the vertical load carrying characteristics. The SGC can also provide data during compaction to help determine reactions of a particular mix while compaction is occurring.

Superpave asphalt mixtures are designed at a specific level of compactive effort, which is a function of the design number of gyrations, N_{design}. N_{design} is a function of climate and traffic level. Climate condition is a function of the average high air temperature determined using the average seven-day maximum air temperature for the project conditions. This is the equivalent temperature at 50% reliability. The traffic level is the design ESALs, taken from actual traffic data or from other nearby data and adapted to the project location. N_{initial} is an estimation of the mixture's ability for compaction. N_{max} is a laboratory density determined by using additional SGC specimens of the selected design as a check to help guard against plastic failure caused by traffic levels that exceed the design level. Table 5.9 shows the range of values for determining N_{initial}, N_{design}, and N_{max}. N_{initial} and N_{max} are determined by Eq. 5.27 and Eq. 5.28.

$$\log_{10} N_{\text{initial}} = 0.45 \log_{10} N_{\text{design}} \qquad 5.27$$

$$\log_{10} N_{\text{max}} = 1.10 \log_{10} N_{\text{design}} \qquad 5.28$$

Design Selection

Individual asphalt and aggregate materials are selected from locally available and approved sources. Using local selections, trial blends are made according to Superpave gradation requirements. Gradation blends for the five Superpave aggregates are shown in App. 5.C.

While there is no set number of trial blends, trials usually begin with three blends as their starting point. Once selected, the initial blend's volumetric analysis is performed on each of the individual three blends. In

Table 5.9 Superpave Design Gyratory Compaction Effort

design ESALs* (millions)	compaction parameters			typical roadway applications
	N_{initial}	N_{design}	N_{max}	
< 0.3	6	50	75	Applications include roadways with very light traffic volumes, such as local roads, county roads, and city streets where truck traffic is prohibited or at a very minimal level. Traffic on these roadways would be considered local in nature, not regional, intrastate, or interstate. Special-purpose roadways serving recreational sites or areas may also be applicable to this level.
0.3 to < 3	7	75	115	Applications include collector roads or access streets. Medium-trafficked city streets and the majority of country roadways may be applicable to this level.
3 to < 30	8	100	160	Applications include many two-lane, multilane, divided, and partially or completely controlled access highways. Among these are medium to heavily trafficked city streets, many state routes, U.S. highways, and some rural intersections.
≥30	9	125	205	Applications include the vast majority of the U.S. Interstate System, both rural and urban in nature. Special applications such as truck-weighing stations or truck-climbing lanes on two-lane highways may also be applicable to this level.

*The significant project traffic level expected on the design lane over a 20 year period. Regardless of the design life of the roadway, determine the design ESALs for 20 years.

Reprinted with permission from the Asphalt Institute, *The Asphalt Handbook, Manual Series No. 4 (MS-4)*, 7th ed., Table 4.4, © 2007.

Construction

order to save time, the Asphalt Institute suggests that the asphalt binder content for aggregates have a bulk specific gravity of 2.65, as shown in Table 5.10. Aggregate with significantly higher bulk specific gravity may need less asphalt, while aggregate with a lower bulk specific gravity may need more asphalt.

Table 5.10 *Typical Asphalt Binder Content for Aggregate with G_{sb} of 2.65*

nominal maximum aggregate size (mm)	trial asphalt binder content (%)
37.5	3.5
25.0	4.0
19.0	4.5
12.5	5.0
9.5	5.5

Once the design aggregate blend is selected from the trial samples, specimens are compacted at varying asphalt binder contents. The mixture properties are evaluated to determine the design asphalt binder content.

To obtain the design asphalt binder content for Superpave design, a minimum of four asphalt contents must be evaluated. A minimum of two specimens are compacted at the trial blend's estimated asphalt content. A minimum of one specimen each is prepared at +0.5%, +1.0%, and −0.5% of the estimated asphalt content. The process is repeated until a satisfactory level of performance is achieved on the aged samples. Table 5.11 shows Superpave design requirements recommended by the Asphalt Institute.

9. ACTIVITY-ON-NODE NETWORKS

The *critical path method* (CPM) is one of several critical path techniques that uses a directed graph to describe the precedence of project activities. The CPM requires that all activity durations be specified by single values. That is, the CPM is a *deterministic method* that does not intrinsically support activity durations that are distributed as random variables.

Another characteristic of the CPM is that each activity (task) is traditionally represented by a node (junction), hence the name *activity-on-node network*. Each node is typically drawn on the graph as a square box and labeled with a capital letter, although these are not absolute or universally observed conventions. Each activity can be thought of as a continuum of work, each with its own implicit "start" and "finish" events. An *event* usually is the beginning of an activity and does not consume time or resources. For example, the activity "grub building site" starts with the event of a bulldozer arriving at the native site, followed by several days of bulldozing, and ending with the bulldozer leaving the cleaned site. An activity, including its start and finish events, occurs completely within its box (node).

Each activity in a CPM diagram is connected by arcs (connecting arrows, lines, etc.). The arcs merely show precedence and dependencies. Events are not represented on the graph, other than, perhaps, as the heads of tails of the arcs. Nothing happens along the arcs, and the arcs have zero durations. Because of this, arcs are not labeled. (See Fig. 5.20.)

For convenience, when a project starts or ends with multiple simultaneous activities, *dummy nodes* with

Table 5.11 *Superpave Volumetric Design Requirements*

	required density (% of theoretical maximum specific gravity)			minimum voids in the mineral aggregate (%)						
				nominal maximum aggregate size (mm)						
design ESALs[a] (million)	$N_{initial}$	N_{design}	N_{max}	37.5[d]	25.0[c]	19.0	12.5	9.5[b]	voids filled with asphalt (%)	dust-to-binder ratio[e]
<0.3	≤91.5	–	–	–	–	–	–	–	70–80	0.6–1.2
0.3 to < 3	≤90.5	96.0	≤98.0	11.0	12.0	13.0	14.0	15.0	65–78	0.6–1.2
3 to < 10	≤89.0	96.0	≤98.0	11.0	12.0	13.0	14.0	15.0	65–75	0.6–1.2
10 to < 30	≤89.0	96.0	≤98.0	11.0	12.0	13.0	14.0	15.0	65–75	0.6–1.2
≥30	≤89.0	96.0	≤98.0	11.0	12.0	13.0	14.0	15.0	65–75	0.6–1.2

(Multiply mm by 3.937×10^{-2} to obtain in.)
[a]Design ESALs are the anticipated project traffic level expected on the design lane over a 20 year period. Regardless of the actual design life of the roadway, determine the design ESALs for 20 years, and choose the appropriate N_{design} level.
[b]For 9.5 mm (0.375 in) nominal maximum size mixtures, the specified VFA range shall be 73% to 76% for design traffic levels ≥3 million ESALs.
[c]For 25.0 mm (1 in) nominal maximum size mixtures, the specified lower limit of the VFA shall be 67% for design traffic levels < 0.3 million ESALs.
[d]For 37.5 mm (1.5 in) nominal maximum size mixtures, the specified lower limit of the VFA shall be 64% for all design traffic levels.
[e]If the aggregate gradation passes beneath the boundaries of the aggregate restricted zone, consideration should be given to increasing the dust-to-binder ratio criteria from 0.6–1.2 to 0.8–1.6.

Reprinted with permission from the Asphalt Institute, *The Asphalt Handbook, Manual Series No. 4, (MS-4)*, 7th ed., Table 4.5, © 2007.

Figure 5.20 *Activity-on-Node Network*

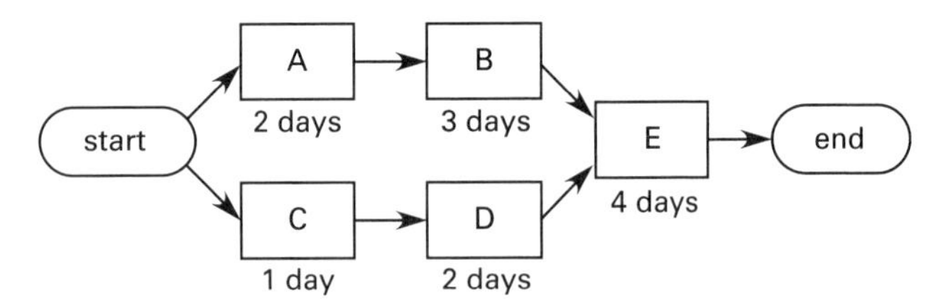

zero durations may be used to specify the start and/or finish of the entire project. Dummy nodes do not add time to the project. They are included for convenience. Since all CPM arcs have zero durations, the arcs connecting dummy nodes are not different from arcs connecting other activities.

A CPM graph depicts the activities required to complete a project and the sequence in which the activities must be completed. No activity can begin until all of the activities with arcs leading into it have been completed. The duration and various dates are associated with each node, and these dates can be written on the CPM graph for convenience. These dates include the earliest possible start date, ES; the latest start date, LS; the earliest finish date, EF; and, the latest finish date, LF. The ES, EF, LS, and LF dates are calculated from the durations and the activity interdependencies.

After the ES, EF, LS, and LF dates have been determined for each activity, they can be used to identify the critical path. The *critical path* is the sequence of activities (the *critical activities*) that must all be started and finished exactly on time in order to not delay the project. Delaying the starting time of any activity on the critical path, or increasing its duration, will delay the entire project. The critical path is the longest path through the network. If it is desired that the project be completed sooner than expected, then one or more activities in the critical path must be shortened. The critical path is generally identified on the network with heavier (thicker) arcs.

Activities not on the critical path are known as *noncritical activities*. Other paths through the network will require less time than the critical path, and hence, will have inherent delays. The noncritical activities can begin or finish earlier or later (within limits) without affecting the overall schedule. The amount of time that an activity can be delayed without affecting the overall schedule is known as the *float* (*float time* or *slack time*). Float can be calculated in two ways, with identical results. The first way uses Eq. 5.29. The second way is described in Sec. 5.10.

$$\text{float} = \text{LS} - \text{ES} = \text{LF} - \text{EF} \qquad 5.29$$

Float is zero along the critical path. This fact can be used to identify the activities along the critical path from their respective ES, EF, LS, and LF dates.

It is essential to maintain a distinction between time, date, length, and duration. Like the timeline for engineering economics problems, all projects start at time = 0, but this corresponds to starting on day = 1. That is, work starts on the first day. Work ends on the last day of the project, but this end rarely corresponds to midnight. So, if a project has a critical path length (duration-to-completion) of 15 days and starts on (at the beginning of) May 1, it will finish on (at the end of) May 15, not May 16.

10. SOLVING A CPM PROBLEM

As previously described, the solution to a critical path method problem reveals the earliest and latest times that an activity can be started and finished, and it also identifies the critical path and generates the float for each activity.

As an alternative to using Eq. 5.29, the following procedure may be used to solve a CPM problem. To facilitate the solution, each node should be replaced by a square that has been quartered. The compartments have the meanings indicated by the key.

ES	EF
LS	LF

key

ES: Earliest Start

EF: Earliest Finish

LS: Latest Start

LF: Latest Finish

step 1: Place the project start time or date in the **ES** and **EF** positions of the start activity. The start time is zero for relative calculations.

step 2: Consider any unmarked activity, all of whose predecessors have been marked in the **EF** and **ES** positions. (Go to step 4 if there are none.) Mark in its **ES** position the largest number marked in the **EF** position of those predecessors.

step 3: Add the activity time to the **ES** time and write this in the **EF** box. Go to step 2.

step 4: Place the value of the latest finish date in the **LS** and **LF** boxes of the finish mode.

step 5: Consider unmarked predecessors whose successors have all been marked. Their **LF** is the smallest **LS** of the successors. Go to step 7 if there are no unmarked predecessors.

step 6: The **LS** for the new node is **LF** minus its activity time. Go to step 5.

step 7: The float for each node is **LS** – **ES** and **LF** – **EF**.

step 8: The critical path encompasses nodes for which the float equals **LS** – **ES** from the start node. There may be more than one critical path.

11. ACTIVITY-ON-ARC NETWORKS

Another variety of deterministic critical path techniques represents project activities as arcs. With *activity-on-arc networks* (also known as *activity-on-branch networks*), the continuum of work occupies the arcs, while the nodes represent instantaneous starting and ending events. The arcs have durations, while the nodes do not. Nothing happens on a node, as it represents an instant in time only. As shown in Fig. 5.21, each node is typically drawn on the graph as a circle and labeled with a number, although this format is not universally adhered to. Since the activities are described by the same precedence table as would be used with an activity-on-node graph, the arcs are labeled with the activity capital letter identifiers or the activity descriptions. The duration of the activity may be written adjacent to its arc. (This is the reason for identifying activities with letters, so that any number appearing with an arc can be interpreted as a duration.)

Although the concepts of ES, EF, LS, and LF dates, critical path, and float are equally applicable to activity-on-arc and activity-on-node networks, the two methods cannot be combined within a single project graph. Calculations for activity-on-arc networks may seem less intuitive, and there are other possible complications.

The activity-on-arc method is complicated by the frequent requirement for dummy activities and nodes to maintain precedence. A *dummy activity* is an artificial activity with a completion time of 0 that is used to show relationships between activities and connect dependent tasks. Dummy activities are often shown using a dashed line and are used when other means of showing precedence are inadequate. Consider the following part of a precedence table.

activity	predecessors
L	–
M	–
N	L, M
P	M

Activity P depends on the completion of only M. Figure 5.22(a) is an activity-on-arc representation of this precedence. However, N depends on the completion of both L and M. It would be incorrect to draw the network as Fig. 5.22(b) since the activity N appears twice. To represent the project, the dummy activity X must be used, as shown in Fig. 5.22(c).

Figure 5.22 *Activity-on-Arc Network with Predecessors*

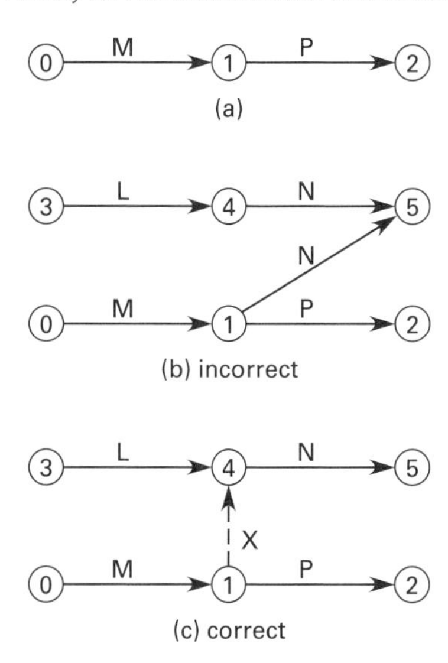

Figure 5.21 *Activity-on-Arc Network*

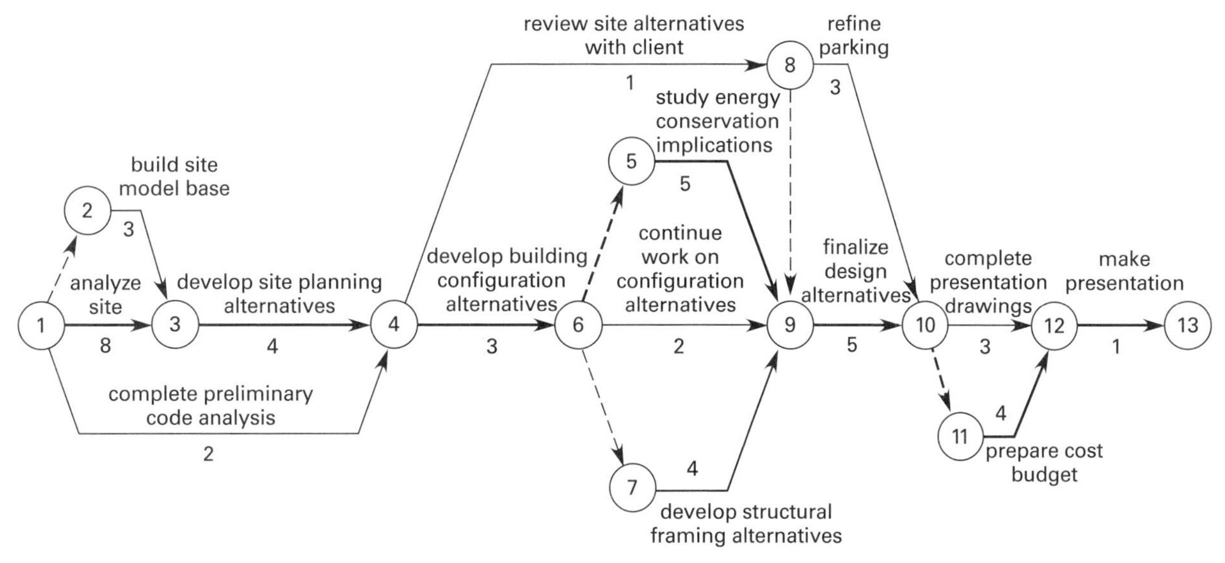

If two activities have the same starting and ending events, a dummy node is required to give one activity a uniquely identifiable completion event. This is illustrated in Fig. 5.23(b).

Figure 5.23 Use of a Dummy Node

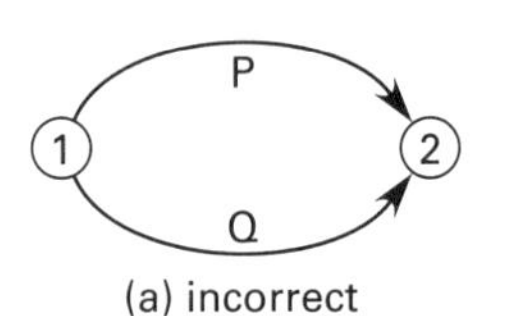

(a) incorrect

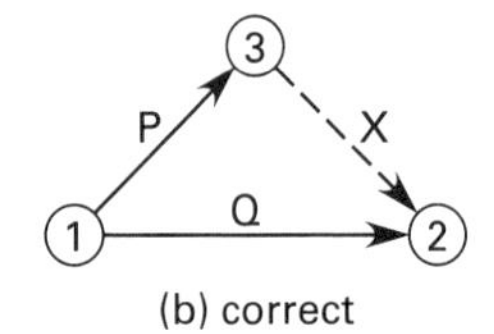

(b) correct

The solution method for an activity-on-arc problem is essentially the same as for the activity-on-node problem, requiring forward and reverse passes to determine earliest and latest dates.

PRACTICE PROBLEMS

1. A project's schedule is to be adjusted in order to minimize cost. Currently, the project will complete on schedule, assuming no delays arise that affect the critical path, but the contract contains a bonus/penalty clause that the contractor would like to take advantage of to achieve maximum benefit-cost. The activity-on-arc network for the project has been simplified by grouping component activities into logical sequence groups using three labor crews. To expedite the work, crew 1 can increase daily output using overtime at 1.5 times the normal daily rate (for simplicity use 1.5 times the total daily rate). Crew 2 and crew 3 must be doubled in size in order to increase daily output, but will not accrue further cost increase from overtime. Crew costs are as follows.

crew	normal daily rate ($)	expedited daily rate ($)
1	800	1200
2	1000	2000
3	650	1300

Supervision and on-site support cost $500 for each day of on-site work activity. The activity schedule is

activity	duration (days)	predecessors	successors	crew
A	3	–	B, F	3
B	6	A	C, E	1
C	21	B	J, D	1
D	12	C, H	L	1
E	0	B	G	–
F	4	A	G	2
G	14	E, F	H, I	2
H	20	G	D,J	2
I	5	G	K	3
J	3	C, H	K	2
K	5	I, J	L	3
L	5	D, K	–	3

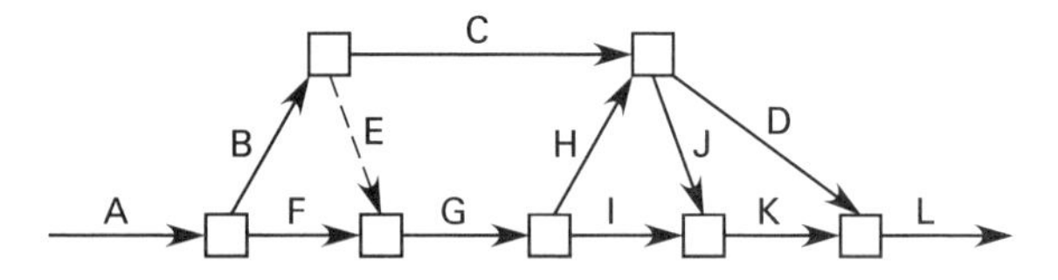

The expedited schedule reduces certain activities to the following durations. These reduced-direction activities use the expedited daily rates to determine the cost of the activities.

activity	normal days	expedited days
B	6	5
C	21	18
D	12	8
G	14	9
H	20	12

Construction

The bonus/penalty clause in the contract states that the contractor will receive a bonus if the project is completed more than three days early and will be penalized if the project is completed more than three days late. The bonus is $1000 per day early the project is completed, not including the first three days, and the penalty is $2500 per day late the project is completed, not including the first three days.

(a) What is the critical path for the normal schedule?

(A) A-B-E-G-H-D-L

(B) A-F-G-H-D-L

(C) A-B-C-D-L

(D) A-F-G-I-K-L

(b) What is the normal, nonexpedited duration of the project?

(A) 45 days

(B) 60 days

(C) 62 days

(D) 77 days

(c) The original labor crew cost, including cost of supervision, is most nearly

(A) $30,000

(B) $84,000

(C) $94,000

(D) $114,000

(d) Comparing total costs, how much more or how much less does the expedited schedule cost than the normal schedule?

(A) $2150 less

(B) $2150 more

(C) $9000 less

(D) $9000 more

(e) Including the project bonus or penalty, what is most nearly the cost of the expedited schedule compared to the cost of the normal schedule?

(A) $12,800 less

(B) $12,800 more

(C) $17,150 less

(D) $17,150 more

(f) Which of the five activities that can be expedited will NOT help decrease the duration of the project if expedited?

(A) B

(B) C

(C) H

(D) none of the above

2. Superpave binder is being selected for a location at 42.2° latitude with a mean seven-day high temperature of 31.9°C and a low air temperature of −21.1°C. The high temperature standard deviation is 1.7°C, and the low temperature standard deviation is 3.9°C. Traffic load rate is projected at 3,000,000 ESALs, with fast traffic operating at greater than 70 kph.

(a) What is the expected pavement high temperature at 20 mm below the pavement surface at the 95% reliability confidence level?

(A) 32°C

(B) 35°C

(C) 54°C

(D) 58°C

(b) What is the binder selection for this location at the 95% reliability level?

(A) PG 52-20

(B) PG 56-20

(C) PG 58-24

(D) PG 58-28

3. Aggregate blends are being prepared for asphalt paving. Coarse aggregate is stored in three cold bins with the following characteristics.

bin	bulk specific gravity, G_{sb}	apparent specific gravity, G_{sa}
1	2.720	2.787
2	2.689	2.755
3	2.450	2.510

A trial blend is prepared with a mix of 25% from bin 1, 35% from bin 2, and 40% from bin 3. The blend has 5% asphalt binder by total mass of mixture. Use 1.030 as the specific gravity of asphalt, G_b. The maximum specific gravity, G_{mm}, of the trial blend is 2.434.

Construction

(a) What is most nearly the bulk specific gravity of the trial blend?

(A) 2.385

(B) 2.595

(C) 2.689

(D) 2.720

(b) What is most nearly the effective specific gravity of the trial blend?

(A) 2.510

(B) 2.595

(C) 2.622

(D) 2.992

(c) What is most nearly the percentage of asphalt absorption?

(A) 0.4%

(B) 0.8%

(C) 4%

(D) 5%

(d) What is most nearly the effective asphalt content of the mixture?

(A) 0.046%

(B) 4.4%

(C) 4.6%

(D) 5.0%

4. An asphalt concrete mix is being prepared with the following material properties.

material	volumetric ratio	oven dry bulk specific gravity, G_{sb}	moisture by weight (%)	absorption by weight (%)
portland cement	1.00	3.14	–	–
fine aggregate	2.27	2.66	3.25	0.41
coarse aggregate	3.65	2.72	2.08	0.25

The effective water content is 5.83 gal/sack of cement. There are 3% air voids in the final mix.

(a) What is most nearly the water-cement ratio of the mix?

(A) 0.016

(B) 0.420

(C) 0.520

(D) 32.0

(b) Most nearly how much water per sack of cement must be added to obtain the effective water content?

(A) 37.7 lbf/sack

(B) 39.9 lbf/sack

(C) 47.7 lbf/sack

(D) 48.5 lbf/sack

(c) What is most nearly the effective yield per sack of cement?

(A) 0.256 yd^3/sack

(B) 0.264 yd^3/sack

(C) 7.13 yd^3/sack

(D) 16.5 yd^3/sack

5. An asphalt mixture has the following properties.

bulk specific gravity of the aggregate mix, $G_{sb} = 2.687$

maximum specific gravity of asphalt mix, $G_{mm} = 2.526$

bulk specific gravity of the compacted mixture, $G_{mb} = 2.438$

asphalt content, $P_b = 5.25\%$

specific gravity of the asphalt, $G_b = 1.035$

volume of water absorbed into the bulk aggregate $= 2.76\%$

volume of water absorbed by the aggregate $= 2.72\%$

asphalt absorption factor, $P_{ba} = 0.7$

(a) What is most nearly the effective asphalt content?

(A) 4.54%

(B) 4.60%

(C) 5.25%

(D) 5.95%

(b) What is most nearly the volume of air voids (VTM) in the mixture?

(A) 2.76%

(B) 3.48%

(C) 3.61%

(D) 5.90%

(c) What is most nearly the volume of voids in the mineral aggregate (VMA)?

(A) 0.91%

(B) 2.76%

(C) 14.0%

(D) 35.3%

(d) What is most nearly the volume of voids filled with asphalt (VFA)?

(A) 10.5%

(B) 14.0%

(C) 28.8%

(D) 75.2%

SOLUTIONS

1. (a) To find the critical path for the project, find the total duration of each path through the CPM network and select the path that has the longest total duration. Possible paths include path A-B-C-D-L, path A-B-C-J-K-L, path A-B-E-G-H-D-L, path A-B-E-G-H-J-K-L, path A-B-E-G-I-K-L, path A-F-G-H-D-L, path A-F-G-H-J-K-L, and path A-F-G-I-K-L.

Path A-B-C-D-L has a total duration of

$$\begin{aligned} D &= 3 \text{ days} + 6 \text{ days} + 21 \text{ days} + 12 \text{ days} + 5 \text{ days} \\ &= 47 \text{ days} \end{aligned}$$

Path A-B-C-J-K-L has a total duration of

$$\begin{aligned} D &= 3 \text{ days} + 6 \text{ days} + 21 \text{ days} + 3 \text{ days} \\ &\quad + 5 \text{ days} + 5 \text{ days} \\ &= 43 \text{ days} \end{aligned}$$

Path A-B-E-G-H-D-L has a total duration of

$$\begin{aligned} D &= 3 \text{ days} + 6 \text{ days} + 0 \text{ days} + 14 \text{ days} \\ &\quad + 20 \text{ days} + 12 \text{ days} + 5 \text{ days} \\ &= 60 \text{ days} \end{aligned}$$

Path A-B-E-G-H-J-K-L has a total duration of

$$\begin{aligned} D &= 3 \text{ days} + 6 \text{ days} + 0 \text{ days} + 14 \text{ days} + 20 \text{ days} \\ &\quad + 3 \text{ days} + 5 \text{ days} + 5 \text{ days} \\ &= 56 \text{ days} \end{aligned}$$

Path A-B-E-G-I-K-L has a total duration of

$$\begin{aligned} D &= 3 \text{ days} + 6 \text{ days} + 0 \text{ days} + 14 \text{ days} + 5 \text{ days} \\ &\quad + 5 \text{ days} + 5 \text{ days} \\ &= 38 \text{ days} \end{aligned}$$

Path A-F-G-H-D-L has a total duration of

$$\begin{aligned} D &= 3 \text{ days} + 4 \text{ days} + 14 \text{ days} + 20 \text{ days} \\ &\quad + 12 \text{ days} + 5 \text{ days} \\ &= 58 \text{ days} \end{aligned}$$

Path A-F-G-H-J-K-L has a total duration of

$$\begin{aligned} D &= 3 \text{ days} + 4 \text{ days} + 14 \text{ days} + 20 \text{ days} + 3 \text{ days} \\ &\quad + 5 \text{ days} + 5 \text{ days} \\ &= 54 \text{ days} \end{aligned}$$

Path A-F-G-I-K-L has a total duration of

$$\begin{aligned} D &= 3 \text{ days} + 4 \text{ days} + 14 \text{ days} + 5 \text{ days} \\ &\quad + 5 \text{ days} + 5 \text{ days} \\ &= 36 \text{ days} \end{aligned}$$

The critical path is path A-B-E-G-H-D-L.

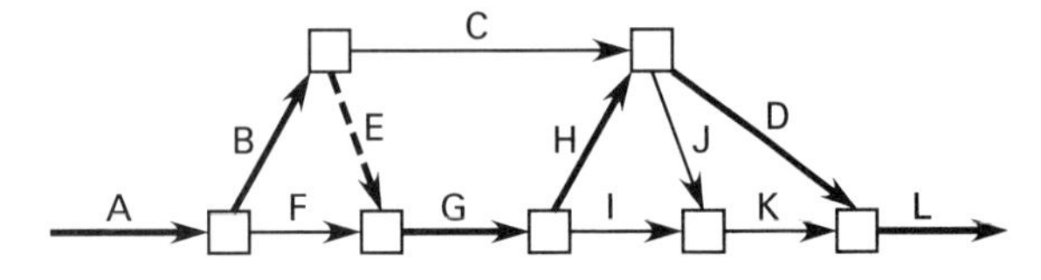

The answer is (A).

(b) The duration of a project is equal to the duration of its critical path, so the duration of this project is 60 days.

The answer is (B).

(c) To find the original labor crew cost, multiply the number of days required for each activity by the normal daily rate of the crew required for that activity, and total up all the costs to find the total project crew cost.

The crew cost for activity A is

$$C_{\text{A},c} = \left(650 \ \frac{\$}{\text{day}}\right)(3 \text{ days}) = \$1950$$

The crew cost for activity B is

$$C_{\text{B},c} = \left(800 \ \frac{\$}{\text{day}}\right)(6 \text{ days}) = \$4800$$

The crew cost for activity C is

$$C_{\text{C},c} = \left(800 \ \frac{\$}{\text{day}}\right)(21 \text{ days}) = \$16{,}800$$

The crew cost for activity D is

$$C_{\text{D},c} = \left(800 \ \frac{\$}{\text{day}}\right)(12 \text{ days}) = \$9600$$

The crew cost for activity E is 0.

The crew cost for activity F is

$$C_{\text{F},c} = \left(1000 \ \frac{\$}{\text{day}}\right)(4 \text{ days}) = \$4000$$

The crew cost for activity G is

$$C_{\text{G},c} = \left(1000 \ \frac{\$}{\text{day}}\right)(14 \text{ days}) = \$14{,}000$$

The crew cost for activity H is

$$C_{\text{H},c} = \left(1000 \ \frac{\$}{\text{day}}\right)(20 \text{ days}) = \$20,000$$

The crew cost for activity I is

$$C_{\text{I},c} = \left(650 \ \frac{\$}{\text{day}}\right)(5 \text{ days}) = \$3250$$

The crew cost for activity J is

$$C_{\text{J},c} = \left(1000 \ \frac{\$}{\text{day}}\right)(3 \text{ days}) = \$3000$$

The crew cost for activity K is

$$C_{\text{K},c} = \left(650 \ \frac{\$}{\text{day}}\right)(5 \text{ days}) = \$3250$$

The crew cost for activity L is

$$C_{\text{L},c} = \left(650 \ \frac{\$}{\text{day}}\right)(5 \text{ days}) = \$3250$$

The total crew cost for the original project schedule is

$$\begin{aligned} C_{t,c} &= \$1950 + \$4800 + \$16{,}800 + \$9600 + \$4000 \\ &\quad + \$14{,}000 + \$20{,}000 + \$3250 + \$3000 \\ &\quad + \$3250 + \$3250 \\ &= \$83{,}900 \end{aligned}$$

The cost of supervision for the project is

$$C_s = \left(500 \ \frac{\$}{\text{day}}\right)(60 \text{ days}) = \$30{,}000$$

The total cost of the project is

$$C_t = \$83{,}900 + \$30{,}000 = \$113{,}900 \quad (\$114{,}000)$$

The answer is (D).

(d) Find the total crew cost of the expedited project. As shown in solution 1(d), only activities B, D, G, and H will be expedited, so new costs only need to be found for these four activities.

The expedited crew cost for activity B is

$$C_{\text{B},c} = \left(1200 \ \frac{\$}{\text{day}}\right)(5 \text{ days}) = \$6000$$

The expedited crew cost for activity D is

$$C_{\text{D},c} = \left(1200 \ \frac{\$}{\text{day}}\right)(8 \text{ days}) = \$9600$$

The expedited crew cost for activity G is

$$C_{\text{G},c} = \left(2000\ \frac{\$}{\text{day}}\right)(9\ \text{days}) = \$18{,}000$$

The expedited crew cost for activity H is

$$C_{\text{H},c} = \left(2000\ \frac{\$}{\text{day}}\right)(12\ \text{days}) = \$24{,}000$$

The total crew cost for the expedited project is

$$\begin{aligned} C_{t,c} &= \$1950 + \$6000 + \$16{,}800 + \$9600 + \$4000 \\ &\quad + \$18{,}000 + \$24{,}000 + \$3250 + \$3000 \\ &\quad + \$3250 + \$3250 \\ &= \$95{,}050 \end{aligned}$$

Find the cost of supervision for the expedited project. The total cost of supervision will decrease, depending on the number of days shorter the expedited schedule is than the normal schedule.

Find the total duration of the expedited schedule. As shown in solution 1(d), the changes in duration for the expedited activities will not cause a change in critical path, so the duration can be found by recalculating the total duration of path A-B-E-G-H-D-L using the expedited durations for activities B, G, H, and D.

The total duration of the expedited schedule is

$$\begin{aligned} D_t &= 3\ \text{days} + 5\ \text{days} + 0\ \text{days} + 9\ \text{days} + 12\ \text{days} \\ &\quad + 8\ \text{days} + 5\ \text{days} \\ &= 42\ \text{days} \end{aligned}$$

The total cost of supervision for the expedited project is

$$C_{t,s} = \left(500\ \frac{\$}{\text{day}}\right)(42\ \text{days}) = \$21{,}000$$

The total cost of the expedited schedule, before bonuses or penalties, is

$$C_t = \$95{,}000 + \$21{,}000 = \$116{,}050$$

This is a cost increase over the normal project. Subtract the total cost of the normal project from the total cost of the expedited project to find the exact amount of the cost increase.

$$\$116{,}050 - \$113{,}900 = \$2150$$

Expediting the project will cost \$2150 more than adhering to the normal schedule.

The answer is (B).

(e) A duration of 42 days means that the project will finish 18 days ahead of schedule. According to the bonus/penalty clause, this means the contractor will receive a bonus of

$$\left(1000\ \frac{\$}{\text{day}}\right)(18\ \text{days} - 3\ \text{days}) = \$15{,}000$$

The total cost of the expedited schedule is

$$C_t = \$116{,}050 - \$15{,}000 = \$101{,}050$$

This is a savings over the normal schedule. Subtract the total cost of the normal schedule from the total cost of the expedited schedule to find the exact amount of the savings.

$$\$113{,}900 - \$101{,}050 = \$12{,}850 \quad (\$12{,}800)$$

The answer is (A).

(f) For the expediting of an activity to decrease the duration of the project, it must decrease the total duration of the critical path. Therefore, only activities that are on the critical path for the expedited schedule should be expedited; expediting other activities will not affect the total duration and will result in an unnecessary increase in cost.

Both activity B and activity H are on the critical path for the normal schedule. For activity B to not be on the critical path for the expedited schedule, its expedited duration must be less than the duration of activity F, the only alternative path from the completion of activity A. For activity H to not be on the critical path for the expedited schedule, either activity C must become a part of the critical path rather than activities E, G, and H, or the expedited duration of activity H must be less than the duration of activity I, the only alternative route from the completion of activity G.

The expedited duration of activity B is 5 days, which is greater than the duration of activity F, so activity B remains on the critical path when expedited, and expediting activity B should help to decrease the project duration.

The expedited duration of activity H is 12 days, which is greater than the duration of activity I, so activity G remains on the critical path unless activity C supplants path E-G-H, and expediting activity G should help to decrease the project duration.

Activity C is not on the critical path for the normal schedule. For activity C to become part of the critical path for the expedited schedule, its expedited duration must be less than the total expedited duration of activities E, G, and H. The expedited duration of activity C is 18 days. The total expedited duration of activities E, G, and H is 21 days. Activity C is not on the critical path for the expedited schedule, meaning activity H is, and expediting activity C will not affect the critical path and would result in an unnecessary increase in cost.

The answer is (B).

2. (a) For 95% reliability, first calculate the high pavement design temperature at a 50% reliability level using Eq. 5.25.

$$\begin{aligned} T_{20\,\text{mm}} &= \left(\begin{array}{l} T_{\text{air}} - (0.00618)(\text{lat})^2 \\ \quad + (0.2289)(\text{lat}) + 42.2^\circ\text{C} \end{array} \right)(0.9545) \\ &\quad - 17.78^\circ\text{C} \\ &= \left(\begin{array}{l} T_{\text{air}} - (0.00618)(42.2^\circ\text{C})^2 \\ \quad + (0.2289)(42.2^\circ\text{C}) + 42.2^\circ\text{C} \end{array} \right)(0.9545) \\ &\quad - 17.78^\circ\text{C} \\ &= 54.3^\circ\text{C} \end{aligned}$$

The 95% reliability level is found from the high pavement design temperature plus two standard deviations.

$$54.3^\circ\text{C} + 2s = 54.3^\circ\text{C} + (2)(1.7)^\circ\text{C} = 57.7^\circ\text{C} \quad (58^\circ\text{C})$$

The answer is (D).

(b) In order to select a binder grade, calculate the low air design temperature at a 95% reliability level. From Eq. 5.26, the low pavement design temperature is

$$\begin{aligned} T_{\min} &= 0.859\,T_{\text{air}} + 1.7^\circ\text{C} = (0.859)(-21.1^\circ\text{C}) + 1.7^\circ\text{C} \\ &= -16.4^\circ\text{C} \end{aligned}$$

The low pavement design temperature at 95% reliability is

$$\begin{aligned} -16.4^\circ\text{C} + 2s &= -16.4^\circ\text{C} + (2)(3.9)^\circ\text{C} \\ &= -24.2^\circ\text{C} \quad (-24^\circ\text{C}) \end{aligned}$$

Using the high pavement design temperature of 58°C calculated in part (a) and the low pavement design temperature of −24°C, the selected grade for 95% reliability would be PG 58-24. According to Table 5.8, no binder grade adjustment is required for standard traffic (i.e., traffic traveling at 70 kph or above) at 3,000,000 ESALs.

The answer is (C).

3. (a) To determine the bulk specific gravity of a combination of aggregates, use Eq. 5.10.

$$\begin{aligned} G_{\text{sb}} &= \frac{P_1 + P_2 + \cdots + P_n}{\frac{P_1}{G_1} + \frac{P_2}{G_2} + \cdots + \frac{P_n}{G_n}} \\ &= \frac{0.25 + 0.35 + 0.40}{\frac{0.25}{2.720} + \frac{0.35}{2.689} + \frac{0.40}{2.450}} \\ &= 2.595 \end{aligned}$$

The answer is (B).

(b) The effective specific gravity is found using Eq. 5.12.

$$\begin{aligned} G_{\text{se}} &= \frac{P_{\text{mm}} - P_b}{\frac{P_{\text{mm}}}{G_{\text{mm}}} - \frac{P_b}{G_b}} \\ &= \frac{100\% - 5.0\%}{\frac{100\%}{2.434} - \frac{5.0\%}{1.030}} \\ &= 2.622 \end{aligned}$$

The answer is (C).

(c) The percentage of asphalt absorption is found using Eq. 5.17.

$$\begin{aligned} P_{\text{ba}} &= \frac{G_b(G_{\text{se}} - G_{\text{sb}})}{G_{\text{sb}}G_{\text{se}}} \times 100\% \\ &= \frac{(1.030)(2.622 - 2.595)}{(2.595)(2.622)} \times 100\% \\ &= 0.409\% \quad (0.4\%) \end{aligned}$$

The answer is (A).

(d) The percentage of aggregate in the mix is

$$\begin{aligned} P_s &= 100\% - P_b \\ &= 100\% - 5\% \\ &= 95.0\% \end{aligned}$$

The effective asphalt content of the mixture is found using Eq. 5.16.

$$\begin{aligned} P_{\text{be}} &= P_b - \frac{P_{\text{ba}}P_s}{100\%} \\ &= 5.0\% - \frac{(0.409\%)(95.0\%)}{100\%} \\ &= 4.61\% \quad (4.6\%) \end{aligned}$$

The answer is (C).

4. (a) The water-cement ratio is found by comparing the unit weights of each. One sack of cement weighs 94 lbf, and water weighs 8.33 lbf/gal.

$$\begin{aligned} w/c &= \frac{\text{weight of water per unit}}{\text{weight of cement per unit}} \\ &= \frac{\left(5.83\ \frac{\text{gal}}{\text{sack}}\right)\left(8.33\ \frac{\text{lbf}}{\text{gal}}\right)}{94\ \frac{\text{lbf}}{\text{sack}}} \\ &= 0.520 \end{aligned}$$

The answer is (C).

Construction

(b) Determine the weight ratio of the components using Eq. 5.21.

The specific weight of cement in the mix is

$$\begin{aligned}\gamma_c &= VG_{\text{sb},c}\gamma_{\text{water}} \\ &= (1.00)(3.14)\left(62.4\ \frac{\text{lbf}}{\text{ft}^3}\right) \\ &= 196\ \text{lbf/ft}^3 \text{ of concrete}\end{aligned}$$

The specific weight of fine aggregate in the mix is

$$\begin{aligned}\gamma_{\text{fa}} &= VG_{\text{sb,fa}}\gamma_{\text{water}} \\ &= (2.27)(2.66)\left(62.4\ \frac{\text{lbf}}{\text{ft}^3}\right) \\ &= 377\ \text{lbf/ft}^3 \text{ of concrete}\end{aligned}$$

The specific weight of coarse aggregate in the mix is

$$\begin{aligned}\gamma_{\text{ca}} &= VG_{\text{sb,ca}}\gamma_{\text{water}} \\ &= (3.65)(2.72)\left(62.4\ \frac{\text{lbf}}{\text{ft}^3}\right) \\ &= 620\ \text{lbf/ft}^3 \text{ of concrete}\end{aligned}$$

Adjust the specific weights to ratios per 94 lbf sack of cement using Eq. 5.22.

The adjusted net quantity of the cement is

$$\begin{aligned}Q_c &= \gamma_c\left(\frac{94\ \frac{\text{lbf}}{\text{sack}}}{\gamma_c}\right) = \left(196\ \frac{\text{lbf}}{\text{ft}^3}\right)\left(\frac{94\ \frac{\text{lbf}}{\text{sack}}}{196\ \frac{\text{lbf}}{\text{ft}^3}}\right) \\ &= 94\ \text{lbf/sack}\end{aligned}$$

The adjusted net quantity of the fine aggregate is

$$\begin{aligned}Q_{\text{fa}} &= \gamma_{\text{fa}}\left(\frac{94\ \frac{\text{lbf}}{\text{sack}}}{\gamma_c}\right) = \left(377\ \frac{\text{lbf}}{\text{ft}^3}\right)\left(\frac{94\ \frac{\text{lbf}}{\text{sack}}}{196\ \frac{\text{lbf}}{\text{ft}^3}}\right) \\ &= 181\ \text{lbf/sack}\end{aligned}$$

The adjusted net quantity of the coarse aggregate is

$$\begin{aligned}Q_{\text{ca}} &= \gamma_{\text{ca}}\left(\frac{94\ \frac{\text{lbf}}{\text{sack}}}{\gamma_c}\right) = \left(620\ \frac{\text{lbf}}{\text{ft}^3}\right)\left(\frac{94\ \frac{\text{lbf}}{\text{sack}}}{196\ \frac{\text{lbf}}{\text{ft}^3}}\right) \\ &- 297\ \text{lbf/sack}\end{aligned}$$

Adjust the aggregate weight to account for water included in the stockpiled aggregate.

The adjusted weight of the fine aggregate is

$$W = \left(181\ \frac{\text{lbf}}{\text{sack}}\right)(1 + 0.0325) = 187\ \text{lbf/sack}$$

The adjusted weight of the coarse aggregate is

$$W = \left(297\ \frac{\text{lbf}}{\text{sack}}\right)(1 + 0.0208) = 303\ \text{lbf/sack}$$

Water absorption is less than the amount of moisture included with each aggregate. Therefore, there will be excess water introduced into the mix along with the aggregate.

$$\text{total water} - \text{absorbed water} = \text{excess water}$$

The excess water included with the fine aggregate is

$$3.25\% - 0.41\% = 2.84\%\ \text{excess water}$$

The excess water included with the coarse aggregate is

$$2.08\% - 0.25\% = 1.83\%\ \text{excess water}$$

Determine the total weight of excess water.

$$\begin{aligned}W_t &= (0.0284)\left(187\ \frac{\text{lbf}}{\text{sack}}\right) + (0.0183)\left(303\ \frac{\text{lbf}}{\text{sack}}\right) \\ &= 10.9\ \text{lbf/sack excess water}\end{aligned}$$

The final adjusted water to be added is

$$\left(5.83\ \frac{\text{gal}}{\text{sack}}\right)\left(8.33\ \frac{\text{lbf}}{\text{gal}}\right) - 10.9\ \frac{\text{lbf}}{\text{sack}} = 37.7\ \text{lbf/sack}$$

The answer is (A).

(c) Using the material quantities adjusted for water content, convert the yield to cubic yard per sack.

$$\begin{aligned}&\left(\frac{\gamma_c}{G_{\text{sb},c}\gamma_w} + \frac{\gamma_{\text{fa}}}{G_{\text{sb,fa}}\gamma_w} + \frac{\gamma_{\text{ca}}}{G_{\text{sb,ca}}\gamma_w}\right)(1 + \text{fraction air}) \\ &= \frac{\left(\frac{196\ \text{lbf}}{(3.14)\left(62.4\ \frac{\text{lbf}}{\text{ft}^3}\right)} + \frac{377\ \text{lbf}}{(2.66)\left(62.4\ \frac{\text{lbf}}{\text{ft}^3}\right)} + \frac{620\ \text{lbf}}{(2.72)\left(62.4\ \frac{\text{lbf}}{\text{ft}^3}\right)}\right)\left(1 + \frac{3\%}{100\%}\right)}{27\ \frac{\text{ft}^3}{\text{yd}^3}} \\ &= 0.264\ \text{yd}^3/\text{sack}\end{aligned}$$

The answer is (B).

Construction

5. (a) Using water absorption to indicate the air voids within the aggregate particles, the apparent specific gravity, G_{sa}, and effective specific gravity, G_{se}, can be estimated.

To find the effective asphalt content, the percentage of aggregates in the mix must be known. The asphalt content is already known, so the percentage of aggregates is

$$\begin{aligned} P_s &= 100\% - P_b = 100\% - 5.25\% \\ &= 94.75\% \end{aligned}$$

Use Eq. 5.16 to find the effective asphalt content.

$$\begin{aligned} P_{be} &= P_b - \frac{P_{ba}P_s}{100\%} = 5.25\% - \frac{(0.7)(94.75\%)}{100\%} \\ &= 4.58\% \quad (4.60\%) \end{aligned}$$

The answer is (B).

(b) The volume of air voids can be found using Eq. 5.18.

$$\begin{aligned} \text{VTM} &= \frac{G_{mm} - G_{mb}}{G_{mm}} \times 100\% \\ &= \frac{2.526 - 2.438}{2.526} \times 100\% \\ &= 3.484\% \quad (3.48\%) \end{aligned}$$

The answer is (B).

(c) The volume of voids in mineral aggregate is found using Eq. 5.14.

$$\begin{aligned} \text{VMA} &= 100\% - \frac{G_{mb}P_s}{G_{sb}} \\ &= 100\% - \frac{(2.438)(94.75\%)}{2.687} \\ &= 14.03\% \quad (14.0\%) \end{aligned}$$

The answer is (C).

(d) The volume of voids filled with asphalt is found using Eq. 5.19.

$$\begin{aligned} \text{VFA} &= \frac{\text{VMA} - \text{VTM}}{\text{VMA}} \times 100\% \\ &= \frac{14.03\% - 3.484\%}{14.03\%} \times 100\% \\ &= 75.17\% \quad (75.2\%) \end{aligned}$$

The answer is (D).

6 Traffic Safety

Nomenclature

A	annual amount	–	–
AADT	average annual daily traffic	veh/day	veh/d
ACC	annual crash cost	$	$
B	benefit	–	–
C	cost	$	$
CZ_c	clear zone on outside curvature	ft	m
d	deceleration	ft/sec^2	m/s^2
EUAC	equivalent uniform annual cost	$	$
F	force	lbf	N
g	annual growth rate	%	%
g	gravitational acceleration, 32.2 (9.81)	ft/sec^2	m/s^2
g_c	gravitational constant, 32.2	lbm-ft/lbf-sec^2	n.a.
i	effective interest rate	decimal	decimal
$i\%$	effective interest rate	%	%
K	constant or factor	–	–
L	length or distance	ft	m
m	mass	lbm	kg
n	number of compounding periods	–	–
N	number	–	–
P	present worth	–	–
P	probability	–	–
R	crash rate	various	various
s	sample standard deviation	various	various
SF	safety factor	–	–
v	velocity	mph	kph
V	traffic volume	vph	vph

Symbols

α	vehicle orientation angle	deg	deg
θ	encroachment angle	deg	deg

Subscripts

a	after
ave	average
A	annualized crash or societal
b	before
cr	crash
cz	clear-zone
D	annualized direct
exp	expected
E	encroachment
i	level i
int	intersection
M	mean
n	year n
PRT	perception-reaction time
s	stopping
seg	segment
SI	injury severity level
th	threshold

1. TRAFFIC SAFETY AND ROADWAY DESIGN ANALYSIS

Traffic safety is a fundamental part of transportation design and affects multiple aspects of planning, including capacity analysis, design standards, construction, operations, and maintenance. Safe design practices include providing adequate roadside clearances, analyzing and planning for traffic management in conflict zones, and providing work zone safety during construction and maintenance activity. Once a roadway network is in place, the emphasis on safety continues in order to mitigate crashes, which often result in property damage, injury, and/or fatalities. Local governments implement *countermeasures* to reduce the severity of crashes or prevent them altogether. When crashes do occur, crash causation analysis should be undertaken to identify factors and mechanisms that increase crash risk and to prioritize changes and improvements that will prevent crashes in the future.

Traffic safety begins with applying highway safety principles during design. The comfort, convenience, and safety of a highway can be greatly improved by designs that work to eliminate surprises, such as sudden sharp curves or rapid lane shifts. One of the greatest past improvements to highway safety was the elimination of at-grade intersection crossings so that all traffic enters and leaves the main roadways via high-speed ramps, and all traffic crossing the roadway is carried on grade-separated bridges. This allows the mainline flow to continue unimpeded (i.e., in "free flow") in all but high-density traffic conditions, hence the term "freeway."

Observations of consistent driver behavior patterns are quantified by AASHTO and in design guides. In particular, AASHTO's *Green Book* provides charts, tables, and graphs for acceleration rates, braking rates, comfortable travel speeds, perception-reaction times, signage visibility, and blending of horizontal and vertical curve geometry. This information is beneficial to designing roadways with safety in mind. AASHTO has adopted the design philosophy that accommodating 85% of the usual driving behavior is a cost-effective goal. Although accommodating the upper or lower 5–10% of behavior beyond the normal range is not necessarily ignored, unusual or erratic behavior and extreme driving patterns are generally not accounted for due to cost or other considerations.

2. TRAFFIC DATA AND CRASH ANALYSIS

The methodologies and time periods used in data collection can have a marked effect on crash analysis. In particular, engineers should consider the various definitions of "average" traffic when selecting and applying traffic data to a location. While the terms *average daily traffic* (ADT) and *annual average daily traffic* (AADT) are frequently used synonymously, the way that average traffic is defined in a study can have a significant impact on the resulting crash statistics.

Strictly speaking, ADT is the average daily traffic over a highway segment measured in a period of 364 days or fewer, and AADT is the average daily traffic measured over the course of a year. If ADT numbers are being used instead of AADT numbers, it is important to establish what the average is based on and how much the traffic varies from this average. For instance, an ADT value may be a weekday daylight average that takes into account the a.m. and p.m. peak flows, and this could vary considerably from a seven-day, 24-hour count. Another possible source of variation is the average daily directional flow and its effect on the average traffic overall. The suitability of the count period and how representative the data are of the prevailing traffic conditions must be considered before proceeding with the analysis using ADT data.

Traffic data that are seasonally adjusted are called *average seasonal daily traffic* (ASDT) data. ASDT data are useful in resort areas with heavy tourist traffic, locations with extreme weather conditions such as large amounts of winter snowfall, and other areas where there are large variations in traffic flow from season to season. For example, ASDT is often recorded for tourist areas where traffic varies substantially within the tourist season, and some jurisdictions use this data to adjust signal timing, parking regulations, lane configuration, or other roadway features for the peak tourist season.

For the most general conditions, and throughout this book, AADT is used instead of ADT. The total annual count can be found using many methods, but the most common is using a continuous counter set up in a semipermanent or permanent location. The total annual traffic count on a roadway segment (i.e., a full 12-month count) can also be extrapolated from a network of counting stations using a variety of count adjustment methods. For instance, counts obtained for seasonal or partial-year periods can be expanded to a full year. Once full-year data are available, the total annual count is divided by 365 days. These numbers become the AADT figures that highway offices use for traffic management, funding applications, crime and crash statistics, and a variety of other purposes. AADT is frequently the most reliable of the various count statistics and is perhaps the most universally accepted as a comparative statistic for describing the traffic on a roadway segment. When a traffic flow is given using only ADT terminology, it can usually be considered as the equivalent AADT if no other conditions are given with the data.

3. ROADWAY ELEMENTS FOR SAFE DESIGN

On-Road Elements of Safe Design

The basis of safe roadway design should be no surprises for the driver so that sudden and erratic moves are unnecessary for maintaining the prevailing speed at a reasonable degree of comfort and safety. Lack of consistency and uniformity of design elements can distract drivers and lead to collisions. All roadway features are selected based on the *design speed*, which is typically set for the 85th percentile driver. The following are the main concepts for designing safe roadways.

Curve radius and *superelevation* must be set according to the minimum design speed or greater.

Curve widening, especially for two-lane roads, is necessary to account for off-tracking of larger vehicles, minimize unnecessary slowing of traffic around curves, and avoid collisions.

Sight distance must be maintained for the design speed, whether the minimum for stopping or a longer distance for passing and making complex decisions.

Cross section elements must be matched to the volume, design speed, and vehicle mix. The number of lanes, provisions for curbs and shoulders, and the width of the roadway must match at least the minimum requirements for the desired service level.

Intersections and *ramps* must provide adequate sight distance for the design speed, provide sufficient room for the vehicles to maneuver, and allow the expected volume of traffic to flow in a normal fashion without requiring the drivers to make extraordinary moves.

Weaving and maneuvering sections, or areas where one-way traffic streams cross at contiguous access points, must be long enough to accommodate the expected flow volumes, especially on freeways and expressways.

Off-Road Elements for Crash Avoidance or Reduction

Once a vehicle leaves the normal roadway, there are a number of design features that can either help the driver avoid a crash or reduce damage or injury should a crash occur. They include the following.

Clearance to obstructions is important because where more clearance is provided for a driver to recover and regain vehicular control, the likelihood of a severe crash is reduced (or a crash may be prevented altogether).

Objects in the clear zone, such as light poles and signs, should be designed with frangible (i.e., breakaway) fittings. These fittings minimize deceleration forces and help prevent hardware from penetrating the vehicle and increasing crash severity.

Ditches, *drain grates*, and *endwalls* should incorporate designs that allow a vehicle to pass over them without snagging or causing the vehicle to vault or overturn.

Earthwork side slopes within the clear zone should be designed to allow a vehicle to traverse the slope with as little potential for overturning as possible.

Guiderails and other protective devices should incorporate impact attenuators and should employ designs known to deflect and redirect vehicles instead of trapping and stopping with blunt force action.

Surrounding clutter and *distractions* can play a large role in traffic safety, as they tend to draw drivers' attention away from the road. Common examples of distractions are animated advertising signs near eye level or within the cone of vision of the driver, lighted signs that distract attention from traffic signals, driveways and store entrances along the roadway in commercial districts, nighttime lighting that glares into oncoming motorists' windshields, and bars and restaurants with busy patronage and night entertainment spilling onto the sidewalks and roadway.

4. CRASH ANALYSIS

Crashes (also known as *accidents*) involve vehicles colliding with other vehicles, stationary objects, or nonstationary objects. There are three categories of crashes: *property damage*, *injury*, or *fatality*. Crashes often involve more than one category and vary in severity. To eliminate small claims, most jurisdictions have a minimum reportable crash threshold for crashes involving property damage or injury. For property damage, this minimum threshold may be as low as $500. Reportable injuries vary by jurisdiction based on nearby emergency medical care, insurance coverage, or other regional differences. Reportable injuries commonly include severe bleeding, head trauma, loss of extremities, broken bones, breathing difficulties, heart attack, stroke, loss of mobility, loss of consciousness, exposure to extreme heat or cold, burns, eye damage, hearing damage, or some other debilitating injury. Because of these reporting thresholds, minor crashes and "near miss" situations are rarely reported.

Crashes are classified according to type and occurrence and are ranked according to exposure. *Exposure* is the number of vehicles that travel over a roadway in a period of time and/or that travel a specific distance. The *crash rate*, R, is the ratio of the number of crashes to the exposure. The general equation for crash rate is Eq. 6.1.

$$R = \frac{N_{\text{cr}}}{\text{exposure}} \tag{6.1}$$

Crash data are commonly grouped into either intersection data or segment data. For intersections, the crash rate is the rate of crashes per number of entering vehicles per year and is found from Eq. 6.2. To make the numbers more manageable, the rate is reported as the rate per million entering vehicles (RMEV) from all directions.

$$R_{\text{int}} = \frac{N_{\text{cr}}(10^6)}{(\text{AADT})(N_{\text{yr}})\left(365\ \frac{\text{days}}{\text{yr}}\right)} \tag{6.2}$$

For a *highway segment*, the number of crashes is reported as the ratio of the number of crashes per year to the AADT per mile of length, L, and is found from Eq. 6.3. To make numbers more manageable, the rate is calculated per 100 million vehicle-miles (HMVM).

$$R_{\text{seg}} = \frac{N_{\text{cr}}(10^8)}{(\text{AADT})(N_{\text{yr}})\left(365\ \frac{\text{days}}{\text{yr}}\right)L_{\text{mi}}} \tag{6.3}$$

Locations with the highest numbers of crashes when compared with other locations of similar design and function are given a high crash rate ranking. Screening processes using classic statistical methods are often performed to select critical locations for study and improvements based on crash rates and/or the severity of the crashes that occur. Once locations are ranked, there may be breaks in the ranking list separated by larger differences in frequency rates. These breaks may indicate a natural grouping of locations by crash type or severity.

Analysis is performed with the assumption that the crash rate follows a standard normal probability distribution as described in the Institute of Transportation Engineers' (ITE's) *Manual of Transportation Engineering Studies*. Logically, the locations with the highest crash rates are likely to benefit the most from a detailed study to find improvements. However, in order to make the best use of available funds for remedial action, there must be a consistent criterion upon which locations are judged and selected. The criterion commonly used is to select locations that appear to have a crash frequency significantly higher than the mean frequency of the rate of occurrence. Locations are ranked according to the rate of occurrence, which is calculated by comparing the crash rate, R, to the threshold rate, R_{th}. If the crash frequency rate is greater than the threshold rate, then the location likely qualifies for improvement funding. If the crash frequency rate is less than the threshold rate, the location does not meet the criterion. Locations where the threshold rate falls below the crash frequency rate are not considered for funding unless special circumstances exist, such as community pressure to make an improvement.

Use Eq. 6.4 to determine how the crash frequency rate compares to the threshold rate. R_M is the mean crash rate, and s is the sample standard deviation. Table 6.1 gives values for the constant, K, at common levels of confidence. The units of the standard deviation will be the same as the units of the crash rate.

$$R_{\text{th}} = R_M + Ks \qquad 6.4$$

Table 6.1 *Values of K at Selected Levels of Confidence Using Two-Tailed Limits*

level of confidence (%)	K
90	1.645
95	1.960
99	2.58

When selecting locations for data analysis, attention must be paid to how broad of an area is being selected, as well as how broad the criteria for selection are. For example, statewide data may exist only in relation to intersection type, but if criteria are narrowed and closer proximity limits to comparison locations are developed, it may become easier to find comparable locations by roadway type or traffic condition, as well. Doing this, however, will also decrease the sample size, which will make statistical analysis more difficult.

The following steps are used to analyze crash data.

step 1: Obtain all crash data for a time period of two years or more.

step 2: Prepare a summary of the crash data, including the time and date of each occurrence, weather conditions, road conditions, crash type, type of vehicles involved, driver actions, and other useful information from the report forms.

step 3: Prepare collision diagrams to illustrate the crash patterns using scale layouts of the intersection or road segment.

step 4: Prepare a diagram of the study location showing the location of relevant physical features such as traffic control devices, utility poles, fire hydrants, building lines, and street furniture. Pavement conditions that could have affected the crash should also be noted.

step 5: Obtain data showing traffic volumes, parking, speeds, driveway access, vehicle classifications, and signal timing.

step 6: Visit the site to become familiar with specific characteristics and conditions of traffic, or other information not readily available from reports or apparent from sketches. Take photographs to supplement and illustrate the report data.

The crash analysis should cover the direct effects of pertinent site features, such as sight distance limits, signal timing, or traffic patterns, to establish a cause and effect relationship. If necessary, statistical analysis can be weighted for the severity of crashes and to account for observed near-miss situations. The report should focus on the effects of possible engineering improvements and the prediction of crash reduction due to these improvements. The effectiveness of the improvements is presented in a before-and-after evaluation using the best approximations of the improvements' effects, obtained from Eq. 6.5.

$$\text{effectiveness} = \frac{N_{\text{cr},b} - N_{\text{cr},a}}{N_{\text{cr},b}} \times 100\% \qquad 6.5$$

Incidents are related to the location and can include crashes with property damage only, crashes with bodily injury, and crashes involving fatalities. The significance of the change in effectiveness should be based on statistical test procedures. This is especially true if there is a small sample of crashes to establish the effectiveness base.

Example 6.1

A highway segment has a crash rate of 257 HMVM. The mean crash rate for all highway sections in the region with similar roadway characteristics is 139 HMVM. The standard deviation of this mean is 68 HMVM. The segment will qualify for funding consideration if the location rate exceeds the threshold criteria using a 90% confidence level. Does this location qualify?

Solution

Use Table 6.1 to determine K, and solve for threshold rate using Eq. 6.4.

$$\begin{aligned} R_{\text{th}} &= R_M + Ks = 139 \text{ HMVM} + (1.645)(68 \text{ HMVM}) \\ &= 251 \text{ HMVM} \end{aligned}$$

The location crash rate (257 HMVM) is in excess of the threshold rate (251 HMVM). Therefore, the location qualifies for funding.

5. ENCROACHMENT CRASH ANALYSIS

Extensive investigation during the early days of interstate highway operations revealed that most off-road vehicle incursions did not result in a crash when the vehicle traveled 30 ft (9 m) or less from the edge of the shoulder. When the landscape adjoining the shoulder had a gentle slope and could support a vehicle, the vehicle often returned to the roadway unharmed. Further analysis of crash data has resulted in clear-zone distance design curves that take into account the design speed, the AADT volume, and the steepness of the slope beyond the paved shoulder. These findings, along with design procedure guidelines, are described in the AASHTO *Roadside Design Guide* (*RDG*). The *RDG* covers detailed procedures to evaluate road design conditions for potential off-pavement crashes and contains guidelines for mitigation procedures to reduce the potential for these crashes.

The National Highway Traffic Safety Association (NHTSA) collects comprehensive data on crashes so that engineers and safety officials can better predict crash occurrence and severity. Figure 6.1, Table 6.2, and Table 6.3 give examples of available data. Approximately 20% of motor vehicle deaths result from a vehicle leaving the roadway and hitting a fixed object (e.g., a tree or utility pole) alongside the road. Generally, fixed-object deaths occur in single-vehicle crashes, as opposed to multiple-vehicle crashes.

Figure 6.1 shows the percentage distribution of fixed-object crashes in 2009 by object struck. Trees are by far the most common fixed object struck, followed by utility poles and traffic barriers. Alcohol is frequently a contributing factor in these crashes, but excessive speed, falling asleep, and inattention also contribute to vehicles leaving the roadway. Table 6.2 can be used with Table 6.3 to predict fatality rates for various off-pavement features based on past occurrences and traffic volume.

Figure 6.1 *Percentage Distribution of Fixed-Object Crash Deaths by Object Struck, 2009*

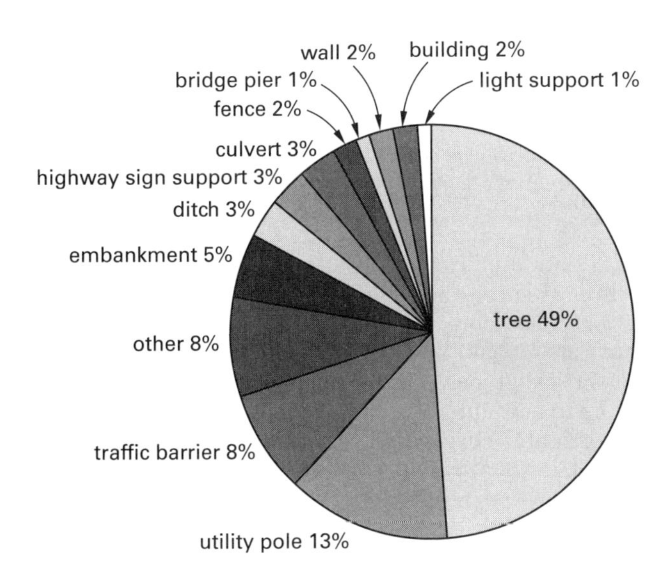

Adapted from Fatality Analysis Reporting System (FARS), U.S. Department of Transportation.

Table 6.2 *Deaths in Fixed-Object Crashes as a Percent of all Motor Vehicle Deaths, 1979–2009*

	fixed object crash deaths		other deaths		all motor vehicle deaths
year	no.	%	no.	%	no.
1979	10,550	21	40,543	79	51,093
1980	10,968	21	40,123	79	51,091
1981	9464	19	39,837	81	49,301
1982	8417	19	35,528	81	43,945
1983	8333	20	34,256	80	42,589
1984	8796	20	35,461	80	44,257
1985	9319	21	34,506	79	43,825
1986	9851	21	36,236	79	46,087
1987	9662	21	36,728	79	46,390
1988	9865	21	37,222	79	47,087
1989	9537	21	36,045	79	45,582
1990	9355	21	35,244	79	44,599
1991	8894	21	32,614	79	41,508
1992	8221	21	31,029	79	39,250
1993	8256	21	31,894	79	40,150
1994	8142	20	32,574	80	40,716
1995	8524	20	33,293	80	41,817
1996	8442	20	33,623	80	42,065
1997	8381	20	33,632	80	42,013
1998	8401	20	33,100	80	41,501
1999	8431	20	33,286	80	41,717
2000	8899	21	33,046	79	41,945
2001	9011	21	33,185	79	42,196
2002	9580	22	33,425	78	43,005
2003	9433	22	33,451	78	42,884
2004	9123	21	33,713	79	42,836
2005	9062	21	34,448	79	43,510
2006	9181	21	33,527	79	42,708
2007	9168	22	32,091	78	41,259
2008	8661	23	28,762	77	37,423
2009	7800	23	26,008	77	33,808

Adapted from Fatality Analysis Reporting System (FARS), U.S. Department of Transportation.

Encroachment Crash Prediction

Crash prediction estimates how many off-road encroachments result in crashes. *Encroachments* refer to the off-pavement travel of vehicles onto the roadside beyond the paved shoulder. Random occurrence and simulations, such as the *Monte Carlo method*,[1] are used based on distributions-derived average data, such as those given in the *RDG*. Included with the *RDG* is the Roadside Safety Analysis Program (RSAP), which can be used to analyze specific highway segments for crash potential. The RSAP crash prediction module simulates one encroachment at a time, randomly assigning characteristics to each encroachment: location along the highway, lane of origination, direction of the encroachment, vehicle type, vehicle speed and angle, vehicle orientation, and lateral extent of the encroachment. The RSAP has

[1]The Monte Carlo method is a class of computational algorithms that make repeated random sampling to calculate results.

Table 6.3 *Deaths, Crashes, and Total Number of Vehicles Involved, 1975–2009*

year	deaths	crashes	motor vehicles
1975	44,525	39,161	55,534
1976	45,523	39,747	56,084
1977	47,878	42,211	60,516
1978	50,331	44,433	64,144
1979	51,093	45,223	64,762
1980	51,091	45,284	63,484
1981	49,301	44,000	62,698
1982	43,945	39,092	56,449
1983	42,589	37,976	55,103
1984	44,257	39,631	57,970
1985	43,825	39,196	58,271
1986	46,087	41,090	60,792
1987	46,390	41,438	61,836
1988	47,087	42,130	62,702
1989	45,582	40,741	60,870
1990	44,599	39,836	59,292
1991	41,508	36,937	54,794
1992	39,250	34,942	52,227
1993	40,150	35,780	53,777
1994	40,716	36,254	54,906
1995	41,817	37,241	56,524
1996	42,065	37,494	57,347
1997	42,013	37,324	57,037
1998	41,501	37,107	56,920
1999	41,717	37,140	56,820
2000	41,945	37,526	57,594
2001	42,196	37,862	57,918
2002	43,005	38,491	58,426
2003	42,884	38,477	58,877
2004	42,836	38,444	58,729
2005	43,510	39,252	59,495
2006	42,708	38,648	58,094
2007	41,059	37,435	55,926
2008	37,423	34,172	49,151
2009	33,808	30,797	44,062

Adapted from Fatality Analysis Reporting System (FARS), U.S. Department of Transportation.

12 different vehicle types that can be used if the data is available, ranging from small passenger cars to tractor-trailers. Weighting may be employed so that rare events, or events with very low frequencies, will be properly represented in the distributions.

In order to predict crashes, traffic growth over the life of the project must be established. The traffic growth adjustment factor averages the traffic volume over the life of the project and is found from Eq. 6.6. g is the annual percentage growth rate and n is the number of years in the project life.

$$\text{average traffic growth adjustment factor} = \sum_{i=1}^{n} \frac{\left(1+\frac{g}{100\%}\right)^n}{n} \quad 6.6$$

The traffic volume entered into the RSAP is for a given year and, to allow for future increases in traffic, the base year AADT is adjusted by Eq. 6.7.

$$\text{AADT}_n = \text{AADT}_1\left(1+\frac{g}{100\%}\right)^n \quad 6.7$$

Figure 6.2 shows the encroachment frequency curves used in the RSAP, which can also be used in manual calculations. The encroachments are expressed as the number of encroachments per mile per year per AADT for undivided and divided highways. The encroachment frequency curves were developed by observing tire tracks in medians and roadsides. The encroachment frequency curves were adjusted upward by a ratio of 2.466 for two-lane, undivided highways and 1.878 for multilane, divided highways to account for under-reporting of encroachments due to paved shoulders. The percentage of uncontrolled encroachments is assumed to be 60% based on a study of reported versus unreported crashes involving longitudinal barriers. Therefore, the encroachment frequency is multiplied by a factor of 0.6 to account for the lack of ability to detect the difference between controlled and uncontrolled encroachments.

Figure 6.2 *Encroachment Frequency Curves Used by the RSAP*

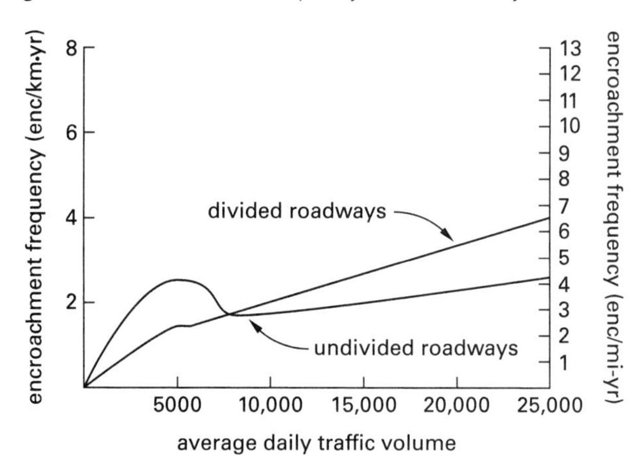

From *Roadside Design Guide*, 2002, by the American Association of State Highway and Transportation Officials, Washington, D.C. Used by permission.

Crash data studies show that crash rates are higher on horizontal curves and vertical grades than on tangent sections. Therefore, it is logical to assume that encroachment rates would also be affected by horizontal curves and vertical grades. The RSAP incorporates adjustment factors to account for an increase in encroachment rates on horizontal curves and vertical grades, as shown in Fig. 6.3.

Vehicle speed, the angle of the encroachment path, and vehicle orientation are determined by distributions estimated from crash data. In the absence of reliable local data, statewide or national data can be used and adjusted for local conditions if they vary from the

Figure 6.3 *Adjustment Factors for Encroachment Rates on Horizontal Curves and Vertical Grades*

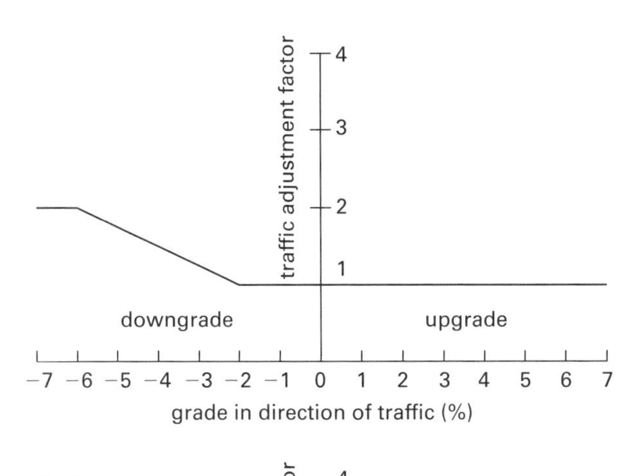

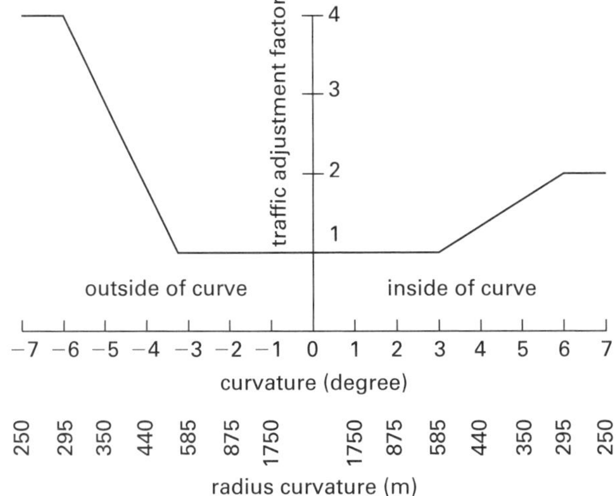

From *Roadside Design Guide*, 2002, by the American Association of State Highway and Transportation Officials, Washington, D.C. Used by permission.

national averages. The assumed vehicle path and impact envelope are illustrated in Fig. 6.4.

The *impact envelope*, also known as the *encroachment path*, is the path of the vehicle once it leaves the roadway and is assumed to be a straight line. The *encroachment angle*, θ, is the angle between the edge of traveled way and the outer edge of the impact envelope, which is measured from the rear tire. The *vehicle orientation angle*, α, describes the orientation of the vehicle in relation to the impact envelope and varies based on the impact envelope and the vehicle dimensions. More information on the determination of the impact envelope is given in the *RDG*.

If no roadside features are within the impact envelope, the encroachment will not result in a crash. If roadside features are within the impact envelope and the probability exists that a vehicle will travel far enough to impact these features, then a crash will occur with a probability determined by the lateral extent of encroachment of the vehicle. This probability is illustrated by Fig. 6.5.

Figure 6.4 *Vehicle Path and Impact Envelope*

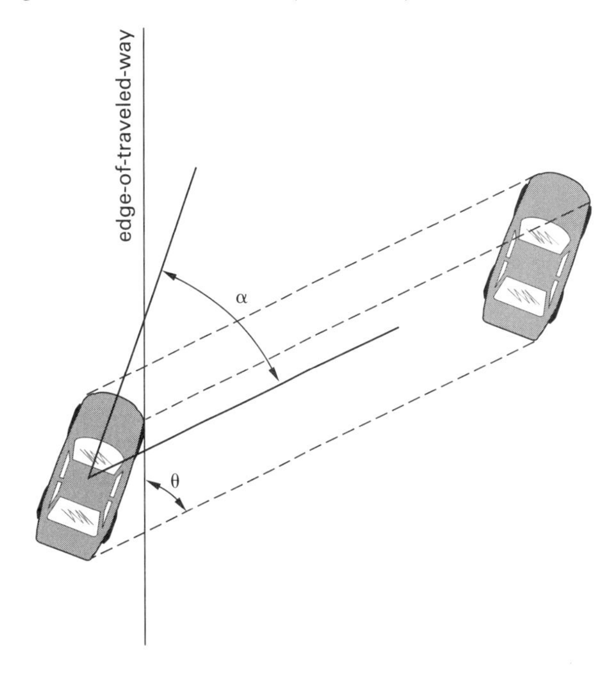

From *Roadside Design Guide*, 2002, by the American Association of State Highway and Transportation Officials, Washington, D.C. Used by permission.

Figure 6.5 *Lateral Extent of Encroachment Distribution*

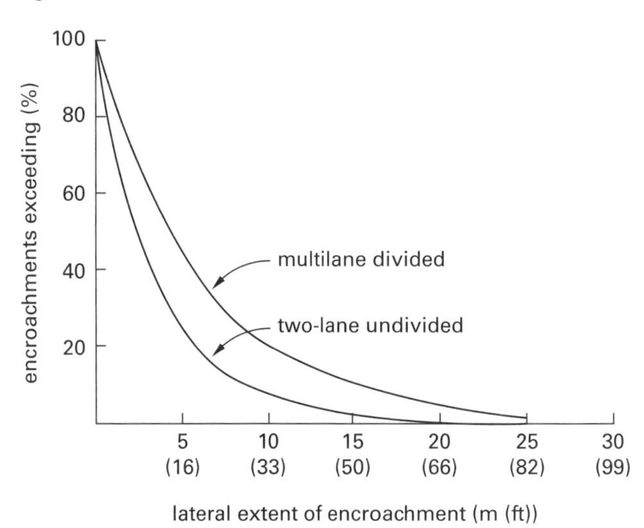

From *Roadside Design Guide*, 2002, by the American Association of State Highway and Transportation Officials, Washington, D.C. Used by permission.

The probability of a crash is related to the encroachment path coverage and the lateral extent of the encroachment. If an encroachment occurs and the vehicle does not impact an object, the encroachment will not result in a crash. The path area that results in an object at a certain lateral distance being impacted, compared with the total roadside area at that same lateral distance in the study zone, can be used to approximate the proportion of encroachments that will result in a crash. The path and impact envelope follow the layout of Fig. 6.4.

The lateral extent of an encroachment can be predicted using Fig. 6.5. It is significant to note that virtually all of the available crash data indicate that nearly 100% of encroachments penetrate no farther than 82 ft (25 m), and only 30–45% of encroachments penetrate farther than 16 ft (5 m). Using the rule of thumb of a 30 ft (10 m) clear zone, 80% of the encroachments penetrate no farther than 16 ft (5 m).

After every 10,000 encroachments, the convergence of a solution is examined. The distributions are compared with the pre-established distributions to check if they are within the specified level of convergence. Outputs that do not meet the convergence criteria are ignored in the final output. The process can be performed manually using a few iterations with general measures of reliability to obtain approximate crash predictions for planning purposes.

Severity Prediction

For every object or feature impacted by a crashing vehicle, the condition of the impact determines the outcome and the severity of the crash. For instance, if a vehicle traveling at a high speed impacts a guiderail at a low angle and is redirected, the severity and probability of injury would be less than if the impact at the same speed were at a high angle and the vehicle snagged or rolled over. The performance of the roadside device would also be taken into account if the device were designed to absorb some of the impact.

The Federal Highway Administration (FHWA) uses a *severity index* (SI) that varies with the type of vehicle involved, the speed and impact angle, and the type of obstacle struck. The intervening severity indices between zero and ten involve varying degrees of property damage coupled with slight, moderate, or severe personal injuries. A crash that results in no probability of injury would be assigned an SI of 0, while a crash that has a 100% probability of fatality would be assigned an SI of 10. Crash severity and probability of injury increase linearly, as shown in Fig. 6.6.

Crashes are also given an injury level rating based on the *KABCO scale*, which is used by police officers when completing crash reports. Police use their own judgment on-site when assigning injury levels. K is used for fatalities, A for disabling injuries, B for evident injuries, C for possible injuries, and O for no apparent injuries or property damage only (PDO). The *RDG* makes a further distinction within the O category. It uses two levels: property damage only level 1 (PDO1) and property damage only level 2 (PDO2), with PDO2 being assigned to crashes with more significant damage. Table 6.4 shows the correlation between SI and the probability of injury.

Figure 6.6 *Example of Relationship Between Severity Index (SI) and Impact Speed*

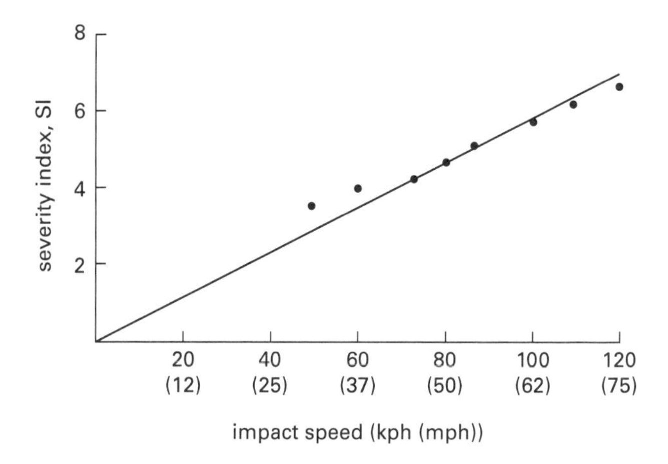

(Multiply kph by 0.621 to obtain mph.)

From *Roadside Design Guide*, 2002, by the American Association of State Highway and Transportation Officials, Washington, D.C. Used by permission.

Table 6.4 *Relationship of Severity Indices (SI) and Probability of Injury*

severity index, SI	injury level (%) K	A	B	C	PDO1	PDO2	none
0.0	–	–	–	–	–	–	100.0
0.5	–	–	–	–	100.0	–	–
1.0	–	–	2.3	7.3	66.7	23.7	–
2.0	–	–	7.0	22.0	–	71.0	–
3.0	1.0	1.0	21.0	34.0	–	43.0	–
4.0	3.0	5.0	32.0	30.0	–	30.0	–
5.0	8.0	10.0	45.0	22.0	–	15.0	–
6.0	18.0	20.0	39.0	16.0	–	7.0	–
7.0	30.0	30.0	28.0	10.0	–	2.0	–
8.0	50.0	27.0	19.0	4.0	–	–	–
9.0	75.0	18.0	7.0	–	–	–	–
10.0	100.0	–	–	–	–	–	–

Reprinted with permission from the American Association of State Highway and Transportation Officials, *Roadside Design Guide*, Table A.1, copyright © 2002.

6. ROADSIDE CLEARANCE ANALYSIS

Clear-Zone Distance

The *RDG* defines a *clear zone* as the total roadside area, beginning at the edge of the roadway, that allows drivers to stop safely or regain control of a vehicle that leaves the roadway. This area may consist of a shoulder, a recoverable slope, a nonrecoverable slope, and/or a clear runout area.

Recoverable slopes are slopes that generally allow a motorist to regain control of a vehicle by slowing or stopping. Slopes flatter than 1V:4H are considered recoverable. *Nonrecoverable slopes* are considered traversable, but the vehicle will continue to the bottom of the slope. Embankment slopes between 1V:3H and 1V:4H are considered traversable (as long as they are smooth and free of obstructing objects) but nonrecoverable. A *clear runout area* is the area at the toe of a nonrecoverable slope that is available for the vehicle's safe use. Slopes steeper than 1V:3H are not considered traversable and are not considered part of the clear zone.

Using design speed and design AADT, the recommended clear-zone distance can be selected from Fig. 6.7 and Fig. 6.8, or from Table 6.5. AASHTO's *Green Book* recommends a clear-zone distance between 7 ft and 10 ft (2 m and 3 m) on local roads and streets without curbs. For collectors without curbs, a 10 ft (3 m) minimum is recommended.

Figure 6.7 and Fig. 6.8 use the term *backslope* to describe an upward slope away from the shoulder and the term *foreslope* for a downward slope away from the shoulder. These are the terms used by the *RDG*, although the terms *cut slope* and *fill slope*, respectively, are more commonly used by design engineers. The *RDG* corrections for curves, or *clear-zone correction factor* values, K_{cz}, are shown in Table 6.6 and can be derived from Eq. 6.8.

$$\text{CZ}_c = L_{cz} K_{cz} \quad \text{6.8}$$

L_{cz} is the clear-zone distance found from either Fig. 6.7 and Fig. 6.8 or Table 6.5, and CZ_c is the clear zone on the outside curvature. The clear-zone correction factor is applied to outside curves only. Curves that are flatter than 2860 ft (900 m) do not need to be adjusted.

For every object or feature impacted by a crashing vehicle, the condition of the impact determines the outcome and the severity of the crash. For instance, if a vehicle traveling at high speed impacts a guiderail at a low angle and is redirected, the probability of injury will be less than if the impact were at a high angle and the vehicle were redirected at the same speed. The performance of the roadside feature must also be taken into account (e.g., whether or not the impact object was designed to absorb impact).

Objects in the Clear Zone

There are several options for how to treat objects in the clear zone. Whenever possible, the best option is to remove the object. However, if the object serves a necessary function, such as a utility pole or sign fixture, the object must be designed so that the lowest degree of personal injury occurs upon impact. For instance, breakaway design or shielding by a longitudinal barrier or crash cushion may be employed to mitigate injury.

Placement of large, fixed objects within the clear zone, such as piers, abutments, or large sign structures, must be protected by impact attenuators, guiderails, or other protective measures. Delineation can also be employed to improve the visibility of the object for oncoming drivers.

Objects should also have adequate shy distance. *Shy distance* is measured beginning at the edge of the traveled way. Objects with adequate shy distance should be far enough from the edge of the traveled way that they will not be perceived as an immediate hazard and cause drivers to slow or change the placement of their vehicle.

Fill slopes, or foreslopes, should be traversable whenever possible. When a traversable slope is not possible, the slope can be steepened with a guiderail placed at the top of the slope. The shoulder can also be widened by 2 ft to 4 ft (0.6 m to 1.2 m) to allow for additional vehicle-runoff recovery. Generally, slopes flatter than 1V:6H do not require guiderails. Slopes as steep as 1V:4H or 1V:3H are recoverable if a suitable runout is provided at the bottom of the slope, whether or not the driver can regain control of the vehicle within the space provided without a collision. Guiderail locations should be analyzed to determine if the potential damage caused by the guiderail would be greater or less than the potential damage if the guiderail were not present. If the guiderail has an equal or greater potential for causing damage, then the guiderail should not be installed.

Cut slopes, or backslopes, follow the same rules of recoverable versus nonrecoverable as fill slopes. However, cut slopes may redirect the vehicle back onto the roadway, which can result in a secondary crash. Both cut and fill slopes are considered nonrecoverable if they are planted with large trees or if vegetation has become so dense that vehicles would not penetrate it upon impact but instead be brought to an abrupt stop.

To reduce construction and right-of-way costs, it may be appropriate to design the cut or fill slope with a *variable-slope design*. This design places a recoverable slope adjacent to the paved shoulder for a certain distance, then allows for a nonrecoverable slope with a recovery zone at the bottom of the steeper slope, as shown in Fig. 6.9. A recoverable slope followed by a nonrecoverable slope is also called a *barn roof section*. This type of design is more cost-effective than a continuous flat foreslope and safer than a continuous steep foreslope, so it provides a good compromise between the two extremes.

Example 6.2

A freeway is designed for a speed of 60 mph (100 kph) and an AADT of 4500. There is an 8 ft (2.4 m) wide paved shoulder with a 0.04 ft/ft (0.04 m/m) cross slope. The distance from the edge of the traveled lane to the right-of-way line is 37 ft (11 m). The ground line at the right-of-way line is 5.5 ft (1.7 m) below the edge of the traveled lane. The design calls for variable slopes, with the section adjacent to the paved shoulder being 1V:8H and as wide as possible. The final slope against the right-of-way line is to be a maximum of 1V:5H. What will the width of the 1V:8H slope be, and what is the equivalent average slope from the edge of the running lane?

Figure 6.7 Clear-Zone Distance Curves (SI units)

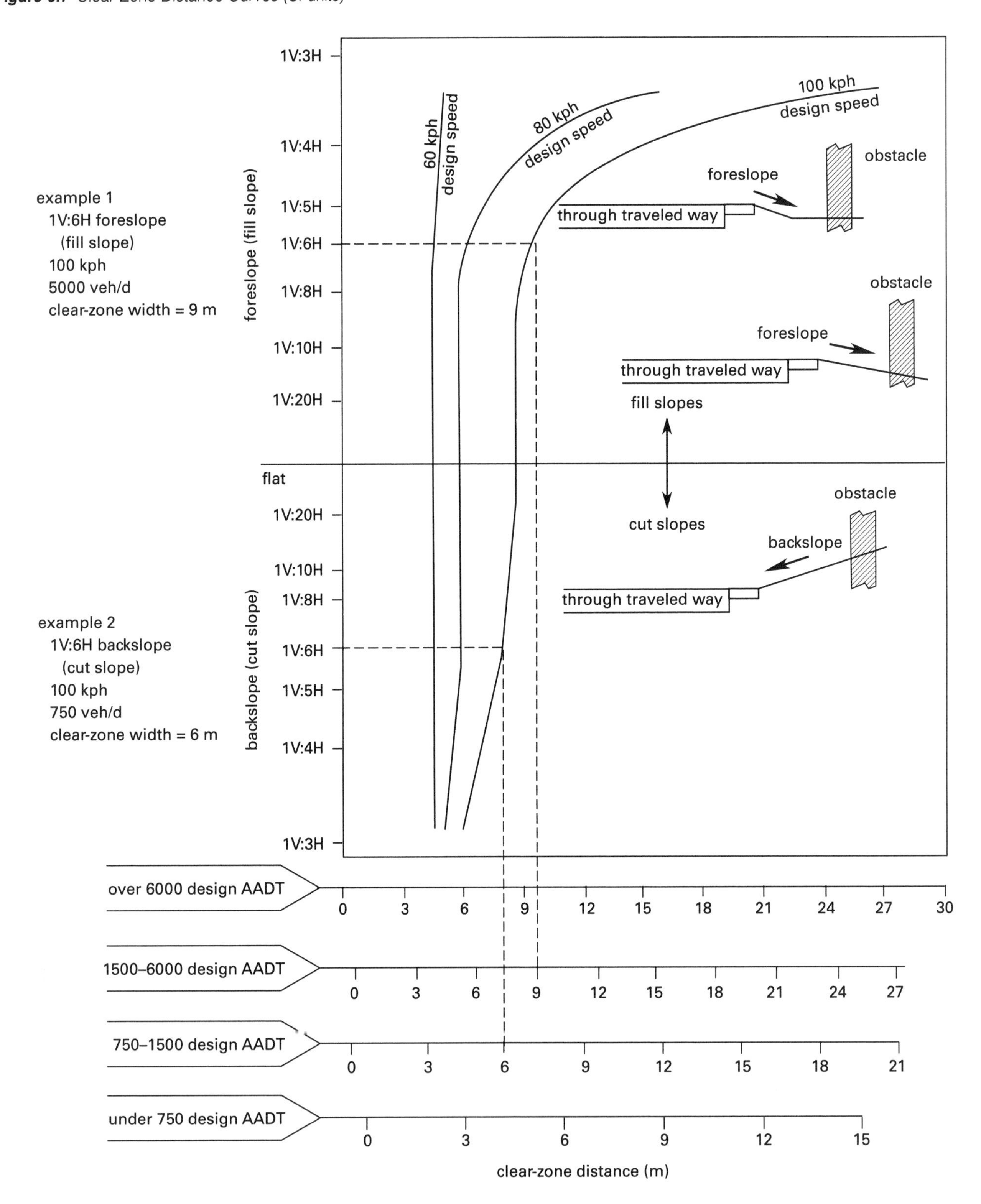

From *Roadside Design Guide*, 2002, by the American Association of State Highway and Transportation Officials, Washington, D.C. Used by permission.

Figure 6.8 *Clear-Zone Distance Curves (customary U.S. units)*

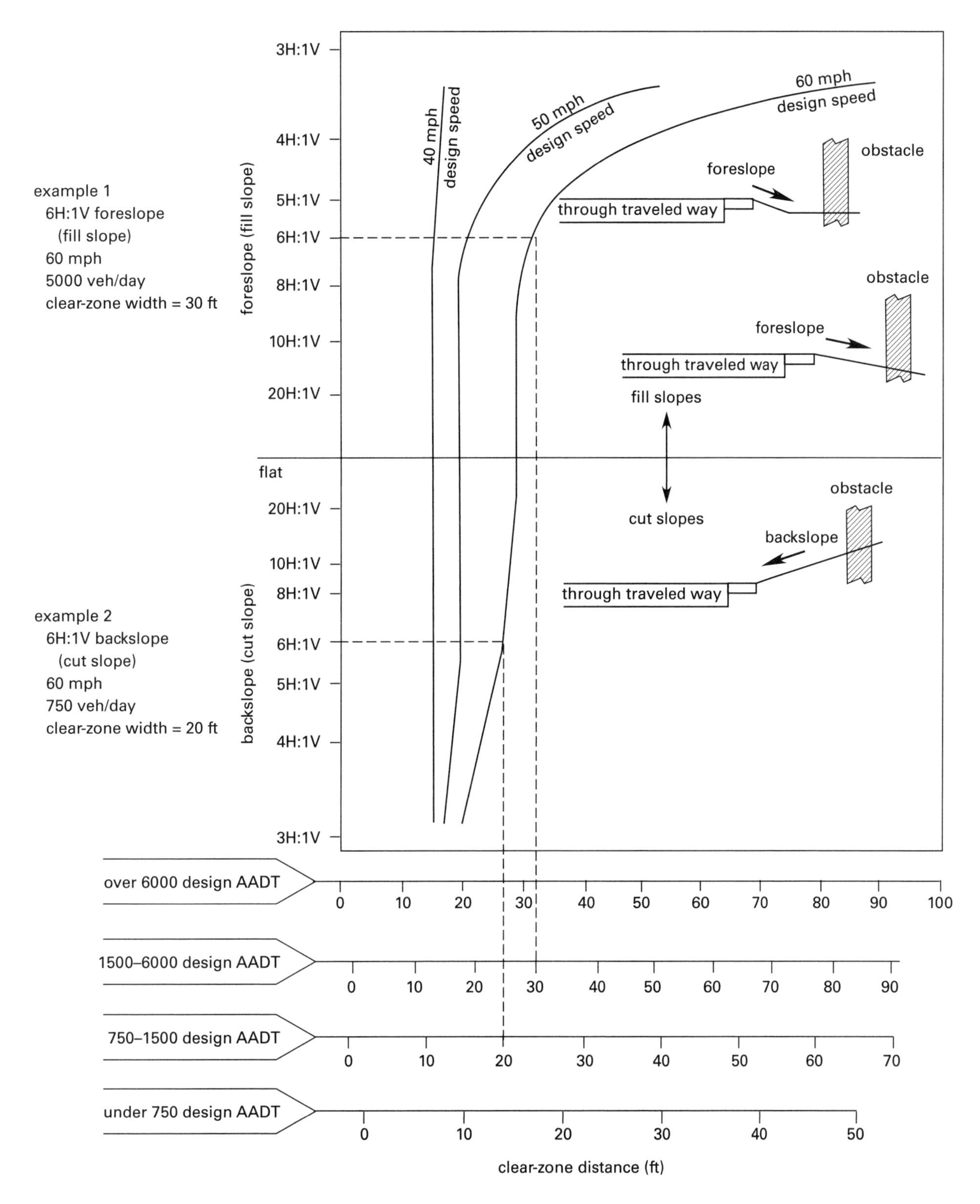

From *Roadside Design Guide*, 2002, by the American Association of State Highway and Transportation Officials, Washington, D.C. Used by permission.

Traffic Safety

Table 6.5 *Clear-Zone Distance from the Edge of Through Traveled Way*

SI units

		foreslopes[b]			backslopes	
design speed	design AADT	1V:6H or flatter	1V:5H to 1V:4H	1V:3H	1V:5H to 1V:4H	1V:6H or flatter
60 kph or less	under 750	2.0–3.0	2.0–3.0	2.0–3.0	2.0–3.0	2.0–3.0
	750–1500	3.0–3.5	3.5–4.5	3.0–3.5	3.0–3.5	3.0–3.5
	1500–6000	3.5–4.5	4.5–5.0	3.5–4.5	3.5–4.5	3.5–4.5
	over 6000	4.5–5.0	5.0–5.5	4.5–5.0	4.5–5.0	4.5–5.0
70–80 kph	under 750	3.0–3.5	3.5–4.5	2.5–3.0	2.5–3.0	3.0–3.5
	750–1500	4.5–5.0	5.0–6.0	3.0–3.5	3.5–4.5	4.5–5.0
	1500–6000	5.0–5.5	6.0–8.0	3.5–4.5	4.5–5.0	5.0–5.5
	over 6000	6.0–6.5	7.5–8.5	4.5–5.0	5.5–6.0	6.0–6.5
90 kph	under 750	3.5–4.5	4.5–5.5	2.5–3.0	3.0–3.5	3.0–3.5
	750–1500	5.0–5.5	6.0–7.5	3.0–3.5	4.5–5.0	5.0–5.5
	1500-6000	6.0–6.5	7.5–9.0	4.5–5.0	5.0–5.5	6.0–6.5
	over 6000	6.5–7.5	8.0–10.0[a]	5.0–5.5	6.0–6.5	6.5–7.5
100 kph	under 750	5.0–5.5	6.0–7.5	3.0–3.5	3.5–4.5	4.5–5.0
	750–1500	6.0–7.5	8.0–10.0[a]	3.5–4.5	5.0–5.5	6.0–6.5
	1500–6000	8.0–9.0	10.0–12.0[a]	4.5–5.5	5.5–6.5	7.5–8.0
	over 6000	9.0–10.0[a]	11.0–13.5[a]	6.0–6.5	7.5–8.0	8.0–8.5
110 kph	under 750	5.5–6.0	6.0–8.0	3.0–3.5	4.5–5.0	4.5–5.0
	750–1500	7.5–8.0	8.5–11.0[a]	3.5–5.0	5.5–6.0	6.0–6.5
	1500–6000	8.5–10.0[a]	10.5–13.0[a]	5.0–6.0	6.5–7.5	8.0–8.5
	over 6000	9.0–10.5[a]	11.5–14.0[a]	6.5–7.5	8.0–9.0	8.5–9.0

customary U.S. units

		foreslopes[b]			backslopes	
design speed	design AADT	1V:6H or flatter	1V:5H to 1V:4H	1V:3H	1V:5H to 1V:4H	1V:6H or flatter
40 mph or less	under 750	7–10	7–10	7–10	7–10	7–10
	750–1500	10–12	12–14	10–12	10–12	10–12
	1500–6000	12–14	14–16	12–14	12–14	12–14
	over 6000	14–16	16–18	14–16	14–16	14–16
45–50 mph	under 750	10–12	12–14	8–10	8–10	10–12
	750–1500	14–16	16–20	10–12	12–14	14–16
	1500–6000	16–18	20–26	12–14	14–16	16–18
	over 6000	20–22	24–28	14–16	18–20	20–22
55 mph	under 750	12–14	14–18	8–10	10–12	10–12
	750–1500	16–18	20–24	10–12	14–16	16–18
	1500–6000	20–22	24–30	14–16	16–18	20–22
	over 6000	22–24	26–32[a]	16–18	20–22	22–24
60 mph	under 750	16–18	20–24	10–12	12–14	14–16
	750–1500	20–24	26–32[a]	12–14	16–18	20–22
	1500–6000	26–30	32–40[a]	14–18	18–22	24–26
	over 6000	30–32[a]	36–44[a]	20–22	24–26	26–28
65–70 mph	under 750	18–20	20–26	10–12	14–16	14–16
	750–1500	24–26	28–36[a]	12–16	18–20	20–22
	1500–6000	28–32[a]	34–42[a]	16–20	22–24	26–28
	over 6000	30–34[a]	38–46[a]	22–24	26–30	28–30

[a]Where a site-specific investigation indicates a high probability of continuing crashes, or such occurrences are indicated by crash history, the design may provide clear-zone distances greater than the clear zone shown in Table 6.5. Clear zones may be limited to 30 ft (9 m) for practicality and to provide a consistent roadway template if previous experience with similar projects or designs indicates satisfactory performance.

[b]Since recovery is less likely on the unshielded, traversable 1V:3H slopes, fixed objects should not be present in the vicinity of the toe of these slopes. Recovery of high-speed vehicles that encroach beyond the edge of the shoulder may be expected to occur beyond the toe of the slope. Determination of the width of the recovery area at the toe of the slope should take into consideration right-of-way availability, environmental concerns, economic factors, safety needs, and crash histories. Also, the distance between the edge of the through traveled lane and the beginning of the 1V:3H slope should influence the recovery area provided at the toe of the slope. While the application may be limited by several factors, the foreslope parameters which may enter into determining a maximum desirable recovery area are illustrated in Fig. 6.5.

SI Solution

Draw the cross section.

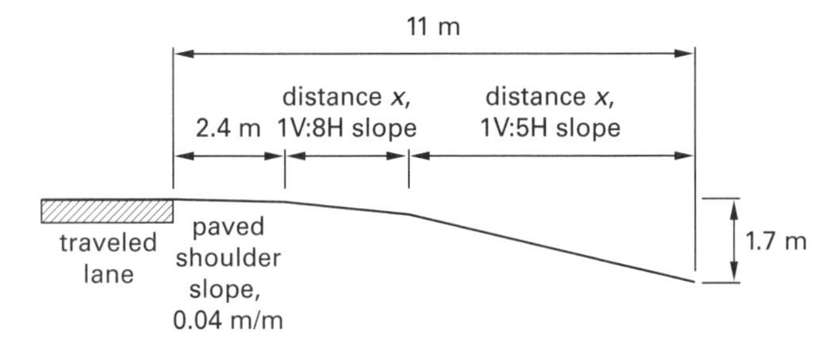

From Table 6.5, for a 100 kph design speed, an AADT of 4500, and a 1V:6H or flatter slope, the clear zone is 8–9 m. For slopes 1V:5H to 1V:4H, the clear zone is 10–12 m. Selecting a 1V:5H slope 6 m wide will gain 1.2 m in elevation. The remaining 0.50 m will be taken in the 2.4 m shoulder and the 1V:8H section. The average slope from the edge of the traveled lane to the right-of-way line is 1V:6.7H, which is within the criteria given in Fig. 6.7 and Table 6.5 for a slope between 1V:6HV and 1V:5H. Rounding the slope ratios to the nearest two significant digits is adequate for the analysis.

For the shoulder, the elevation change is

$$(2.4 \text{ m})\left(0.04 \ \frac{\text{m}}{\text{m}}\right) = 0.096 \text{ m}$$

For the 1V:5H slope, the elevation change is

$$(6 \text{ m})\left(0.2 \ \frac{\text{m}}{\text{m}}\right) = 1.20 \text{ m}$$

For the 1V:8H slope, the width is

$$11 \text{ m} - 6 \text{ m} - 2.4 \text{ m} = 2.6 \text{ m}$$

The elevation change of the 1V:8H slope is

$$(2.6 \text{ m})\left(0.125 \ \frac{\text{m}}{\text{m}}\right) = 0.325 \text{ m}$$

The elevation gain calculated by the slope ratio is

$$0.096 \text{ m} + 1.20 \text{ m} + 0.325 \text{ m} = 1.621 \text{ m}$$

Check the elevation difference.

$$1.7 \text{ m} - 1.621 \text{ m} = 0.079 \text{ m}$$

The elevation difference of 0.079 m is made by slightly adjusting either of the foreslopes, as the earthwork elevation in the clear zone will be within 0.03 m, which is a reasonable tolerance for this location.

The 11 m width has an elevation change of 1.7 m. The average slope is

$$\frac{11 \text{ m}}{1.7 \text{ m}} = 6.47 \quad (6.5)$$

Therefore, the slope ratio is 1V:6.5H.

Customary U.S. Solution

Draw the cross section.

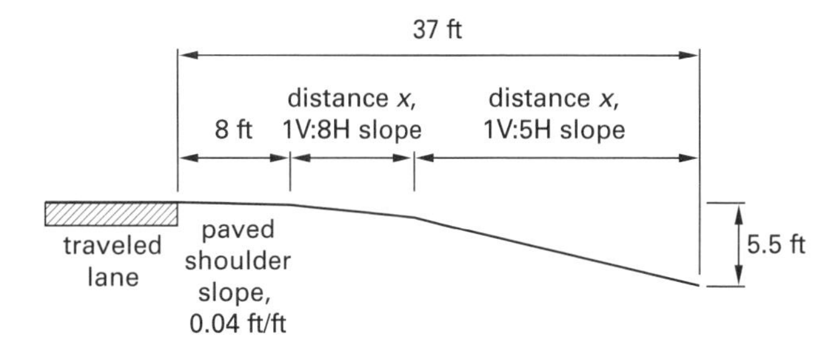

From Table 6.5, for a design speed of 60 mph, an AADT of 4500, and a 1V:6H or flatter slope, the clear zone is 26–30 ft. For slopes 1V:5H to 1V:4H, the clear zone is 32–40 ft. Selecting a 1V:5H slope 20 ft wide will gain 4.0 ft in elevation. The remaining 1.5 ft will be taken in the 8 ft shoulder and the 1V:8H section. The average slope from the edge of the traveled lane to the right-of-way line is 1V:6.7, which is within the criteria shown in Fig. 6.8 and Table 6.5 for a slope between 1V:6H and 1V:5H. Rounding the slope ratios to the nearest two significant digits is adequate for the analysis.

For the shoulder, the elevation change is

$$(8 \text{ ft})\left(0.04 \ \frac{\text{ft}}{\text{ft}}\right) = 0.32 \text{ ft}$$

For the 1V:5H slope, the elevation change is

$$(20 \text{ ft})\left(0.20 \ \frac{\text{ft}}{\text{ft}}\right) = 4.0 \text{ ft}$$

For the 1V:8H slope, the width is

$$37 \text{ ft} - 20 \text{ ft} - 8 \text{ ft} = 9 \text{ ft}$$

The elevation change of the 1V:8H slope is

$$(9 \text{ ft})\left(0.125 \ \frac{\text{ft}}{\text{ft}}\right) = 1.125 \text{ ft}$$

The elevation gain calculated by the slope ratio is

$$0.32 \text{ ft} + 4.0 \text{ ft} + 1.125 \text{ ft} = 5.445 \text{ ft}$$

Check the elevation difference.

$$5.5 \text{ ft} - 5.445 \text{ ft} = 0.055 \text{ ft}$$

Figure 6.9 Example of Parallel Foreslope Design

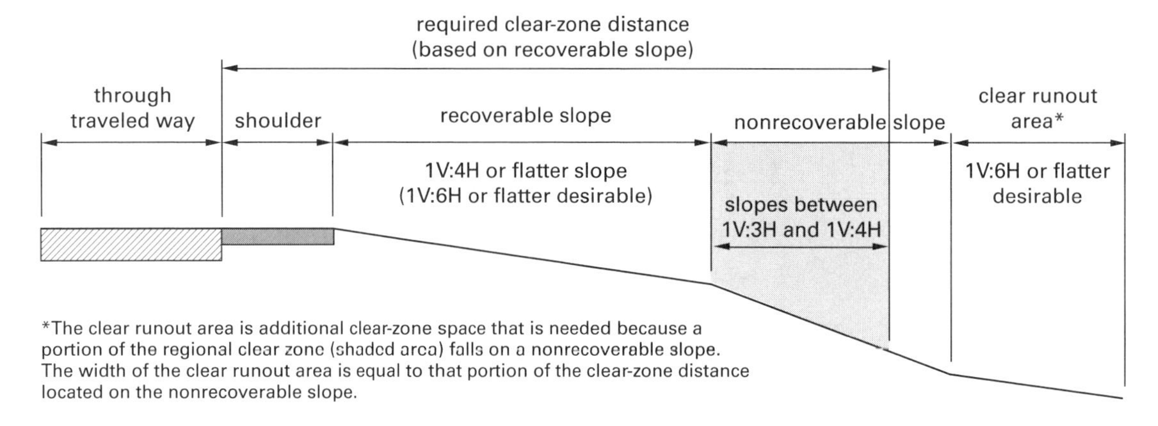

From *Roadside Design Guide*, 2002, by the American Association of State Highway and Transportation Officials, Washington, D.C. Used by permission.

The elevation difference of 0.055 ft is made by slightly adjusting either of the foreslopes, as the earthwork elevation in the clear zone will be within 0.1 ft, which is a reasonable tolerance for this location.

The 37 ft width has an elevation change of 5.5 ft. The average slope is

$$\frac{37 \text{ ft}}{5.5 \text{ ft}} = 6.73 \qquad (6.7)$$

Therefore, the slope ratio is 1V:6.7H.

7. ECONOMIC ANALYSIS

After determining the severity index of a crash, the costs associated with the crash can be calculated. Benefit-cost and engineering economic analysis can be applied to roadside safety-design features to determine a project's priority ranking. Economic analysis is most often used to compare different strategies to find the most economical alternative to implement on the jobsite. Economic evaluation of roadside safety using benefit-cost analysis requires data related to

- encroachments
- roadside geometry
- crash costs

Encroachments are primarily affected by traffic volume, roadway alignment, and lane widths. The number of encroachments that result in crashes is estimated from actual crash data (*crash statistics*) from the study location or from data derived from similar locations. The crash rates may be adjusted using the angle of departure from the roadway and the types of vehicles involved, as well as engineering judgment.

Roadside geometry encompasses the physical characteristics of the surface and appurtenances that a stray

Table 6.6 Horizontal Curve Adjustments*

K_{cz}, SI units

radius (m)	design speed (kph) 60	70	80	90	100	110
900	1.1	1.1	1.1	1.2	1.2	1.2
700	1.1	1.1	1.2	1.2	1.2	1.3
600	1.1	1.2	1.2	1.2	1.3	1.4
500	1.1	1.2	1.2	1.3	1.3	1.4
450	1.2	1.2	1.3	1.3	1.4	1.5
400	1.2	1.2	1.3	1.3	1.4	–
350	1.2	1.2	1.3	1.4	1.5	–
300	1.2	1.3	1.4	1.5	1.5	–
250	1.3	1.3	1.4	1.5	–	–
200	1.3	1.4	1.5	–	–	–
150	1.4	1.5	–	–	–	–
100	1.5	–	–	–	–	–

K_{cz}, customary U.S. units

radius (ft)	design speed (mph) 40	45	50	55	60	65	70
2860	1.1	1.1	1.1	1.2	1.2	1.2	1.3
2290	1.1	1.1	1.2	1.2	1.2	1.3	1.3
1910	1.1	1.2	1.2	1.2	1.3	1.3	1.4
1640	1.1	1.2	1.2	1.3	1.3	1.4	1.5
1430	1.2	1.2	1.3	1.3	1.4	1.4	–
1270	1.2	1.2	1.3	1.3	1.4	1.5	–
1150	1.2	1.2	1.3	1.4	1.5	–	–
950	1.2	1.3	1.4	1.5	1.5	–	–
820	1.3	1.3	1.4	1.5	–	–	–
720	1.3	1.4	1.5	–	–	–	–
640	1.3	1.4	1.5	–	–	–	–
570	1.4	1.5	–	–	–	–	–
380	1.5	–	–	–	–	–	–

*The clear-zone correction factor is applied to the outside of curves only. Curves flatter than 2860 ft (900 m) do not require an adjusted clear zone.

Reprinted with permission from the American Association of State Highway and Transportation Officials, *Roadside Design Guide*, Table 3.2, copyright © 2002.

vehicle is likely to traverse. It is important to take roadside geometry into account when performing an economic analysis of a roadway. Objects in the clear zone deserve special consideration. There are several options for how to treat objects in the clear zone, including relocating them or eliminating them entirely, mounting them on breakaway devices, shielding them, modifying them, or giving them a greater degree of visibility. The decision for how to treat an object in the clear zone should be based not only on the safety and feasibility of any option, but also on economic considerations.

Crash costs are a function of the number of crashes and the severity of crashes. Dollar values are applied to the severity indices using actual crash cost data from sources such as National Security Council (NSC), NHTSA, and FHWA databases, and insurance company information, as shown in Table 6.7. This information must be applied with careful engineering judgment to be sure of a good fit with the study location characteristics.

Table 6.7 *Crash Cost Figures*

crash severity	*RDG*	FHWA comprehensive cost
fatal crash (K)	$1,000,000	$2,600,000
severe injury crash (A)	$200,000	$180,000
moderate injury crash (B)	$12,500	$36,000
slight injury crash (C)	$3750	$19,000
PDO2	$3125	$2000
PDO1	$625	$2000

Measures of Cost-Effectiveness

Benefit-cost analysis can be performed using available data and a prediction model developed for the study location. The AASHTO computer-based RSAP can help reduce the tedium of repetitive evaluations of locations and alternatives. The program is adaptable to local data and site-specific applications for the selection of alternative treatments during the planning and design process of a project. Predictions can also be performed manually using readily available traffic crash data from the Bureau of Transportation Statistics or from individual state transportation department sources.

Crash cost is estimated using an encroachment probability model, which is unique to roadside safety cost-effectiveness procedures. Crash-cost predictions are based on the concept that the "run-off-the-road" crash frequency can be directly related to the encroachment frequency.

Encroachment Cost

The expected crash cost, C_{exp}, can be calculated using Eq. 6.9.

$$C_{\text{exp}} = \sum_{i=1}^{N_{\text{SI}}} VP_E P_{\text{cr}/E} P_{\text{SI},i/\text{cr}} C_{\text{SI},i} \qquad 6.9$$

V is the traffic volume, P_E is the probability of an encroachment, $P_{\text{cr}/E}$ is the probability of a crash given an encroachment, $P_{\text{SI},i/\text{cr}}$ is the probability of injury severity level i given a crash, $C_{\text{SI},i}$ is the cost associated with injury severity level i, and N_{SI} is the number of injury severity levels. The term VP_E is the expected encroachment frequency.

Benefit-Cost Determination

Using the predicted severity of a crash, the annualized crash costs, C_A, are calculated from Eq. 6.10 by multiplying the probability of each level of injury by the cost associated with that level of injury.

$$C_A = \sum_{i=1}^{N_{\text{SI}}} P_{\text{SI},i} C_{\text{SI},i} \qquad 6.10$$

$P_{\text{SI},i}$ is the probability of injury severity level i, $C_{\text{SI},i}$ is the cost associated with injury severity i, and N_{SI} is the total number of injury severity levels. Examples of crash cost figures are shown in Table 6.7.

The concept of benefit-cost analysis is that public funds should be invested in projects in proportion to the benefits exceeding the direct costs of the project. Benefits are considered the reduction in crash costs or societal costs due to the decrease in number and severity of crashes. Direct project costs are the costs of installation, annual maintenance, and crash repairs. Comparing the benefit-cost ratio of one alternative to another is generally used for the primary ranking of mitigation measures and determining whether or not a public investment in the improvement is appropriate. Incrementing benefit-cost (B/C) is determined by comparing alternatives as shown in Eq. 6.11.

$$B/C_{2\text{-}1} = \frac{C_{A,1} - C_{A,2}}{C_{D,1} - C_{D,1}} \qquad 6.11$$

$C_{A,1}$ and $C_{A,2}$ are the annualized crash or societal costs of alternatives 1 and 2, and $C_{D,1}$ and $C_{D,2}$ are the annualized direct costs of alternatives 1 and 2.

8. DRIVER BEHAVIOR AND PERFORMANCE

Driving is an acquired skill that relies on auditory and visual awareness, motor skill coordination, and the ability to judge situations and react accordingly. Therefore, driver education plays an important role in expected on-road driver behavior. Skills are taught and demonstrated, then studied and practiced. Although the quality of instruction is important in driver training, the driver's attitude toward the rules of the road has a greater effect on how well the driver's actions will conform to general traffic behavior. Attempts to change nonconforming behavior can range from passive measures, such as road designs and signs, to active measures, such as law enforcement.

Transportation systems should be designed to accommodate the *design driver*, a term that encompasses a range of drivers whose differences and limitations are accounted for during road, vehicle, and traffic control designs. Historically, designers have used the 85th percentile of driver behavior as the cutoff to determine transportation system designs and controls. The 85th percentile is used to represent the average driver, or the *reasonable worst case driver*, though this percentile may need to be adjusted depending on regional population characteristics. For instance, regions with large elderly populations may require more careful consideration of sign placement and sign lettering, as well as an increase in standard parking stall size and signal timing design.

Driver behavior is also influenced by vehicle design. Drivers of large commercial vehicles tend to drive conservatively due to the size of the vehicle, limited maneuverability, and the nature and value of the carried cargo. Similarly, bus drivers generally make driving decisions based on the comfort and safety of the passengers. Of standard passenger cars, the faster acceleration and maneuverability of performance cars allow and even encourage their drivers to engage in higher-risk driving behaviors than drivers of sedan-like cars.

Of the three major elements making up a highway system—the driver, the vehicle, and the road—the driver is perhaps the most variable and least predictable. Therefore, it is important to understand how the following internal and external factors influence driver performance so that they can be accounted for in a transportation system's design.

Perception-Reaction Responses

Perception-reaction refers to how quickly a driver can react to a situation. Most studies divide the perception-reaction response into four subprocesses—perception, identification, emotion, and volition—also known as PIEV. *Perception* first occurs when the driver sees a control device, sign, object, or another vehicle on the road. *Identification* follows perception when the driver classifies the nature of the perceived object. *Emotion* encompasses a driver's decision-making process to determine what action needs to be taken. Finally, *volition*, or *reaction*, occurs at the point at which the driver executes the action determined to be necessary.

These subprocesses take small but measurable amounts of time, together known as the *perception-reaction time* (PRT). The average necessary time to react and execute a response to a roadway condition or object message must be accounted for in the design process, such as placement of warning signs at the proper distance ahead of the condition being warned of.

PRT is important in determining the necessary sight distance on curves and for setting the amber phase of traffic signals. It is also a necessary factor to include when determining the safe following distance between vehicles. PRT increases with age, but is not uniform across the driver population at any age group. Experienced and alert drivers can react in 1.5 sec or less, while more passive drivers may take as long as 3 sec to respond to the same situation. AASHTO design guides generally use 2.5 sec as an average reaction time for determining stopping distances. However, PRT can vary significantly depending on roadway types and conditions. 2.5 sec may not be adequate under adverse or very complex conditions. For instance, an intersection approach may be complicated enough that the amber phase must be increased by 1–2 sec to eliminate running the red phase. Other instances where a longer PRT may exist include confusing or incorrect information on warning or directional signs. Drivers have generally learned to expect that adequate warning time will be given in order for them to react to a situation and make adjustments in vehicle operation. When drivers come across a situation that deviates from this norm (i.e., detour signs or unanticipated construction), they will likely need more time to react than usual, as the situation will come as a surprise. A *surprise* is when the time allowed for adjustment to a change in road condition, traffic condition, or visibility is less than the required normal perception-reaction time. The goal of good highway design is to eliminate surprises whenever possible.

Visual and Recognition Acuity

Visual acuity refers to the sharpness with which a person can see an object, while *recognition acuity* requires the viewer to not only see an object, but recognize the meaning of the object as well. For a normal eye, visual acuity is greatest near the visual center of the retina, which is along the visual axis of the eye. The cone of greatest visual acuity is about 3°, or about 1.5° in each direction from the center axis of vision. *Peripheral vision* can extend as far as 180° or more in some people, but 160° is the accepted normal limit of peripheral vision. In the *peripheral zone*, objects are visible and important for orientation and awareness, but may be blurred compared to the central cone of vision. Traffic signs and critical road markings should be placed so that they are within the normal central cone of vision in order to be clearly read by drivers.

The standard *Snellen chart* is used universally for checking visual acuity during an eye examination. A person reads letters of different heights on the chart from a specified distance. Charts used to determine visual acuity are designed so that a person with normal vision can read lettering 1/3 in (0.85 cm) in height from a distance of 20 ft (6.1 m). This ratio is called *20/20 vision*. When a person needs an object twice as large to be visible at 20 ft (6.1 m), the ratio is given as 20/40.

Visual and recognition acuity can deteriorate because of age, fatigue, eye disease, medication side effects, and/or alcohol and drug use.

The criterion for letter height on signs is often cited as 1 in (25 mm) for each 50 ft (15 m) of viewing distance. However, this criterion is incomplete without including letter spacing and letter shape. The letter shapes used on interstate highway signs were developed after extensive visibility research during the development of the early interstate highway system. The FHWA has developed the *federal alphabet* with respect to within letter spaces and between letter spaces. *Within letter spaces* are the openings in loop letters such as a, p, and so on. *Between letter spaces* are the spaces that exist between letters when they are put together to form words. Each letter in the federal alphabet has been created with an optimum amount of white space around, and, if necessary, within it. By this design, when letters are put together, the white space between them optically balances with the white space within them, producing optimum readability and legibility. These guidelines allow traffic signs to be viewed from a distance of 100 ft (30 m) for each 1 in (25 mm) of letter height, or twice the viewing distance when using other lettering styles. This standard has not been replicated in most current electronic digital lettering signs, meaning the viewing distance of many of these changeable lettering signs is much less than the fixed lettering version. Standards for a sign's height, color, letter spacing, message content, panels, layout, and location are given in further detail in the *Manual on Uniform Traffic Control Devices* (*MUTCD*).

Color Vision

The colors chosen for traffic control devices and informational signs were selected to minimize the effects of color blindness on their readability. *Color blindness* refers to a difficulty in perceiving differences between colors. Driver color blindness exists in various types and degrees that range from a difficulty in distinguishing between minor hue differences to a complete lack of color definition, in which the driver sees everything as varying intensities of gray. Background and letter coloring have been researched to provide the greatest amount of contrast so that lettering does not disappear when viewed by someone with color-impaired vision. For example, the green background with white lettering on roadway direction signs has become a standard on U.S. federal highways, eliminating the effect of color versus contrast sensitivities. Similarly, traffic signals place the red, amber, and green lenses in standardized positions so that a driver with total or red/green color blindness will still be able to determine signal phases.

Glare Vision

Glare can be caused by a direct shaft of light in the field of vision or by reflected light. When too much light enters the eye, the pupil closes to reduce the amount of incoming light. *Glare recovery* can take 6 sec or more for a normal eye, though factors such as age, congenital disabilities, or drug and alcohol use can increase the eye's recovery time.

Glare also affects the eye's visual contrast ratio between light and dark objects. The human eye cannot discern a light-intensity-contrast ratio greater than 3 to 1. *Glare blindness* occurs when the presence of an intense light source at one part of the visual plane causes images in the rest of the visual plane to disappear. For example, glare blindness can occur when driving at night along a dark road. Objects are illuminated by the headlight's radius, but objects appear invisible outside that radius.

When designing transportation systems, designers must account for changes between lit and unlit sections of the roadway and allow sufficient time for drivers' eyes to acclimate to new conditions before introducing critical movement, such as exit ramps or abrupt geometry conditions. Sources of light off of, but near to, roadways, such as advertising lighting and lighting in shopping center parking lots, can also create glare and affect the safety of traffic.

Depth Perception

Depth perception allows a person to estimate the distance to an object by using either stereo vision for objects closer to the eye or by estimating the relative size, position, and movement of objects farther away. The human eye has difficulty measuring the speed and the relative distances of objects that are more than a few feet away. This is especially true of approaching objects that are moving in a straight line, such as two oncoming vehicles occupying the same lane. Traffic signs and marking devices have standardized locations to account for the driver's viewing distance, which helps maintain uniformity and predictability of approaching roadway conditions. One example is the placement of a stop sign on the corner of a minor street intersecting a major street.

Hearing

Sounds received by the human ear can help a driver adjust speed, vehicle location, and other factors in relation to certain circumstances, such as an oncoming vehicle or the sound of a car horn. They can also warn drivers of unusual occurrences, such as an approaching emergency vehicle. Drivers with limited or no hearing are generally able to drive quite well, as long as they take extra precautions by using more visual feedback to compensate for the lack of hearing.

Driver Error and Unavoidable Situations

When a driver encounters traffic or road conditions that are outside of the range of normal conditions, collisions, off-road excursions, or other out-of-the-ordinary travel may occur. A pattern of similar crashes, particularly if involving drivers unfamiliar with the local conditions, may indicate a specific need for countermeasure development.

For example, if drivers frequently miss a turn at an intersection and drive off the side of the road, there may be a sight-limiting condition that is blocking the view at the driver's eye level. If the condition is primarily weather-related, such as icing, there may be a need to redirect drainage or recontour the intersection.

When a specific countermeasure cannot be developed to prevent a crash, or when crashes are random, countermeasures may include providing a safer environment for the most damaging or injury-prone type of occurrence.

Crash Attenuators

The chance that a vehicle will strike an object, such as a bridge pier or light pole, increases when the object is placed nearer to the edge of a high-speed roadway. Objects can be designed to yield under impact (e.g., light poles), or they can be designed to withstand considerable impact (e.g., bridge piers). In order to reduce the number of injuries or deaths of vehicle occupants, protective devices can be installed ahead of fixed objects to attenuate, or weaken, the decelerating force on the vehicle. Protective devices include barrels of sand, water-filled tubes, plastic foam cartridges, lightweight concrete, and a variety of sliding metal shapes.

The objective for all protective systems is to provide a stopping or deflecting action to the vehicle at a predetermined *maximum g-force* (usually 8.0 g). The *g*-force associated with an object is its acceleration in standard earth gravities, 32.2 ft/sec^2 (9.81 m/s^2). The maximum survivable *g*-force is also dependent on the duration of the force being applied. Higher *g*-forces can be tolerated for a shorter duration, while lower *g*-forces can be endured longer. Personal injury, vehicle control, rebound, damage to other vehicles, reusability, reparability, installation cost, maintenance cost, and liability are some of the factors considered in the design and installation of protective devices. To determine the minimum length of deceleration, an engineer selects the maximum *g*-force and the probable impact speed along with other predetermined criteria, such as the design vehicle weight, and calculates the required working length of the selected device. The loss of kinetic energy is equated to the work in deforming the attenuator, or crash cushion. Some energy is also expended in the deformation of the vehicle. However, this can vary from vehicle to vehicle and is not always included in the calculations. Equation 6.12 can be used for deceleration of a vehicle during a crash barrier impact. v is the design speed, and L is the unit length of the attenuator.

$$d = \frac{\mathrm{v}^2}{2L(0.75)} \qquad 6.12$$

The 0.75 factor in the denominator is the attenuation efficiency factor, which takes into account the portion of the length of the attenuator that is occupied by the energy absorption mechanism. Efficiency can be set at any level, but 75% is commonly used.

Once the engineer determines the required deceleration distance needed, a unit is selected from the vendor's catalog based on the specific requirements of the application. The engineer must then determine the design force on the backer wall or anchor system for the attenuator. A safety factor, SF, is also commonly applied to the anchor system or backer wall. Therefore, the stopping force, F_s, is calculated from Eq. 6.13 as

$$F_s = (\mathrm{SF})md \qquad \text{[SI]} \quad 6.13(a)$$

$$F_s = \frac{(\mathrm{SF})md}{g_c} \qquad \text{[U.S.]} \quad 6.13(b)$$

Example 6.3

An energy cushion unit is being considered for a design speed of 50 mph (80 kph) and a maximum deceleration of 8.0 g. The design vehicle weighs 4500 lbm (2040 kg). The unit is 14.5 ft (4.4 m) long, and 75% efficiency is to be used. (a) Determine whether the cushion unit length is sufficient, and (b) using a factor of safety of 1.5, determine the force requirement for the backer wall supporting the unit.

SI Solution

(a) From Eq. 6.12, the deceleration in m/s^2 is

$$\begin{aligned} d &= \frac{\mathrm{v}^2}{2L(0.75)} \\ &= \frac{\left(\left(80\ \frac{\mathrm{km}}{\mathrm{h}}\right)\left(\frac{1000\ \frac{\mathrm{m}}{\mathrm{km}}}{3600\ \frac{\mathrm{s}}{\mathrm{h}}}\right)\right)^2}{(2)(4.4\ \mathrm{m})(0.75)} \\ &= 74.6\ \mathrm{m/s^2} \end{aligned}$$

Determine the *g*-force.

$$\begin{aligned} d_g &= \frac{d}{g} = \frac{74.6\ \frac{\mathrm{m}}{\mathrm{s}^2}}{9.81\ \frac{\mathrm{m}}{\mathrm{s}^2\cdot\mathrm{g}}} \\ &= 7.6\ \mathrm{g} \quad \left[\begin{array}{c} < 8.0\ \mathrm{g}.\ \text{The cushion unit} \\ \text{length is sufficient.} \end{array}\right] \end{aligned}$$

(b) Determine the stopping force of a 2040 kg vehicle on the backup wall from Eq. 6.13(a), using a factor of safety of 1.5.

$$\begin{aligned} F_s &= (\mathrm{SF})md = (1.5)(2040\ \mathrm{kg})\left(74.6\ \frac{\mathrm{m}}{\mathrm{s}^2}\right) \\ &= 228\,000\ \mathrm{N} \end{aligned}$$

Customary U.S. Solution

(a) From Eq. 6.12, the deceleration in ft/sec² is

$$d = \frac{v^2}{2L(0.75)}$$

$$= \frac{\left(\left(50\ \frac{\text{mi}}{\text{hr}}\right)\left(\frac{5280\ \frac{\text{ft}}{\text{mi}}}{3600\ \frac{\text{sec}}{\text{hr}}}\right)\right)^2}{(2)(14.5\ \text{ft})(0.75)}$$

$$= 247\ \text{ft/sec}^2$$

Determine the g-force.

$$d_g = \frac{d}{g}$$

$$= \frac{247\ \frac{\text{ft}}{\text{sec}^2}}{32.2\ \frac{\text{ft}}{\text{sec}^2\text{-g}}}$$

$$= 7.7\ \text{g} \quad \left[\begin{array}{c} < 8.0\ \text{g. The cushion unit} \\ \text{length is sufficient.} \end{array}\right]$$

(b) Determine the stopping force of a 4500 lbm vehicle on the backup wall from Eq. 6.13(b), using a factor of safety of 1.5.

$$F_s = \frac{(\text{SF})md}{g_c}$$

$$= \frac{(1.5)(4500\ \text{lbm})\left(247\ \frac{\text{ft}}{\text{sec}^2}\right)}{32.2\ \frac{\text{lbm-ft}}{\text{lbf-sec}^2}}$$

$$= 52{,}000\ \text{lbf}$$

Example 6.4

A directional sign is needed before a highway turnoff. The highway has an 85th percentile speed of 50 mph. Vehicles normally make the turn at the design speed of 25 mph, and cars travel at a constant speed of 50 mph during the perception-reaction time. Using a perception-reaction time of 1.5 sec to read and understand the sign and a deceleration (or braking) rate of $0.3g$, what is the minimum distance the directional sign should be placed in advance of the turnoff?

Solution

The total distance traveled from the moment when the sign is observed includes the perception-reaction distance plus the braking distance needed to decelerate to the turnoff speed of 25 mph. Add the perception-reaction time to the deceleration distance. Since the car travels at a constant speed of 50 mph during the perception-reaction time, t_{PRT}, the total distance is

$$s_d = \frac{v_{\text{full}}^2 - v_{\text{final}}^2}{2d} + v_{\text{full}} t_{\text{PRT}}$$

$$= \left(\frac{\left(50\ \frac{\text{mi}}{\text{hr}}\right)^2 - \left(25\ \frac{\text{mi}}{\text{hr}}\right)^2}{(2)(0.3)\left(32.2\ \frac{\text{ft}}{\text{sec}^2}\right)}\right)\left(\frac{5280\ \frac{\text{ft}}{\text{mi}}}{3600\ \frac{\text{sec}}{\text{hr}}}\right)^2 + \left(50\ \frac{\text{mi}}{\text{hr}}\right)(1.5\ \text{sec})\left(\frac{5280\ \frac{\text{ft}}{\text{mi}}}{3600\ \frac{\text{sec}}{\text{hr}}}\right)$$

$$= 319\ \text{ft}$$

The car will travel 110 ft before braking begins, and the car requires 209 ft to brake from 50 mph to 25 mph. Therefore, the sign should be placed at least 209 ft before the turnoff.

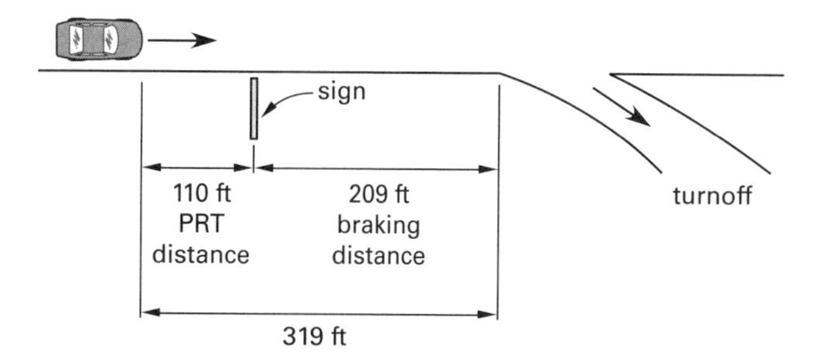

9. WORK ZONE SAFETY

Work zones increase crashes and congestion, and account for approximately 2% of all roadway fatalities each year. With the majority of the fatalities being motorists, the FHWA created the National Highway Work Zone Safety Program, which focuses on standardization and evaluation to improve work zone safety. Work zone standardization is set by the FHWA in both traffic control and work zone safety devices. All safety standards are contained in the *MUTCD*.

The National Work Zone Safety Information Clearinghouse is a public outreach organization that provides information and resources about work zone safety. Managed by the FHWA Office of Safety, it contains the largest online database of accident and crash rate data and statistics, training programs and materials, research services, latest technologies and equipment, best practices, safety engineer contact information, laws and regulations, and public education.

The FHWA is also responsible for National Cooperative Highway Research Program *Report 350* (NCHRP 350), which contains the federal standards and guidelines for all work zone safety devices, including traffic barriers, end treatments, crash cushions, and breakaway devices. NCHRP 350 provides a wide range of test procedures to evaluate these devices, with different test levels defined for various classes of roadside safety features. It provides enhanced measurement techniques related to occupant risk, as well as guidelines for device installation and test instrumentation. Evaluation criteria to assess the performance of work zone safety devices include occupant risk, occupant compartment integrity, test article debris, and vehicle stability.

10. CONFLICT ANALYSIS

Conflicts in traffic streams occur at any location where two traffic flows cross, merge, or diverge. Conflicts can be within the same type of traffic, such as automobile traffic, or between different types of traffic, such as between auto traffic and pedestrian traffic. Conflicts involve gaps in the traffic flow and the driver's judgment on whether the gap is acceptable or not. Diverging moves, such as for turning vehicles at intersections, are the least disruptive conflicts as long as the diverging paths operate at about the same speed. If one path is considerably slower than the other, a conflict occurs with the faster flow. Figure 6.10 shows vehicle streams at a three-way intersection.

Crossing and merging flows require larger gaps and have the greatest potential for collision. At a typical intersection of a two-lane street crossing another two-lane street (see Fig. 6.11), there are 32 conflict points. 16 points are crossing conflicts, 8 points are merging conflicts, and 8 points are diverging conflicts. Pedestrian crossings add another 20 conflict points to the intersection.

Figure 6.10 *Vehicle Streams at a Three-Way Intersection*

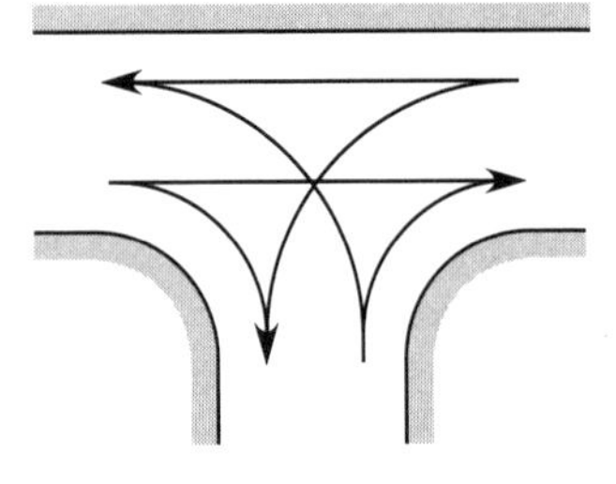

Adapted with permission of C. Jotin Khisty and B. Kent Lall, *Transportation Engineering: An Introduction*, copyright © 2003, by Prentice Hall.

Figure 6.11 illustrates where vehicle conflict points occur for a four-way intersection and two three-way intersections. Replacing a four-way intersection with two three-way intersections may reduce some conflict points for certain moves. However, this improvement creates two intersections. The delay caused by eliminating one of the through-movements is more than offset by the extra maneuvering required to perform an offset through-movement. For a three-way intersection, right turns and through-movements along the lanes opposite the entering traffic are the least impeded by other vehicle movements.

Conflict analysis involves reviewing data on conflicting traffic stream movements that show the flow density of each movement, acceptable gaps in following distance, the capacity limits of conflicting movements, and the delays caused by conflicting movements. Conflict analysis can be used to develop strategies to reduce delay, improve average speed, increase capacity, and improve safety.

Figure 6.11 *Vehicle Intersection Conflict Points*

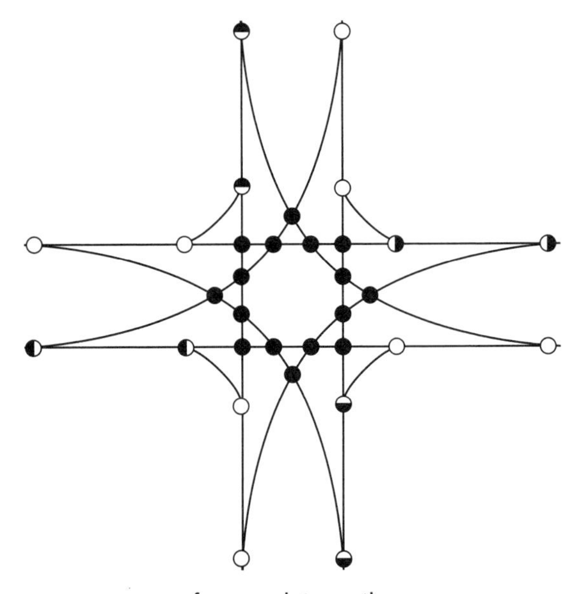

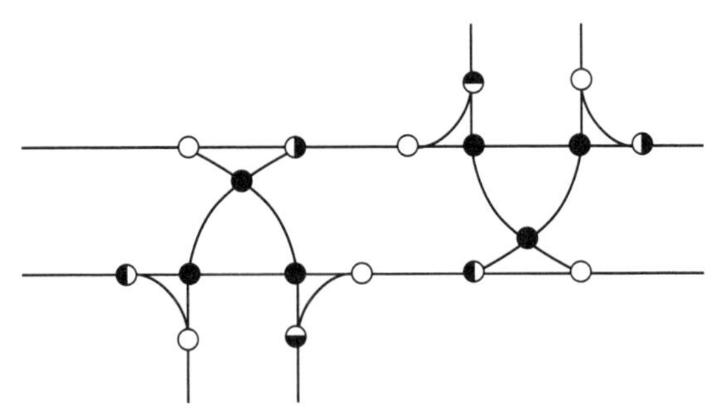

Adapted with permission of C. Jotin Khisty and B. Kent Lall, *Transportation Engineering: An Introduction*, copyright © 2003, by Prentice Hall.

PRACTICE PROBLEMS

1. A 5 mi segment of highway is analyzed for inclusion in a funding program. Included in the analysis is a statistical test to determine if the crash rate is significantly greater than the statewide average. The crash statistics over the previous four years are as follows.

year	fatal crashes	injury crashes	total crashes	AADT (veh/day)
1	4	162	320	38,800
2	3	176	340	40,000
3	2	190	365	41,200
4	5	185	352	42,400

The statewide average total property damage crash rate for similar roadway types is 106 crashes/HMVM for the 4 yr period. The standard deviation corresponding to this mean rate is 56 crashes/HMVM.

(a) What is most nearly the average crash rate per HMVM involving property damage only for the 4 yr period?

(A) 8.0 HMVM

(B) 44 HMVM

(C) 220 HMVM

(D) 650 HMVM

(b) At what confidence level do the statistics exceed the statewide average crash rate for the 4 yr period?

(A) 90%

(B) 95%

(C) 99%

(D) more than 99%

2. A four-lane urban freeway with a 9 m wide median and a 95 kph design speed is being studied for construction of additional lanes for capacity. The alternative proposed is to construct an additional 3.6 m lane on the left side in each direction with a concrete barrier, leaving a left shoulder of 1.3 m between the edge of the left lane and the median barrier. The freeway is 16 km long, lies in flat terrain, and has an AADT of 10 000. Construction cost of the alternative will be $30,000,000, with a life expectancy of 15 years. The annualized maintenance cost will be $150,000/km·yr for the project life. The discount rate is 4% per year, and the traffic growth factor is projected at 2.5% per year. Crash statistics are projected as follows.

crash severity	current roadway crash rate per 10^6 veh·km	alternate crash rate per 10^6 veh·km	crash cost
K—fatal crash	0.0172	0.0258	$1,000,000
A—disabling injury crash	0.0487	0.0688	$200,000
B—evident injury crash	0.0602	0.0860	$12,500
C—possible injury crash	0.119	0.150	$3750
PDO2	0.150	0.173	$625
PDO1	0.0989	0.120	$3125

(a) The number of additional crashes per year projected for the proposed alternative, averaged over the life of the project, is most nearly

(A) 9.0 crashes/yr

(B) 34 crashes/yr

(C) 44 crashes/yr

(D) 78 crashes/yr

(b) The average annualized cost of construction and maintenance for the life of the project is most nearly

(A) $2,850,000

(B) $2,890,000

(C) $215,000,000

(D) $235,000,000

(c) What is most nearly the average projected crash cost per year for the proposed alternative?

(A) $57,000

(B) $69,000

(C) $89,000

(D) $110,000

(d) Using benefit-cost analysis, compare the annualized cost of implementing the proposed alternative with the additional crash cost.

(A) 0.23:1

(B) 0.43:1

(C) 2.3:1

(D) 2.8:1

(e) What information would best complete the benefit-cost analysis for this study?

(A) data on right shoulder crashes

(B) effects of changing design speed

(C) number of trucks and buses

(D) benefits of added lane for increased capacity

3. An existing 24 ft wide two-lane highway is to be widened to four lanes of 12 ft each. The shoulders are 8 ft wide, and the retaining wall is projected above the extended shoulder slope 26 ft from the edge of the traveled lane. The proposed cross section will have 6 ft shoulders. The wall is to be reconstructed farther from the edge of pavement with a foreslope of 8H:1V between the wall and the shoulder. Currently, the highway has a design speed of 60 mph. The proposed design speed will be reduced to 50 mph, and the traffic projection is for more than 22,000 AADT.

What should be the shy distance from the edge of the pavement for the new wall in order to maintain the AASHTO recommended clear zone?

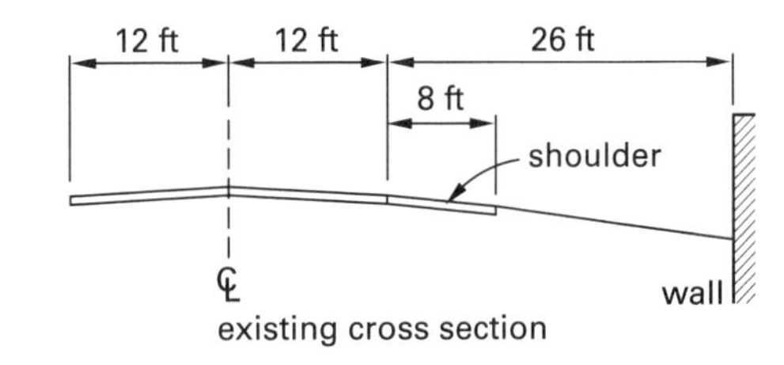

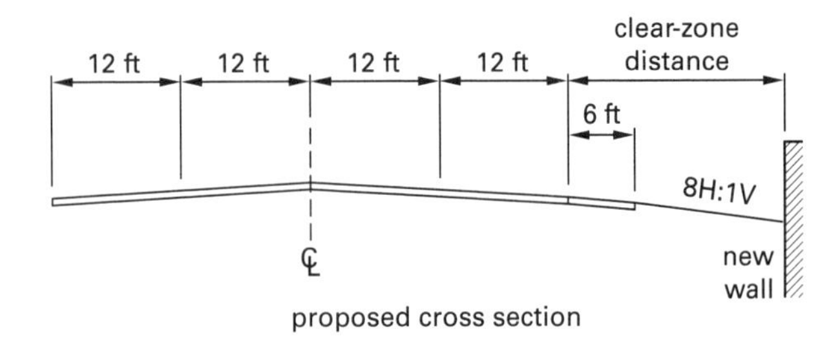

(A) 14 ft

(B) 18 ft

(C) 22 ft

(D) 30 ft

4. A 5 mi long, nearly straight highway segment through flat lowlands has experienced a high number of off-road crashes due to very heavy traffic, frequent icing conditions, and poor visibility caused by fog. The AADT of 55,000 operates at LOS C during the midday, and a peak period flow of about 8000 vph is at LOS E for the prevailing direction in the morning and evening periods. The cross section consists of a four-lane divided freeway with a wide median and a 10 ft paved right shoulder. The clear zone measures 20 ft wide from the traveled lane edge to the top slope of a drainage ditch running parallel to the roadway. The clear zone beyond the shoulder has a foreslope of 12H:1V, and the drainage ditch has 3H:1V side slopes along its 2 ft average depth. Typical crashes involve a vehicle being forced off the roadway and into the clear zone, traversing the drainage ditch and landing in an adjoining pasture. About 35% of the ditch crossings involve vehicle upset, and about 2% of heavy vehicle crashes involve at least one fatality, primarily due to vehicle upset while crossing the drainage ditch. The posted speed limit is 65 mph for passenger cars and 55 mph for trucks. Several alternative countermeasures have been proposed.

Alternative 1 is to lower the speed limit to 50 mph for all vehicles, along with strict enforcement. No other changes to the roadway cross section are proposed.

Alternative 2 is to place a guiderail behind the shoulder 10 ft from the edge of the traveled lanes.

Alternative 3 is to place a guiderail near the top of the slope of the drainage ditch, about 18 ft from the edge of the traveled lanes.

Alternative 4 is to change the slope of the drainage ditch sides to 6H:1V, narrowing the ditch bottom to 3 ft. This alternative also involves narrowing the clear zone adjacent to the shoulder from 20 ft wide to 16 ft wide and increasing its slope to 8H:1V. In certain limited areas, the entire 16 ft wide clear zone may be as steep as 6H:1V in order to maintain the needed ditch invert. The ditch bottom would be 31 ft from the traveled lanes, versus 30 ft for the current design. Right-of-way is available to perform the ditch and slope changes.

Which alternative would be the preferred countermeasure based on AASHTO guidelines?

(A)

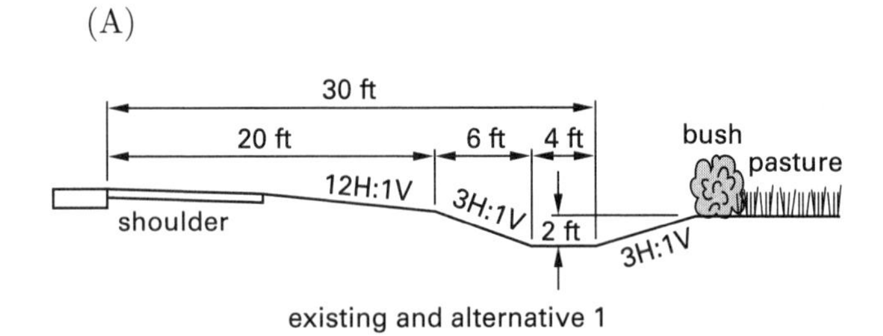

(B)

30 ft
10 ft
10 ft
6 ft
4 ft
12H:1V
3H:1V
2 ft
shoulder
3H:1V

alternative 2

(C)

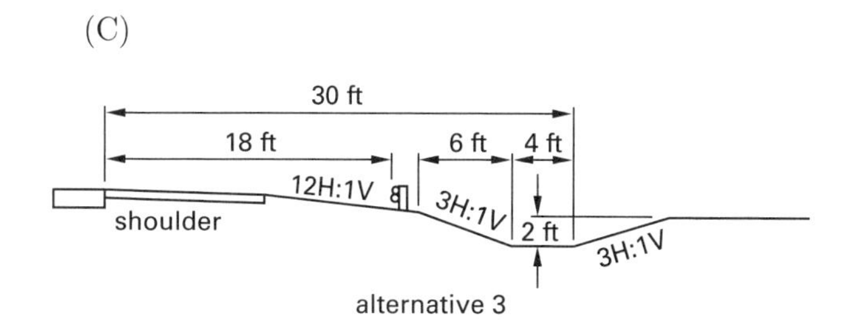

(D)

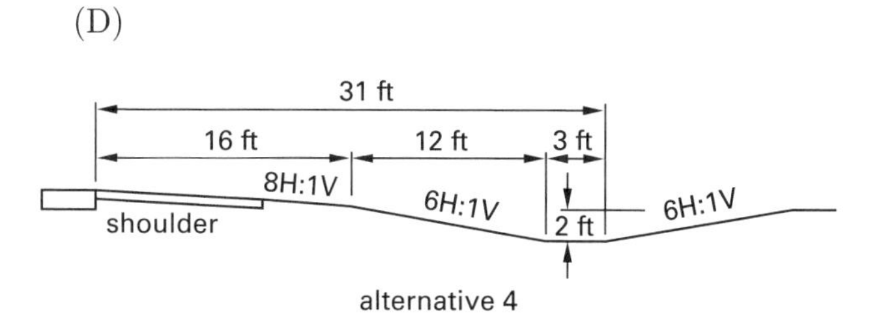

5. An intersection near a school with several conflict points has an increasing number of crashes. The intersection has stop sign control, and a school guard is stationed at the major school crossing for 3 hr each school day. An additional problem is the intersection flooding at least four times a year due to clogged and inadequately sized drainage pipes, which requires cleanup after every flood. Community groups are demanding improvements, including a traffic signal and increased school guard presence. Crash costs are given in the following table.

crash type	average cost per incident
K	$1,000,000
A	$200,000
PDO1	$3000

Crash type K involves a fatality, crash type A involves a disabling injury, and crash type PDO1 involves property damage only, including the result of swerving to avoid pedestrians.

Several alternatives have been proposed to reduce the conflict between vehicles, pedestrians, and/or other vehicles.

Alternative 1 is to do nothing with the intersection geometry. The drain pipes would be cleaned, but not repaired or replaced. The intersection would be given a maintenance overlay and re-striped in the existing traffic lane configuration. The historical crash rate is expected to continue for the next 2 yr. This alternative is used for comparison purposes.

Alternative 2 provides drainage improvements and a grade separation for the major cross traffic flows, with left-turning bays and slip ramps to accommodate the traffic movements. A pedestrian bridge is provided to eliminate the conflict with young children crossing the main traffic flow. However, a few older children cross at other places. The school guard would be eliminated with this alternative.

Alternative 3 is to provide drainage improvements, perform intersection geometric improvements, and install a traffic signal. A school guard would be stationed at the main school crossing for a total of 3 hr per school day.

Alternative 4 is to perform drainage improvements, make modest geometric improvements, and install a blinking traffic beacon, which would flash amber for the main movement and red for all other movements. A school guard would be stationed at the intersection for a total of 5 hr per school day.

Funding is available from the FHWA, the state transportation department, and local sources, sharing costs as shown in the following table. Crash occurrence and annual maintenance costs are also shown. School guard costs are entirely local. The projected crash statistics are for a 2 yr period following construction.

	alternative 1	alternative 2	alternative 3	alternative 4
crashes				
K	5	0	0	0
A	45	4	12	16
PDO1	160	14	28	40
crash years (total)	5 yr	2 yr	2 yr	2 yr
total construction cost	$100,000	$2,500,000	$650,000	$450,000
construction cost share				
federal	0%	50%	0%	0%
state	0%	30%	80%	80%
local	100%	20%	20%	20%
total annual maintenance	$50,000	$50,000	$45,000	$30,000
maintenance cost share				
state	0%	60%	20%	20%
local	100%	40%	80%	80%
school guard annual cost (local)	$30,000	$0	$30,000	$45,000
design life	3 yr	20 yr	12 yr	7 yr

Cost escalation is projected to be 5%/yr for construction and maintenance activities and 3%/yr for school guard wages.

(a) Using equivalent uniform annual cost (EUAC), which alternative will have the lowest local annual cost?

(A) alternative 1

(B) alternative 2

(C) alternative 3

(D) alternative 4

(b) Using a benefit-cost analysis of annual crash reduction cost compared with EUAC construction cost, what is most nearly the highest benefit-cost ratio?

(A) 17:1 for alternative 2

(B) 23:1 for alternative 3

(C) 27:1 for alternative 4

(D) 51:1 for alternative 3

SOLUTIONS

1. (a) The number of crashes that cause property damage for each year is found by subtracting the number of fatal crashes and the number of injury crashes from the total number of crashes.

$$N_{\text{property}} = N_t - N_{\text{fatal}} - N_{\text{injury}}$$

$$\begin{aligned}
\text{year 1: } N_{\text{property},1} &= 320 - 4 - 162 = 154 \\
\text{year 2: } N_{\text{property},2} &= 340 - 3 - 176 = 161 \\
\text{year 3: } N_{\text{property},3} &= 365 - 2 - 190 = 173 \\
\text{year 4: } N_{\text{property},4} &= 352 - 5 - 185 = 162
\end{aligned}$$

The crash rate is determined by Eq. 6.3.

$$\begin{aligned}
R_{\text{seg}} &= \frac{N_{\text{cr}}(10^8)}{(\text{AADT})(N_{\text{yr}})\left(365\ \frac{\text{days}}{\text{yr}}\right)L_{\text{mi}}} \\
&= \frac{(154 + 161 + 173 + 162)(10^8)}{\left(\begin{array}{l} 38{,}800\ \frac{\text{veh}}{\text{day}} + 40{,}000\ \frac{\text{veh}}{\text{day}} \\ \quad + 41{,}200\ \frac{\text{veh}}{\text{day}} + 42{,}400\ \frac{\text{veh}}{\text{day}} \end{array}\right) \times (1\ \text{yr})\left(365\ \frac{\text{days}}{\text{yr}}\right)(5\ \text{mi})} \\
&= 219\ \text{HMVM} \quad (220\ \text{HMVM})
\end{aligned}$$

The answer is (C).

(b) Test the data for statistical significance. Choose a value of K that corresponds to the level of confidence desired (see Table 6.1). Calculate the threshold rate for each interval and compare to the crash frequency rate using Eq. 6.4.

For a 90% confidence interval,

$$\begin{aligned}
R_{\text{th}} &= R_M + Ks = 106\ \text{HMVM} + (1.645)(56\ \text{HMVM}) \\
&= 198\ \text{HMVM}
\end{aligned}$$

For a 95% confidence interval,

$$\begin{aligned}
R_{\text{th}} &= R_M + Ks = 106\ \text{HMVM} + (1.960)(56\ \text{HMVM}) \\
&= 216\ \text{HMVM}
\end{aligned}$$

For a 99% confidence interval,

$$\begin{aligned}
R_{\text{th}} &= R_M + Ks = 106\ \text{HMVM} + (2.58)(56\ \text{HMVM}) \\
&= 250\ \text{HMVM}
\end{aligned}$$

The crash rate exceeds the threshold rate at confidence intervals of 90% and 95%, but not at 99%.

The answer is (B).

2. (a) Determine the total life of project average annual vehicle kilometers using the projected traffic growth.

$$\begin{aligned}\text{average traffic} &= (\text{AADT})\left(365\ \frac{\text{d}}{\text{yr}}\right)(F/A, i\%, n)\\ &\quad\times\left(\frac{\text{distance traveled}}{\text{study years}}\right)\\ &= \left(10\,000\ \frac{\text{veh}}{\text{d}}\right)\left(365\ \frac{\text{d}}{\text{yr}}\right)\\ &\quad\times\left(\frac{(1+0.025)^{15}-1}{0.025}\right)\left(\frac{16\ \text{km}}{15\ \text{yr}}\right)\\ &= 69\,814\,000\ \text{veh·km/yr}\end{aligned}$$

Determine the projected number of crashes for the current roadway and for the proposed alternative, and the annual cost of each.

For the current roadway, the average fatalities are

$$\begin{aligned}\text{average fatalities} &= \left(69\,814\,000\ \frac{\text{veh·km}}{\text{yr}}\right)\left(\frac{0.0172\ \text{fatalities}}{10^6\ \text{veh·km}}\right)\\ &= 1.20\ \text{fatalities/yr}\end{aligned}$$

Using the $(F/A, i\%, n)$ factor, the average annual cost is

$$\begin{aligned}&\left(1.20\ \frac{\text{fatalities}}{\text{yr}}\right)(\$1{,}000{,}000)\left(\frac{(1+0.04)^{15}-1}{0.04}\right)\left(\frac{1}{15\ \text{yr}}\right)\\ &\quad= \$1{,}602{,}000/\text{yr}\end{aligned}$$

Extend the data table to determine the average projected crashes and average projected crash cost over the life of the project.

crash severity	current roadway (crashes/yr)	current crash ($/yr)	proposed alternative (crashes/yr)	proposed alternative ($/yr)
K	1.20	1,602,000	1.80	2,404,000
A	3.40	908,000	4.80	1,282,000
B	4.20	70,100	6.00	100,100
C	8.31	41,600	10.47	52,400
PDO2	6.90	28,800	8.38	34,900
PDO1	10.47	8800	12.08	10,100
total	34.49	2,659,300	43.5	3,883,500

The average annual increase in the number of crashes per year for the alternative is

$$43.5\ \frac{\text{crashes}}{\text{yr}} - 34.49\ \frac{\text{crashes}}{\text{yr}} = 9.01\ \text{crashes/yr}\quad (9.0\ \text{crashes/yr})$$

The answer is (A).

(b) Using the $(A/P, i\%, n)$ factor, the annualized cost of construction is

$$C = (\$30{,}000{,}000)\left(\frac{(0.04)(1+0.04)^{15}}{(1+0.04)^{15}-1}\right) = \$2{,}698{,}000$$

The annualized cost of construction and maintenance is

$$C = \$2{,}698{,}000 + \$150{,}000 = \$2{,}848{,}000\quad (\$2{,}850{,}000)$$

The answer is (A).

(c) The average projected crash cost is the total crash cost divided by the number of crashes projected per year.

$$\begin{aligned}C_{\text{ave}} &= \frac{C_{\text{cr}}}{N_{\text{cr}}} = \frac{3{,}883{,}500\ \frac{\$}{\text{yr}}}{43.5\ \frac{1}{\text{yr}}}\\ &= \$89{,}276\quad (\$89{,}000)\end{aligned}$$

The answer is (C).

(d) The additional crash cost of implementing the proposed alternative is

$$C_{\text{cr}} = \$3{,}883{,}500 - \$2{,}659{,}300 = \$1{,}224{,}200$$

The benefit-cost ratio of the proposed alternative is

$$B/C = \frac{\$1{,}224{,}200}{\$2{,}848{,}000} = 0.43{:}1$$

The benefit-cost ratio is less than one, meaning there is more cost in predicted crashes than the cost of constructing the proposed alternative.

The answer is (B).

(e) Benefit-cost analysis is best used when comparing all direct benefits and all direct costs. The analysis docs not include the benefit of adding the extra lane to increase capacity of the roadway.

The answer is (D).

3. Refer to Fig. 6.7 for recommended clear-zone distances. Projecting down from the 50 mph line in the vicinity of 8H:1V foreslopes to the horizontal graph for over 6000 AADT, the intersect is near 18 ft offset.

The answer is (B).

4. Alternative 1 proposes lowering the speed limit with strict enforcement. With a peak period flow of 8000 vph, all lanes in both directions are operating at near capacity. Referring to Table 2.15, for a free flow speed of 65 mph, the minimum speed to maintain the flow rate at LOS E is 52.2 mph. From this observation, the traffic is already operating well below the posted speed limit of 65 mph. Therefore, lowering the speed limit would have a negligible effect on the peak period flow. However, lowering the average travel speed limit would reduce the severity of crashes during off-peak periods.

Alternative 2 places the guiderail close to the edge of the shoulder. This placement would reduce the number of vehicles that cross the drainage ditch. The guiderail may not have a pronounced effect on large vehicle crashes as large vehicles frequently vault guiderails, which are primarily designed for automobile protection. Also, the 12 ft clear zone remaining between the guiderail and the ditch slope could no longer be used as a recovery area. Referring to Fig. 6.4, about 60% of vehicles are able to recover from lateral excursions within 20 ft of an off-road encroachment, as opposed to less than 35% recovering from a 10 ft encroachment. Placing the guiderail 10 ft from the edge-of-traveled-way is not preferred by AASHTO guidelines when a greater offset is available.

Alternative 3 places the guiderail 18 ft from the edge of the traveled way. While this allows a greater proportion of vehicles to recover and would prevent many vehicles from traversing the drainage ditch, it would have a negligible effect on large vehicles for the reason outlined in alternative 2. If there were no other countermeasures available, this alternative would be more effective than alternative 1 or alternative 2.

Alternative 4 alters the clear-zone slopes by flattening the ditch side slopes and slightly increasing the recovery area foreslope. The increased slope could be considered traversable because the roadway is flat and nearly straight. Therefore, most of the lateral excursions can be assumed to be nearly straight or nearly parallel with the lanes of traffic. Based on the problem statement, the steep side slopes of the drainage ditch have caused vehicle upsets. Reducing the ditch side slopes to 6H:1V would reduce the number of vehicle upsets as well as the fatality rate. Alternative 4 would be the most effective countermeasure.

The answer is (D).

5. (a) The equivalent uniform annual cost is the present value of the construction cost amortized over the life of the project, A/P, plus the present annual cost of maintenance and the school guard.

Alternative 1

To calculate the local construction cost, multiply the cost by the percentage that must be paid by local sources.

$$\begin{aligned}\text{local construction cost} &= (1.0)(\$100{,}000)\\ &= \$100{,}000\end{aligned}$$

The A/P factor can be found using economic factor tables.

$$(A/P, i\%, n) = (A/P, 5\%, 3) = 0.3672$$

$$\begin{aligned}\text{EUAC} &= (\$100{,}000)(A/P, i\%, n)\\ &\quad + \text{annual maintenance cost}\\ &\quad + \text{school guard cost}\\ &= (\$100{,}000)(0.3672) + \$50{,}000 + \$30{,}000\\ &= \$117{,}000\end{aligned}$$

Alternative 2

The equivalent uniform annual cost for alternative 2 is

$$\begin{aligned}\text{local construction cost} &= (0.2)(\$2{,}500{,}000)\\ &= \$500{,}000\end{aligned}$$

$$(A/P, i\%, n) = (A/P, 5\%, 20) = 0.0802$$

$$\begin{aligned}\text{EUAC} &= (\$500{,}000)(A/P, i\%, n)\\ &\quad + \text{annual maintenance cost}\\ &\quad + \text{school guard cost}\\ &= (\$500{,}000)(0.0802) + (0.4)(\$50{,}000) + 0\\ &= \$60{,}100\end{aligned}$$

Alternative 3

The equivalent uniform annual cost for alternative 3 is

$$\begin{aligned}\text{local construction cost} &= (0.2)(\$650{,}000)\\ &= \$130{,}000\end{aligned}$$

$$(A/P, i\%, n) = (A/P, 5\%, 12) = 0.1128$$

Traffic Safety

$$\begin{aligned}\text{EUAC} &= (\$130{,}000)(A/P, i\%, n)\\ &\quad + \text{annual maintenance cost}\\ &\quad + \text{school guard cost}\\ &= (\$130{,}000)(0.1128) + (0.8)(\$45{,}000) + \$30{,}000\\ &= \$80{,}700\end{aligned}$$

Alternative 4

The equivalent uniform annual cost for alternative 4 is

$$\begin{aligned}\text{local construction cost} &= (0.2)(\$450{,}000)\\ &= \$90{,}000\end{aligned}$$

$$(A/P, i\%, n) = (A/P, 5\%, 7) = 0.1728$$

$$\begin{aligned}\text{EUAC} &= (\$90{,}000)(A/P, i\%, n)\\ &\quad + \text{annual maintenance cost}\\ &\quad + \text{school guard cost}\\ &= (\$90{,}000)(0.1728) + (0.8)(\$30{,}000) + \$45{,}000\\ &= \$84{,}600\end{aligned}$$

Alternative 2 has the lowest equivalent uniform annual cost.

The answer is (B).

(b) The format to determine annual crash cost (ACC) is

$$\text{ACC}_i = \left(\frac{1}{\text{no. crash data years}}\right) \times \begin{pmatrix} C_{\text{cr,K}}N_{\text{cr}} \\ + C_{\text{cr,A}}N_{\text{cr}} \\ + C_{\text{cr,PDO1}}N_{\text{cr}}\end{pmatrix}$$

For each alternative,

$$\begin{aligned}\text{ACC}_1 &= \left(\tfrac{1}{5}\right)\begin{pmatrix}(\$1{,}000{,}000)(5) + (\$200{,}000)(45)\\ + (\$3000)(160)\end{pmatrix}\\ &= \$2{,}896{,}000\\ \text{ACC}_2 &= \left(\tfrac{1}{2}\right)\begin{pmatrix}(\$1{,}000{,}000)(0) + (\$200{,}000)(4)\\ + (\$3000)(14)\end{pmatrix}\\ &= \$421{,}000\\ \text{ACC}_3 &= \left(\tfrac{1}{2}\right)\begin{pmatrix}(\$1{,}000{,}000)(0) + (\$200{,}000)(12)\\ + (\$3000)(28)\end{pmatrix}\\ &= \$1{,}242{,}000\\ \text{ACC}_4 &= \left(\tfrac{1}{2}\right)\begin{pmatrix}(\$1{,}000{,}000)(0) + (\$200{,}000)(16)\\ + (\$3000)(40)\end{pmatrix}\\ &= \$1{,}660{,}000\end{aligned}$$

The EUAC of construction for each alternative is

$$\begin{aligned}\text{EUAC}_2 &= (\text{construction cost})(A/P, i\%, n)\\ &= (\text{construction cost})(A/P, 5\%, 20)\\ &= (\$2{,}500{,}000)(0.0802)\\ &= \$201{,}000\\ \text{EUAC}_3 &= (\text{construction cost})(A/P, i\%, n)\\ &= (\text{construction cost})(A/P, 5\%, 12)\\ &= (\$650{,}000)(0.1128)\\ &= \$73{,}300\\ \text{EUAC}_4 &= (\text{construction cost})(A/P, i\%, n)\\ &= (\text{construction cost})(A/P, 5\%, 7)\\ &= (\$450{,}000)(0.1728)\\ &= \$77{,}800\end{aligned}$$

The benefit-cost ratio (B/C) is found by

$$\begin{aligned}B/C_i &= \frac{\text{ACC}_1 - \text{ACC}_i}{\text{EUAC}_i}\\ B/C_2 &= \frac{\$2{,}896{,}000 - \$421{,}000}{\$201{,}000} = 12.3 \quad (12{:}1)\\ B/C_3 &= \frac{\$2{,}896{,}000 - \$1{,}242{,}000}{\$73{,}300} = 22.6 \quad (23{:}1)\\ B/C_4 &= \frac{\$2{,}896{,}000 - \$1{,}660{,}000}{\$77{,}800} = 15.9 \quad (16{:}1)\end{aligned}$$

The answer is (B).

Appendices Table of Contents

APPENDIX 2.A

Travel Time, Speed, and Delay Study Field Sheet Using the Moving Vehicle Method

Travel Time and Delay Study
Moving Vehicle Method
Field Sheet

route ______________________ date ______________________

start point ______________________ end point ______________________

weather ______________________

run	start time	finish time	travel time	vehicles met	vehicles overtaking	vehicles passed
__bound						
1						
2						
3						
4						
5						
6						
7						
8						
total						
average						
__bound						
1						
2						
3						
4						
5						
6						
7						
8						
total						
average						

comments ______________________

recorder(s) ______________________

Appendices

APPENDIX 2.B
Passenger Car Equivalents for Upgrades and Downgrades on Multilane Highways

for trucks and buses on uniform upgrades, E_T

upgrade (%)	length (mi)	percentage of trucks and buses								
		2%	4%	5%	6%	8%	10%	15%	20%	25%
<2	all	1.5	1.5	1.5	1.5	1.5	1.5	1.5	1.5	1.5
≥2–3	0.00–0.25	1.5	1.5	1.5	1.5	1.5	1.5	1.5	1.5	1.5
	>0.25–0.50	1.5	1.5	1.5	1.5	1.5	1.5	1.5	1.5	1.5
	>0.50–0.75	1.5	1.5	1.5	1.5	1.5	1.5	1.5	1.5	1.5
	>0.75–1.00	2.0	2.0	2.0	2.0	1.5	1.5	1.5	1.5	1.5
	>1.00–1.50	2.5	2.5	2.5	2.5	2.0	2.0	2.0	2.0	2.0
	>1.50	3.0	3.0	2.5	2.5	2.0	2.0	2.0	2.0	2.0
>3–4	0.00–0.25	1.5	1.5	1.5	1.5	1.5	1.5	1.5	1.5	1.5
	>0.25–0.50	2.0	2.0	2.0	2.0	2.0	2.0	1.5	1.5	1.5
	>0.50–0.75	2.5	2.5	2.0	2.0	2.0	2.0	2.0	2.0	2.0
	>0.75–1.00	3.0	3.0	2.5	2.5	2.5	2.5	2.0	2.0	2.0
	>1.00–1.50	3.5	3.5	3.0	3.0	3.0	3.0	2.5	2.5	2.5
	>1.50	4.0	3.5	3.0	3.0	3.0	3.0	2.5	2.5	2.5
>4–5	0.00–0.25	1.5	1.5	1.5	1.5	1.5	1.5	1.5	1.5	1.5
	>0.25–0.50	3.0	2.5	2.5	2.5	2.0	2.0	2.0	2.0	2.0
	>0.50–0.75	3.5	3.0	3.0	3.0	2.5	2.5	2.5	2.5	2.5
	>0.75–1.00	4.0	3.5	3.5	3.5	3.0	3.0	3.0	3.0	3.0
	>1.00	5.0	4.0	4.0	4.0	3.5	3.5	3.0	3.0	3.0
>5–6	0.00–0.25	2.0	2.0	1.5	1.5	1.5	1.5	1.5	1.5	1.5
	>0.25–0.30	4.0	3.0	2.5	2.5	2.0	2.0	2.0	2.0	2.0
	>0.30–0.50	4.5	4.0	3.5	3.0	2.5	2.5	2.5	2.5	2.5
	>0.50–0.75	5.0	4.5	4.0	3.5	3.0	3.0	3.0	3.0	3.0
	>0.75–1.00	5.5	5.0	4.5	4.0	3.0	3.0	3.0	3.0	3.0
	>1.00	6.0	5.0	5.0	4.5	3.5	3.5	3.5	3.5	3.5
>6	0.00–0.25	4.0	3.0	2.5	2.5	2.5	2.5	2.0	2.0	2.0
	>0.25–0.30	4.5	4.0	3.5	3.5	3.5	3.0	2.5	2.5	2.5
	>0.30–0.50	5.0	4.5	4.0	4.0	3.5	3.0	2.5	2.5	2.5
	>0.50–0.75	5.5	5.0	4.5	4.5	4.0	3.5	3.0	3.0	3.0
	>0.75–1.00	6.0	5.5	5.0	5.0	4.5	4.0	3.5	3.5	3.5
	>1.00	7.0	6.0	5.5	5.5	5.0	4.5	4.0	4.0	4.0

(Multiply mi by 1.609 to obtain km.)

Highway Capacity Manual 2000. Copyright, National Academy of Sciences, Washington, D.C. Exhibit 21.9. Reproduced with permission of the Transportation Research Board.

(continued)

APPENDIX 2.B *(continued)*

Passenger Car Equivalents for Upgrades and Downgrades on Multilane Highways

for RVs on uniform upgrades, E_R

grade (%)	length (mi)	percentage of RVs								
		2%	4%	5%	6%	8%	10%	15%	20%	25%
<2	all	1.2	1.2	1.2	1.2	1.2	1.2	1.2	1.2	1.2
≥2–3	0.00–0.50	1.2	1.2	1.2	1.2	1.2	1.2	1.2	1.2	1.2
	>0.50	3.0	1.5	1.5	1.5	1.5	1.5	1.2	1.2	1.2
>3–4	0.00–0.25	1.2	1.2	1.2	1.2	1.2	1.2	1.2	1.2	1.2
	>0.25–0.50	2.5	2.5	2.0	2.0	2.0	2.0	1.5	1.5	1.5
	>0.50	3.0	2.5	2.5	2.5	2.0	2.0	2.0	1.5	1.5
>4–5	0.00–0.25	2.5	2.0	2.0	2.0	1.5	1.5	1.5	1.5	1.5
	>0.25–0.50	4.0	3.0	3.0	3.0	2.5	2.5	2.0	2.0	2.0
	>0.50	4.5	3.5	3.0	3.0	3.0	2.5	2.5	2.0	2.0
>5	0.00–0.25	4.0	3.0	2.5	2.5	2.5	2.0	2.0	2.0	1.5
	>0.25–0.50	6.0	4.0	4.0	3.5	3.0	3.0	2.5	2.5	2.0
	>0.50	6.0	4.5	4.0	4.0	3.5	3.0	3.0	2.5	2.0

(Multiply mi by 1.609 to obtain km.)

Highway Capacity Manual 2000. Copyright, National Academy of Sciences, Washington, D.C. Exhibit 21.9. Reproduced with permission of the Transportation Research Board.

for trucks on downgrades, E_T

downgrade	length	percentage of trucks			
(%)	(mi)	5%	10%	15%	20%
<4	all	1.5	1.5	1.5	1.5
4–5	≤4	1.5	1.5	1.5	1.5
4–5	>4	2.0	2.0	2.0	1.5
>5–6	≤4	1.5	1.5	1.5	1.5
>5–6	>4	5.5	4.0	4.0	3.0
>6	≤4	1.5	1.5	1.5	1.5
>6	>4	7.5	6.0	5.5	4.5

(Multiply mi by 1.609 to obtain km.)

Highway Capacity Manual 2000. Copyright, National Academy of Sciences, Washington, D.C. Exhibit 21.9. Reproduced with permission of the Transportation Research Board.

APPENDIX 2.C
Warrant 1, Eight-Hour Vehicular Volume

condition A—minimum vehicular volume

number of lanes for moving traffic on each approach		vehicles per hour on major street (total of both approaches)				vehicles per hour on higher-volume minor-street approach (one direction only)			
major street	minor street	100%[a]	80%[b]	70%[c]	56%[d]	100%[a]	80%[b]	70%[c]	56%[d]
1	1	500	400	350	280	150	120	105	84
2 or more	1	600	480	420	336	150	120	105	84
2 or more	2 or more	600	480	420	336	200	160	140	112
1	2 or more	500	400	350	280	200	160	140	112

condition B—interruption of continuous traffic

number of lanes for moving traffic on each approach		vehicles per hour on major street (total of both approaches)				vehicles per hour on higher-volume minor-street approach (one direction only)			
major street	minor street	100%[a]	80%[b]	70%[c]	56%[d]	100%[a]	80%[b]	70%[c]	56%[d]
1	1	750	600	525	420	75	60	53	42
2 or more	1	900	720	630	504	75	60	53	42
2 or more	2 or more	900	720	630	504	100	80	70	56
1	2 or more	750	600	525	420	100	80	70	56

[a]basic minimum hourly volume
[b]used for combination of conditions A and B after adequate trial of other remedial measures
[c]may be used when the major-street speed exceeds 40 mph (65 kph) or in an isolated community with a population of less than 10,000
[d]may be used for combination of conditions A and B after adequate trial of other remedial measures when the major-street speed exceeds 40 mph (65 kph) or in an isolated community with a population of less than 10,000

Reprinted from the *Manual on Uniform Traffic Control Devices*, 2009 ed., Table 4C-1, U.S. Department of Transportation, Federal Highway Administration, 2009.

Appendices

APPENDIX 2.D
Warrant 2

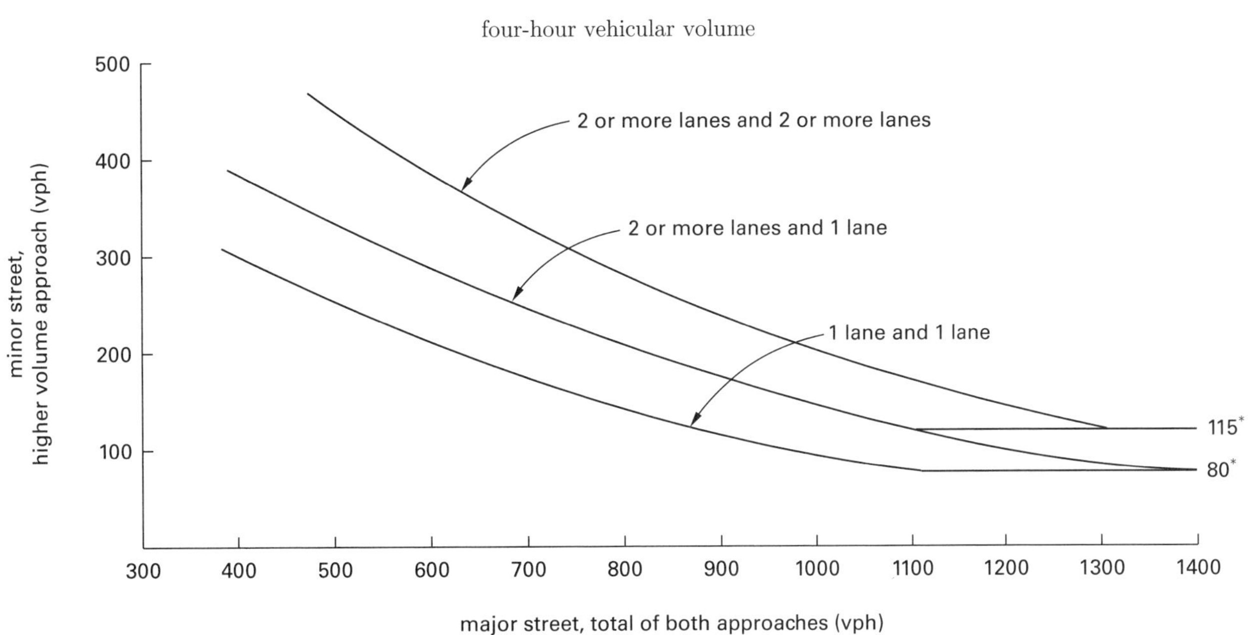

*Note: 115 vph applies as the lower threshold volume for a minor-street approach with two or more lanes and 80 vph applies as the lower threshold volume for a minor-street approach with one lane.

Reprinted from the *Manual on Uniform Traffic Control Devices*, 2009 ed., Fig. 4C-1, U.S. Department of Transportation, Federal Highway Administration, 2009.

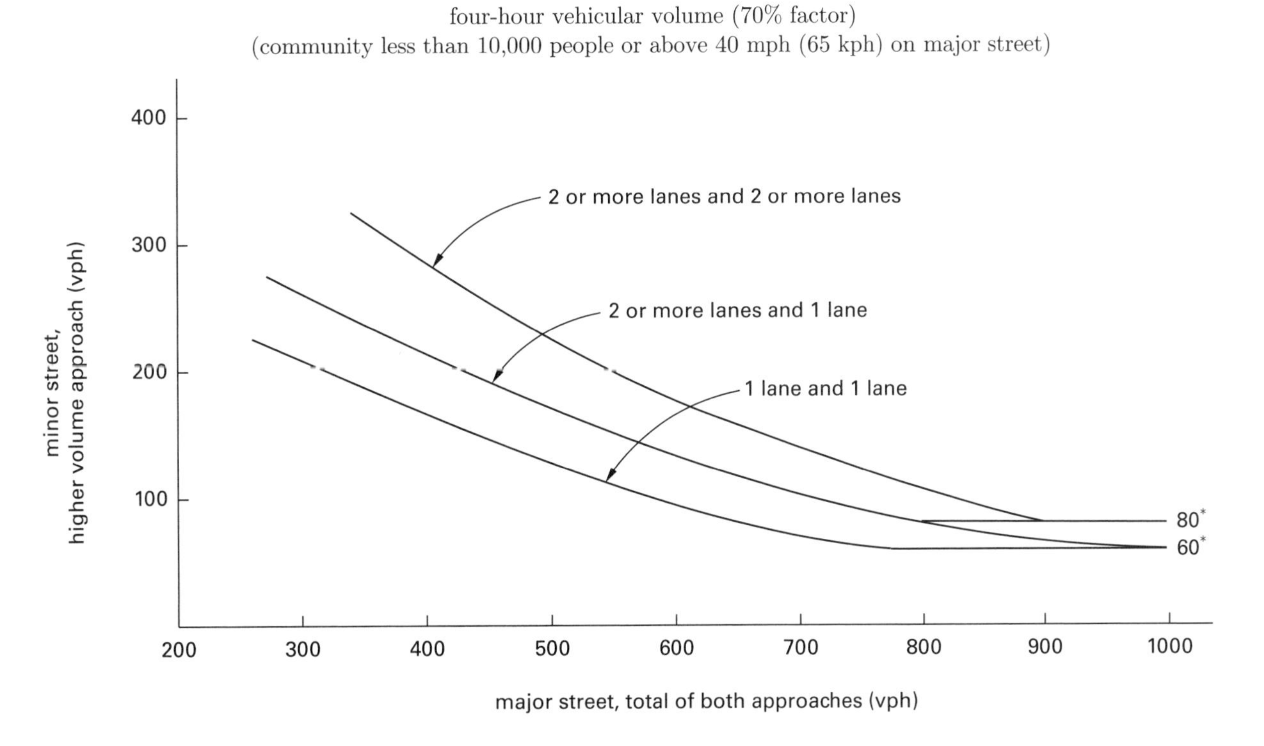

*Note: 80 vph applies as the lower threshold volume for a minor-street approach with two or more lanes and 60 vph applies as the lower threshold volume for a minor-street approach with one lane.

Reprinted from the *Manual on Uniform Traffic Control Devices*, 2009 ed., Fig. 4C-2, U.S. Department of Transportation, Federal Highway Administration, 2009.

APPENDIX 2.E
Warrant 3

peak hour

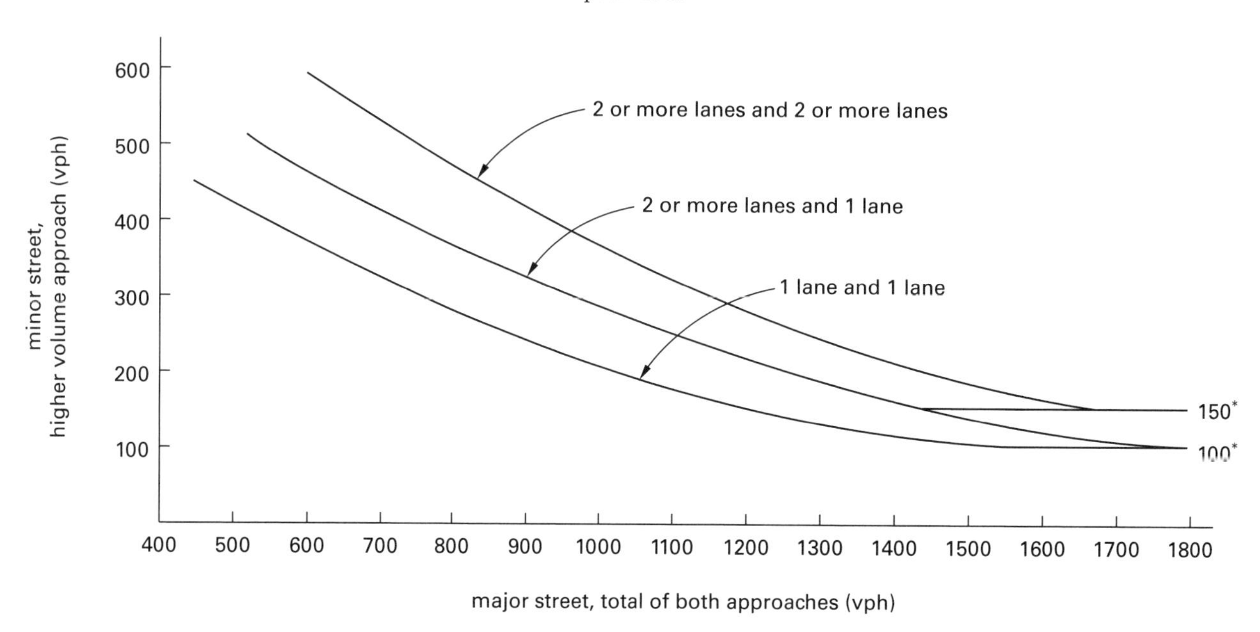

*Note: 150 vph applies as the lower threshold volume for a minor-street approach with two or more lanes and 100 vph applies as the lower threshold volume for a minor-street approach with one lane.

Reprinted from the *Manual on Uniform Traffic Control Devices*, 2009 ed., Fig. 4C-3, U.S. Department of Transportation, Federal Highway Administration, 2009.

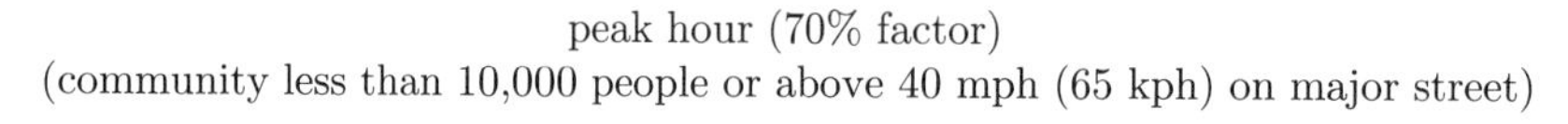
peak hour (70% factor)
(community less than 10,000 people or above 40 mph (65 kph) on major street)

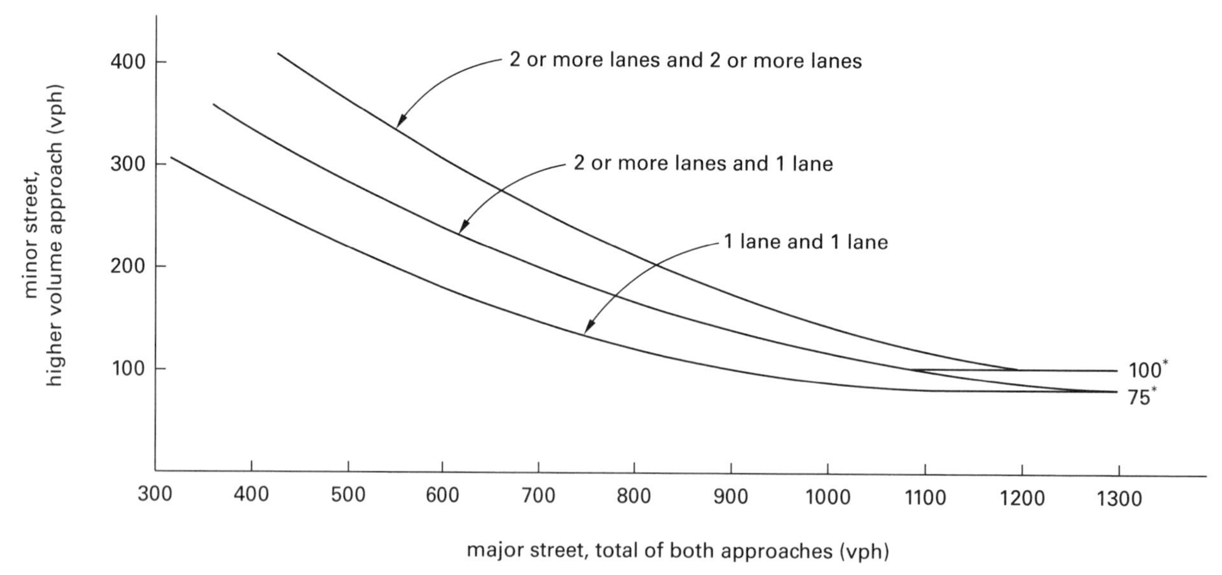

*Note: 100 vph applies as the lower threshold volume for a minor-street approach with two or more lanes and 75 vph applies as the lower threshold volume for a minor-street approach with one lane.

Reprinted from the *Manual on Uniform Traffic Control Devices*, 2009 ed., Fig. 4C-4, U.S. Department of Transportation, Federal Highway Administration, 2009.

APPENDIX 2.F
Warrant 4

pedestrian four-hour volume

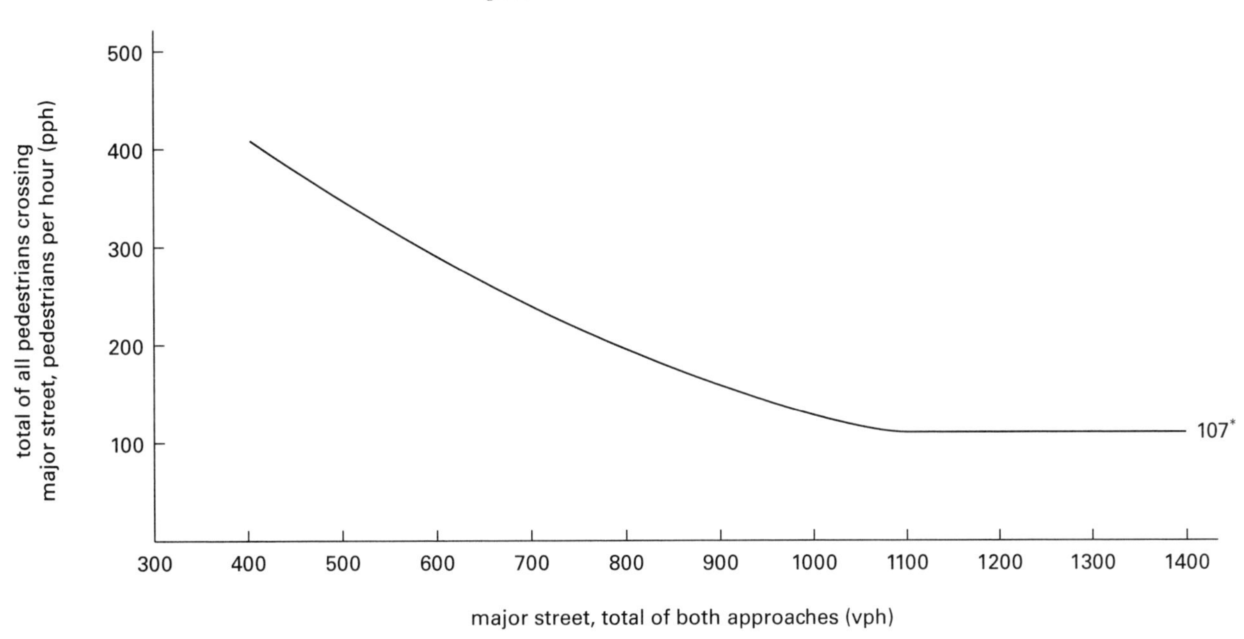

*Note: 107 pph applies as the lower threshold volume.

Reprinted from the *Manual on Uniform Traffic Control Devices*, 2009 ed., Fig. 4C-5, U.S. Department of Transportation, Federal Highway Administration, 2009.

pedestrian four-hour volume (70% factor)

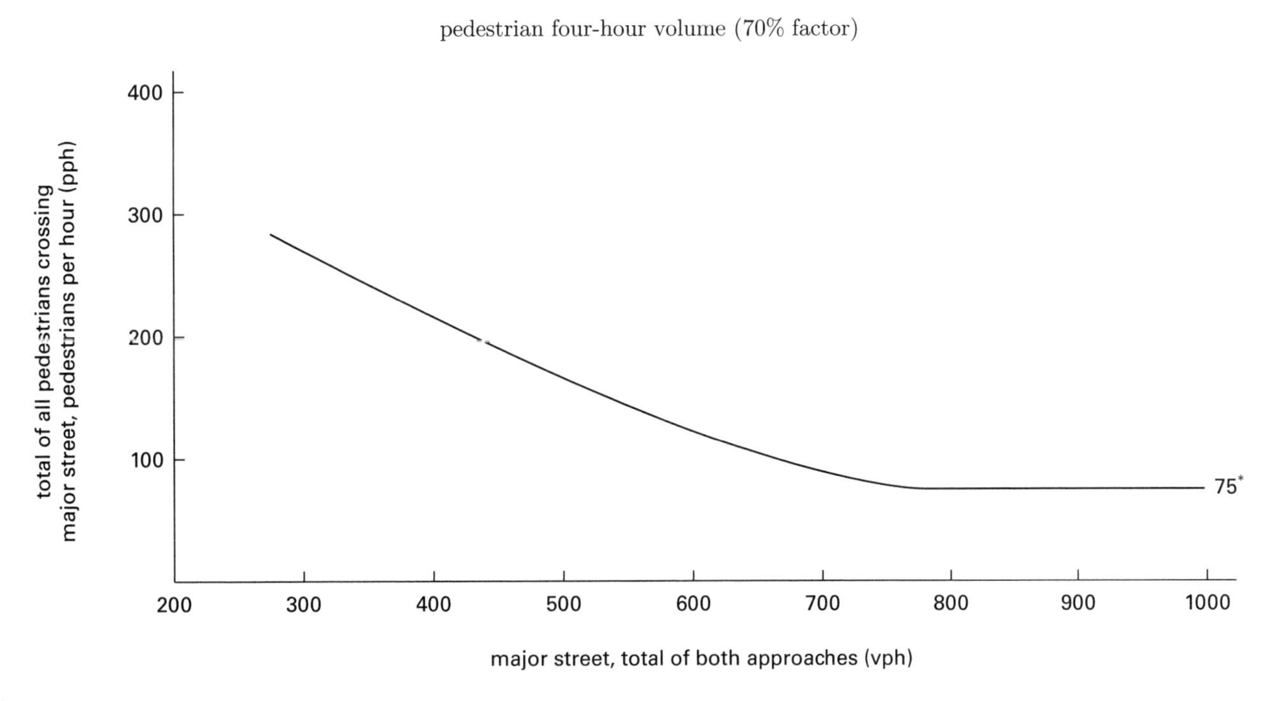

*Note: 75 pph applies as the lower threshold volume.

Reprinted from the *Manual on Uniform Traffic Control Devices*, 2009 ed., Fig. 4C-6, U.S. Department of Transportation, Federal Highway Administration, 2009.

(continued)

APPENDIX 2.F *(continued)*
Warrant 4

pedestrian peak hour

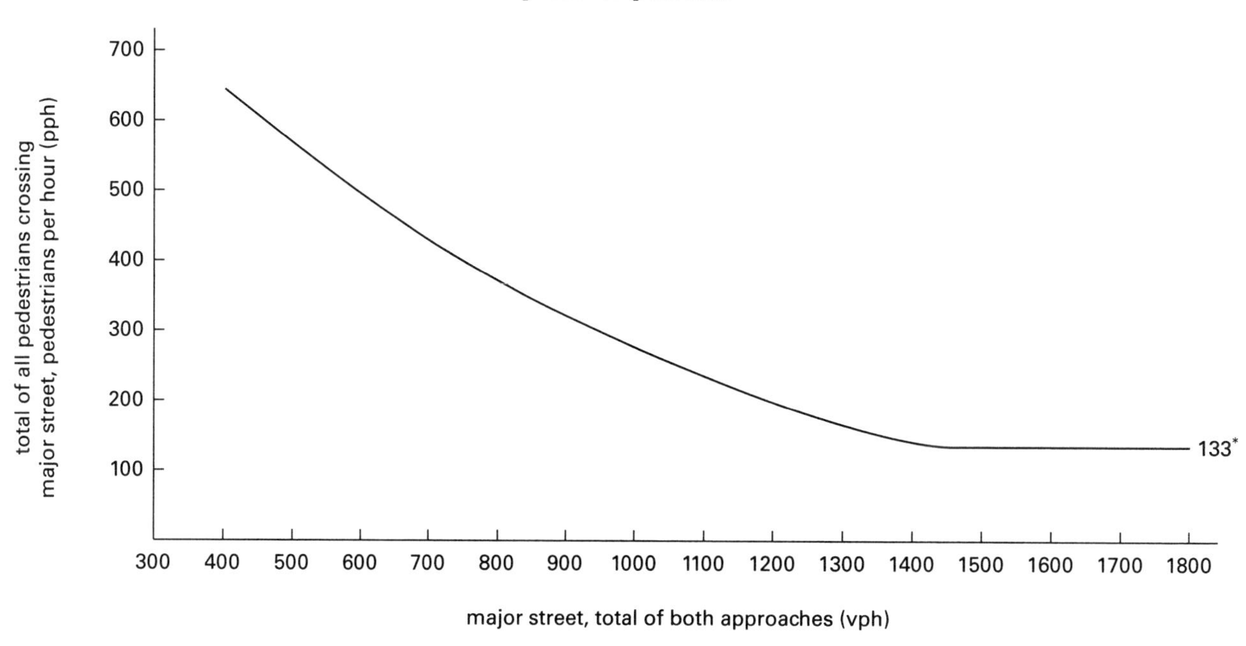

*Note: 133 pph applies as the lower threshold volume.

Reprinted from the *Manual on Uniform Traffic Control Devices*, 2009 ed., Fig. 4C-7, U.S. Department of Transportation, Federal Highway Administration, 2009.

pedestrian peak hour (70% factor)

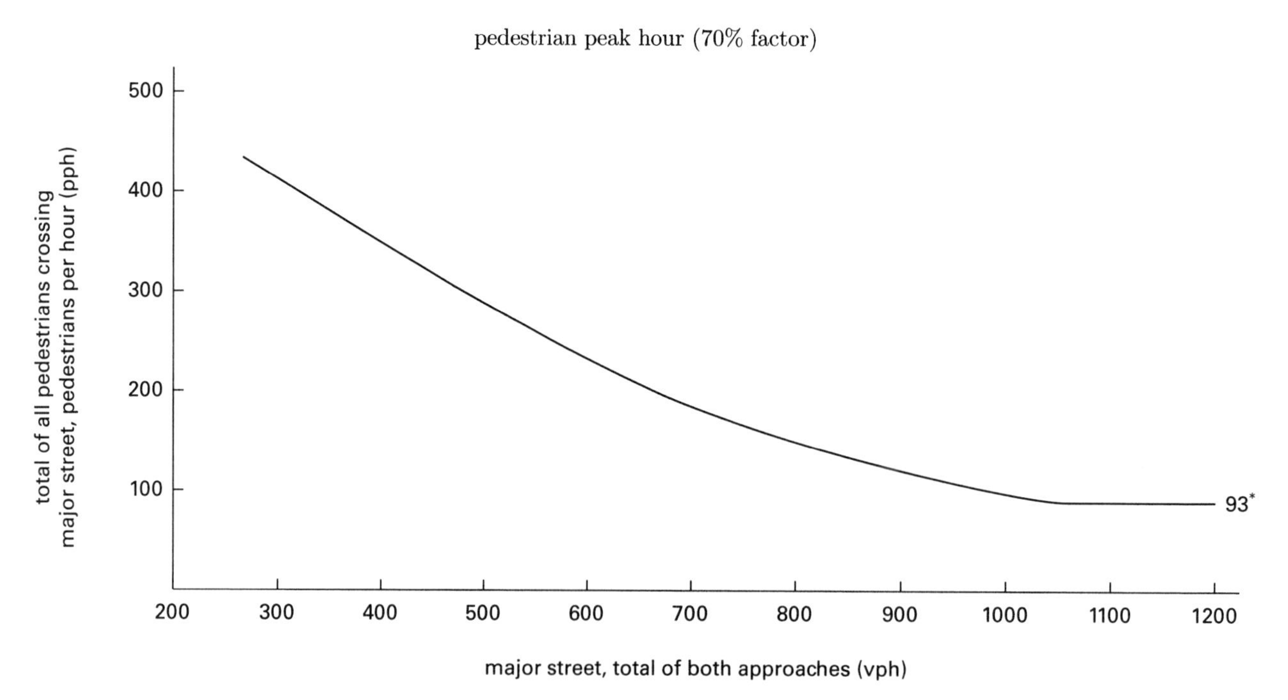

*Note: 93 pph applies as the lower threshold volume.

Reprinted from the *Manual on Uniform Traffic Control Devices*, 2009 ed., Fig. 4C-8, U.S. Department of Transportation, Federal Highway Administration, 2009.

APPENDIX 2.G
Warrant 9

intersection near a grade crossing
(one approach lane at the track crossing)

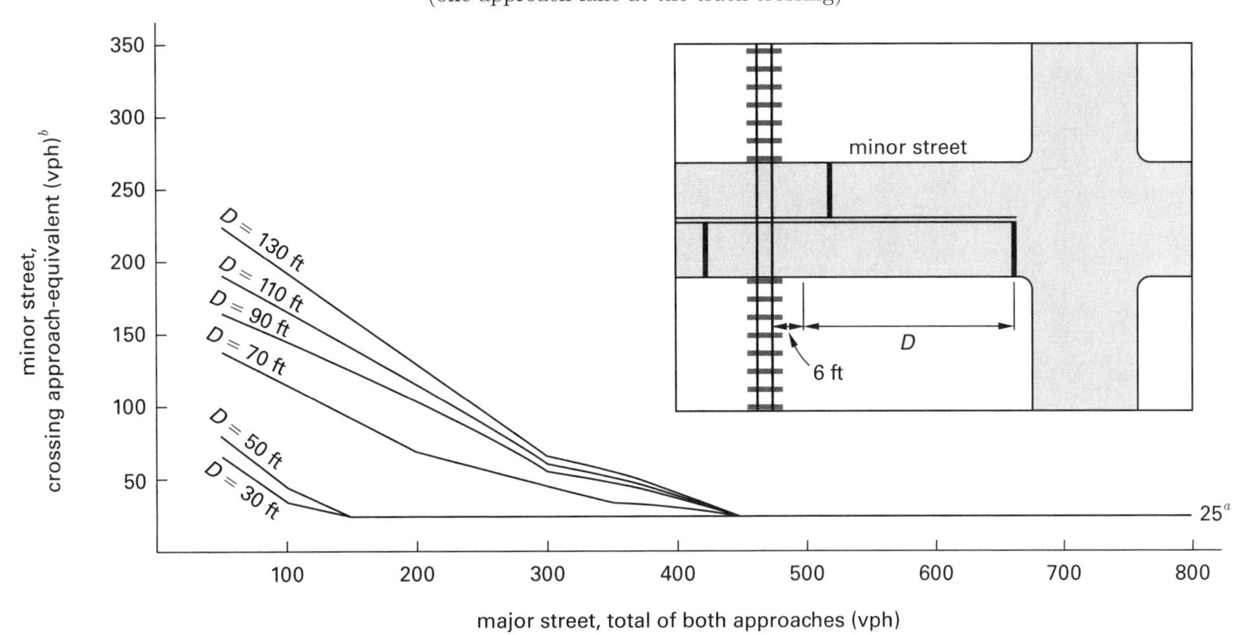

[a]25 vph applies as the lower threshold volume.
[b]vph after applying the adjustment factors, if appropriate

Reprinted from the *Manual on Uniform Traffic Control Devices*, 2009 ed., Fig. 4C-9, U.S. Department of Transportation, Federal Highway Administration, 2009.

intersection near a grade crossing
(two or more approach lanes at the track crossing)

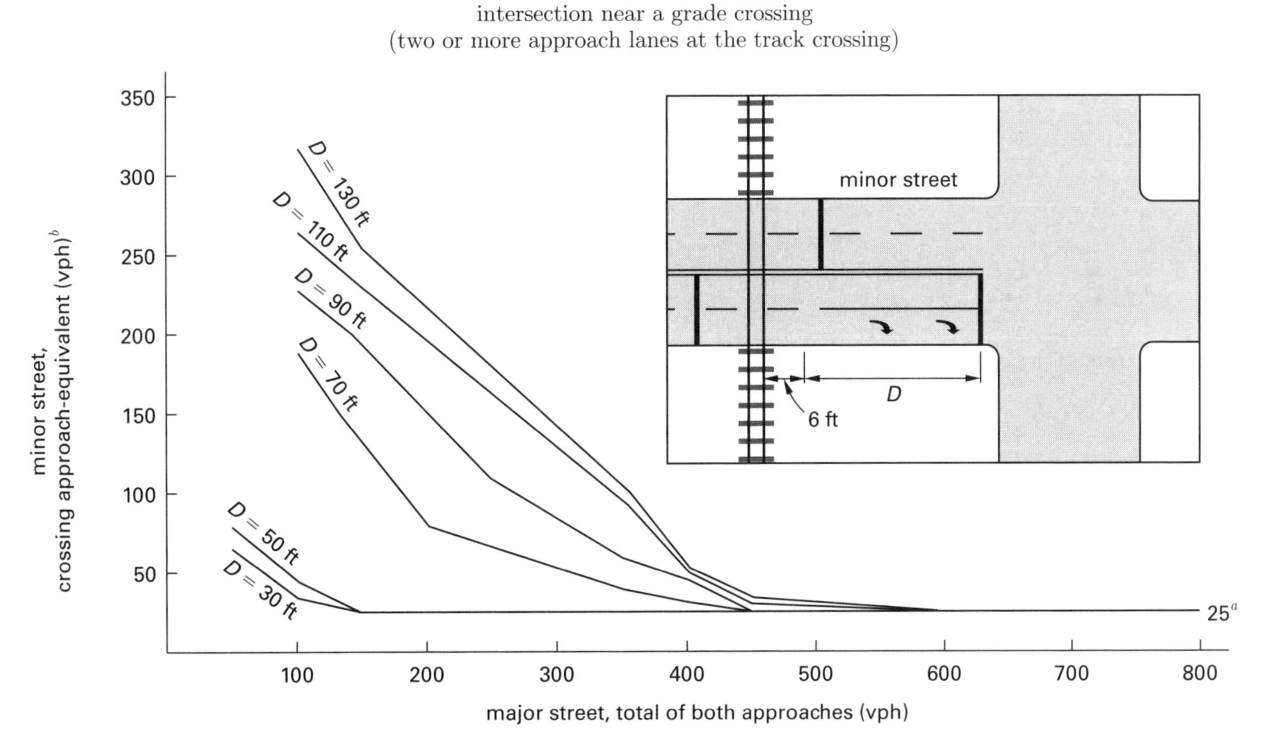

[a]25 vph applies as the lower threshold volume.
[b]vph after applying the adjustment factors, if appropriate

Reprinted from the *Manual on Uniform Traffic Control Devices*, 2009 ed., Fig. 4C-10, U.S. Department of Transportation, Federal Highway Administration, 2009.

Appendices

APPENDIX 2.G
Warrant 9 *(continued)*

adjustment factor for daily frequency of rail traffic

rail traffic per day	adjustment factor
1	0.67
2	0.91
3 to 5	1.00
6 to 8	1.18
9 to 11	1.25
12 or more	1.33

adjustment factor for percentage of high-occupancy buses

high-occupancy buses* on minor-street approach	adjustment factor
0%	1.00
2%	1.09
4%	1.19
6% or more	1.32

adjustment factor for percentage of tractor-trailer trucks

	adjustment factor	
tractor-trailer trucks on minor-street approach	clear storage distance, $D < 70$ ft	clear storage distance, $D \geq 70$ ft
0% to 2.5%	0.50	0.50
2.6% to 7.5%	0.75	0.75
7.6% to 12.5%	1.00	1.00
12.6% to 17.5%	2.30	1.15
17.6% to 22.5%	2.70	1.35
22.6% to 27.5%	3.28	1.64
more than 27.5%	4.18	2.09

*A high-occupancy bus is defined as a bus occupied by at least 20 people.

Reprinted from the *Manual on Uniform Traffic Control Devices*, 2009 ed., Table 4C-2, Table 4C-3, and Table 4C-4, U.S. Department of Transportation, Federal Highway Administration, 2009.

APPENDIX 2.H

Left-Turn Adjustment Factors for Permitted Phasing of Signalized Intersections

case	type of lane group	left-turn adjustment factor (f_{LT})
1	exclusive LT lane; protected phasing	0.95
2	exclusive LT lane; permitted phasing	special procedure; see *HCM* Exh. C16-9 or C16-10
3	exclusive LT lane; protected-plus-permitted phasing	apply case 1 to protected phase; apply case 2 to permitted phase
4	shared LT lane; protected phasing	$f_{LT} = 1.0/(1.0 + 0.05p_{LT})$
5	shared LT lane; permitted phasing	special procedure; see *HCM* Exh. C16-9 or C16-10
6	shared LT lane; protected-plus-permitted phasing	apply case 4 to protected phase; apply case 5 to permitted phase

Appendices

APPENDIX 4.A
Design Superelevation Rates

4% maximum (SI units)

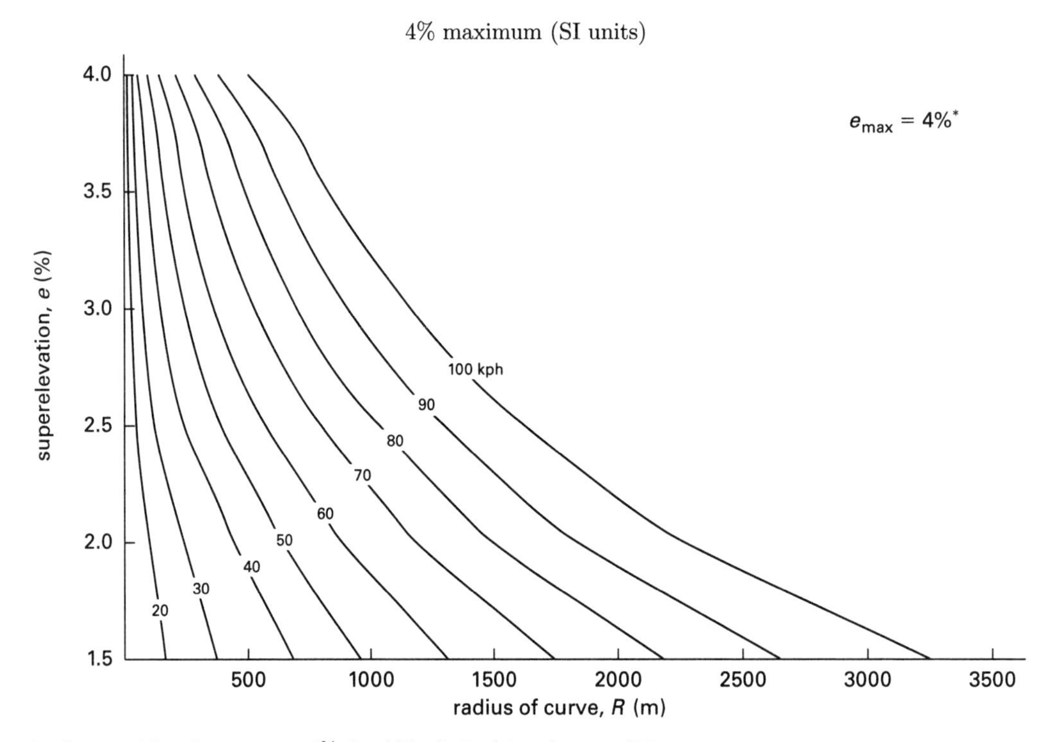

*In recognition of safety considerations, e_{max} = 4% should be limited to urban conditions.

From *A Policy on Geometric Design of Highways and Streets*, 2004, by the American Association of State Highway and Transportation Officials, Washington, D.C. Used by permission.

4% maximum (customary U.S. units)

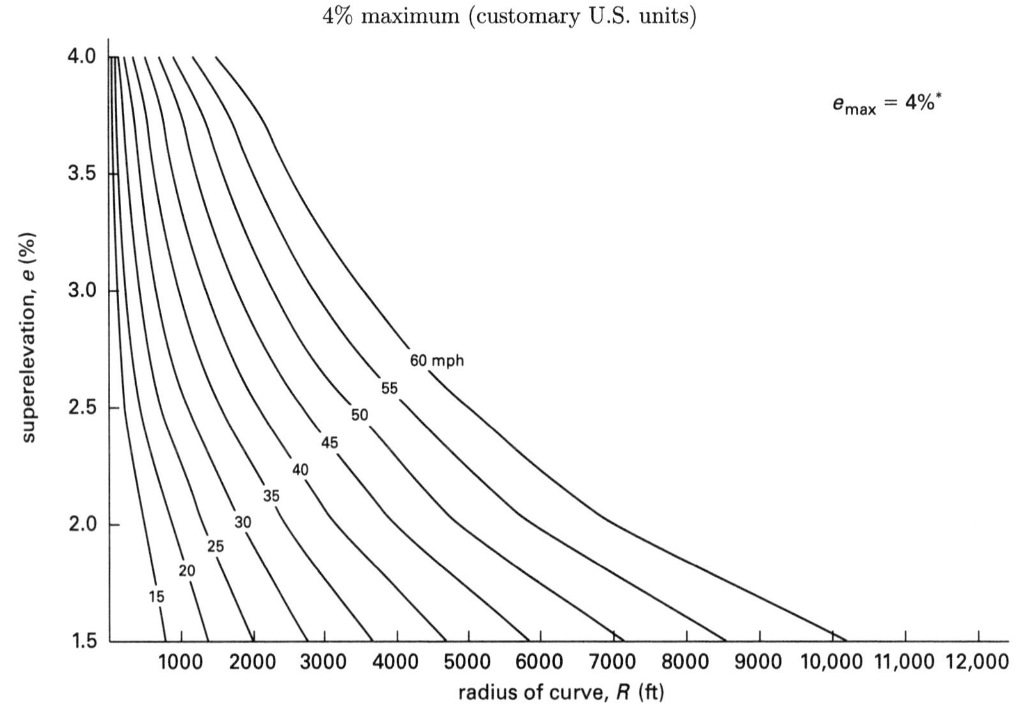

*In recognition of safety considerations, e_{max} = 4% should be limited to urban conditions.

From *A Policy on Geometric Design of Highways and Streets*, 2004, by the American Association of State Highway and Transportation Officials, Washington, D.C. Used by permission.

(continued)

Appendices

APPENDIX 4.A *(continued)*
Design Superelevation Rates

6% maximum (SI units)

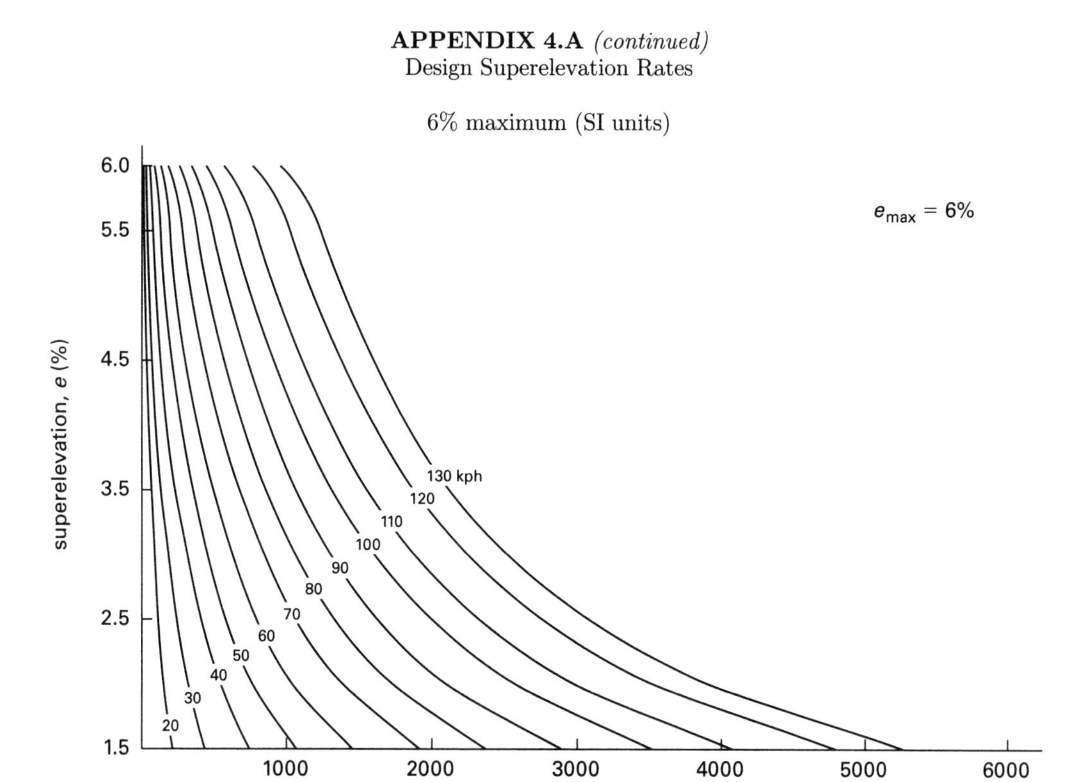

From *A Policy on Geometric Design of Highways and Streets*, 2004, by the American Association of State Highway and Transportation Officials, Washington, D.C. Used by permission.

6% maximum (customary U.S. units)

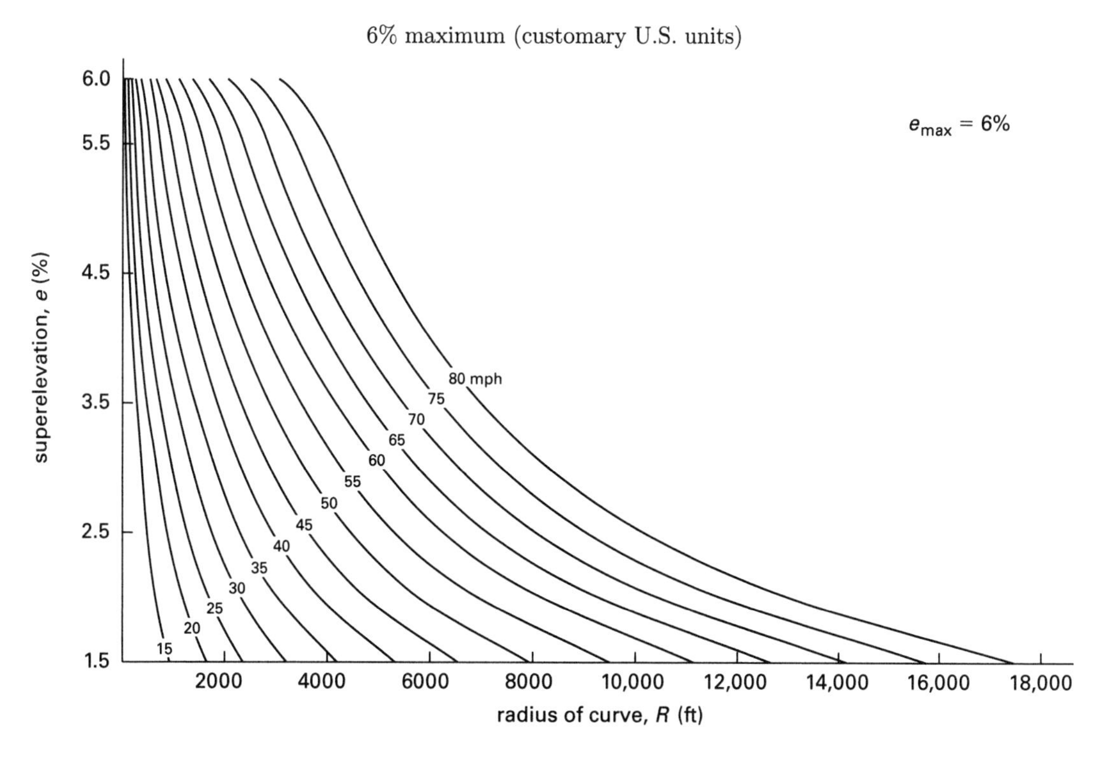

From *A Policy on Geometric Design of Highways and Streets*, 2004, by the American Association of State Highway and Transportation Officials, Washington, D.C. Used by permission.

(continued)

APPENDIX 4.A *(continued)*
Design Superelevation Rates

8% maximum (SI units)

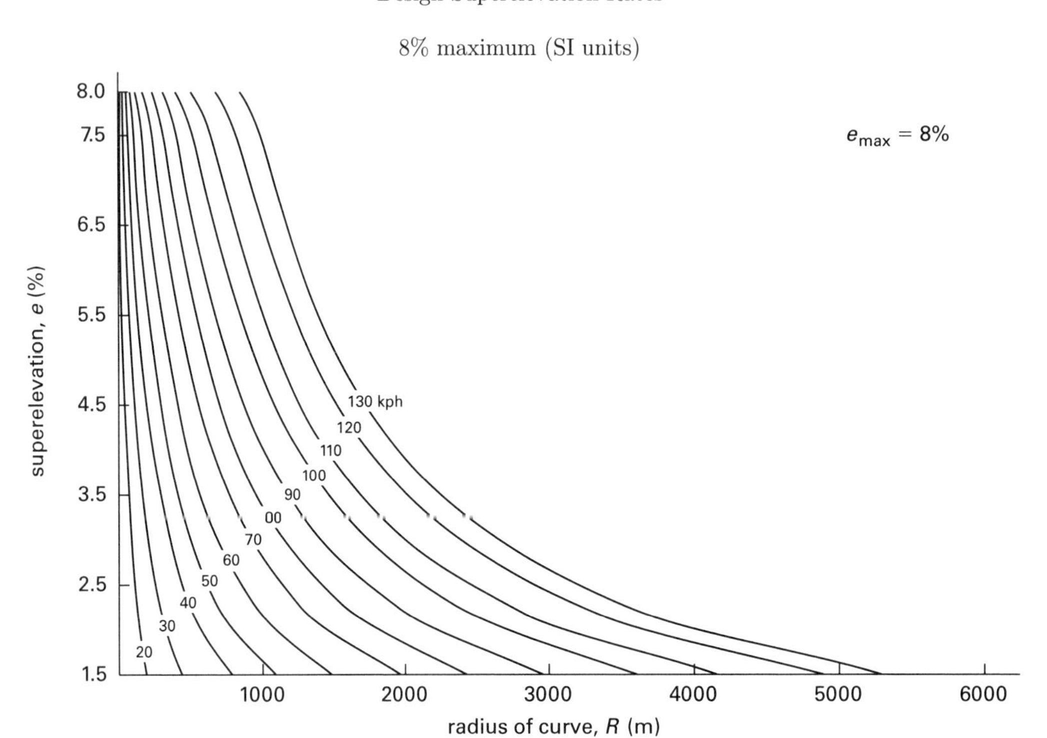

From *A Policy on Geometric Design of Highways and Streets*, 2004, by the American Association of State Highway and Transportation Officials, Washington, D.C. Used by permission.

8% maximum (customary U.S. units)

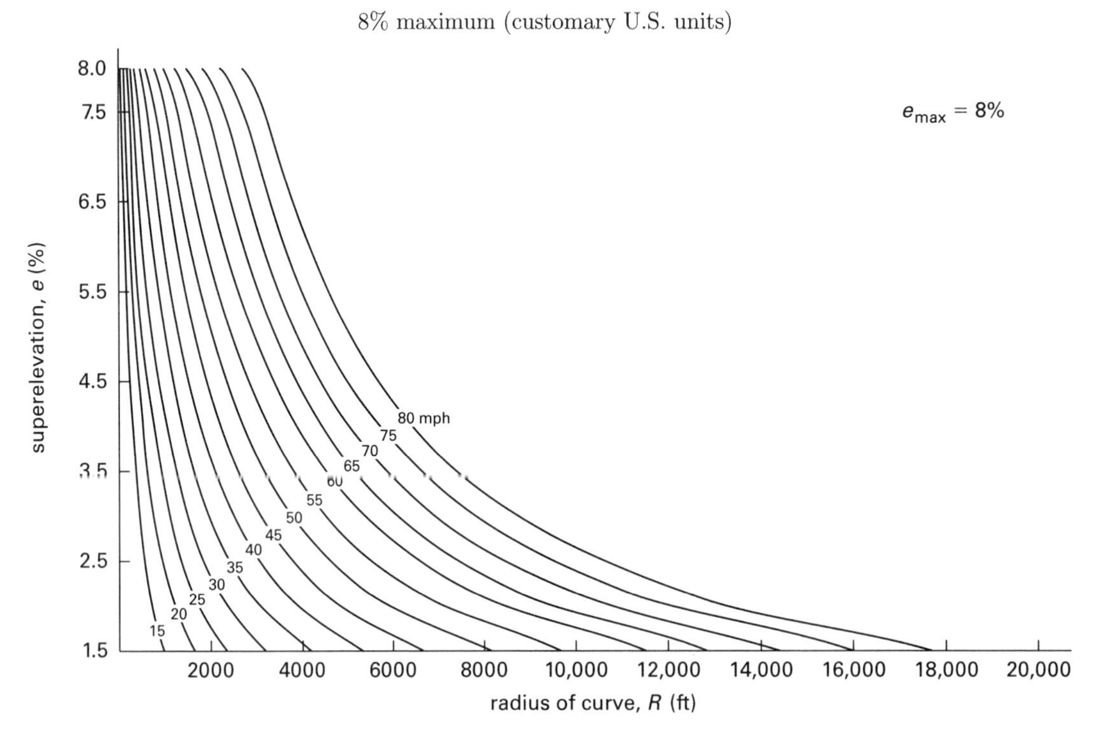

From *A Policy on Geometric Design of Highways and Streets*, 2004, by the American Association of State Highway and Transportation Officials, Washington, D.C. Used by permission.

(continued)

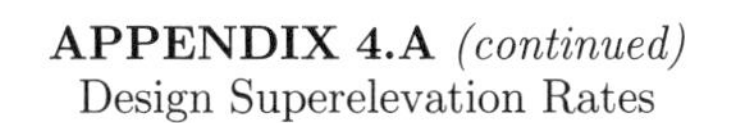

APPENDIX 4.A *(continued)*
Design Superelevation Rates

10% maximum (SI units)

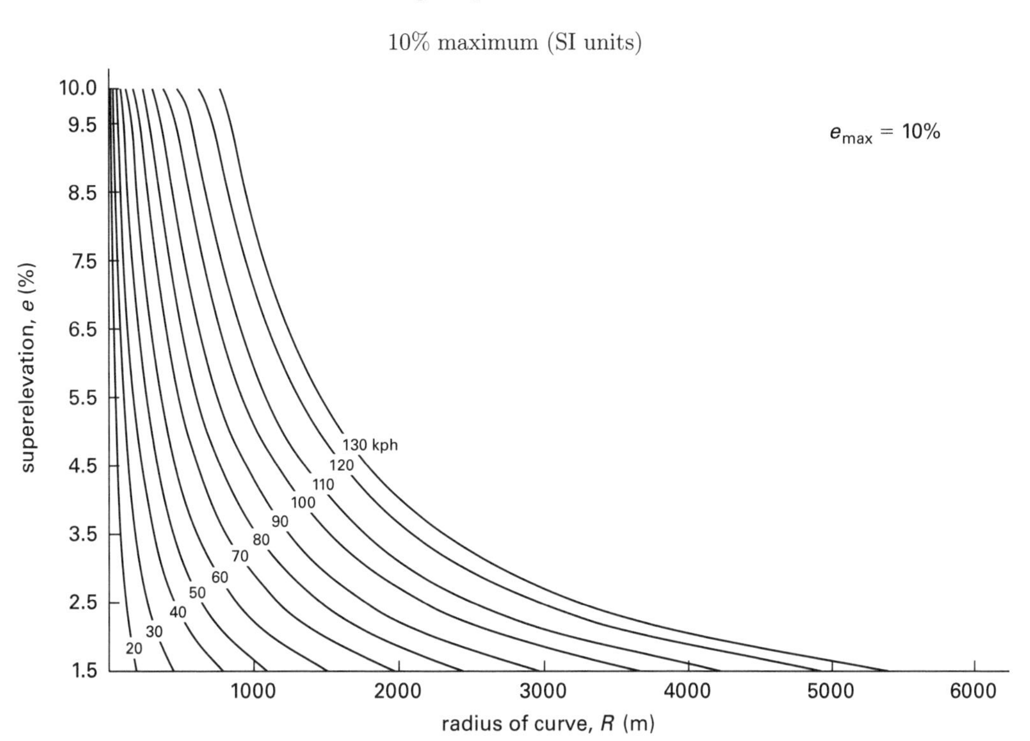

From *A Policy on Geometric Design of Highways and Streets*, 2004, by the American Association of State Highway and Transportation Officials, Washington, D.C. Used by permission.

10% maximum (customary U.S. units)

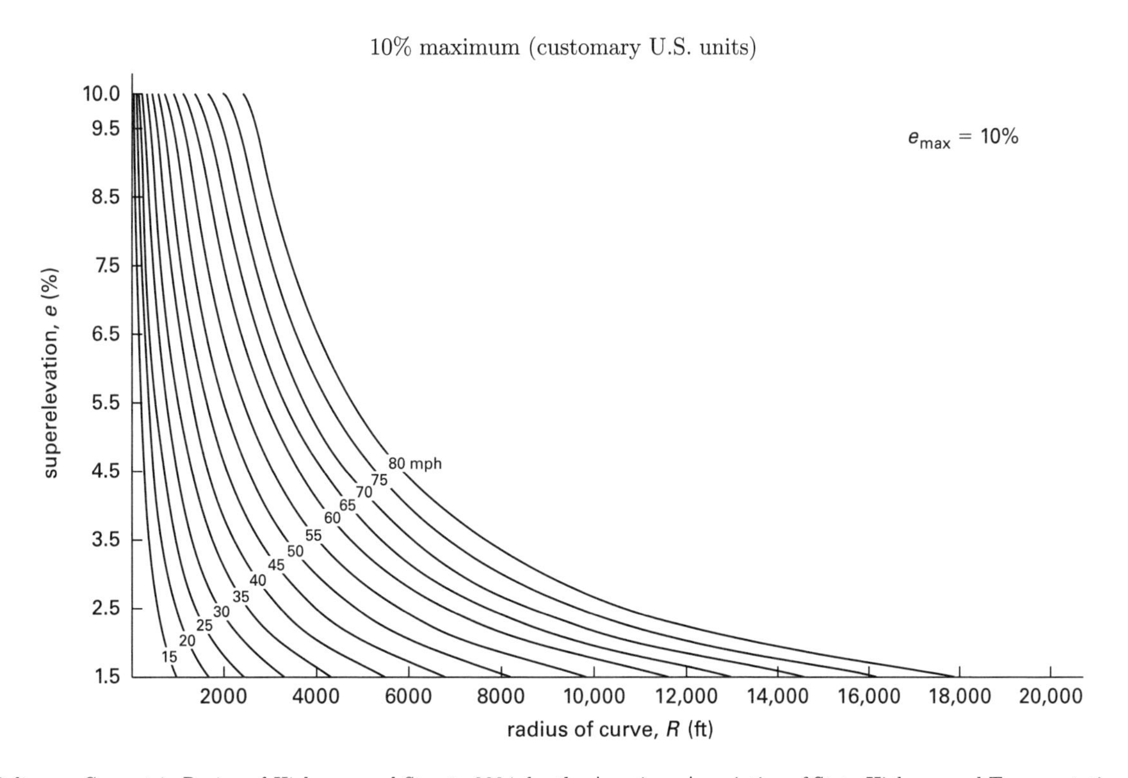

From *A Policy on Geometric Design of Highways and Streets*, 2004, by the American Association of State Highway and Transportation Officials, Washington, D.C. Used by permission.

(continued)

Appendices

APPENDIX 4.A *(continued)*
Design Superelevation Rates

12% maximum (SI units)

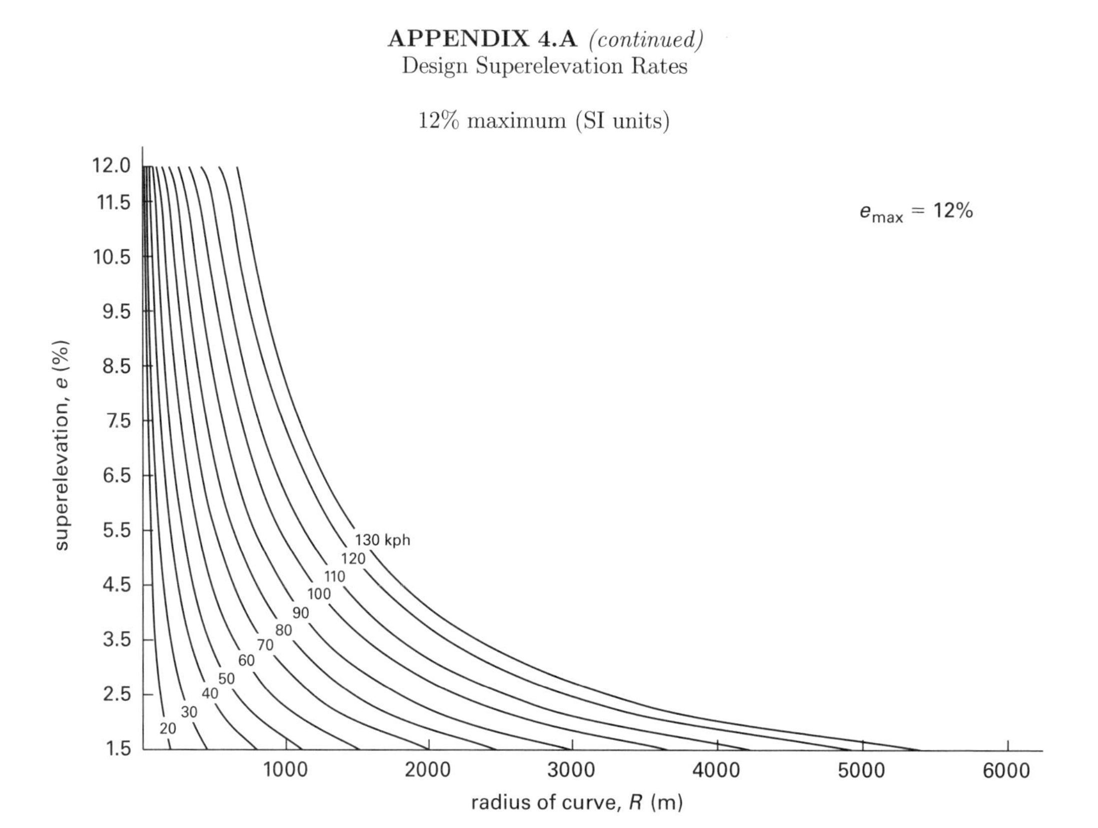

From *A Policy on Geometric Design of Highways and Streets*, 2004, by the American Association of State Highway and Transportation Officials, Washington, D.C. Used by permission.

12% maximum (customary U.S. units)

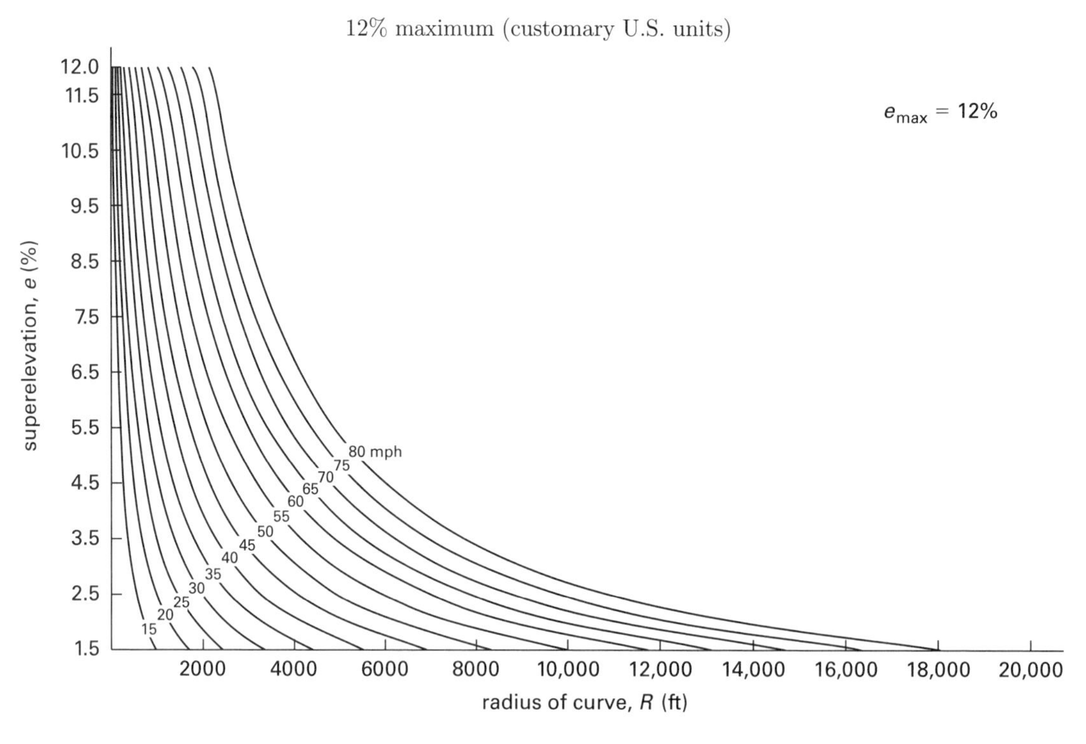

From *A Policy on Geometric Design of Highways and Streets*, 2004, by the American Association of State Highway and Transportation Officials, Washington, D.C. Used by permission.

APPENDIX 4.B

Minimum Radii for Design Superelevation Rates and Design Speeds

4% maximum (SI units)*

	v_d (kph)								
	20	30	40	50	60	70	80	90	100
e (%)	R (m)								
1.5	163	371	679	951	1310	1740	2170	2640	3250
2.0	102	237	441	632	877	1180	1490	1830	2260
2.2	75	187	363	534	749	1020	1290	1590	1980
2.4	51	132	273	435	626	865	1110	1390	1730
2.6	38	99	209	345	508	720	944	1200	1510
2.8	30	79	167	283	422	605	802	1030	1320
3.0	24	64	137	236	356	516	690	893	1150
3.2	20	54	114	199	303	443	597	779	1010
3.4	17	45	96	170	260	382	518	680	879
3.6	14	38	81	144	222	329	448	591	767
3.8	12	31	67	121	187	278	381	505	658
4.0	8	22	47	86	135	203	280	375	492

4% maximum (customary U.S. units)*

	v_d (mph)									
	15	20	25	30	35	40	45	50	55	60
e (%)	R (ft)									
1.5	796	1410	2050	2830	3730	4770	5930	7220	8650	10,300
2.0	506	902	1340	1680	2490	3220	4040	4940	5950	7080
2.2	399	723	1110	1580	2120	2760	3480	4280	5180	6190
2.4	271	513	836	1270	1760	2340	2980	3690	4500	5410
2.6	201	388	650	1000	1420	1930	2490	3130	3870	4700
2.8	157	308	524	817	1170	1620	2100	2660	3310	4060
3.0	127	251	433	681	982	1370	1800	2290	2860	3530
3.2	105	209	363	576	835	1180	1550	1980	2490	3090
3.4	88	175	307	490	714	1010	1340	1720	2170	2700
3.6	73	147	259	416	610	865	1150	1480	1880	2350
3.8	61	122	215	348	512	730	970	1260	1600	2010
4.0	42	86	154	250	371	533	711	926	1190	1500

*Use of $e_{max} = 4\%$ should be limited to urban conditions.

From *A Policy on Geometric Design of Highways and Streets*, 2004, by the American Association of State Highway and Transportation Officials, Washington, D.C. Used by permission.

(continued)

APPENDIX 4.B *(continued)*

Minimum Radii for Design Superelevation Rates and Design Speeds

6% maximum (SI units)

	v_d (kph)											
	20	30	40	50	60	70	80	90	100	110	120	130
e (%)	R (m)											
1.5	194	421	738	1050	1440	1910	2360	2880	3510	4060	4770	5240
2.0	138	299	525	750	1030	1380	1710	2090	2560	2970	3510	3880
2.2	122	265	465	668	919	1230	1530	1880	2300	2670	3160	3500
2.4	109	236	415	599	825	1110	1380	1700	2080	2420	2870	3190
2.6	97	212	372	540	746	1000	1260	1540	1890	2210	2630	2930
2.8	87	190	334	488	676	910	1150	1410	1730	2020	2420	2700
3.0	78	170	300	443	615	831	1050	1290	1590	1870	2240	2510
3.2	70	152	269	402	561	761	959	1190	1470	1730	2080	2330
3.4	61	133	239	364	511	697	882	1100	1360	1600	1940	2180
3.6	51	113	206	329	465	640	813	1020	1260	1490	1810	2050
3.8	42	96	177	294	422	586	749	939	1170	1390	1700	1930
4.0	36	82	155	261	380	535	690	870	1090	1300	1590	1820
4.2	31	72	136	234	343	488	635	806	1010	1220	1500	1720
4.4	27	63	121	210	311	446	584	746	938	1140	1410	1630
4.6	24	56	108	190	283	408	538	692	873	1070	1330	1540
4.8	21	50	97	172	258	374	496	641	812	997	1260	1470
5.0	19	45	88	156	235	343	457	594	755	933	1190	1400
5.2	17	40	79	142	214	315	421	549	701	871	1120	1330
5.4	15	36	71	128	195	287	386	506	648	810	1060	1260
5.6	13	32	63	115	176	260	351	463	594	747	980	1190
5.8	11	28	56	102	156	232	315	416	537	679	900	1110
6.0	8	21	43	79	123	184	252	336	437	560	756	951

6% maximum (customary U.S. units)

	v_d (mph)													
	15	20	25	30	35	40	45	50	55	60	65	70	75	80
e (%)	R (ft)													
1.5	868	1580	2290	3130	4100	5230	6480	7870	9410	11,100	12,600	14,100	15,700	17,400
2.0	614	1120	1630	2240	2950	3770	4680	5700	6820	8060	9130	10,300	11,500	12,900
2.2	543	991	1450	2000	2630	3370	4190	5100	6110	7230	8200	9240	10,400	11,600
2.4	482	884	1300	1790	2360	3030	3770	4600	5520	6540	7430	8380	9420	10,600
2.6	430	791	1170	1610	2130	2740	3420	4170	5020	5950	6770	7660	8620	9670
2.8	384	709	1050	1460	1930	2490	3110	3800	4580	5440	6200	7030	7930	8910
3.0	341	635	944	1320	1760	2270	2840	3480	4200	4990	5710	6490	7330	8260
3.2	300	566	850	1200	1600	2080	2600	3200	3860	4600	5280	6010	6810	7680
3.4	256	498	761	1080	1460	1900	2390	2940	3560	4250	4890	5580	6340	7180
3.6	209	422	673	972	1320	1740	2190	2710	3290	3940	4540	5210	5930	6720
3.8	176	358	583	864	1190	1590	2010	2490	3040	3650	4230	4860	5560	6320
4.0	151	309	511	766	1070	1440	1840	2300	2810	3390	3950	4550	5220	5950
4.2	131	270	452	684	960	1310	1680	2110	2590	3140	3680	4270	4910	5620
4.4	116	238	402	615	868	1190	1540	1940	2400	2920	3440	4010	4630	5320
4.6	102	212	360	555	788	1090	1410	1780	2210	2710	3220	3770	4380	5040
4.8	91	189	324	502	718	995	1300	1640	2050	2510	3000	3550	4140	4790
5.0	82	169	292	456	654	911	1190	1510	1890	2330	2800	3330	3910	4550
5.2	73	152	264	413	595	833	1090	1390	1750	2160	2610	3120	3690	4320
5.4	65	136	237	373	540	759	995	1280	1610	1990	2420	2910	3460	4090
5.6	58	121	212	335	487	687	903	1160	1470	1830	2230	2700	3230	3840
5.8	51	106	186	296	431	611	806	1040	1320	1650	2020	2460	2970	3560
6.0	39	81	144	231	340	485	643	833	1060	1330	1660	2040	2500	3050

From *A Policy on Geometric Design of Highways and Streets*, 2004, by the American Association of State Highway and Transportation Officials, Washington, D.C. Used by permission.

(continued)

APPENDIX 4.B *(continued)*
Minimum Radii for Design Superelevation Rates and Design Speeds

8% maximum (SI units)

	v_d (kph)											
	20	30	40	50	60	70	80	90	100	110	120	130
e (%)	R (m)											
1.5	184	443	784	1090	1490	1970	2440	2970	3630	4180	4900	5360
2.0	133	322	571	791	1090	1450	1790	2190	2680	3090	3640	4000
2.2	119	288	512	711	976	1300	1620	1960	2420	2790	3290	3620
2.4	107	261	463	644	885	1190	1470	1800	2200	2550	3010	3310
2.6	97	237	421	587	808	1080	1350	1650	2020	2340	2760	3050
2.8	88	216	385	539	742	992	1240	1520	1860	2160	2550	2830
3.0	81	199	354	496	684	916	1150	1410	1730	2000	2370	2630
3.2	74	183	326	458	633	849	1060	1310	1610	1870	2220	2460
3.4	68	169	302	425	588	790	988	1220	1500	1740	2080	2310
3.6	62	156	279	395	548	738	924	1140	1410	1640	1950	2180
3.8	57	144	259	368	512	690	866	1070	1320	1540	1840	2060
4.0	52	134	241	344	479	648	813	1010	1240	1450	1740	1950
4.2	48	124	224	321	449	608	766	948	1180	1380	1650	1850
4.4	43	115	208	301	421	573	722	895	1110	1300	1570	1760
4.6	38	106	192	281	395	540	682	847	1050	1240	1490	1680
4.8	33	96	178	263	371	509	645	803	996	1180	1420	1610
5.0	30	87	163	246	349	480	611	762	947	1120	1360	1540
5.2	27	78	148	229	328	454	579	724	901	1070	1300	1480
5.4	24	71	136	213	307	429	549	689	859	1020	1250	1420
5.6	22	65	125	198	288	405	521	656	819	975	1200	1360
5.8	20	59	115	185	270	382	494	625	781	933	1150	1310
6.0	19	55	106	172	253	360	469	595	746	894	1100	1260
6.2	17	50	98	161	238	340	445	567	713	857	1060	1220
6.4	16	46	91	151	224	322	422	540	681	823	1020	1180
6.6	15	43	85	141	210	304	400	514	651	789	982	1140
6.8	14	40	79	132	198	287	379	489	620	757	948	1100
7.0	13	37	73	123	185	270	358	464	591	724	914	1070
7.2	12	34	68	115	174	254	338	440	561	691	879	1040
7.4	11	31	62	107	162	237	318	415	531	657	842	998
7.6	10	29	57	99	150	221	296	389	499	621	803	962
7.8	9	26	52	90	137	202	273	359	462	579	757	919
8.0	7	20	41	73	113	168	229	304	394	501	667	832

(continued)

APPENDIX 4.B *(continued)*
Minimum Radii for Design Superelevation Rates and Design Speeds

8% maximum (customary U.S. units)

	v_d (mph)													
	15	20	25	30	35	40	45	50	55	60	65	70	75	80
e (%)	R (ft)													
1.5	932	1640	2370	3240	4260	5410	6710	8150	9720	11,500	12,900	14,500	16,100	17,800
2.0	676	1190	1720	2370	3120	3970	4930	5990	7150	8440	9510	10,700	12,000	13,300
2.2	605	1070	1550	2130	2800	3570	4440	5400	6450	7620	8600	9660	10,800	12,000
2.4	546	959	1400	1930	2540	3240	4030	4910	5870	6930	7830	8810	9850	11,000
2.6	496	872	1280	1760	2320	2960	3690	4490	5370	6350	7180	8090	9050	10,100
2.8	453	796	1170	1610	2130	2720	3390	4130	4950	5850	6630	7470	8370	9340
3.0	415	730	1070	1480	1960	2510	3130	3820	4580	5420	6140	6930	7780	8700
3.2	382	672	985	1370	1820	2330	2900	3550	4250	5040	5720	6460	7260	8130
3.4	352	620	911	1270	1690	2170	2700	3300	3970	4700	5350	6050	6800	7620
3.6	324	572	845	1180	1570	2020	2520	3090	3710	4400	5010	5680	6400	7180
3.8	300	530	784	1100	1470	1890	2360	2890	3480	4140	4710	5350	6030	6780
4.0	277	490	729	1030	1370	1770	2220	2720	3270	3890	4450	5050	5710	6420
4.2	255	453	678	955	1280	1660	2080	2560	3080	3670	4200	4780	5410	6090
4.4	235	418	630	893	1200	1560	1960	2410	2910	3470	3980	4540	5140	5800
4.6	215	384	585	834	1130	1470	1850	2280	2750	3290	3770	4310	4890	5530
4.8	193	349	542	779	1060	1390	1750	2160	2610	3120	3590	4100	4670	5280
5.0	172	314	499	727	991	1310	1650	2040	2470	2960	3410	3910	4460	5050
5.2	154	284	457	676	929	1230	1560	1930	2350	2820	3250	3740	4260	4840
5.4	139	258	420	627	870	1160	1480	1830	2230	2680	3110	3570	4090	4640
5.6	126	236	387	582	813	1090	1390	1740	2120	2550	2970	3420	3920	4460
5.8	115	216	358	542	761	1030	1320	1650	2010	2430	2840	3280	3760	4290
6.0	105	199	332	506	713	965	1250	1560	1920	2320	2710	3150	3620	4140
6.2	97	184	308	472	669	909	1180	1480	1820	2210	2600	3020	3480	3990
6.4	89	170	287	442	628	857	1110	1400	1730	2110	2490	2910	3360	3850
6.6	82	157	267	413	590	808	1050	1330	1650	2010	2380	2790	3240	3720
6.8	76	146	248	386	553	761	990	1260	1560	1910	2280	2690	3120	3600
7.0	70	135	231	360	518	716	933	1190	1480	1820	2180	2580	3010	3480
7.2	64	125	214	336	485	672	878	1120	1400	1720	2070	2470	2900	3370
7.4	59	115	198	312	451	628	822	1060	1320	1630	1970	2350	2780	3250
7.6	54	105	182	287	417	583	765	980	1230	1530	1850	2230	2650	3120
7.8	48	94	164	261	380	533	701	901	1140	1410	1720	2090	2500	2970
8.0	38	76	134	214	314	444	587	758	960	1200	1480	1810	2210	2670

From *A Policy on Geometric Design of Highways and Streets*, 2004, by the American Association of State Highway and Transportation Officials, Washington, D.C. Used by permission.

(continued)

APPENDIX 4.B *(continued)*
Minimum Radii for Design Superelevation Rates and Design Speeds

10% maximum (SI units)

	v_d (kph)											
	20	30	40	50	60	70	80	90	100	110	120	130
e (%)	R (m)											
1.5	197	454	790	1110	1520	2000	2480	3010	3690	4250	4960	5410
2.0	145	333	580	815	1120	1480	1840	2230	2740	3160	3700	4050
2.2	130	300	522	735	1020	1340	1660	2020	2480	2860	3360	3680
2.4	118	272	474	669	920	1220	1520	1840	2260	2620	3070	3370
2.6	108	249	434	612	844	1120	1390	1700	2080	2410	2830	3110
2.8	99	229	399	564	778	1030	1290	1570	1920	2230	2620	2880
3.0	91	211	368	522	720	952	1190	1460	1790	2070	2440	2690
3.2	85	196	342	485	670	887	1110	1360	1670	1940	2280	2520
3.4	79	182	318	453	626	829	1040	1270	1560	1820	2140	2370
3.6	73	170	297	424	586	777	974	1200	1470	1710	2020	2230
3.8	68	159	278	398	551	731	917	1130	1390	1610	1910	2120
4.0	64	149	261	374	519	690	866	1060	1310	1530	1810	2010
4.2	60	140	245	353	490	652	820	1010	1240	1450	1720	1910
4.4	56	132	231	333	464	617	777	953	1180	1380	1640	1820
4.6	53	124	218	315	439	586	738	907	1120	1310	1560	1740
4.8	50	117	206	299	417	557	703	864	1070	1250	1490	1670
5.0	47	111	194	283	396	530	670	824	1020	1200	1430	1600
5.2	44	104	184	269	377	505	640	788	975	1150	1370	1540
5.4	41	98	174	256	359	482	611	754	934	1100	1320	1480
5.6	39	93	164	243	343	461	585	723	896	1060	1270	1420
5.8	36	88	155	232	327	441	561	693	860	1020	1220	1370
6.0	33	82	146	221	312	422	538	666	827	976	1180	1330
6.2	31	77	138	210	298	404	516	640	795	941	1140	1280
6.4	28	72	130	200	285	387	496	616	766	907	1100	1240
6.6	26	67	121	191	273	372	476	593	738	876	1060	1200
6.8	24	62	114	181	261	357	458	571	712	846	1030	1170
7.0	22	58	107	172	249	342	441	551	688	819	993	1130
7.2	21	55	101	164	238	329	425	532	664	792	963	1100
7.4	20	51	95	156	228	315	409	513	642	767	934	1070
7.6	18	48	90	148	218	303	394	496	621	743	907	1040
7.8	17	45	85	141	208	291	380	479	601	721	882	1010
8.0	16	43	80	135	199	279	366	463	582	699	857	981
8.2	15	40	76	128	190	268	353	448	564	679	834	956
8.4	14	38	72	122	182	257	339	432	546	660	812	932
8.6	14	36	68	116	174	246	326	417	528	641	790	910
8.8	13	34	64	110	166	236	313	402	509	621	770	888
9.0	12	32	61	105	158	225	300	386	491	602	751	867
9.2	11	30	57	99	150	215	287	371	472	582	731	847
9.4	11	28	54	94	142	204	274	354	453	560	709	828
9.6	10	26	50	88	133	192	259	337	432	537	685	809
9.8	9	24	46	81	124	179	242	316	407	509	656	786
10.0	7	19	38	68	105	154	210	277	358	454	597	739

(continued)

APPENDIX 4.B *(continued)*
Minimum Radii for Design Superelevation Rates and Design Speeds

10% maximum (customary U.S. units)

	v_d (mph)													
	15	20	25	30	35	40	45	50	55	60	65	70	75	80
e (%)	R (ft)													
1.5	947	1680	2420	3320	4350	5520	6830	8280	9890	11,700	13,100	14,700	16,300	18,000
2.0	694	1230	1780	2440	3210	4080	5050	6130	7330	8630	9720	10,900	12,200	13,500
2.2	625	1110	1600	2200	2900	3680	4570	5540	6630	7810	8800	9860	11,000	12,200
2.4	567	1010	1460	2000	2640	3350	4160	5050	6050	7130	8040	9010	10,100	11,200
2.6	517	916	1330	1840	2420	3080	3820	4640	5550	6550	7390	8290	9260	10,300
2.8	475	841	1230	1690	2230	2840	3520	4280	5130	6050	6840	7680	8580	9550
3.0	438	777	1140	1570	2060	2630	3270	3970	4760	5620	6360	7140	7990	8900
3.2	406	720	1050	1450	1920	2450	3040	3700	4440	5250	5930	6680	7480	8330
3.4	377	670	978	1360	1790	2290	2850	3470	4160	4910	5560	6260	7020	7830
3.6	352	625	913	1270	1680	2150	2670	3250	3900	4620	5230	5900	6620	7390
3.8	329	584	856	1190	1580	2020	2510	3060	3680	4350	4940	5570	6260	6990
4.0	308	547	804	1120	1490	1900	2370	2890	3470	4110	4670	5270	5930	6630
4.2	289	514	756	1060	1400	1800	2240	2740	3290	3900	4430	5010	5630	6300
4.4	271	483	713	994	1330	1700	2120	2590	3120	3700	4210	4760	5370	6010
4.6	255	455	673	940	1260	1610	2020	2460	2970	3520	4010	4540	5120	5740
4.8	240	429	636	890	1190	1530	1920	2340	2830	3360	3830	4340	4900	5490
5.0	226	404	601	844	1130	1460	1830	2240	2700	3200	3660	4150	4690	5270
5.2	213	381	569	802	1080	1390	1740	2130	2580	3060	3500	3980	4500	5060
5.4	200	359	539	762	1030	1330	1660	2040	2460	2930	3360	3820	4320	4860
5.6	188	339	511	724	974	1270	1590	1950	2360	2810	3220	3670	4160	4680
5.8	176	319	484	689	929	1210	1520	1870	2260	2700	3090	3530	4000	4510
6.0	164	299	458	656	886	1160	1460	1790	2170	2590	2980	3400	3860	4360
6.2	152	280	433	624	846	1110	1400	1720	2090	2490	2870	3280	3730	4210
6.4	140	260	409	594	808	1060	1340	1650	2010	2400	2760	3160	3600	4070
6.6	130	242	386	564	772	1020	1290	1590	1930	2310	2670	3060	3480	3940
6.8	120	226	363	536	737	971	1230	1530	1860	2230	2570	2960	3370	3820
7.0	112	212	343	509	704	931	1190	1470	1790	2150	2490	2860	3270	3710
7.2	105	199	324	483	671	892	1140	1410	1730	2070	2410	2770	3170	3600
7.4	98	187	306	460	641	855	1100	1360	1670	2000	2330	2680	3070	3500
7.6	92	176	290	437	612	820	1050	1310	1610	1940	2250	2600	2990	3400
7.8	86	165	274	416	585	786	1010	1260	1550	1870	2180	2530	2900	3310
8.0	81	156	260	396	558	754	968	1220	1500	1810	2120	2450	2820	3220
8.2	76	147	246	377	533	722	930	1170	1440	1750	2050	2380	2750	3140
8.4	72	139	234	359	509	692	893	1130	1390	1690	1990	2320	2670	3060
8.6	68	131	221	341	486	662	856	1080	1340	1630	1930	2250	2600	2980
8.8	64	124	209	324	463	633	820	1040	1290	1570	1870	2190	2540	2910
9.0	60	116	198	307	440	604	784	992	1240	1520	1810	2130	2470	2840
9.2	56	109	186	291	418	574	748	948	1190	1460	1740	2060	2410	2770
9.4	52	102	175	274	395	545	710	903	1130	1390	1670	1990	2340	2710
9.6	48	95	163	256	370	513	671	854	1080	1320	1600	1910	2260	2640
9.8	44	87	150	236	343	477	625	798	1010	1250	1510	1820	2160	2550
10.0	36	72	126	200	292	410	540	694	877	1090	1340	1630	1970	2370

From *A Policy on Geometric Design of Highways and Streets*, 2004, by the American Association of State Highway and Transportation Officials, Washington, D.C. Used by permission.

(continued)

APPENDIX 4.B *(continued)*
Minimum Radii for Design Superelevation Rates and Design Speeds

12% maximum (SI units)

	v_d (kph)											
	20	30	40	50	60	70	80	90	100	110	120	130
e (%)	R (m)											
1.5	210	459	804	1130	1540	2030	2510	3040	3720	4280	4990	5440
2.0	155	338	594	835	1150	1510	1870	2270	2770	3190	3740	4080
2.2	139	306	536	755	1040	1360	1690	2050	2510	2900	3390	3710
2.4	127	278	488	688	942	1250	1550	1880	2300	2650	3110	3400
2.6	116	255	448	631	865	1140	1420	1730	2110	2440	2860	3140
2.8	107	235	413	583	799	1060	1320	1600	1960	2260	2660	2910
3.0	99	218	382	541	742	980	1220	1490	1820	2110	2480	2720
3.2	92	202	356	504	692	914	1140	1390	1700	1970	2320	2550
3.4	86	189	332	472	648	856	1070	1300	1600	1850	2180	2400
3.6	81	177	312	443	609	805	1010	1230	1510	1750	2060	2270
3.8	76	166	293	417	573	759	947	1160	1420	1650	1950	2150
4.0	71	157	276	393	542	718	896	1100	1350	1560	1850	2040
4.2	67	148	261	372	513	680	850	1040	1280	1490	1760	1940
4.4	64	140	247	353	487	646	808	988	1220	1420	1680	1850
4.6	60	132	234	335	463	615	770	941	1160	1350	1600	1770
4.8	57	126	222	319	441	586	734	899	1110	1290	1530	1700
5.0	54	119	211	304	421	560	702	860	1060	1240	1470	1630
5.2	52	114	201	290	402	535	672	824	1020	1190	1410	1570
5.4	49	108	192	277	384	513	644	790	973	1140	1360	1510
5.6	47	103	183	265	368	492	618	759	936	1100	1310	1460
5.8	45	98	175	254	353	472	594	730	900	1060	1260	1410
6.0	43	94	167	244	339	454	572	703	867	1020	1220	1360
6.2	41	90	159	234	326	436	551	678	837	981	1180	1310
6.4	39	86	153	225	313	420	531	654	808	948	1140	1270
6.6	37	82	146	216	302	405	512	632	781	917	1100	1230
6.8	35	78	140	208	290	391	494	611	755	888	1070	1200
7.0	34	75	134	200	280	377	478	591	731	860	1040	1160
7.2	32	71	128	192	270	364	462	572	708	834	1010	1130
7.4	30	68	122	185	260	352	447	554	686	810	974	1100
7.6	29	65	117	178	251	340	433	537	666	786	947	1070
7.8	27	61	112	172	243	329	420	521	646	764	921	1040
8.0	26	58	107	165	235	319	407	506	628	743	897	1020
8.2	24	55	102	159	227	309	395	491	610	723	874	989
8.4	23	52	97	154	219	299	383	477	593	704	852	965
8.6	22	50	93	148	212	290	372	464	577	686	831	942
8.8	20	47	88	142	205	281	361	451	562	668	811	921
9.0	19	45	85	137	198	273	351	439	547	652	792	900
9.2	18	43	81	132	191	264	341	428	533	636	774	880
9.4	18	41	77	127	185	256	332	416	520	621	756	861
9.6	17	39	74	123	179	249	323	406	507	606	739	843
9.8	16	37	71	118	173	241	314	395	494	592	723	826
10.0	15	36	68	114	167	234	305	385	482	579	708	809
10.2	14	34	65	110	161	226	296	375	471	566	693	793
10.4	14	33	62	105	155	219	288	365	459	553	679	778
10.6	13	31	59	101	150	212	279	355	448	541	665	763
10.8	12	30	57	97	144	204	270	345	436	529	652	749
10.0	12	28	54	93	139	197	261	335	423	516	639	735
11.2	11	27	51	89	133	189	252	324	411	503	626	722
11.4	11	25	49	85	127	182	242	312	397	488	613	709
11.6	10	24	46	80	120	173	232	300	382	472	598	697
11.8	9	22	43	75	113	163	219	285	364	453	579	685
12.0	7	18	36	64	98	143	194	255	328	414	540	665

(continued)

Appendices

APPENDIX 4.B *(continued)*
Minimum Radii for Design Superelevation Rates and Design Speeds

12% maximum (customary U.S. units)

	v_d (mph)													
	15	20	25	30	35	40	45	50	55	60	65	70	75	80
e (%)	*R* (ft)													
1.5	950	1690	2460	3370	4390	5580	6910	8370	9990	11,800	13,200	14,800	16,400	18,100
2.0	700	1250	1820	2490	3260	4140	5130	6220	7430	8740	9840	11,000	12,300	13,600
2.2	631	1130	1640	2250	2950	3750	4640	5640	6730	7930	8920	9980	11,200	12,400
2.4	574	1030	1500	2060	2690	3420	4240	5150	6150	7240	8160	9130	10,200	11,300
2.6	526	936	1370	1890	2470	3140	3900	4730	5660	6670	7510	8420	9380	10,500
2.8	484	863	1270	1740	2280	2910	3600	4380	5240	6170	6960	7800	8700	9660
3.0	448	799	1170	1620	2120	2700	3350	4070	4870	5740	6480	7270	8110	9010
3.2	417	743	1090	1510	1970	2520	3130	3800	4550	5370	6060	6800	7600	8440
3.4	389	693	1020	1410	1850	2360	2930	3560	4270	5030	5690	6390	7140	7940
3.6	364	649	953	1320	1730	2220	2750	3350	4020	4740	5360	6020	6740	7500
3.8	341	610	896	1250	1630	2090	2600	3160	3790	4470	5060	5700	6380	7100
4.0	321	574	845	1180	1540	1980	2460	2990	3590	4240	4800	5400	6050	6740
4.2	303	542	798	1110	1460	1870	2330	2840	3400	4020	4560	5130	5750	6420
4.4	286	512	756	1050	1390	1780	2210	2700	3240	3830	4340	4890	5490	6120
4.6	271	485	717	997	1320	1690	2110	2570	3080	3650	4140	4670	5240	5850
4.8	257	460	681	948	1260	1610	2010	2450	2940	3480	3960	4470	5020	5610
5.0	243	437	648	904	1200	1540	1920	2340	2810	3330	3790	4280	4810	5380
5.2	231	415	618	862	1140	1470	1840	2240	2700	3190	3630	4110	4620	5170
5.4	220	395	589	824	1090	1410	1760	2150	2590	3060	3490	3950	4440	4980
5.6	209	377	563	788	1050	1350	1690	2060	2480	2940	3360	3800	4280	4800
5.8	199	359	538	754	1000	1300	1620	1980	2390	2830	3230	3660	4130	4630
6.0	190	343	514	723	960	1250	1560	1910	2300	2730	3110	3530	3990	4470
6.2	181	327	492	694	922	1200	1500	1840	2210	2630	3010	3410	3850	4330
6.4	172	312	471	666	886	1150	1440	1770	2140	2540	2900	3300	3730	4190
6.6	164	298	452	639	852	1110	1390	1710	2060	2450	2810	3190	3610	4060
6.8	156	284	433	615	820	1070	1340	1650	1990	2370	2720	3090	3500	3940
7.0	148	271	415	591	790	1030	1300	1590	1930	2290	2630	3000	3400	3820
7.2	140	258	398	568	762	994	1250	1540	1860	2220	2550	2910	3300	3720
7.4	133	246	382	547	734	960	1210	1490	1810	2150	2470	2820	3200	3610
7.6	125	234	366	527	708	928	1170	1440	1750	2090	2400	2740	3120	3520
7.8	118	222	351	507	684	897	1130	1400	1700	2020	2330	2670	3030	3430
8.0	111	210	336	488	660	868	1100	1360	1650	1970	2270	2600	2950	3340
8.2	105	199	321	470	637	840	1070	1320	1600	1910	2210	2530	2880	3260
8.4	100	190	307	452	615	813	1030	1280	1550	1860	2150	2460	2800	3180
8.6	95	180	294	435	594	787	997	1240	1510	1810	2090	2400	2740	3100
8.8	90	172	281	418	574	762	967	1200	1470	1760	2040	2340	2670	3030
9.0	85	164	270	403	554	738	938	1170	1430	1710	1980	2280	2610	2960
9.2	81	156	259	388	535	715	910	1140	1390	1660	1940	2230	2550	2890
9.4	77	149	248	373	516	693	883	1100	1350	1620	1890	2180	2490	2830
9.6	74	142	238	359	499	671	857	1070	1310	1580	1840	2130	2440	2770
9.8	70	136	228	346	481	650	832	1040	1280	1540	1800	2080	2380	2710
10.0	67	130	219	333	465	629	806	1010	1250	1500	1760	2030	2330	2660
10.2	64	124	210	320	448	608	781	980	1210	1460	1720	1990	2280	2600
10.4	61	118	201	308	432	588	757	951	1180	1430	1680	1940	2240	2550
10.6	58	113	192	296	416	568	732	922	1140	1390	1640	1900	2190	2500
10.8	55	108	184	284	400	548	707	892	1110	1350	1600	1860	2150	2460
11.0	52	102	175	272	384	527	682	862	1070	1310	1560	1820	2110	2410
11.2	49	97	167	259	368	506	656	831	1040	1270	1510	1780	2070	2370
11.4	47	92	158	247	351	485	629	799	995	1220	1470	1730	2020	2320
11.6	44	86	149	233	333	461	600	763	953	1170	1410	1680	1970	2280
11.8	40	80	139	218	312	434	566	722	904	1120	1350	1620	1910	2230
12.0	34	68	119	188	272	381	500	641	807	1000	1220	1480	1790	2130

From *A Policy on Geometric Design of Highways and Streets*, 2004, by the American Association of State Highway and Transportation Officials, Washington, D.C. Used by permission.

APPENDIX 4.C
Superelevation, Radius, and Design Speed for Low-Speed Urban Streets*

SI units

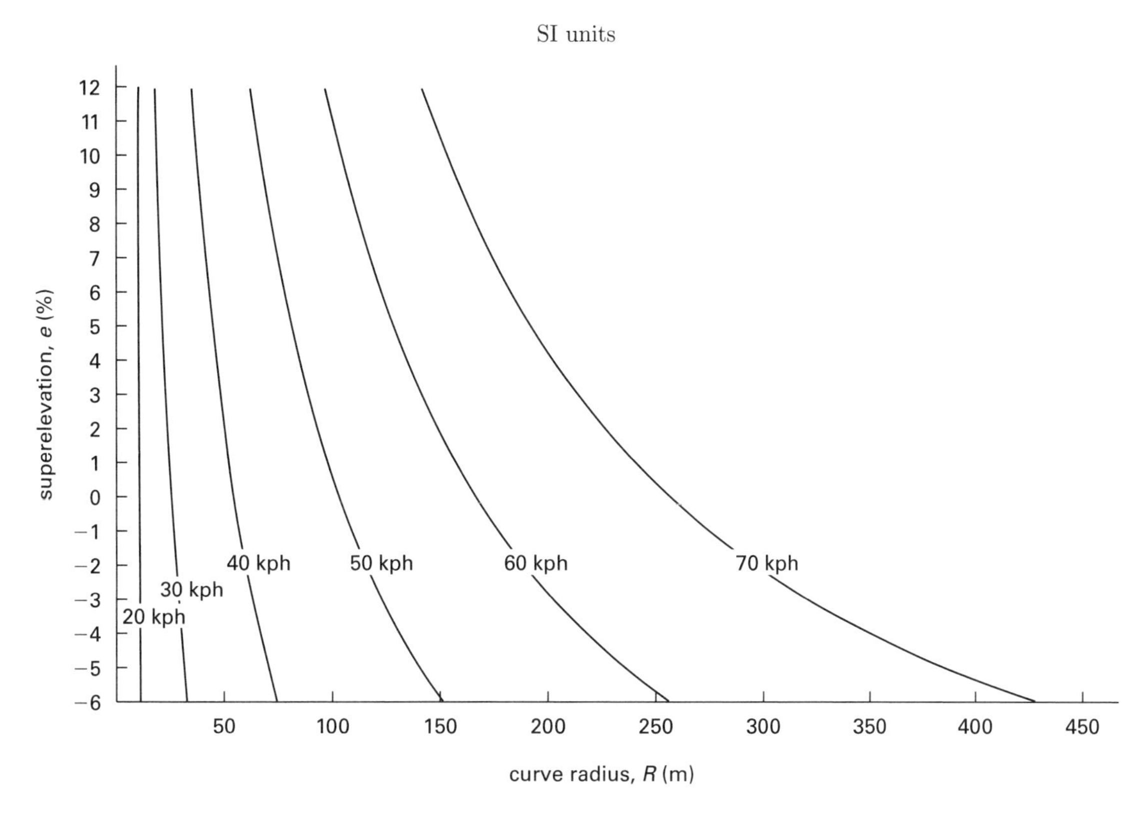

*Negative superelevation values beyond −2.0% should be used for low type surfaces such as gravel, crushed stone, and earth. However, areas with intense rainfall may use normal cross slopes of −2.5% on high type surfaces.

(continued)

Appendices

APPENDIX 4.C *(continued)*
Superelevation, Radius, and Design Speed for Low-Speed Urban Streets*

customary U.S. units

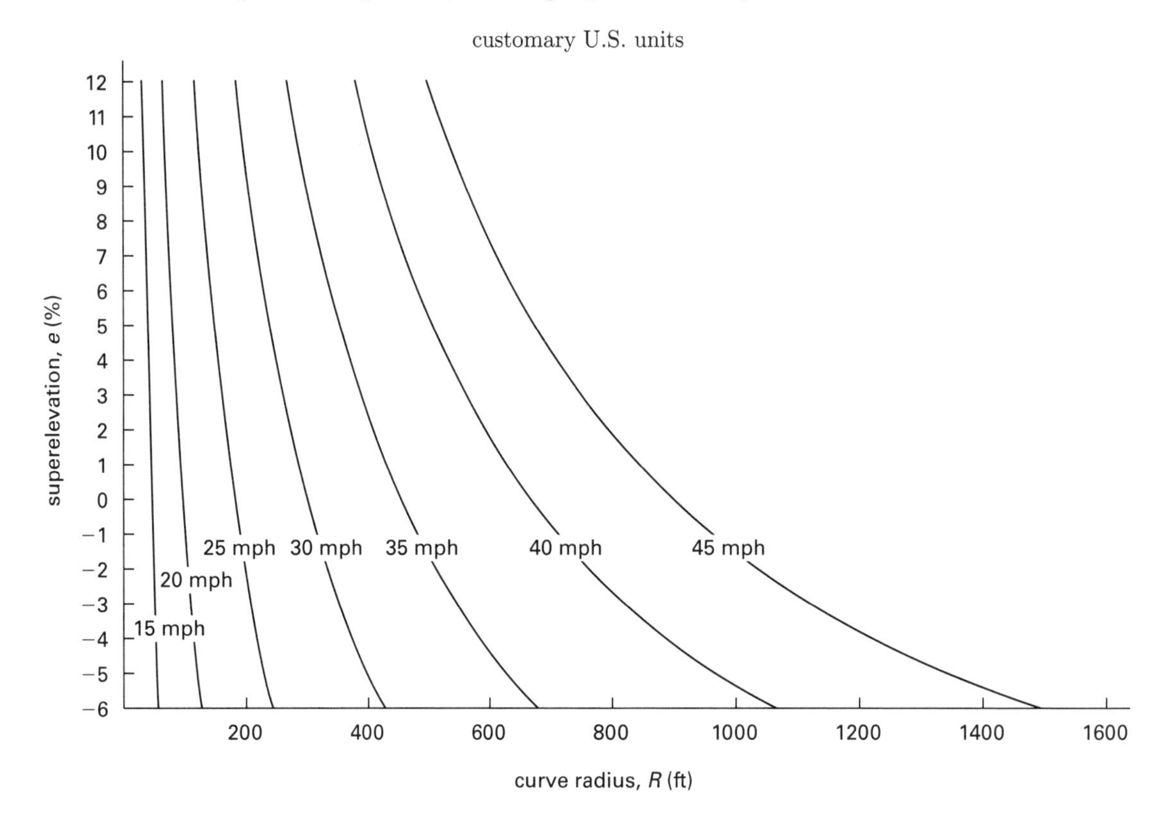

*Negative superelevation values beyond −2.0% should be used for low type surfaces such as gravel, crushed stone, and earth. However, areas with intense rainfall may use normal cross slopes of −2.5% on high type surfaces.

From *A Policy on Geometric Design of Highways and Streets*, 2004, by the American Association of State Highway and Transportation Officials, Washington, D.C. Used by permission.

Appendices

APPENDIX 4.D

Minimum Radii and Superelevation for Low-Speed Urban Streets[a,b]

SI units

	v_d (kph)					
	20	30	40	50	60	70
e (%)	R (m)					
–6.0	11	32	74	151	258	429
–5.0	10	31	70	141	236	386
–4.0	10	30	66	131	218	351
–3.0	10	28	63	123	202	322
–2.8	10	28	62	122	200	316
–2.6	10	28	62	120	197	311
–2.4	10	28	61	119	194	306
–2.2	10	27	61	117	192	301
–2.0	10	27	60	116	189	297
–1.5	9	27	59	113	183	286
0	9	25	55	104	167	257
1.5	9	24	51	96	153	234
2.0	9	24	50	94	149	227
2.2	8	23	50	93	148	224
2.4	8	23	50	92	146	222
2.6	8	23	49	91	145	219
2.8	8	23	49	90	143	217
3.0	8	23	48	89	142	214
3.2	8	23	48	89	140	212
3.4	8	23	48	88	139	210
3.6	8	22	47	87	138	207
3.8	8	22	47	86	136	205
4.0	8	22	47	86	135	203
4.2	8	22	46	85	134	201
4.4	8	22	46	84	132	199
4.6	8	22	46	83	131	197
4.8	8	22	45	83	130	195
5.0	8	21	45	82	129	193
5.2	8	21	45	81	128	191
5.4	8	21	44	81	127	189
5.6	8	21	44	80	125	187
5.8	8	21	44	79	124	185
6.0	8	21	43	79	123	184
6.2	8	21	43	78	122	182
6.4	8	21	43	78	121	180
6.6	8	20	43	77	120	179
6.8	8	20	42	76	119	177
7.0	7	20	42	76	118	175
7.2	7	20	42	75	117	174
7.4	7	20	41	75	116	172
7.6	7	20	41	74	115	171
7.8	7	20	41	73	114	169
8.0	7	20	41	73	113	168
8.2	7	20	40	72	112	166
8.4	7	19	40	72	112	165
8.6	7	19	40	71	111	163
8.8	7	19	40	71	110	162
9.0	7	19	39	70	109	161
9.2	7	19	39	70	108	159
9.4	7	19	39	69	107	158
9.6	7	19	39	69	107	157
9.8	7	19	38	68	106	156
10.0	7	19	38	68	105	154

(continued)

APPENDIX 4.D *(continued)*
Minimum Radii and Superelevation for Low-Speed Urban Streets[a,b]

SI units

	v_d (kph)					
	20	30	40	50	60	70
e (%)			R (m)			
10.2	7	19	38	67	104	153
10.4	7	18	38	67	103	152
10.6	7	18	37	67	103	151
10.8	7	18	37	66	102	150
11.0	7	18	37	66	101	148
11.2	7	18	37	65	101	147
11.4	7	18	37	65	100	146
11.6	7	18	36	64	99	145
11.8	7	18	36	64	98	144
12.0	7	18	36	64	98	143

[a]calculated using Superelevation Distribution Method 2
[b]Superelevation may be optional on low-speed urban streets.

From *A Policy on Geometric Design of Highways and Streets*, 2004, by the American Association of State Highway and Transportation Officials, Washington, D.C. Used by permission.

Appendices

APPENDIX 4.E
Minimum Superelevation Runoff and Tangent Runout Lengths

SI units

	v_d (kph)																							
	20		30		40		50		60		70		80		90		100		110		120		130	
	no. of lanes rotated, L_R (m)																							
e (%)	1	2	1	2	1	2	1	2	1	2	1	2	1	2	1	2	1	2	1	2	1	2	1	2
1.5	0	0	0	0	0	0	0	0	0	0	0	0	0	0	0	0	0	0	0	0	0	0	0	0
2.0	9	14	10	14	10	15	11	17	12	18	13	20	14	22	15	23	16	25	18	26	19	28	21	31
2.2	10	15	11	16	11	17	12	18	13	20	14	22	16	24	17	25	18	27	19	29	21	31	23	34
2.4	11	18	12	17	12	19	13	20	14	22	16	24	17	26	18	28	20	29	21	32	23	34	25	37
2.6	12	18	12	19	13	20	14	22	16	23	17	26	19	28	20	30	21	32	23	34	25	37	27	40
2.8	13	19	13	20	14	22	16	23	17	25	18	27	20	30	21	32	23	34	25	37	27	40	29	43
3.0	14	20	14	22	15	23	17	25	18	27	20	29	22	32	23	34	25	37	26	40	28	43	31	46
3.2	14	22	15	23	16	25	18	27	19	29	21	31	23	35	25	37	26	39	28	42	30	45	33	49
3.4	15	23	16	24	17	26	19	28	20	31	22	33	24	37	26	39	28	42	30	45	32	48	35	52
3.6	16	24	17	26	19	26	20	30	22	32	24	35	26	39	28	41	29	44	32	47	34	51	37	56
3.8	17	26	18	27	20	29	21	32	23	34	25	37	27	41	29	44	31	47	33	50	36	54	39	59
4.0	18	27	19	29	21	31	22	33	24	36	26	39	29	43	31	46	33	49	35	53	38	57	41	62
4.2	19	28	20	30	22	32	23	35	25	38	27	41	30	45	32	48	34	52	37	55	40	60	43	65
4.4	20	30	21	32	23	34	24	37	26	40	29	43	32	48	34	51	36	54	39	58	42	63	45	68
4.6	21	31	22	33	24	35	25	38	28	41	30	45	33	50	35	53	38	56	40	61	44	65	47	71
4.8	22	32	23	35	25	37	27	40	29	43	31	47	35	52	37	55	39	59	42	63	45	68	49	74
5.0	23	34	24	36	26	39	28	42	30	45	33	49	36	54	38	57	41	61	44	66	47	71	51	77
5.2	23	35	25	37	27	40	29	43	31	47	34	51	37	56	40	60	43	64	46	68	49	74	53	80
5.4	24	36	26	39	28	42	30	45	32	49	35	53	39	58	41	62	44	66	47	71	51	77	56	83
5.6	25	38	27	40	29	43	31	47	34	50	37	55	40	60	43	64	46	69	49	74	53	80	58	86
5.8	26	39	28	42	30	45	32	48	35	52	38	57	42	63	44	67	47	71	51	76	55	82	60	89
6.0	27	41	29	43	31	46	33	50	36	54	39	59	43	65	46	69	49	74	53	79	57	85	62	93
6.2	28	42	30	45	32	48	34	52	37	56	41	61	45	67	47	71	51	76	54	82	59	88	64	96
6.4	29	43	31	46	33	49	35	53	38	58	42	63	46	69	49	74	52	79	56	84	61	91	66	99
6.6	30	45	32	48	34	51	37	55	40	59	43	65	48	71	51	76	54	81	58	87	63	94	68	102
6.8	31	46	33	49	35	52	38	56	41	61	45	67	49	73	52	78	56	83	60	90	64	97	70	105
7.0	31	47	34	50	36	54	39	58	42	63	46	69	50	76	54	80	57	86	61	92	66	99	72	108
7.2	32	49	35	52	37	56	40	60	43	65	47	71	52	78	55	83	59	88	63	95	68	102	74	111
7.4	33	50	36	53	38	57	41	61	44	67	48	73	53	80	57	85	61	91	65	97	70	105	76	114
7.6	34	51	36	55	39	59	42	63	46	68	50	75	55	82	58	87	62	93	67	100	72	108	78	117
7.8	35	53	37	56	40	60	43	65	47	70	51	77	56	84	60	90	64	96	68	103	74	111	80	120
8.0	36	54	38	58	41	62	44	65	48	72	52	79	58	86	61	92	65	98	70	105	76	114	82	123
8.2	37	55	39	59	42	63	45	68	49	74	54	81	59	89	63	94	67	101	72	108	78	117	84	127
8.4	38	57	40	60	43	65	47	70	50	76	55	82	60	91	64	97	69	103	74	111	80	119	86	130
8.6	39	58	41	62	44	66	48	71	52	77	56	84	62	93	66	99	70	106	76	113	81	122	88	133
8.8	40	59	42	63	45	68	49	73	53	79	58	86	63	95	67	101	72	108	77	116	83	125	91	136
9.0	40	61	43	65	48	69	50	75	54	81	59	88	65	97	69	103	74	110	79	119	85	128	93	139
9.2	41	62	44	66	47	71	51	76	55	83	60	90	66	99	70	106	75	113	81	121	87	131	95	142
9.4	42	63	45	68	48	73	52	78	56	85	62	92	68	102	72	108	77	115	83	124	89	134	97	145
9.6	43	65	46	69	49	74	53	80	58	86	63	94	69	104	74	110	79	118	84	126	91	136	99	148
9.8	44	66	47	71	50	76	54	81	59	88	64	96	71	106	75	113	80	120	86	129	93	139	101	151
10.0	45	68	48	72	51	77	55	83	60	90	65	98	72	108	77	115	82	123	88	132	95	142	103	154
10.2	46	69	49	73	52	79	56	85	61	92	67	100	73	110	78	117	83	125	90	134	97	145	105	157
10.4	47	70	50	75	53	80	58	86	62	94	68	102	75	112	80	119	85	128	91	137	99	148	107	160
10.6	48	72	51	76	55	82	59	88	64	95	89	104	76	114	81	122	87	130	93	140	100	151	109	164
10.8	49	73	52	78	56	83	60	90	65	97	71	106	78	117	83	124	88	133	95	142	102	153	111	167
11.0	50	74	53	79	57	85	61	91	66	99	72	108	79	119	84	126	90	135	97	145	104	158	113	170
11.2	50	76	54	81	58	86	62	93	67	101	73	110	81	121	86	129	92	137	98	148	106	159	115	173
11.4	51	77	55	82	59	88	63	95	68	103	75	112	82	123	87	131	93	140	100	150	108	162	117	176
11.6	52	78	56	84	60	89	64	96	70	104	76	114	84	125	89	133	95	142	102	153	110	165	119	179
11.8	53	80	57	85	61	91	65	98	71	106	77	116	85	127	90	136	97	145	104	155	112	168	121	182
12.0	54	81	58	86	62	93	66	100	72	108	79	118	86	130	92	138	98	147	105	158	114	171	123	185

*One lane rotated is typical for a two-lane highway, two lanes rotated for a four-lane highway, and so on. (See Fig. 4.27.)

(continued)

APPENDIX 4.E *(continued)*

Minimum Superelevation Runoff and Tangent Runout Lengths

(customary U.S. units)

	v_d (mph)																											
	15		20		25		30		35		40		45		50		55		60		65		70		75		80	
	no. of lanes rotated, L_R (ft)																											
e (%)	1	2	1	2	1	2	1	2	1	2	1	2	1	2	1	2	1	2	1	2	1	2	1	2	1	2	1	2
1.5	0	0	0	0	0	0	0	0	0	0	0	0	0	0	0	0	0	0	0	0	0	0	0	0	0	0	0	0
2.0	31	46	32	49	34	51	36	55	39	58	41	62	44	67	48	72	51	77	53	80	56	84	60	90	63	95	59	103
2.2	34	51	36	54	38	57	40	60	43	64	46	68	49	73	53	79	56	84	59	88	61	92	66	99	69	104	75	113
2.4	37	55	39	58	41	62	44	65	46	70	50	74	53	80	58	86	61	92	64	96	67	100	72	108	76	114	82	123
2.6	40	60	42	63	45	67	47	71	50	75	54	81	58	87	62	94	66	100	69	104	73	109	78	117	82	123	89	134
2.8	43	65	45	68	48	72	51	76	54	81	58	87	62	93	67	101	71	107	75	112	78	117	84	126	88	133	96	144
3.0	46	69	49	73	51	77	55	82	58	87	62	93	67	100	72	108	77	115	80	120	84	126	90	135	95	142	103	154
3.2	49	74	52	78	55	82	58	87	62	93	66	99	71	107	77	115	82	123	85	128	89	134	96	144	101	152	110	165
3.4	52	78	55	83	58	87	62	93	66	99	70	106	76	113	82	122	87	130	91	138	95	142	102	153	107	161	117	175
3.6	55	83	58	88	62	93	65	98	70	105	74	112	80	120	86	130	92	138	96	144	100	151	108	162	114	171	123	185
3.8	58	88	62	92	65	98	69	104	74	110	79	118	84	127	91	137	97	146	101	152	106	159	114	171	120	180	130	195
4.0	62	92	65	97	69	103	73	109	77	116	83	124	89	133	96	144	102	153	107	160	112	167	120	180	126	189	137	206
4.2	65	97	68	102	72	108	76	115	81	122	87	130	93	140	101	151	107	161	112	168	117	176	126	189	133	199	144	216
4.4	68	102	71	107	75	113	80	120	85	128	91	137	98	147	106	158	112	169	117	176	123	184	132	198	139	208	151	226
4.6	71	106	75	112	79	118	84	125	89	134	95	143	102	153	110	166	117	176	123	184	128	193	138	207	145	218	158	237
4.8	74	111	78	117	82	123	87	131	93	139	99	149	107	160	115	173	123	184	128	192	134	201	144	216	152	227	165	247
5.0	77	115	81	122	86	129	91	136	97	145	103	155	111	167	120	180	128	191	133	200	140	209	150	225	158	237	171	257
5.2	80	120	84	126	89	134	95	142	101	151	108	161	116	173	125	187	133	199	139	208	145	218	156	234	164	246	178	267
5.4	83	125	88	131	93	139	98	147	105	157	112	168	120	180	130	194	138	207	144	216	151	226	162	243	171	256	185	278
5.6	86	129	91	136	96	144	102	153	108	163	116	174	124	187	134	202	143	214	149	224	156	234	168	252	177	265	192	288
5.8	89	134	94	141	99	149	105	158	112	168	120	180	129	193	139	209	146	222	155	232	162	243	174	261	183	275	199	298
6.0	92	138	97	146	103	154	109	164	116	174	124	186	133	200	144	216	153	230	160	240	167	251	180	270	189	284	206	309
6.2	95	143	101	151	106	159	113	169	120	180	128	192	138	207	149	223	158	237	165	248	173	260	186	279	196	294	213	319
6.4	98	148	104	156	110	165	116	175	124	186	132	199	142	213	154	230	163	245	171	256	179	268	192	288	202	303	219	329
6.6	102	152	107	161	113	170	120	180	128	192	137	205	147	220	158	238	169	253	176	264	184	276	196	297	208	313	226	339
6.8	105	157	110	165	117	175	124	185	132	197	141	211	151	227	163	245	174	260	181	272	190	285	204	306	215	322	233	350
7.0	108	162	114	170	120	180	127	191	135	203	145	217	156	233	168	252	179	266	187	280	195	293	210	315	221	332	240	360
7.2	111	166	117	175	123	185	131	196	139	209	149	223	160	240	173	259	184	276	192	288	201	301	216	324	227	341	247	370
7.4	114	171	120	180	127	190	135	202	143	215	153	230	164	247	178	266	189	283	197	296	207	310	222	333	234	351	254	381
7.6	117	175	123	185	130	195	138	207	147	221	157	236	169	253	182	274	194	291	203	304	212	318	228	342	240	350	261	391
7.8	120	180	126	190	134	201	142	213	151	226	161	242	173	260	187	281	199	299	208	312	218	327	234	351	246	369	267	401
8.0	123	185	130	195	137	206	145	218	155	232	166	248	178	267	192	288	204	306	213	320	223	335	240	360	253	379	274	411
8.2	126	189	133	199	141	211	149	224	159	238	170	254	182	273	197	295	209	314	219	328	229	343	246	369	259	388	281	422
84	129	194	136	204	144	216	153	229	163	244	174	261	187	280	202	302	214	322	224	336	234	352	252	378	265	398	288	432
85	132	198	139	209	147	221	156	235	166	250	178	267	191	287	206	310	220	329	229	344	240	360	258	387	272	407	295	442
88	135	203	143	214	151	226	160	240	170	255	182	273	196	293	211	317	225	337	235	352	246	368	264	396	278	417	302	453
90	138	208	146	219	154	231	164	245	174	261	186	279	200	300	216	324	230	345	240	360	251	377	270	405	284	426	309	463
92	142	212	149	224	158	237	167	251	178	267	190	286	204	307	221	331	235	352	245	368	257	385	276	414	291	436	315	473
94	145	217	152	229	161	242	171	256	182	273	194	292	209	313	226	338	240	360	251	376	262	393	282	423	297	445	322	483
96	148	222	156	234	165	247	175	262	166	279	199	298	213	320	230	346	245	368	256	384	268	402	268	432	303	455	329	494
98	151	226	159	238	168	252	178	267	190	285	203	304	218	327	235	353	250	375	261	392	273	410	294	441	309	464	336	504
100	154	231	162	243	171	257	182	273	194	290	207	310	222	333	240	360	255	383	267	400	279	419	300	450	316	474	343	514
102	157	235	165	248	175	262	185	278	197	296	211	317	227	340	245	367	260	391	272	408	285	427	306	459	322	483	350	525
104	160	240	169	253	178	267	189	284	201	302	215	323	231	347	250	374	266	398	277	415	290	435	312	468	328	493	357	535
106	163	245	172	258	182	273	193	289	205	308	219	329	236	353	254	382	271	406	283	424	296	444	318	477	335	502	363	545
108	166	249	175	263	185	278	196	295	209	314	223	335	240	360	259	389	276	414	288	432	301	452	324	486	341	512	370	555
110	169	254	178	268	189	283	200	300	213	319	228	341	244	367	264	396	281	421	293	440	307	450	330	495	347	521	377	566
112	172	258	182	272	192	288	204	305	217	325	232	348	249	373	269	403	286	429	299	448	313	469	336	504	354	531	384	576
114	175	263	185	277	195	293	207	311	221	331	236	354	253	380	274	410	291	437	304	456	318	477	342	513	360	540	391	586
116	178	268	188	282	199	298	211	316	225	337	240	360	258	387	278	418	296	444	309	464	324	486	348	522	366	549	398	597
118	182	272	191	287	202	303	215	322	228	343	244	366	262	393	283	425	301	452	315	472	329	494	354	531	373	559	405	607
120	185	277	195	292	208	309	218	327	232	348	248	372	267	400	288	432	306	460	320	480	335	502	360	540	379	569	411	617

*One lane rotated is typical for a two-lane highway, two lanes rotated for a four-lane highway, and so on. (See Fig. 4.27.)

From *A Policy on Geometric Design of Highways and Streets*, 2004, by the American Association of State Highway and Transportation Officials, Washington, D.C. Used by permission.

APPENDIX 4.F
Acceleration Distances for Passenger Cars, Level Conditions

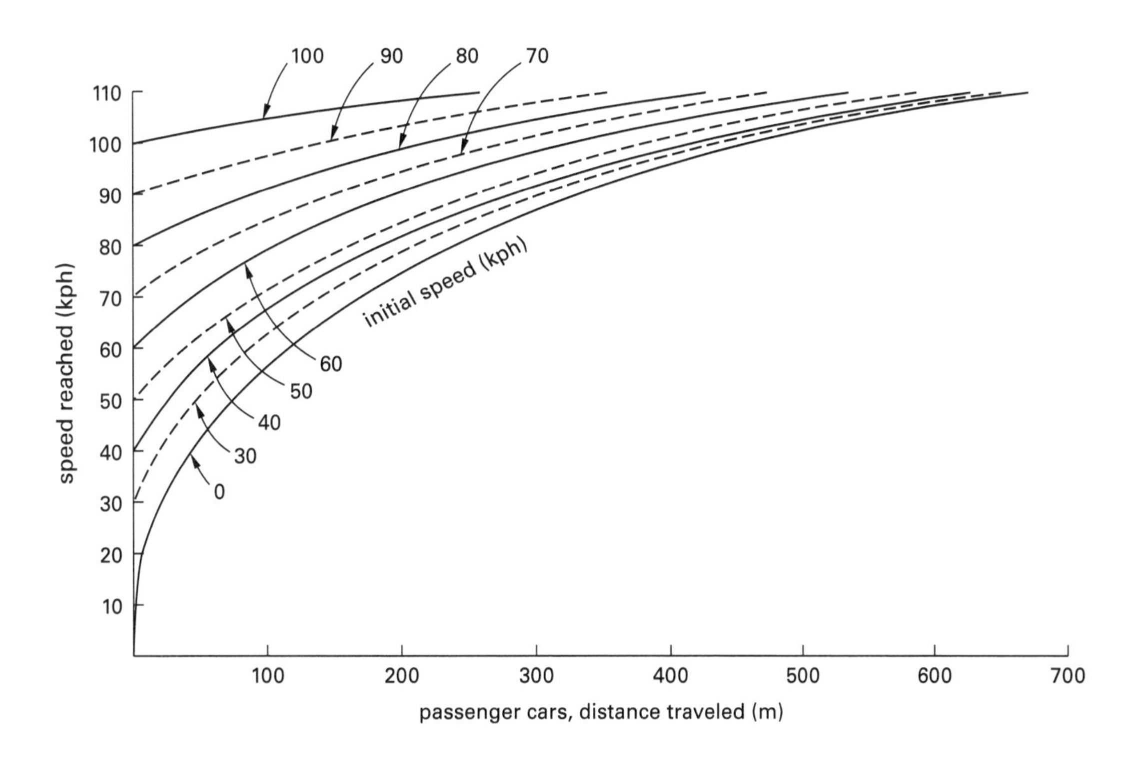

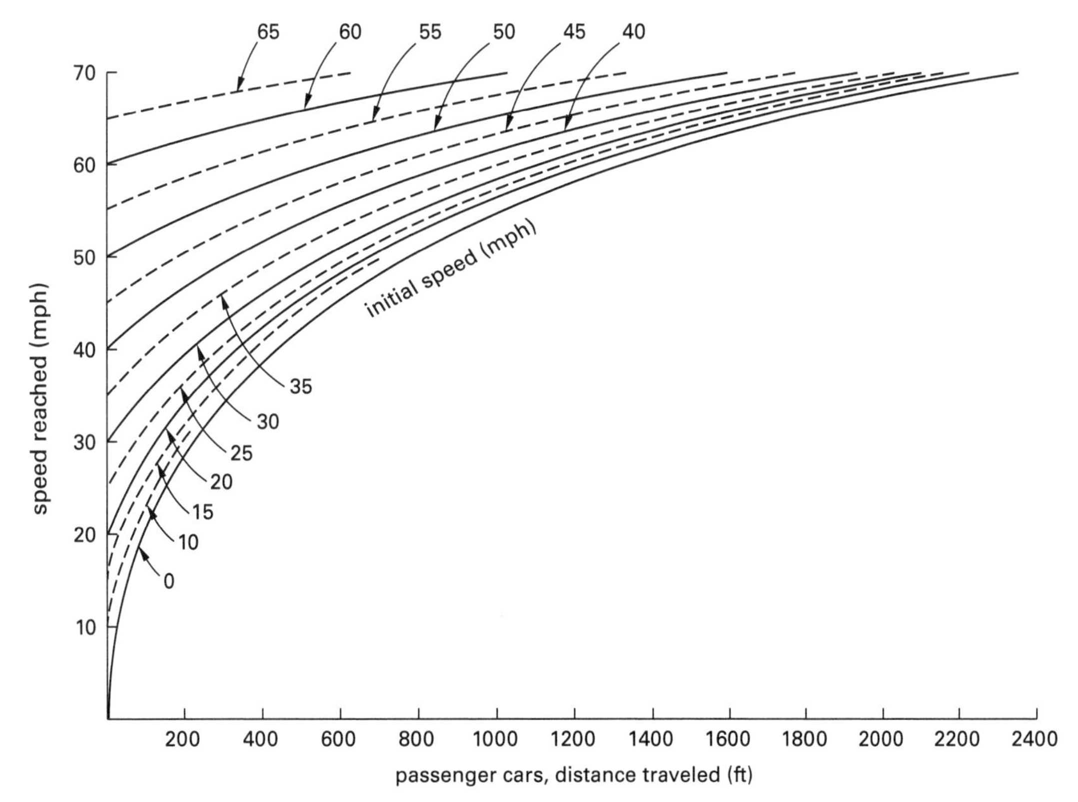

From *A Policy on Geometric Design of Highways and Streets*, 2004, by the American Association of State Highway and Transportation Officials, Washington, D.C. Used by permission.

Appendices

APPENDIX 4.G
Deceleration Distances for Passenger Vehicles Approaching Intersections

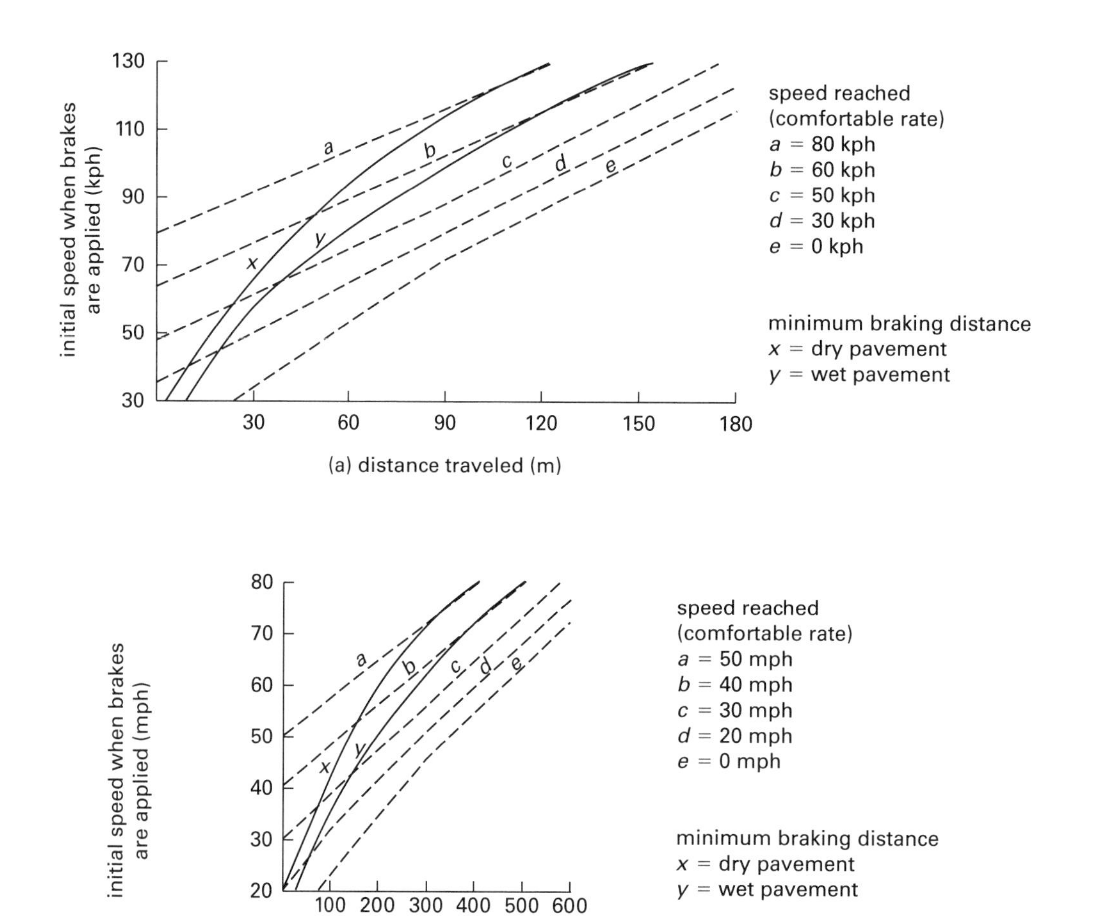

Appendices

APPENDIX 4.H
AASHTO Design Vehicle Dimensions

SI units[a]

design vehicle type	symbol	dimensions (m) overall: height	width	length	overhang: front	rear	WB_1	WB_2	S	T	WB_3	WB_4	typical kingpin to center of rear axle
passenger car	P	1.3	2.1	5.8	0.9	1.5	3.4	–	–	–	–	–	–
single-unit truck	SU	3.4–4.1	2.4	9.2	1.2	1.8	6.1	–	–	–	–	–	–
buses													
intercity bus (motor coaches)	BUS-12	3.7	2.6	12.2	1.8	1.9[d]	7.3	1.1	–	–	–	–	–
	BUS-14	3.7	2.6	13.7	1.8	2.6[d]	8.1	1.2	–	–	–	–	–
city transit bus	CITY-BUS	3.2	2.6	12.2	2.1	2.4	7.6	–	–	–	–	–	–
conventional school bus (65 pass.)	S-BUS 11	3.2	2.4	10.9	0.8	3.7	6.5	–	–	–	–	–	–
large school bus (84 pass.)	S-BUS 12	3.2	2.4	12.2	2.1	4.0	6.1	–	–	–	–	–	–
articulated bus	A-BUS	3.4	2.6	18.3	2.6	3.1	6.7	5.9	1.9[e]	4.0[e]	–	–	–
trucks													
intermediate semitrailer	WB-12	4.1	2.4	13.9	0.9	0.8[d]	3.8	8.4	–	–	–	–	8.4
intermediate semitrailer	WB-15	4.1	2.6	16.8	0.9	0.6[d]	4.5	10.8	–	–	–	–	11.4
interstate semitrailer	WB-19[b]	4.1	2.6	20.9	1.2	0.8[d]	6.6	12.3	–	–	–	–	13.0
interstate semitrailer	WB-20[c]	4.1	2.6	22.4	1.2	1.4–0.8[d]	6.6	13.2–13.8	–	–	–	–	13.9–14.5
"double-bottom" semitrailer/trailer	WB-20D	4.1	2.6	22.4	0.7	0.9	3.4	7.0	0.9[f]	2.1[f]	7.0	–	7.0
triple semitrailer/trailer	WB-30T	4.1	2.6	32.0	0.7	0.9	3.4	6.9	0.9[g]	2.1[g]	7.0	7.0	7.0
turnpike double semitrailer/trailer	WB-33D[b]	4.1	2.6	34.8	0.7	0.8[a]	4.4	12.2	0.8[h]	3.1[h]	13.6	–	13.0
recreational vehicles													
motor home	MH	3.7	2.4	9.2	1.2	1.8	6.1	–	–	–	–	–	–
car and camper trailer	P/T	3.1	2.4	14.8	0.9	3.1	3.4	–	1.5	5.8	–	–	–
car and boat trailer	P/B	–	2.4	12.8	0.9	2.4	3.4	–	1.5	4.6	–	–	–
motor home and boat trailer	MH/B	3.7	2.4	16.2	1.2	2.4	6.1	–	1.8	4.6	–	–	–
farm tractor[i]	TR	3.1	2.4–3.1	4.9[j]	–	–	3.1	2.7	0.9	2.0	–	–	–

[a]Since vehicles are manufactured in customary U.S. dimensions to provide only one physical size for each design vehicle, the values shown in the design drawings have been soft converted from numbers listed in feet, and then the numbers in this table have been rounded to the nearest tenth of a meter.
[b]Design vehicle with 14.63 m trailer as adopted in the 1982 Surface Transportation Assistance Act (STAA).
[c]Design vehicle with 16.16 m trailer as grandfathered in with the 1982 Surface Transportation Assistance Act (STAA).
[d]This is overhang from the back axle of the tandem axle assembly.
[e]Combined dimension is 5.91 m and articulating section is 1.22 m wide.
[f]Combined dimension is typically 3.05 m.
[g]Combined dimension is typically 3.05 m.
[h]Combined dimension is typically 3.81 m.
[i]Dimensions are for a 150–200 hp tractor excluding any wagon length.
[j]To obtain the total length of tractor and one wagon, add 5.64 m to the tractor length. Wagon length is measured from front of drawbar to rear of wagon, and drawbar is 1.98 m long.

- WB_1, WB_2, and WB_4 are the effective vehicle wheelbases, or distances between axle groups, starting at the front and working toward the back of each unit.
- *S* is the distance from the rear effective axle to the hitch point or point of articulation.
- *T* is the distance from the hitch point of articulation measured back to the center of the next axle or center of the tandem axle assembly.

(continued)

APPENDIX 4.H *(continued)*
AASHTO Design Vehicle Dimensions

customary U.S. units

		dimensions (ft)											
		overall			overhang								
design vehicle type	symbol	height	width	length	front	rear	WB_1	WB_2	S	T	WB_3	WB_4	typical kingpin to center of rear axle
passenger car	P	4.25	7	19	3	5	11	–	–	–	–	–	–
single-unit truck	SU	11–13.5	8.0	30	4	6	20	–	–	–	–	–	–
buses													
intercity bus (motor coaches)	BUS-40	12.0	8.5	40	6	6.3[c]	24	3.7	–	–	–	–	–
	BUS-45	12.0	8.5	45	6	8.5[c]	26.5	4.0	–	–	–	–	–
city transit bus	CITY-BUS	10.5	8.5	40	7	8	25	–	–	–	–	–	–
conventional school bus (65 pass.)	S-BUS 36	10.5	8.0	35.8	2.5	12	21.3	–	–	–	–	–	–
large school bus (84 pass.)	S-BUS 40	10.5	8.0	40	7	13	20	–	–	–	–	–	–
articulated bus	A-BUS	11.0	8.5	60	8.6	10	22.0	19.4	6.2[d]	13.2[d]	–	–	–
trucks													
intermediate semitrailer	WB-40	13.5	8.0	45.5	3	2.5[c]	12.5	27.5	–	–	–	–	27.5
intermediate semitrailer	WB-50	13.5	8.5	55	3	2[c]	14.6	35.4	–	–	–	–	37.5
interstate semitrailer	WB-62[a]	13.5	8.5	68.5	4	2.5[c]	21.6	40.4	–	–	–	–	42.5
interstate semitrailer	WB-65[b] or WB-67	13.5	8.5	73.5	4	4.5–2.5[c]	21.6	43.4–45.4	–	–	–	–	45.5–47.5
"double-bottom" semitrailer/trailer	WB-67D	13.5	8.5	73.5	2.33	3	11.0	23.0	3.0[e]	7.0[e]	23.0	–	23.0
triple semitrailer/trailer	WB-100T	13.5	8.5	104.8	2.33	3	11.0	22.5	3.0[f]	7.0[f]	23.0	23.0	23.0
turnpike double semitrailer/trailer	WB-109D[a]	13.5	8.5	114	2.33	2.5[g]	14.3	39.9	2.5[g]	10.0[g]	44.5	–	42.5
recreational vehicles													
motor home	MH	12	8	30	4	6	20	–	–	–	–	–	–
car and camper trailer	P/T	10	8	48.7	3	10	11	–	5	19	–	–	–
car and boat trailer	P/B	–	8	42	3	8	11	–	5	15	–	–	–
motor home and boat trailer	MH/B	12	8	53	4	8	20	–	6	15	–	–	–
farm tractor[h]	TR	10	8–10	16[i]	–	–	10	9	3	6.5	–	–	–

[a]Design vehicle with 48 ft trailer as adopted in the 1982 Surface Transportation Assistance Act (STAA).
[b]Design vehicle with 53 ft trailer as grandfathered in with the 1982 Surface Transportation Assistance Act (STAA).
[c]This is overhang from the back axle of the tandem axle assembly.
[d]Combined dimension is 19.4 ft and articulating section is 4 ft wide.
[e]Combined dimension is typically 10.0 ft.
[f]Combined dimension is typically 10.0 ft.
[g]Combined dimension is typically 12.5 ft.
[h]Dimensions are for a 150–200 hp tractor excluding any wagon length.
[i]To obtain the total length of tractor and one wagon, add 18.5 ft to the tractor length. Wagon length is measured from front of drawbar to rear of wagon, and drawbar is 6.5 ft long.

- WB_1, WB_2, and WB_4 are the effective vehicle wheelbases, or distances between axle groups, starting at the front and working toward the back of each unit.
- S is the distance from the rear effective axle to the hitch point or point of articulation.
- T is the distance from the hitch point of articulation measured back to the center of the next axle or center of the tandem axle assembly.

From *A Policy on Geometric Design of Highways and Streets*, 2004, by the American Association of State Highway and Transportation Officials, Washington, D.C. Used by permission.

APPENDIX 4.I
Minimum Turning Radii of Design Vehicles

SI units[a]

design vehicle type	passenger car	single-unit truck	intercity bus (motor coach)		city transit bus	conventional school bus (65 pass.)	large[e] school bus (84 pass.)	articulated bus	intermediate semitrailer	intermediate semitrailer
symbol	P	SU	BUS-12	BUS-14	CITY-BUS	S-BUS11	S-BUS12	A-BUS	WB-12	WB-15
minimum design turning radius (m)	7.3	12.8	13.7	13.7	12.8	11.9	12.0	12.1	12.2	13.7
centerline[d] turning radius, CTR (m)	6.4	11.6	12.4	12.4	11.5	10.6	10.8	10.8	11.0	12.5
minimum inside radius (m)	4.4	8.6	8.4	7.8	7.5	7.3	7.7	6.5	5.9	5.2

design vehicle type	interstate semitrailer		"double-bottom" combination	triple semitrailer/ trailer	turnpike double semitrailer/ trailer	motor home	car and camper trailer	car and boat trailer	motor home and boat trailer	farm tractor[f] with one wagon
symbol	WB-19[b]	WB-20[c]	WB-20D	WB-30T	WB-33D[b]	MH	P/T	P/B	MH/B	TR/W
minimum design turning radius (m)	13.7	13.7	13.7	13.7	18.3	12.2	10.1	7.3	15.2	5.5
centerline[d] turning radius, CTR (m)	12.5	12.5	12.5	12.5	17.1	11.0	9.1	6.4	14.0	4.3
minimum inside radius (m)	2.4	1.3	5.9	3.0	4.5	7.9	5.3	2.4	10.7	3.2

[a]Numbers in table have been rounded to the nearest tenth of a meter.
[b]Design vehicle with 14.63 m trailer as adopted in the 1982 Surface Transportation Assistance Act (STAA).
[c]Design vehicle with 16.16 m trailer as grandfathered in with the 1982 Surface Transportation Assistance Act (STAA).
[d]The turning radius assumed by a designer when investigating possible turning paths and is set at the centerline of the front axle of the vehicle. If the minimum turning path is assumed, the CTR approximately equals the minimum design turning radius minus one-half the front width of the vehicle.
[e]School buses are manufactured from 42 passenger to 84 passenger sizes. This corresponds to wheelbase lengths of 3.35 m to 6.1 m, respectively. For these different sizes, the minimum design turning radii vary from 8.78 m to 12.01 m, and the minimum inside radii vary from 4.27 m to 7.74 m.
[f]The turning radius is for 150–200 hp tractor with one 5.64 m long wagon attached to the hitch point. Front wheel drive is disengaged and brakes are not applied.

customary U.S. units

design vehicle type	passenger car	single-unit truck	intercity bus (motor coach)		city transit bus	conventional school bus (65 pass.)	large[d] school bus (84 pass.)	articulated bus	intermediate semitrailer	intermediate semitrailer
symbol	P	SU	BUS-40	BUS-45	CITY-BUS	S-BUS36	S-BUS40	A-BUS	WB-40	WB-50
minimum design turning radius (ft)	24	42	45	45	42.0	38.9	39.4	39.8	40	45
centerline[c] turning radius, CTR (ft)	21	38	40.8	40.8	37.8	34.9	35.4	35.5	36	41
minimum inside radius (ft)	14.4	28.3	27.6	25.5	24.5	23.8	25.4	21.3	19.3	17.0

design vehicle type	interstate semitrailer		"double-bottom" combination	triple semitrailer/ trailer	turnpike double semitrailer/ trailer	motor home	car and camper trailer	car and boat trailer	motor home and boat trailer	farm tractor[e] with one wagon
symbol	WB-62[a]	WB-65[b] or WB-67	WB-67D	WB-100T	WB-109D[a]	MH	P/T	P/B	MH/B	TR/W
minimum design turning radius (ft)	45	45	45	45	60	40	33	24	50	18
centerline[c] turning radius, CTR (ft)	41	41	41	41	56	36	30	21	46	14
minimum inside radius (ft)	7.9	4.4	19.3	9.9	14.9	25.9	17.4	8.0	35.1	10.5

[a]Design vehicle with 48 ft trailer as adopted in the 1982 Surface Transportation Assistance Act (STAA).
[b]Design vehicle with 53 ft trailer as grandfathered in with the 1982 Surface Transportation Assistance Act (STAA).
[c]The turning radius assumed by a designer when investigating possible turning paths and is set at the centerline of the front axle of the vehicle. If the minimum turning path is assumed, the CTR approximately equals the minimum design turning radius minus one-half the front width of the vehicle.
[d]School buses are manufactured from 42 passenger to 84 passenger sizes. This corresponds to wheelbase lengths of 11.0 ft to 20.0 ft, respectively. For these different sizes, the minimum design turning radii vary from 28.8 ft to 39.4 ft, and the minimum inside radii vary from 14.0 ft to 25.4 ft.
[e]The turning radius is for 150–200 hp tractor with one 18.5 ft long wagon attached to the hitch point. Front wheel drive is disengaged and brakes are not applied.

From *A Policy on Geometric Design of Highways and Streets*, 2004, by the American Association of State Highway and Transportation Officials, Washington, D.C. Used by permission.

APPENDIX 4.J
Minimum Turning Path for Passenger Car (P) Design Vehicle[a,b]

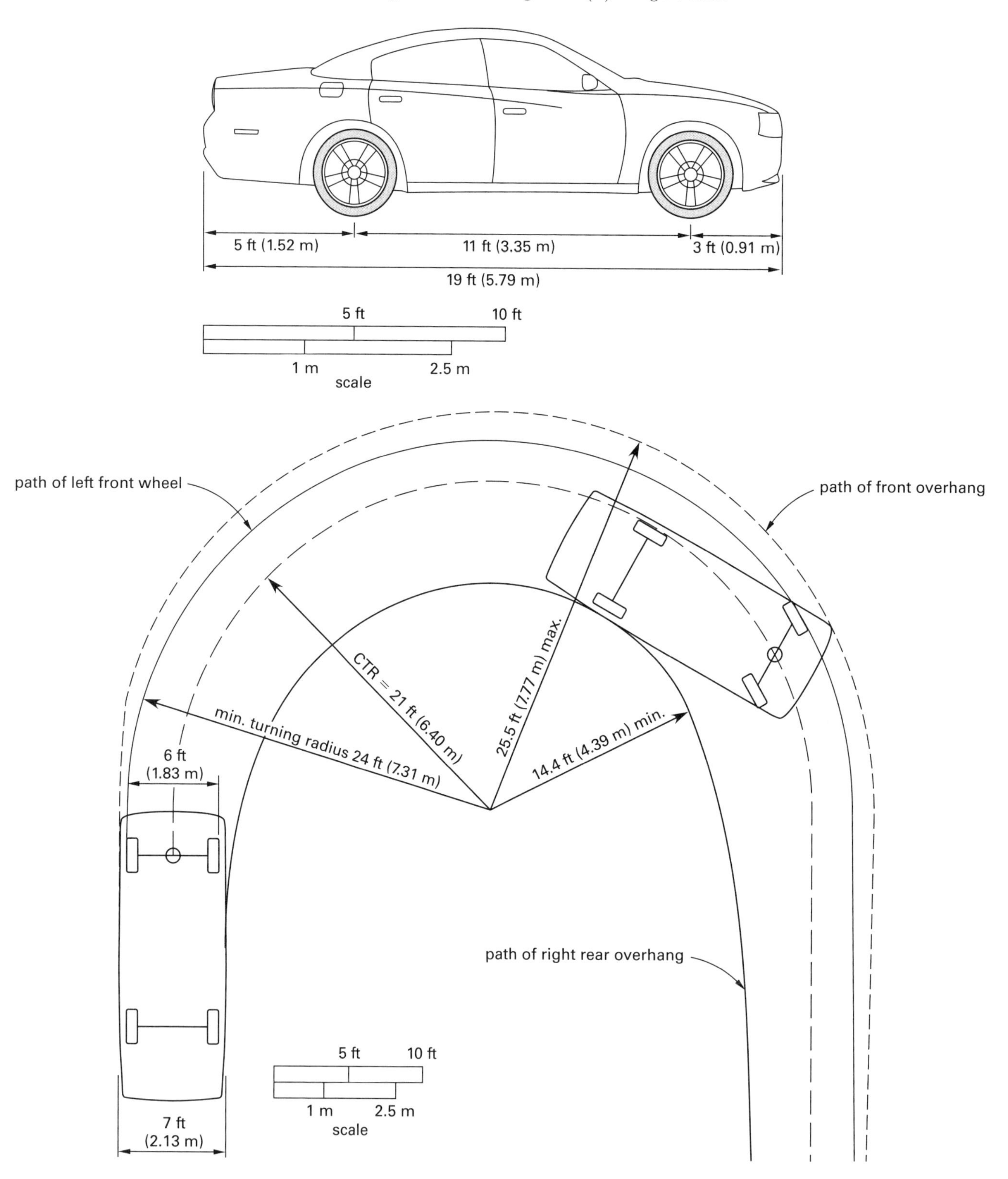

[a]assumed steering angle is 31.6°
[b]CTR = centerline turning radius at front axle

From *A Policy on Geometric Design of Highways and Streets*, 2004, by the American Association of State Highway and Transportation Officials, Washington, D.C. Used by permission.

APPENDIX 4.K

Turning Characteristics of a Typical Tractor/Semitrailer Combination Truck

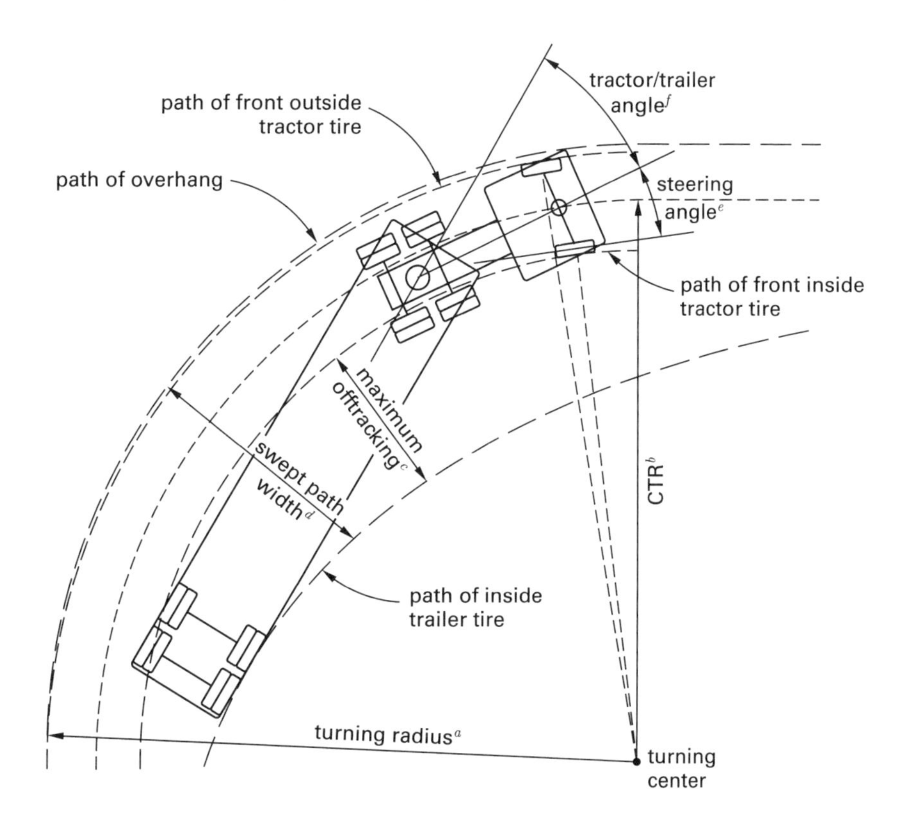

[a]*Turning radius:* The circular arc formed by the turning path radius of the front outside tire of a vehicle. This radius is also described by vehicle manufacturers as the *turning curb radius.*
[b]*CTR:* The turning radius of the centerline of the front axle of a vehicle.
[c]*Offtracking:* The difference in the paths of the front and rear wheels of a tractor/semitrailer as it negotiates a turn. The path of the rear tires of the turning truck does not coincide with that of the front tires, and this effect is shown in the drawing above.
[d]*Swept path width:* The amount of roadway width that a truck covers in negotiating a turn. The swept path width is equal to the amount of offtracking plus the width of the tractor unit. The most significant dimension affecting the swept path width of a tractor/semitrailer is the distance from the kingpin to the rear trailer axle or axles. The greater this distance is, the greater the swept path width.
[e]*Steering angle:* The maximum angle of turn built into the steering mechanism of the front wheels of a vehicle. This maximum angle controls the minimum turning radius of the vehicle.
[f]*Tractor/trailer angle:* The angle between adjoining units of a tractor/semitrailer when the combination unit is placed into a turn. This angle is measured between the longitudinal axes of the tractor and trailer as the vehicle turns. The maximum tractor/trailer angle occurs when a vehicle makes a 180° turn at the minimum turning radius. This angle is reached slightly beyond the point where the maximum swept path width is achieved.

From *A Policy on Geometric Design of Highways and Streets*, 2004, by the American Association of State Highway and Transportation Officials, Washington, D.C. Used by permission.

Appendices

APPENDIX 4.L

Minimum Turning Path for Single-Unit Truck (SU) Design Vehicle[a,b]

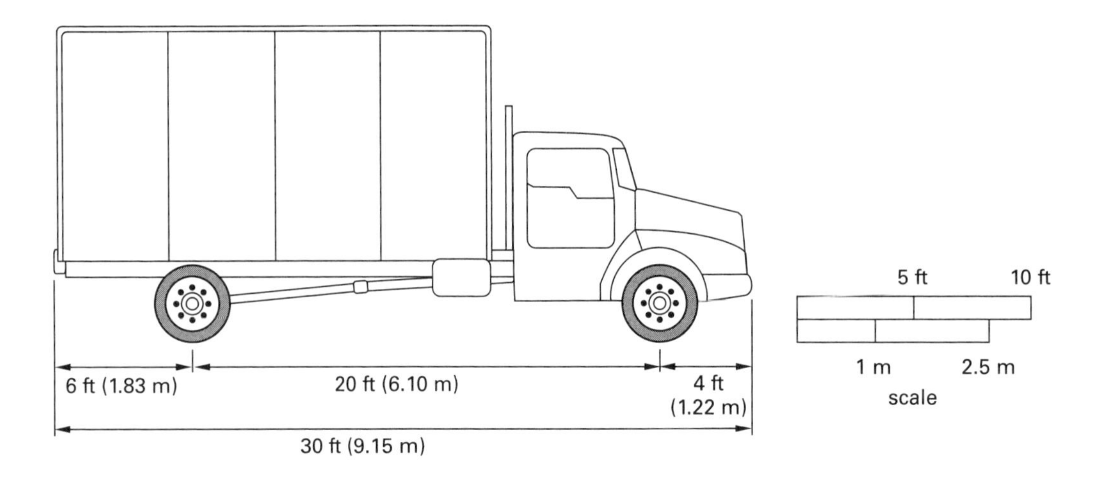

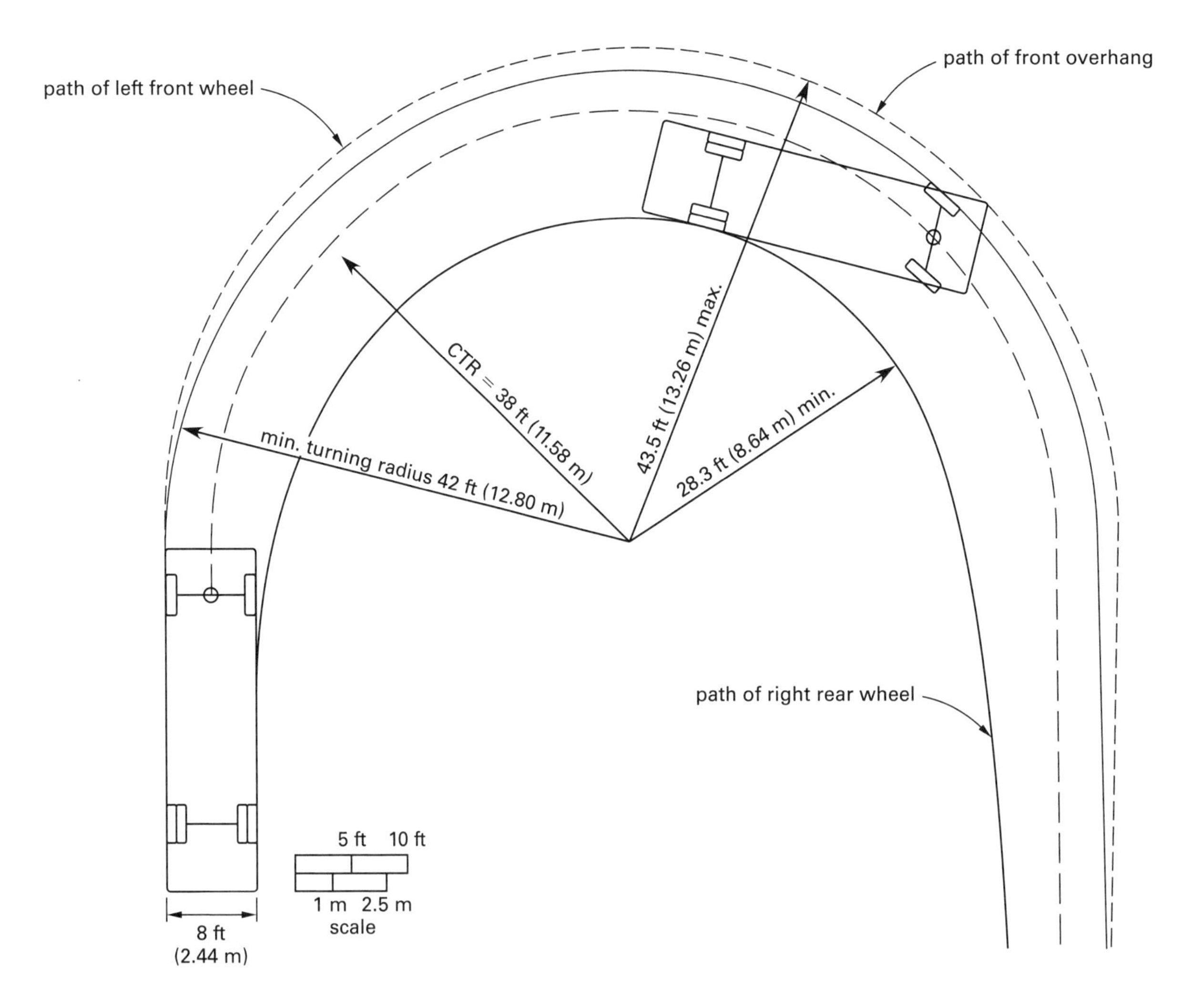

[a]assumed steering angle is 31.7°

[b]CTR = centerline turning radius at front axle

From *A Policy on Geometric Design of Highways and Streets*, 2004, by the American Association of State Highway and Transportation Officials, Washington, D.C. Used by permission.

APPENDIX 4.M

Minimum Turning Path for Intermediate Semitrailer WB-50 (WB-15) Design Vehicle[a,b,c]

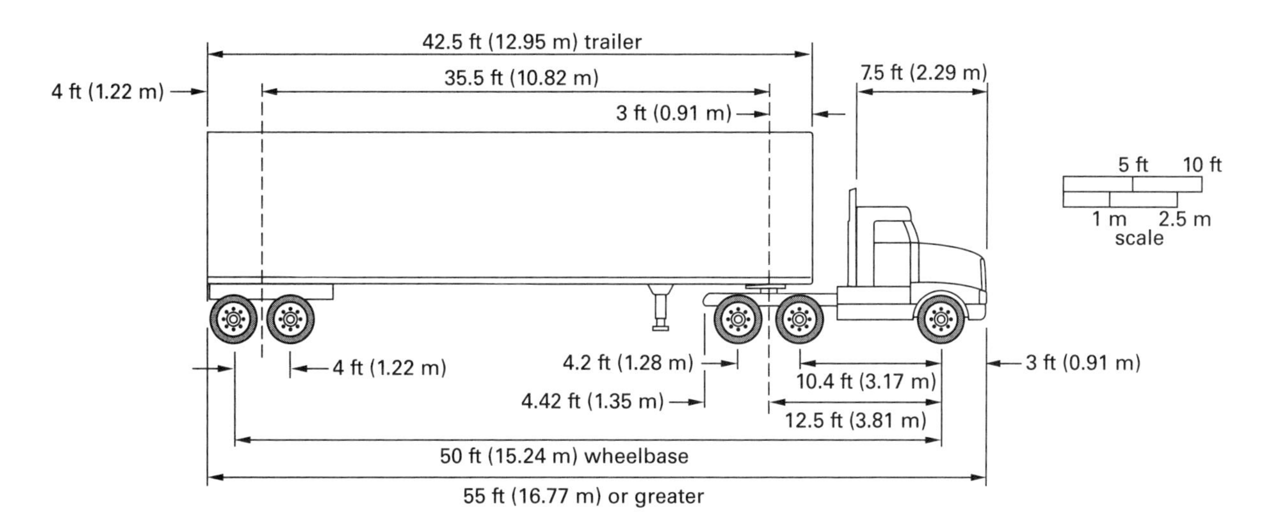

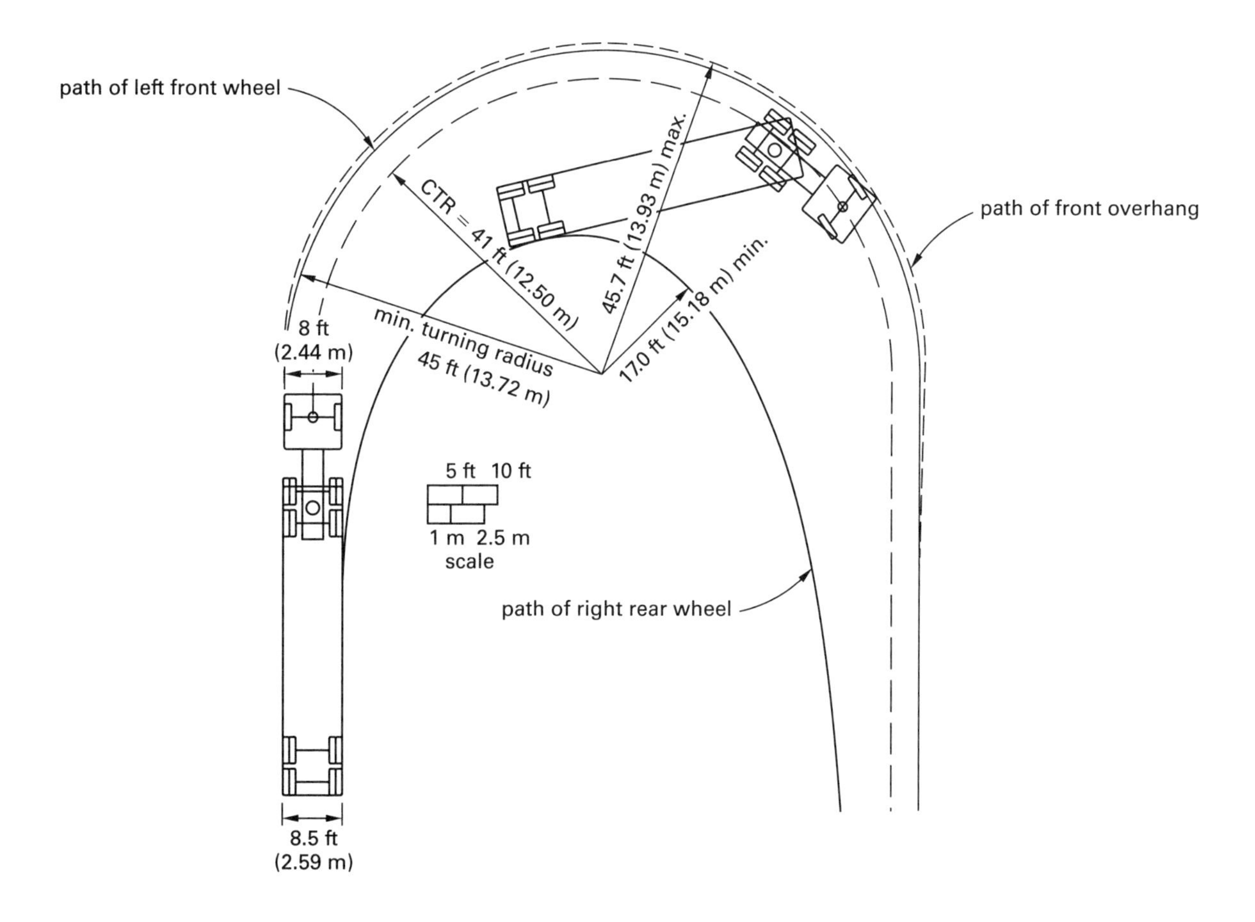

[a]assumed steering angle is 17.9°
[b]assumed tractor/trailer angle is 56°
[c]CTR = centerline turning radius at front axle

From *A Policy on Geometric Design of Highways and Streets*, 2004, by the American Association of State Highway and Transportation Officials, Washington, D.C. Used by permission.

APPENDIX 4.N

Minimum Turning Path for Interstate Semitrailer WB-65 and WB-67 (WB-20) Design Vehicle[a,b,c,d]

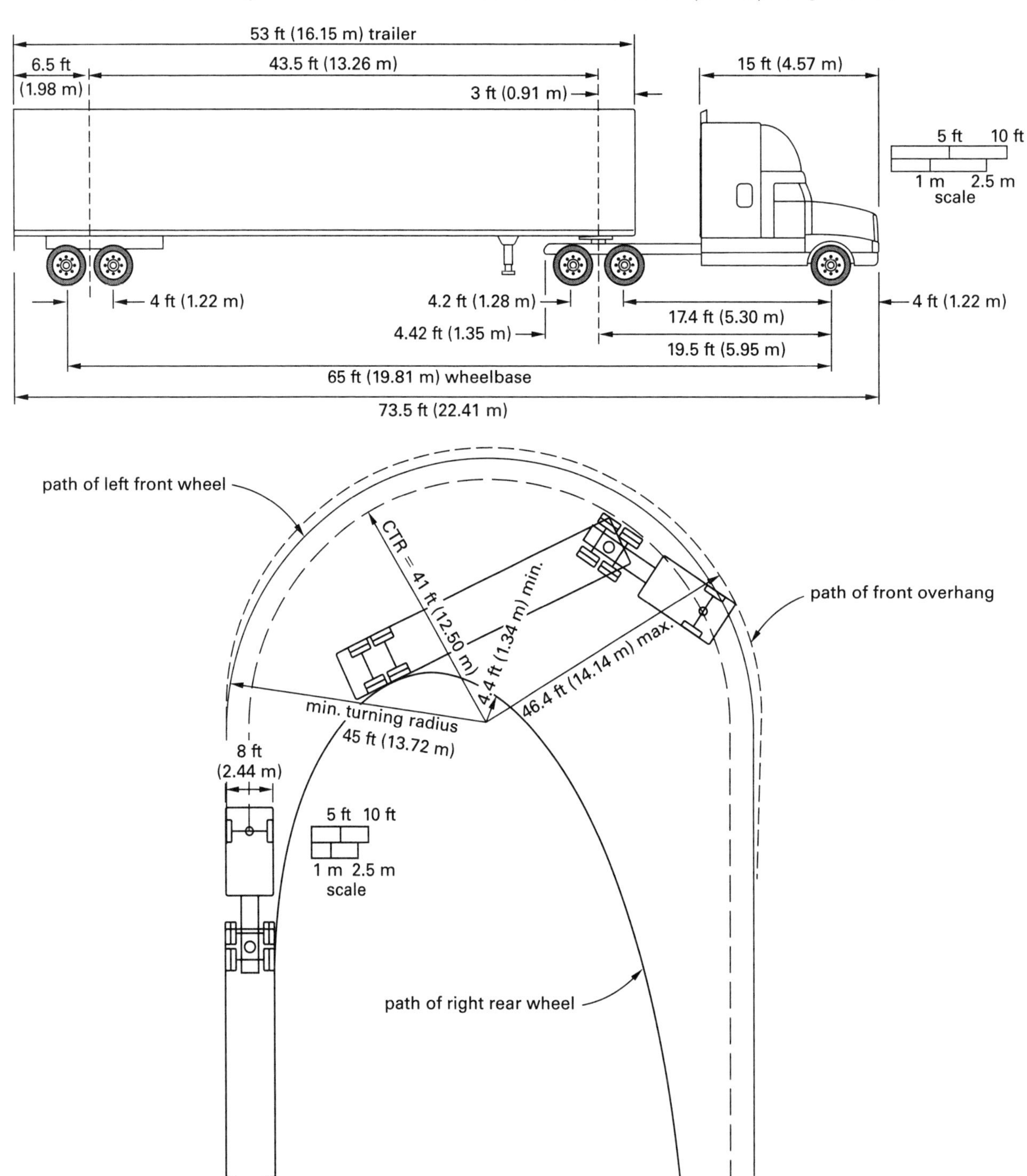

[a]The WB-65 (WB-20) is shown. A longer wheelbase vehicle, the WB-67 (WB-20), can be created by moving the tandem wheel assembly on the trailer back by 2 ft (0.61 m).
[b]assumed steering angle is 28.4°
[c]assumed tractor/trailer angle is 68.5°
[d]CTR = centerline turning radius at front axle

From *A Policy on Geometric Design of Highways and Streets*, 2004, by the American Association of State Highway and Transportation Officials, Washington, D.C. Used by permission.

APPENDIX 4.O

Minimum Turning Path for City Transit Bus (CITY-BUS) Design Vehicle[a,b]

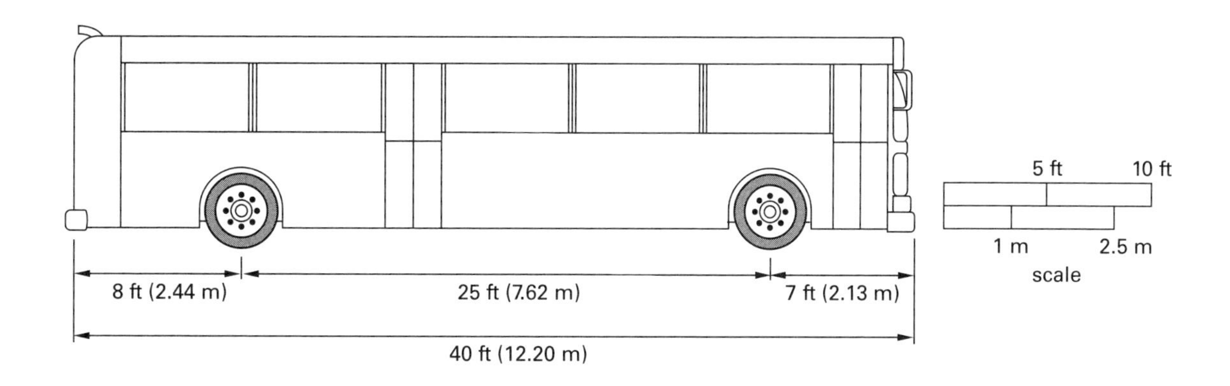

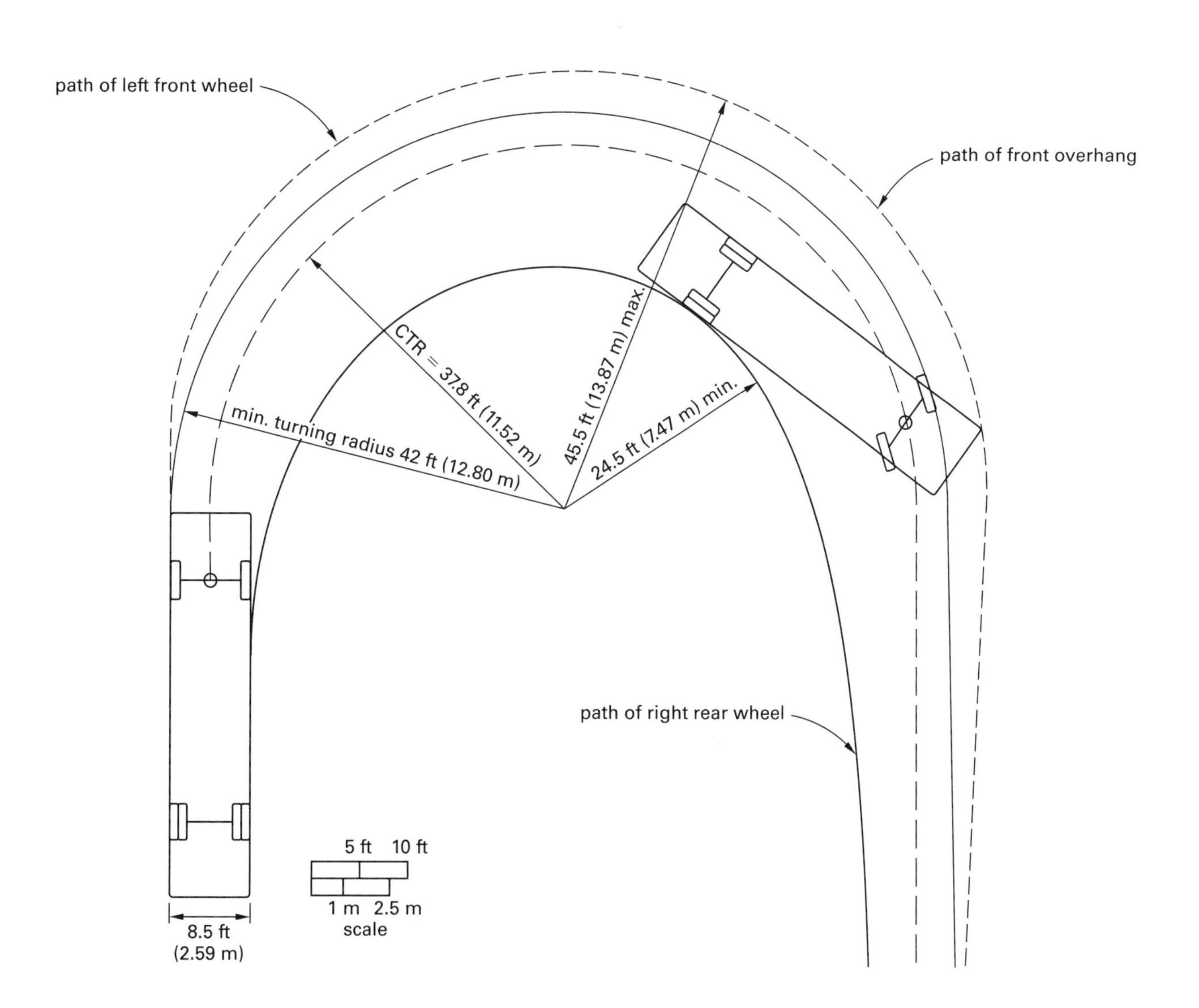

[a]assumed steering angle is 41°

[b]CTR = centerline turning radius at front axle

From *A Policy on Geometric Design of Highways and Streets*, 2004, by the American Association of State Highway and Transportation Officials, Washington, D.C. Used by permission.

APPENDIX 4.P

Minimum Turning Path for Conventional School Bus S-BUS36 (S-BUS11) Design Vehicle[a,b,c]

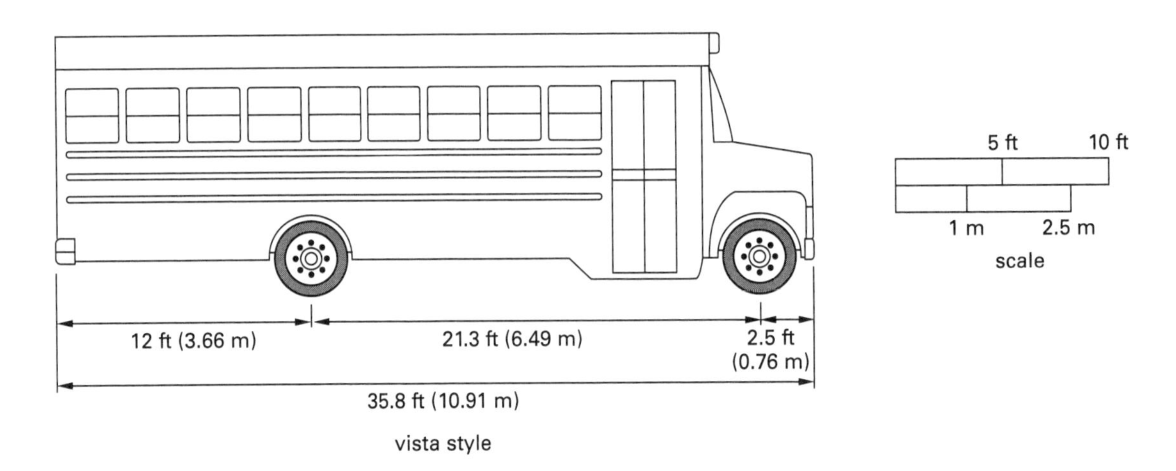

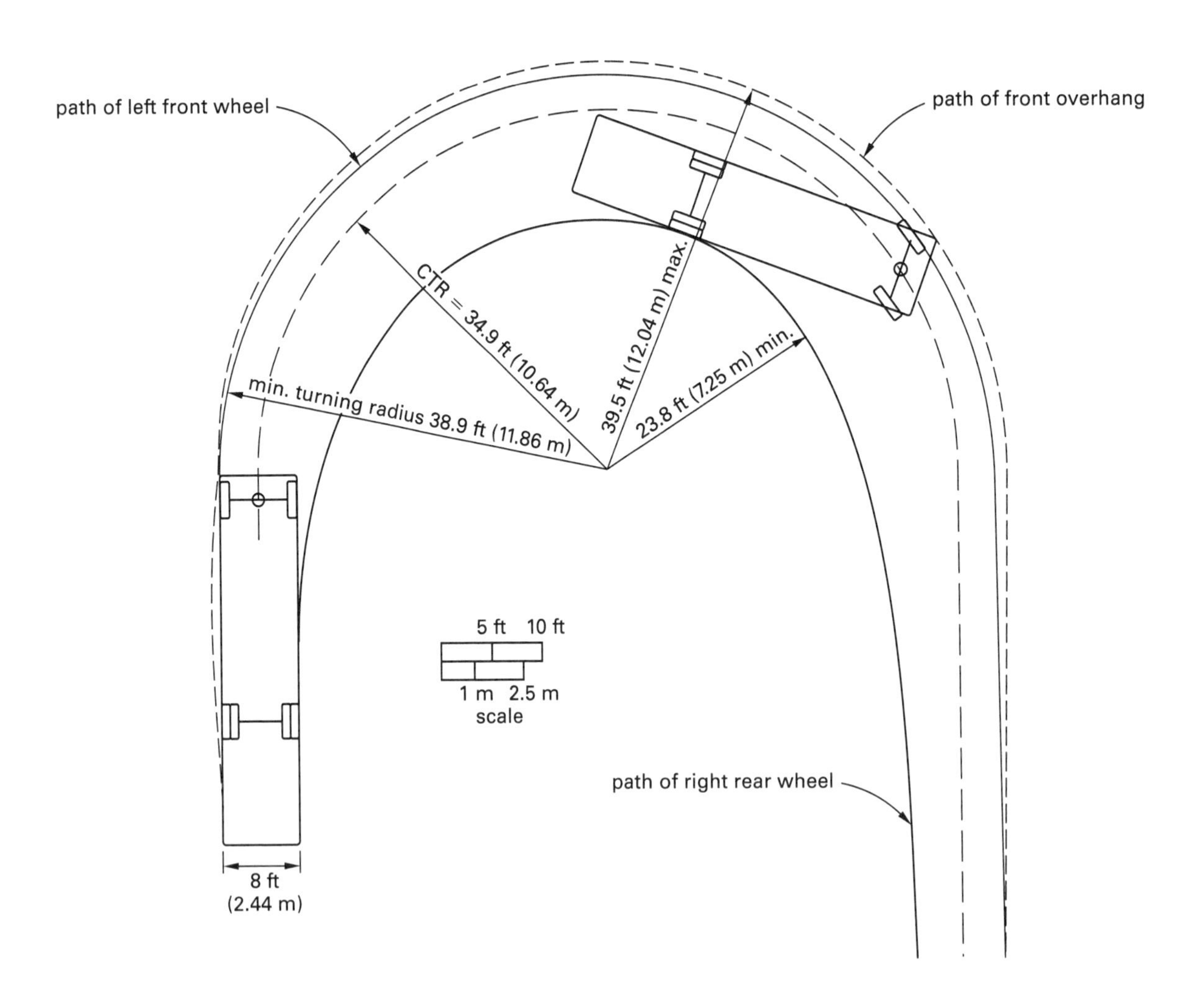

[a]assumed steering angle is 37.2°
[b]CTR = centerline turning radius at front axle
[c]65 passenger bus

From *A Policy on Geometric Design of Highways and Streets*, 2004, by the American Association of State Highway and Transportation Officials, Washington, D.C. Used by permission.

APPENDIX 4.Q
Minimum Traveled Way, Passenger (P) Design Vehicle Path

SI units

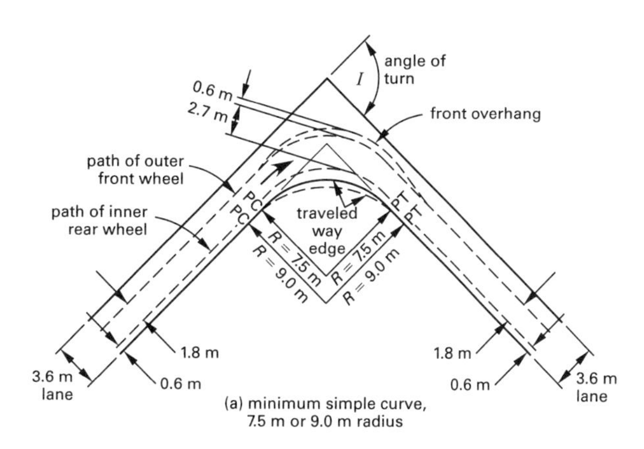

(a) minimum simple curve, 7.5 m or 9.0 m radius

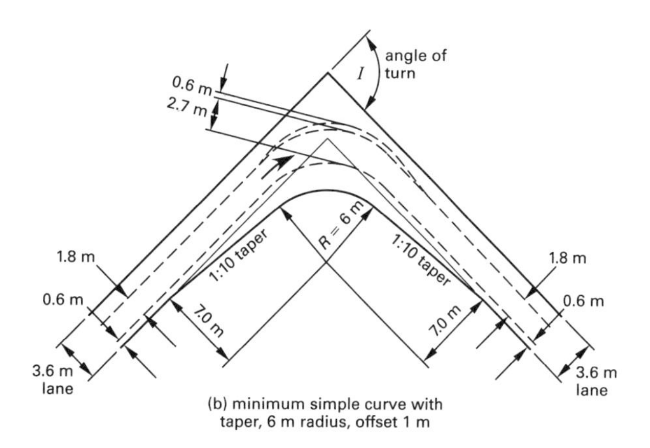

(b) minimum simple curve with taper, 6 m radius, offset 1 m

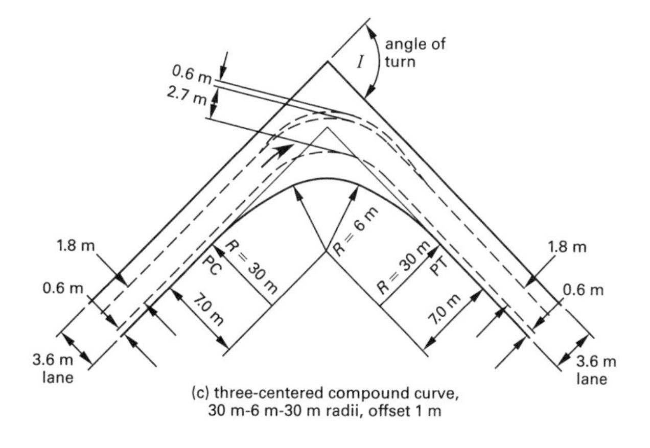

(c) three-centered compound curve, 30 m-6 m-30 m radii, offset 1 m

(continued)

APPENDIX 4.Q *(continued)*
Minimum Traveled Way, Passenger (P) Design Vehicle Path

customary U.S. units

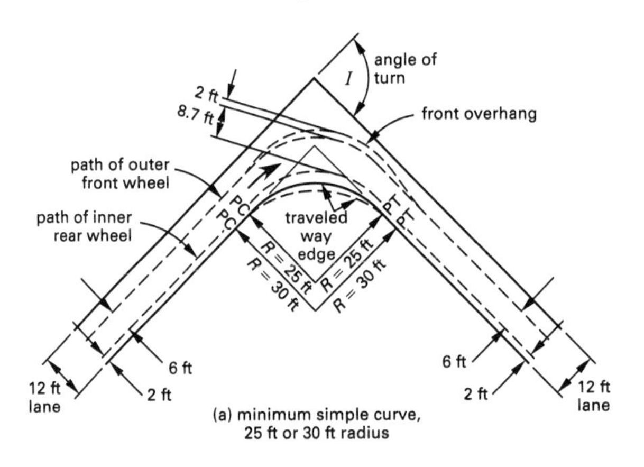

(a) minimum simple curve, 25 ft or 30 ft radius

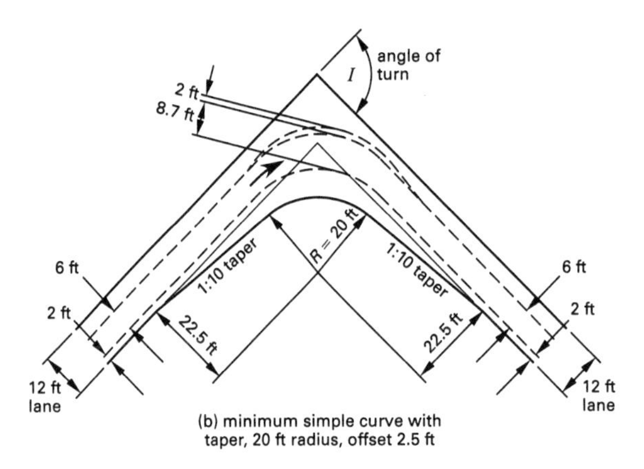

(b) minimum simple curve with taper, 20 ft radius, offset 2.5 ft

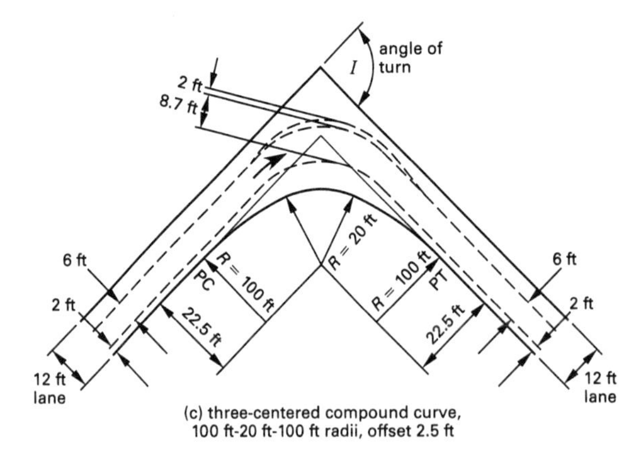

(c) three-centered compound curve, 100 ft-20 ft-100 ft radii, offset 2.5 ft

From *A Policy on Geometric Design of Highways and Streets*, 2004, by the American Association of State Highway and Transportation Officials, Washington, D.C. Used by permission.

APPENDIX 4.R
Minimum Traveled Way Designs, Single-Unit Trucks and City Transit Buses

SI units

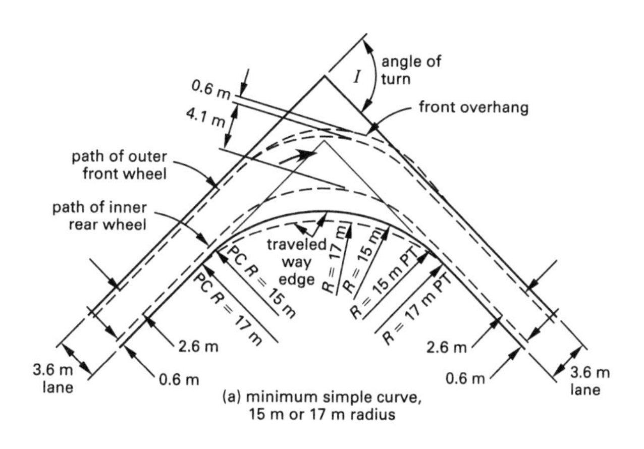

(a) minimum simple curve, 15 m or 17 m radius

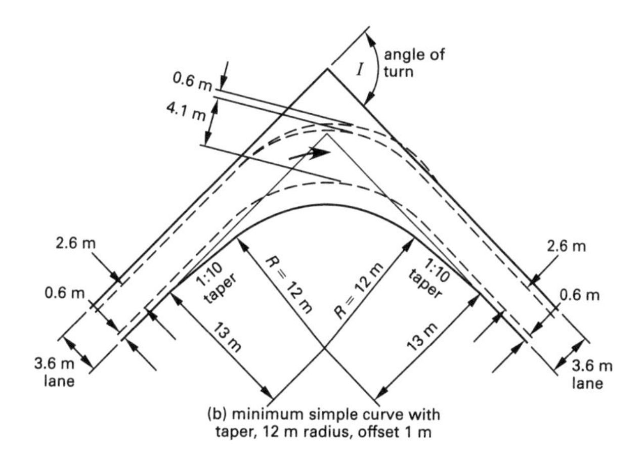

(b) minimum simple curve with taper, 12 m radius, offset 1 m

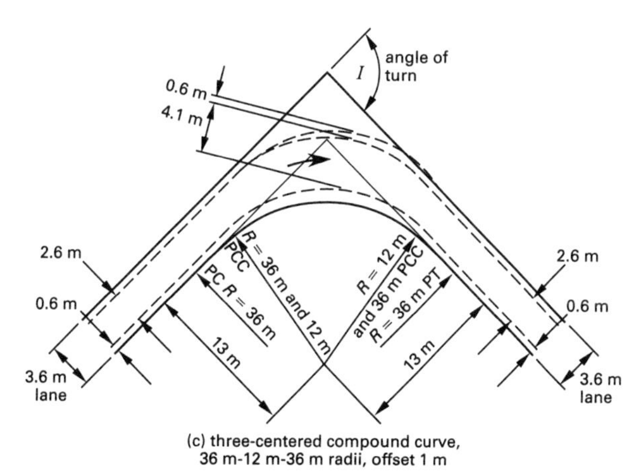

(c) three-centered compound curve, 36 m-12 m-36 m radii, offset 1 m

(continued)

APPENDIX 4.R *(continued)*
Minimum Traveled Way Designs, Single-Unit Trucks and City Transit Buses

customary U.S. units

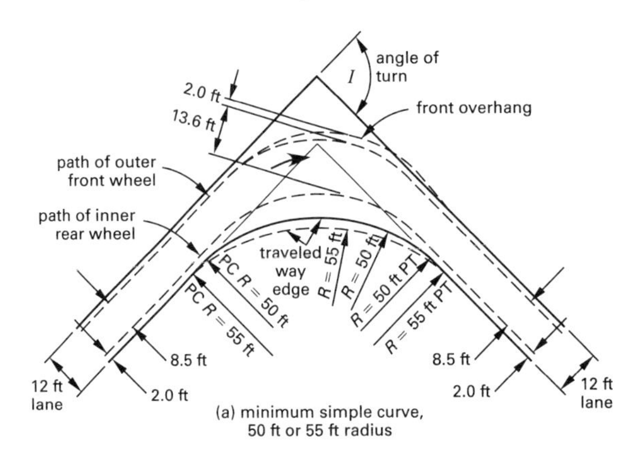

(a) minimum simple curve, 50 ft or 55 ft radius

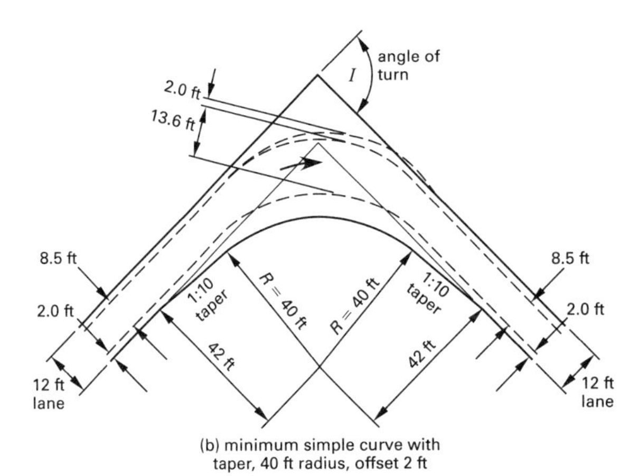

(b) minimum simple curve with taper, 40 ft radius, offset 2 ft

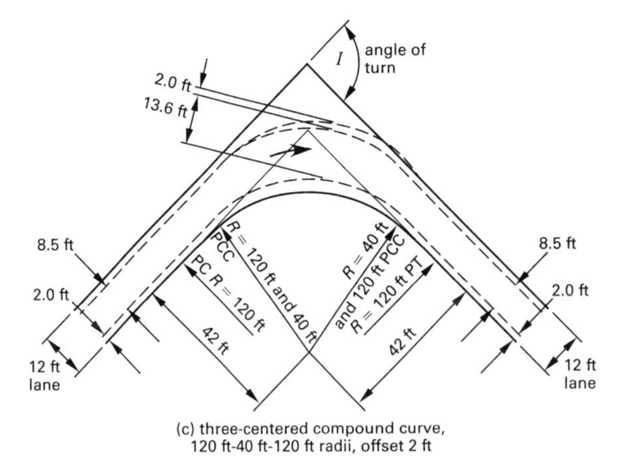

(c) three-centered compound curve, 120 ft-40 ft-120 ft radii, offset 2 ft

From *A Policy on Geometric Design of Highways and Streets*, 2004, by the American Association of State Highway and Transportation Officials, Washington, D.C. Used by permission.

APPENDIX 4.S
Edge-of-Traveled-Way for Turns at Intersections

SI units

angle of turn	design vehicle	simple curve radius (m)	simple curve radius with taper		
			radius (m)	offset (m)	taper $L:T$
30°	P	18	–	–	–
	SU	30	–	–	–
	WB-12	45	–	–	–
	WB-15	60	–	–	–
	WB-19	110	67	1.0	15:1
	WB-20	116	67	1.0	15:1
	WB-30T	77	37	1.0	15:1
	WB-33D	145	77	1.1	20:1
45°	P	15	–	–	–
	SU	23	–	–	–
	WB-12	36	–	–	–
	WB-15	53	36	1.0	15:1
	WB-19	70	43	1.2	15:1
	WB-20	76	43	1.3	15:1
	WB-30T	60	35	0.8	15:1
	WB-33D	–	60	1.3	20:1
60°	P	12	–	–	–
	SU	18	–	–	–
	WB-12	28	–	–	–
	WB-15	45	29	1.0	15:1
	WB-19	50	43	1.2	15:1
	WB-20	60	43	1.3	15:1
	WB-30T	46	29	0.8	15:1
	WB-33D	–	54	1.3	20:1
75°	P	11	8	0.6	10:1
	SU	17	14	0.6	10:1
	WB-12	–	18	0.6	15:1
	WB-15	–	20	1.0	15:1
	WB-19	–	43	1.2	20:1
	WB-20	–	43	1.3	20:1
	WB-30T	–	26	1.0	15:1
	WB-33D	–	42	1.7	20:1
90°	P	9	6	0.8	10:1
	SU	15	12	0.6	10:1
	WB-12	–	14	1.2	10:1
	WB-15	–	18	1.2	15:1
	WB-19	–	36	1.3	30:1
	WB-20	–	37	1.3	30:1
	WB-30T	–	25	0.8	15:1
	WB-33D	–	35	0.9	15:1
105°	P	–	6	0.8	8:1
	SU	–	11	1.0	10:1
	WB-12	–	12	1.2	10:1
	WB-15	–	17	1.2	15:1
	WB-19	–	35	1.0	15:1
	WB-20	–	35	1.0	15:1
	WB-30T	–	22	1.0	15:1
	WB-33D	–	28	2.8	20:1

(continued)

Appendices

APPENDIX 4.S *(continued)*
Edge-of-Traveled-Way for Turns at Intersections

SI units

angle of turn	design vehicle	simple curve radius (m)	simple curve radius with taper		
			radius (m)	offset (m)	taper $L:T$
120°	P	–	6	0.6	10:1
	SU	–	9	1.0	10:1
	WB-12	–	11	1.5	8:1
	WB-15	–	14	1.2	15:1
	WB-19	–	30	1.5	15:1
	WB-20	–	31	1.6	15:1
	WB-30T	–	20	1.1	15:1
	WB-33D	–	26	2.8	20:1
135°	P	–	6	0.5	10:1
	SU	–	9	1.2	10:1
	WB-12	–	9	2.5	15:1
	WB-15	–	12	2.0	15:1
	WB-19	–	24	1.5	20:1
	WB-20	–	25	1.6	20:1
	WB-30T	–	19	1.7	15:1
	WB-33D	–	25	2.6	20:1
150°	P	–	6	0.6	10:1
	SU	–	9	1.2	8:1
	WB-12	–	9	2.0	8:1
	WB-15	–	11	2.1	6:1
	WB-19	–	18	3.0	10:1
	WB-20	–	19	3.1	10:1
	WB-30T	–	19	2.2	10:1
	WB-33D	–	20	4.6	10:1
180°	P	–	5	0.2	20:1
	SU	–	9	0.5	10:1
	WB-12	–	6	3.0	5:1
	WB-15	–	8	3.0	5:1
	WB-19	–	17	3.0	15:1
	WB-20	–	16	4.2	10:1
	WB-30T	–	17	3.1	10:1
	WB-33D	–	17	6.1	10:1

From *A Policy on Geometric Design of Highways and Streets*, 2004, by the American Association of State Highway and Transportation Officials, Washington, D.C. Used by permission.

(continued)

APPENDIX 4.S *(continued)*
Edge-of-Traveled-Way for Turns at Intersections

customary U.S. units

angle of turn	design vehicle	simple curve radius (ft)	simple curve radius with taper: radius (ft)	offset (ft)	taper $L:T$
30°	P	60	–	–	–
	SU	100	–	–	–
	WB-40	150	–	–	–
	WB-50	200	–	–	–
	WB-62	360	220	3.0	15:1
	WB-67	380	220	3.0	15:1
	WB-100T	260	125	3.0	15:1
	WB-109D	475	260	3.5	20:1
45°	P	50	–	–	–
	SU	75	–	–	–
	WB-40	120	–	–	–
	WB-50	175	120	2.0	15:1
	WB-62	230	145	4.0	15:1
	WB-67	250	145	4.5	15:1
	WB-100T	200	115	2.5	15:1
	WB-109D	–	200	4.5	20:1
60°	P	40	–	–	–
	SU	60	–	–	–
	WB-40	90	–	–	–
	WB-50	150	120	3.0	15:1
	WB-62	170	140	4.0	15:1
	WB-67	200	140	4.5	15:1
	WB-100T	150	95	2.5	15:1
	WB-109D	–	180	4.5	20:1
75°	P	35	20	2.0	10:1
	SU	55	45	2.0	10:1
	WB-40	–	60	2.0	15:1
	WB-50	–	65	3.0	15:1
	WB-62	–	145	4.0	20:1
	WB-67	–	145	4.5	20:1
	WB-100T	–	85	3.0	15:1
	WB-109D	–	140	5.5	20:1
90°	P	30	20	2.5	10:1
	SU	50	40	2.0	10:1
	WB-40	–	45	4.0	10:1
	WB-50	–	60	4.0	15:1
	WB-62	–	120	4.5	30:1
	WB-67	–	125	4.5	30:1
	WB-100T	–	85	2.5	15:1
	WB-109D	–	115	2.9	15:1
105°	P	–	20	2.5	8:1
	SU	–	35	3.0	10:1
	WB-40	–	40	4.0	10:1
	WB-50	–	55	4.0	15:1
	WB-62	–	115	3.0	15:1
	WB-67	–	115	3.0	15:1
	WB-100T	–	75	3.0	15:1
	WB-109D	–	90	9.2	20:1

(continued)

APPENDIX 4.S *(continued)*
Edge-of-Traveled-Way for Turns at Intersections

customary U.S. units

angle of turn	design vehicle	simple curve radius (ft)	simple curve radius with taper		
			radius (ft)	offset (ft)	taper $L:T$
120°	P	–	20	2.0	10:1
	SU	–	30	3.0	10:1
	WB-40	–	35	5.0	8:1
	WB-50	–	45	4.0	15:1
	WB-62	–	100	5.0	15:1
	WB-67	–	105	5.2	15:1
	WB-100T	–	65	3.5	15:1
	WB-109D	–	85	9.2	20:1
135°	P	–	20	1.5	10:1
	SU		30	4.0	10:1
	WB-40	–	30	8.0	15:1
	WB-50	–	40	6.0	15:1
	WB-62	–	80	5.0	20:1
	WB-67	–	85	5.2	20:1
	WB-100T	–	65	5.5	15:1
	WB-109D	–	85	8.5	20:1
150°	P	–	18	2.0	10:1
	SU	–	30	4.0	8:1
	WB-40	–	30	6.0	8:1
	WB-50	–	35	7.0	6:1
	WB-62	–	60	10.0	10:1
	WB-67	–	65	10.2	10:1
	WB-100T	–	65	7.3	10:1
	WB-109D	–	65	15.1	10:1
180°	P	–	15	0.5	20:1
	SU	–	30	1.5	10:1
	WB-40	–	20	9.5	5:1
	WB-50	–	25	9.5	5:1
	WB-62	–	55	10.0	15:1
	WB-67	–	55	13.8	10:1
	WB-100T	–	55	10.2	10:1
	WB-109D	–	55	20.0	10:1

From *A Policy on Geometric Design of Highways and Streets*, 2004, by the American Association of State Highway and Transportation Officials, Washington, D.C. Used by permission.

(continued)

APPENDIX 4.S *(continued)*
Edge-of-Traveled-Way for Turns at Intersections

SI units

		three-centered compound		three-centered compound	
angle of turn	design vehicle	curve radii (m)	symmetric offset (m)	curve radii (m)	asymmetric offset (m)
30°	P	–	–	–	–
	SU	–	–	–	–
	WB-12	–	–	–	–
	WB-15	–	–	–	–
	WB-19	–	–	–	–
	WB-20	140-53-140	1.2	91-53-168	0.6–1.4
	WB-30T	67-24-67	1.4	61-24-91	0.8–1.5
	WB-33D	168-76-168	1.5	76-61-198	0.5–2.1
45°	P	–	–	–	–
	SU	–	–	–	–
	WB-12	–	–	–	–
	WB-15	60-30-60	1.0	–	–
	WB-19	140-72-140	0.6	36-43-150	1.0–2.6
	WB-20	140-53-140	1.2	76-38-183	0.3–1.8
	WB-30T	76-24-76	1.4	67-24-91	0.8–1.7
	WB-33D	168-61-168	1.5	61-52-198	0.5–2.1
60°	P	–	–	–	–
	SU	–	–	–	–
	WB-12	–	–	–	–
	WB-15	60-23-60	1.7	60-23-84	0.6–2.0
	WB-19	120-30-120	4.5	34-30-67	3.0–3.7
	WB-20	122-30-122	2.4	76-38-183	0.3–1.8
	WB-30T	76-24-76	1.4	61-24-91	0.6–1.7
	WB-33D	198-46-198	1.7	61-43-183	0.5–2.4
75°	P	30-8-30	0.6	–	–
	SU	36-14-36	0.6	–	–
	WB-12	36-14-36	1.5	36-14-60	0.6–2.0
	WB-15	45-15-45	2.0	45-15-69	0.6–3.0
	WB-19	134-23-134	4.5	43-30-165	1.5–3.6
	WB-20	128-23-128	3.0	61-24-183	0.3–0.3
	WB-30T	76-24-76	1.4	30-24-91	0.5–1.5
	WB-33D	213-38-213	2.0	46-34-168	0.5–3.5
90°	P	30-6-30	0.8	–	–
	SU	36-12-36	0.6	–	–
	WB-12	36-12-36	1.5	36-12-60	0.6–2.0
	WB-15	55-18-55	2.0	36-12-60	0.6–3.0
	WB-19	120-21-120	3.0	48-21-110	2.0–3.0
	WB-20	134-20-134	3.0	61-21-183	0.3–3.4
	WB-30T	76-21-76	1.4	61-21-91	0.3–1.5
	WB-33D	213-34-213	2.0	30-29-168	0.6–3.5
105°	P	30-6-30	0.8	–	–
	SU	30-11-30	1.0	–	–
	WB-12	30-11-30	1.5	30-17-60	0.6–2.5
	WB-15	55-14-55	2.5	45-12-64	0.6–3.0
	WB-19	160-15-160	4.5	110-23-180	1.2–3.2
	WB-20	152-15-152	4.0	61-20-183	0.3–3.4
	WB-30T	76-18-76	1.5	30-18-91	0.5–1.8
	WB-33D	213-29-213	2.4	46-24-152	0.9–4.6

(continued)

Appendices

APPENDIX 4.S *(continued)*
Edge-of-Traveled-Way for Turns at Intersections

SI units

		three-centered compound		three-centered compound	
angle of turn	design vehicle	curve radii (m)	symmetric offset (m)	curve radii (m)	asymmetric offset (m)
120°	P	30-6-30	0.6	–	–
	SU	30-9-30	1.0	–	–
	WB-12	36-9-36	2.0	30-9-55	0.6–2.7
	WB-15	55-12-55	2.6	45-11-67	0.6–3.6
	WB-19	160-21-160	3.0	24-17-160	5.2–7.3
	WB-20	168-14-168	4.6	61-18-183	0.6–3.8
	WB-30T	76-18-76	1.5	30-18-91	0.5–1.8
	WB-33D	213-26-213	2.7	46-21-152	2.0–5.3
135°	P	30-6-30	0.5	–	–
	SU	30-9-30	1.2	–	–
	WB-12	36-9-36	2.0	30-8-55	1.0–4.0
	WB-15	48-11-48	2.7	40-9-55	1.0-4.3
	WB-19	180-18-180	3.6	30-18-195	2.1–4.3
	WB-20	168-14-168	5.0	61-18-183	0.6–3.8
	WB-30T	76-18-76	1.7	30-18-91	0.8–2.0
	WB-33D	213-21-213	3.8	46-20-152	2.1–5.6
150°	P	23-6-23	0.6	–	–
	SU	30-9-30	1.2	–	–
	WB-12	30-9-30	2.0	28-8-48	0.3–3.6
	WB-15	48-11-48	2.1	36-9-55	1.0–4.3
	WB-19	145-17-145	4.5	43-18-170	2.4–3.0
	WB-20	168-14-168	5.8	61-17-183	2.0–5.0
	WB-30T	76-18-76	2.1	30-18-91	1.5–2.4
	WB-33D	213-20-213	4.6	61-20-152	2.7–5.6
180°	P	15-5-15	0.2	–	–
	SU	30-9-30	0.5	–	–
	WB-12	30-6-30	3.0	26-6-45	2.0–4.0
	WB-15	40-8-40	3.0	30-8-55	2.0–4.0
	WB-19	245-14-245	6.0	30-17-275	4.5–4.5
	WB-20	183-14-183	6.2	30-17-122	1.8–4.6
	WB-30T	76-17-76	2.9	30-17-91	2.6–3.2
	WB-33D	213-17-213	6.1	61-18-152	3.0–6.4

From *A Policy on Geometric Design of Highways and Streets*, 2004, by the American Association of State Highway and Transportation Officials, Washington, D.C. Used by permission.

(continued)

APPENDIX 4.S *(continued)*
Edge-of-Traveled-Way for Turns at Intersections

customary U.S. units

angle of turn	design vehicle	three-centered compound curve radii (ft)	three-centered compound symmetric offset (ft)	three-centered compound curve radii (ft)	three-centered compound asymmetric offset (ft)
30°	P	–	–	–	–
	SU	–	–	–	–
	WB-40	–	–	–	–
	WB-50	–	–	–	–
	WB-62	–	–	–	–
	WB-67	460-175-460	4.0	300-175-550	2.0–4.5
	WB-100T	220-80-220	4.5	200-80-300	2.5–5.0
	WB-109D	550-250-550	5.0	250-200-650	1.5–7.0
45°	P	–	–	–	–
	SU	–	–	–	–
	WB-40	–	–	–	–
	WB-50	200-100-200	3.0	–	–
	WB-62	460-240-460	2.0	120-140-500	3.0–8.5
	WB-67	460-175-460	4.0	250-125-600	1.0–6.0
	WB-100T	250-80-250	4.5	200-80-300	2.5–5.5
	WB-109D	550-200-550	5.0	200-170-650	1.5–7.0
60°	P	–	–	–	–
	SU	–	–	–	–
	WB-40	–	–	–	–
	WB-50	200-75-200	5.5	200-75-275	2.0–7.0
	WB-62	400-100-400	15.0	110-100-220	10.5–12.5
	WB-67	400-100-400	8.0	250-125-600	1.0–6.0
	WB-100T	250-80-250	4.5	200-80-300	2.0–5.5
	WB-109D	650-150-650	5.5	200-140-600	1.5–8.0
75°	P	100-25-100	2.0	–	–
	SU	120-45-120	2.0	–	–
	WB-40	120-45-120	5.0	120-45-195	2.0–6.5
	WB-50	150-50-150	6.5	150-50-225	2.0–10.0
	WB-62	440-75-440	15.0	140-100-540	5.0–12.0
	WB-67	420-75-420	10.0	200-80-600	1.0–10.0
	WB-100T	250-80-250	4.5	100-80-300	1.5–5.0
	WB-109D	700-125-700	6.5	150-110-550	1.5–11.5
90°	P	100-20-100	2.5	–	–
	SU	120-40-120	2.0	–	–
	WB-40	120-40-120	5.0	120-40-200	2.0–6.5
	WB-50	180-60-180	6.5	120-40-200	2.0–10.0
	WB-62	400-70-400	10.0	160-70-360	6.0–10.0
	WB-67	440-65-440	10.0	200-70-600	1.0–11.0
	WB-100T	250-70-250	4.5	200-70-300	1.0–5.0
	WB-109D	700-110-700	6.5	100-95-550	2.0–11.5
105°	P	100-20-100	2.5	–	–
	SU	100-35-100	3.0	–	–
	WB-40	100-35-100	5.0	100-55-200	2.0–8.0
	WB-50	180-45-180	8.0	150-40-210	2.0–10.0
	WB-62	520-50-520	15.0	360-75-600	4.0–10.5
	WB-67	500-50-500	13.0	200-65-600	1.0–11.0
	WB-100T	250-60-250	5.0	100-60-300	1.5–6.0
	WB-109D	700-95-700	8.0	150-80-500	3.0–15.0

(continued)

APPENDIX 4.S *(continued)*
Edge-of-Traveled-Way for Turns at Intersections

customary U.S. units

		three-centered compound		three-centered compound	
angle of turn	design vehicle	curve radii (ft)	symmetric offset (ft)	curve radii (ft)	asymmetric offset (ft)
120°	P	100-20-100	2.0	–	–
	SU	100-30-100	3.0	–	–
	WB-40	120-30-120	6.0	100-30-180	2.0–9.0
	WB-50	180-40-180	8.5	150-35-220	2.0–12.0
	WB-62	520-70-520	10.0	80-55-520	24.0–17.0
	WB-67	550-45-550	15.0	200-60-600	2.0–12.5
	WB-100T	250-60-250	5.0	100-60-300	1.5–6.0
	WB-109D	700-85-700	9.0	150-70-500	7.0–17.4
135°	P	100-20-100	1.5	–	–
	SU	100-30-100	4.0	–	–
	WB-40	120-30-120	6.5	100-25-180	3.0–13.0
	WB-50	160-35-160	9.0	130-30-185	3.0–14.0
	WB-62	600-60-600	12.0	100-60-640	14.0–7.0
	WB-67	550-45-550	16.0	200-60-600	2.0–12.5
	WB-100T	250-60-250	5.5	100-60-300	2.5–7.0
	WB-109D	700-70-700	12.5	150-65-500	14.0–18.4
150°	P	75-20-75	2.0	–	–
	SU	100-30-100	4.0	–	–
	WB-40	100-30-100	6.0	90-20-160	1.0–12.0
	WB-50	160-35-160	7.0	120-30-180	3.0–14.0
	WB-62	480-55-480	15.0	140-60-560	8.0–10.0
	WB-67	550-45-550	19.0	200-55-600	7.0–16.4
	WB-100T	250-60-250	7.0	100-60-300	5.0–8.0
	WB-109D	700-65-700	15.0	200-65-500	9.0–18.4
180°	P	50-15-50	0.5	–	–
	SU	100-30-100	1.5	–	–
	WB-40	100-20-100	9.5	85-20-150	6.0–13.0
	WB-50	130-25-130	9.5	100-25-180	6.0–13.0
	WB-62	800-45-800	20.0	100-55-900	15.0–15.0
	WB-67	600-45-600	20.5	100-55-400	6.0–15.0
	WB-100T	250-55-250	9.5	100-55-300	8.5–10.5
	WB-109D	700-55-700	20.0	200-60-500	10.0–21.0

From *A Policy on Geometric Design of Highways and Streets*, 2004, by the American Association of State Highway and Transportation Officials, Washington, D.C. Used by permission.

Appendices

APPENDIX 4.T
Effect of Curb Radii on Right-Turning Paths of Various Vehicles

SI units

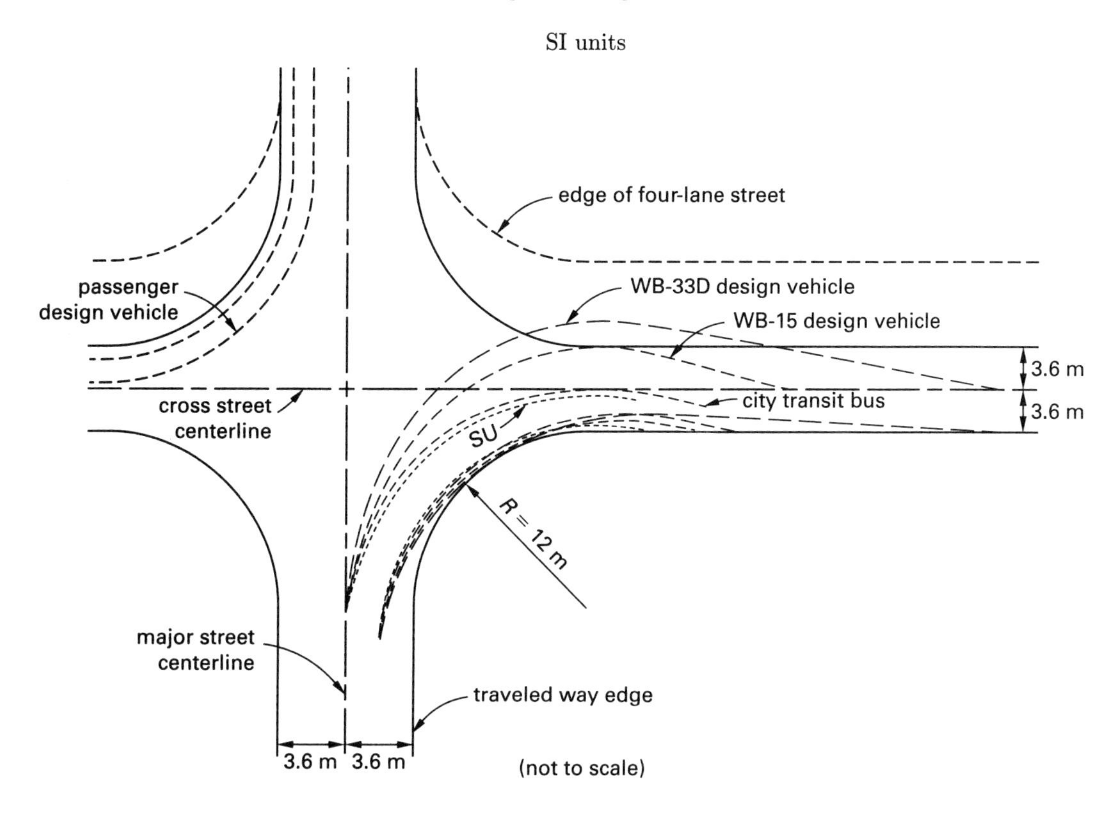

customary U.S. units

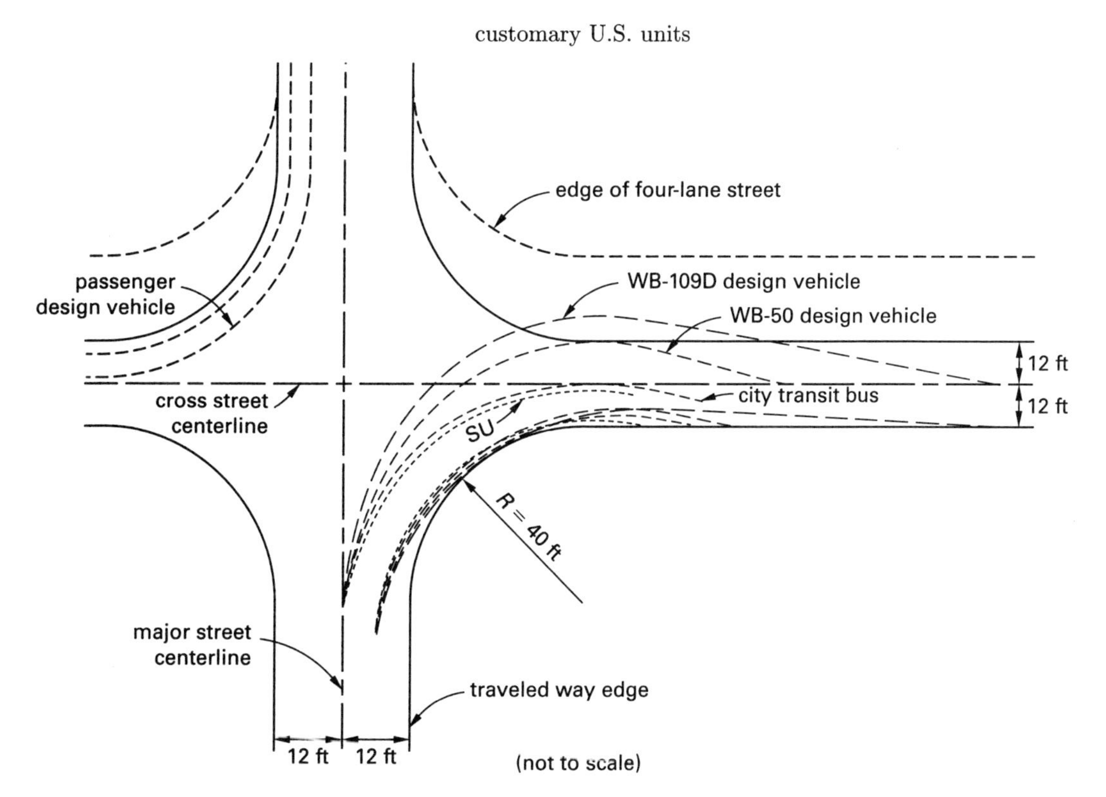

From *A Policy on Geometric Design of Highways and Streets*, 2004, by the American Association of State Highway and Transportation Officials, Washington, D.C. Used by permission.

APPENDIX 4.U

Cross Street Widths Occupied by Turning Vehicles*

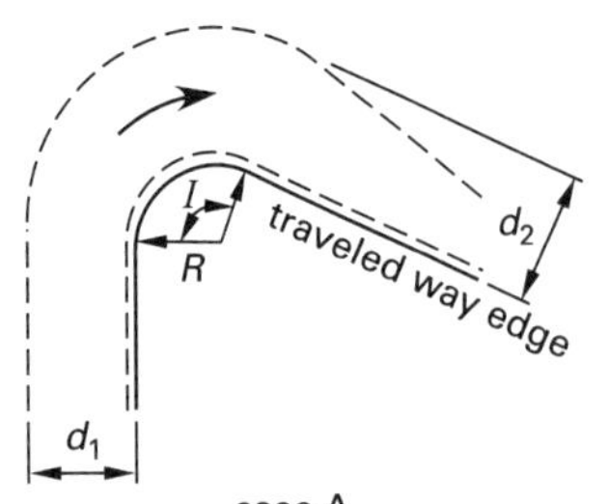

case A
Vehicle turns from proper lane and swings wide on cross street.*
d_1 is 12 ft (3.6 m), and d_2 is variable.

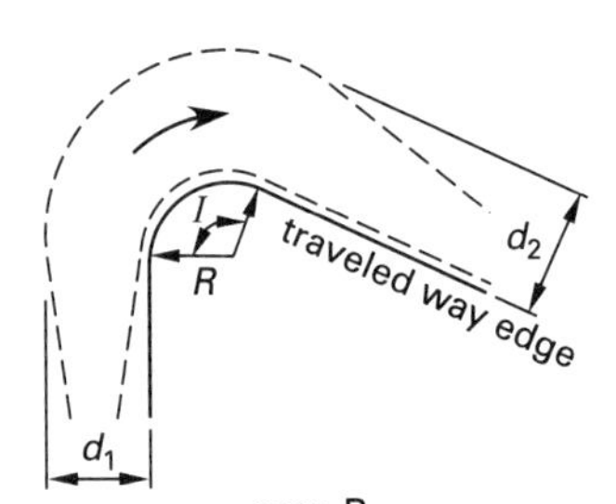

case B
Turning vehicle swings equally wide on both streets.*
d_1 and d_2 are both variable.

SI units

		d_2 for cases A and B where									
		$R = 4.5$ m		$R = 6$ m		$R = 7.5$ m		$R = 9$ m		$R = 12$ m	
angle of intersection (I)	design vehicle	A (m)	B (m)	A (m)	B (m)	A (m)	B (m)	A (m)	B (m)	A (m)	B (m)
30°	SU	4.3	4.0	4.3	4.0	4.0	4.0	4.0	4.0	4.0	4.0
	BUS	6.7	5.2	5.8	5.2	5.8	5.2	5.8	5.2	5.5	5.2
	WB-12	4.3	4.3	4.3	4.3	4.3	4.3	4.3	4.3	4.3	4.3
	WB-15	6.1	5.2	6.1	5.2	6.1	5.2	5.8	4.9	5.5	4.9
	WB-19	–	–	–	–	–	–	–	–	8.2	5.2
	WB-20	–	–	–	–	–	–	–	–	8.5	5.5
60°	SU	5.8	4.9	5.8	4.9	5.2	4.6	4.9	4.6	4.3	4.3
	BUS	8.5	6.4	7.9	6.1	7.3	6.1	7.0	5.8	6.7	5.5
	WB-12	7.3	5.8	6.7	5.8	6.4	5.8	5.8	5.5	5.2	4.9
	WB-15	9.4	6.7	8.2	6.4	8.5	6.1	7.6	5.8	6.7	5.5
	WB-19	–	–	–	–	–	–	–	–	9.1	6.7
	WB-20	–	–	–	–	–	–	–	–	11.3	7.3
90°	SU	7.9	6.1	7.0	5.5	5.8	4.9	5.2	4.6	4.0	4.0
	BUS	11.6	7.0	10.0	6.7	9.1	6.7	7.6	6.4	6.7	5.5
	WB-12	9.4	6.7	8.2	6.4	7.0	6.4	5.8	5.5	5.2	4.9
	WB-15	12.8	6.7	11.3	7.3	9.8	6.7	8.8	6.4	6.7	5.5
	WB-19	–	–	–	–	–	–	–	–	11.9	7.0
	WB-20	–	–	–	–	–	–	–	–	11.9	7.6
120°	SU	10.4	6.7	8.2	5.8	6.4	5.5	5.2	4.9	4.0	4.0
	BUS	14.0	8.5	12.2	7.6	9.8	7.0	7.9	5.8	5.8	5.5
	WB-12	11.3	7.0	8.8	6.7	7.3	6.7	5.8	5.5	5.2	4.9
	WB-15	15.2	8.8	13.1	8.5	11.0	8.2	9.1	7.9	6.7	5.5
	WB-19	–	–	–	–	–	–	–	–	7.9	6.7
	WB-20	–	–	–	–	–	–	–	–	9.1	7.0
150°	SU	12.2	7.6	9.8	6.4	6.7	5.8	5.2	4.9	3.6	3.6
	BUS	14.6	8.5	12.2	7.6	9.8	7.0	6.7	5.5	5.2	4.9
	WB-12	11.9	7.3	8.8	6.7	7.0	6.7	5.8	5.5	5.2	4.9
	WB-15	16.2	9.4	14.0	8.5	11.0	8.2	8.5	7.9	6.7	5.5
	WB-19	–	–	–	–	–	–	–	–	6.1	5.5
	WB-20	–	–	–	–	–	–	–	–	8.2	5.5

*P design vehicle turns within 12 ft (3.6 m) where $R = 15$ ft (4.5 m) or more. No parking on either street.

(continued)

APPENDIX 4.U *(continued)*
Cross Street Widths Occupied by Turning Vehicles*

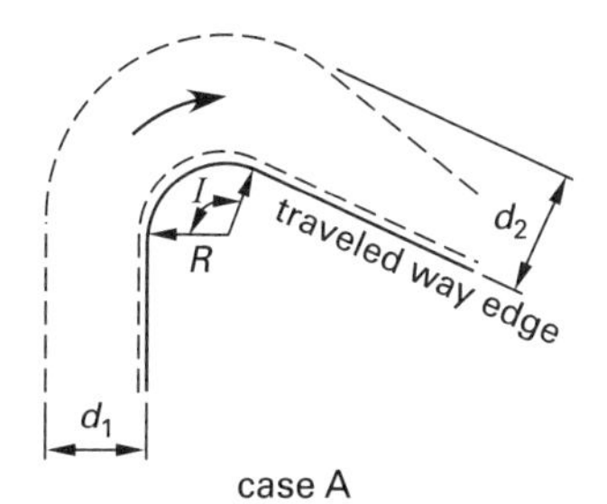

case A
Vehicle turns from proper lane and swings wide on cross street.*
d_1 is 12 ft (3.6 m), and d_2 is variable.

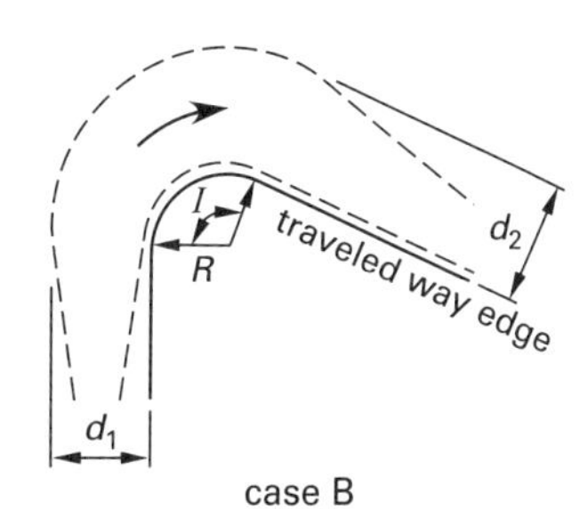

case B
Turning vehicle swings equally wide on both streets.*
d_1 and d_2 are both variable.

customary U.S. units

		d_2 for cases A and B where									
		R = 15 ft		R = 20 ft		R = 25 ft		R = 30 ft		R = 40 ft	
angle of intersection (I)	design vehicle	A (ft)	B (ft)	A (ft)	B (ft)	A (ft)	B (ft)	A (ft)	B (ft)	A (ft)	B (ft)
30°	SU	14	13	14	13	13	13	13	13	13	13
	BUS	22	17	19	17	19	17	19	17	18	17
	WB-40	14	14	14	14	14	14	14	14	14	14
	WB-50	20	17	20	17	20	17	19	16	18	16
	WB-62	–	–	–	–	–	–	–	–	27	17
	WB-67	–	–	–	–	–	–	–	–	28	18
60°	SU	19	16	19	16	17	15	16	15	14	14
	BUS	28	21	26	20	24	20	23	19	22	18
	WB-40	24	19	22	19	21	19	19	18	17	16
	WB-50	31	22	27	21	28	20	25	19	22	18
	WB-62	–	–	–	–	–	–	–	–	30	22
	WB-67	–	–	–	–	–	–	–	–	37	24
90°	SU	26	20	23	18	19	16	17	15	13	13
	BUS	38	23	33	22	30	22	25	21	22	18
	WB-40	31	22	27	21	23	21	19	18	17	16
	WB-50	42	22	37	24	32	22	29	21	22	18
	WB-62	–	–	–	–	–	–	–	–	39	23
	WB-67	–	–	–	–	–	–	–	–	39	25
120°	SU	34	22	27	19	21	18	17	16	13	13
	BUS	46	28	40	25	32	23	26	19	19	18
	WB-40	37	23	29	22	24	22	19	18	17	16
	WB-50	50	29	43	28	36	27	30	26	22	18
	WB-62	–	–	–	–	–	–	–	–	26	22
	WB-67	–	–	–	–	–	–	–	–	30	23
150°	SU	40	25	32	21	22	19	17	16	12	12
	BUS	48	28	40	25	32	23	22	18	17	16
	WB-40	39	24	29	22	23	22	19	18	17	16
	WB-50	53	31	46	28	36	27	28	26	22	18
	WB-62	–	–	–	–	–	–	–	–	20	18
	WB-67	–	–	–	–	–	–	–	–	27	18

*P design vehicle turns within 12 ft (3.6 m) where R = 15 ft (4.5 m) or more. No parking on either street.

From *A Policy on Geometric Design of Highways and Streets*, 2004, by the American Association of State Highway and Transportation Officials, Washington, D.C. Used by permission.

APPENDIX 4.V

Crosswalk Length Variations with Different Curb Radii and Width of Borders

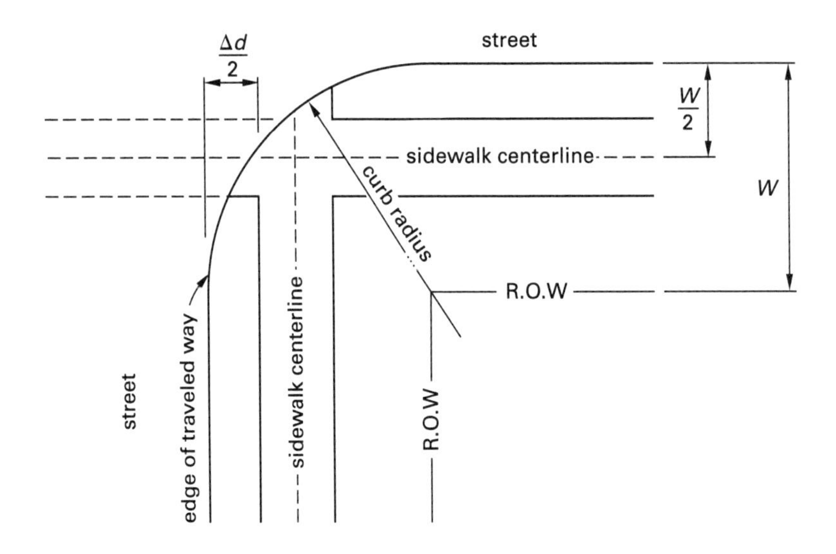

curb radius, R		added crosswalk distance, Δd $W = 10$ ft (3 m)		added crosswalk distance, Δd $W = 20$ ft (6 m)	
ft	m	ft	m	ft	m
10	3	3	0.8	0	0.0
20	6	14	4.0	5	1.6
30	9	27	8.0	15	4.6
40	12	42	12.4	27	8.1
50	15	57	16.9	40	12.0

From *A Policy on Geometric Design of Highways and Streets*, 2004, by the American Association of State Highway and Transportation Officials, Washington, D.C. Used by permission.

APPENDIX 4.W
Corner Setbacks with Different Curb Radii and Width of Borders

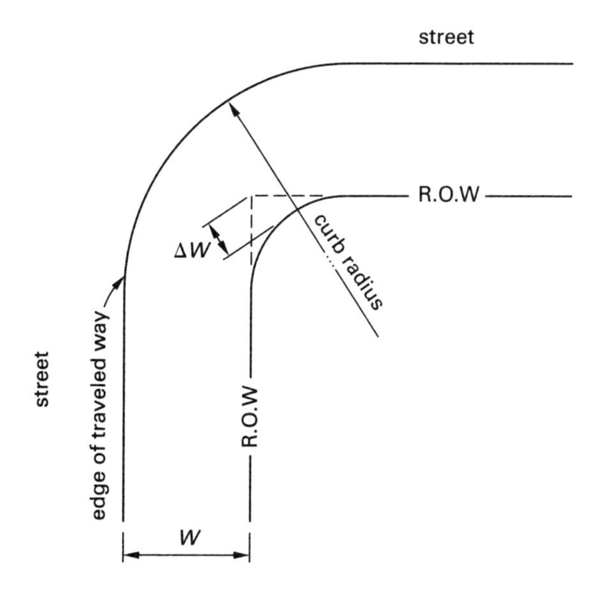

curb radius, R		additional corner setback, ΔW			
		$W = 10$ ft (3 m)		$W = 20$ ft (6 m)	
ft	m	ft	m	ft	m
10	3	0	0.0	0	0.0
20	6	4	1.3	0	0.0
30	9	8	2.5	4	1.3
40	12	13	4.0	8	2.5
50	15	17	5.0	13	4.0

From *A Policy on Geometric Design of Highways and Streets*, 2004, by the American Association of State Highway and Transportation Officials, Washington, D.C. Used by permission.

Appendices

APPENDIX 5.A
AASHTO Soil Classification System

	granular materials (35% or less passing no. 200 sieve)							silt-clay materials (more than 35% passing no. 200 sieve)				
	A-1		A-3	A-2				A-4	A-5	A-6	A-7	A-8
	A-1-a	A-1-b		A-2-4	A-2-5	A-2-6	A-2-7				A-7-5 or A-7-6	
sieve analysis: % passing												
no. 10	50 max											
no. 40	30 max	50 max	51 min									
no. 200	15 max	25 max	10 max	35 max	35 max	35 max	35 max	36 min	36 min	36 min	36 min	
characteristics of fraction passing no. 40: LL: liquid limit				40 max	41 min	40 max	41 min	40 max	41 min	40 max	41 min	
PI: plasticity index	6 max		NP	10 max	10 max	11 min	11 min	10 max	10 max	11 min	11 min	
usual types of significant constituents	stone fragments gravel and sand		fine sand	silty or clayey gravel and sand				silty soils		clayey soils		peat, highly organic soils
general subgrade rating	excellent to good					fair to poor						unsatisfactory

APPENDIX 5.B
USDA Soil Triangle

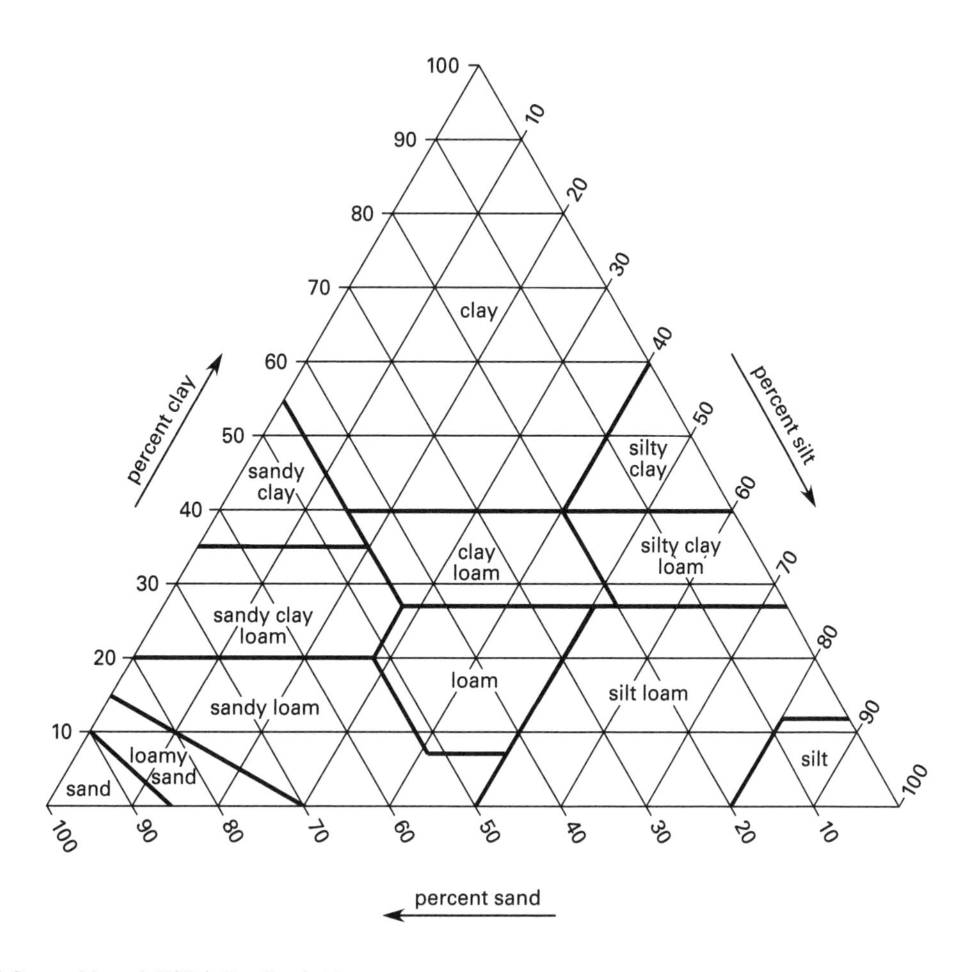

Reprinted from *Soil Survey Manual*, USDA Handbook 18.

APPENDIX 5.C
Performance-Graded Asphalt Binder Specification

performance grade	PG 46			PG 52							PG 58					PG 64					
	34	40	46	10	16	22	28	34	40	46	16	22	28	34	40	10	16	22	28	34	40
average 7 day max. pavement design temp. (°C)	< 46			< 52							< 58					< 64					
min. pavement design temp. (°C)	−34	−40	−46	−10	−16	−22	−28	−34	−40	−46	−16	−22	−28	−34	−40	−10	−16	−22	−28	−34	−40
original binder																					
flash point temp., T 48, min. (°C)	230																				
viscosity, T 316: max. 3 Pa·s, test temp. (°C)	135																				
dynamic shear, T 315: $G^*/\sin\delta$, min. 1.10 kPa test temp. at 10 rad/s (°C)	46			52							58					64					
rolling thin-film oven residue (T 240)																					
mass change, max. (%)	1.00																				
dynamic shear, T 315: $G^*/\sin\delta$, min. 2.20 kPa test temp. at 10 rad/s (°C)	46			52							58					64					
pressure aging vessel (R 28)																					
PAV aging temp. (°C)	90			90							100					100					
dynamic shear, T 315: $G^*/\sin\delta$, max. 5000 kPa test temp. at 10 rad/s (°C)	10	7	4	25	22	19	16	13	10	7	25	22	19	16	13	31	28	25	22	19	16
creep stiffness, T 313: S, max. 300 MPa m-value, min. 0.300 test temp. at 60 s (°C)	−24	−30	−36	0	−6	−12	−18	−24	−30	−36	−6	−12	−18	−24	−30	0	−6	−12	−18	−24	−30
direct tension, T 314: failure strain, min. 0.300 test temp. at 60 s (°C)	−24	−30	−36	0	−6	−12	−18	−24	−30	−36	−6	−12	−18	−24	−30	0	−6	−12	−18	−24	−30
critical low cracking temp., PP 42: critical low cracking temp. determined by PP 42, test temp. (°C)	−24	−30	−36	0	−6	−12	−18	−24	−30	−36	−6	−12	−18	−24	−30	0	−6	−12	−18	−24	−30

(continued)

APPENDIX 5.C *(continued)*
Performance-Graded Asphalt
Binder Specification

performance grade	PG 70						PG 76					PG 82				
	10	16	22	28	34	40	10	16	22	28	34	10	16	22	28	34
average 7 day max. pavement design temp. (°C)	< 70						< 76					< 82				
min. pavement design temp. (°C)	−10	−16	−22	−28	−34	−40	−10	−16	−22	−28	−34	−10	−16	−22	−28	−34
original binder																
flash point temp., T 48, min. (°C)	230															
viscosity, ASTM D 4402: max. 3 Pa·s, test temp. (°C)	135															
dynamic shear, TP 5: $G^*/\sin\delta$, min. 1.10 kPa test temp. at 10 rad/s (°C)	70						76					82				
rolling thin film oven residue (T 240)																
mass change, max. (%)	1.00															
dynamic shear, TP 5: $G^*/\sin\delta$, min. 2.20 kPa test temp. at 10 rad/s (°C)	70						76					82				
pressure aging vessel residue (PP 1)																
PAV aging temp. (°C)	100						100					100				
dynamic shear, TP 5: $G^*/\sin\delta$, max. 5000 kPa test temp. at 10 rad/s (°C)	34	31	28	25	22	19	37	34	31	28	25	40	37	34	31	28
creep stiffness, T 313: S, max. 300 MPa m-value, min. 0.300 test temp. at 60 s (°C)	0	−6	−12	−18	−24	−30	0	−6	−12	−18	−24	0	−6	−12	−18	−24
direction tension, T 314: failure strain, min. 0.300 test temp. at 60 s (°C)	0	−6	−12	−18	−24	−30	0	−6	−12	−18	−24	0	−6	−12	−18	−24
critical low cracking temp., PP 42: critical cracking temp. determined by PP 42, test temp. (°C)	0	−6	−12	−18	−24	−30	0	−6	−12	−18	−24	0	−6	−12	−18	−24

From *Standard Specification for Performance-Grade Asphalt Binder*, 2010, by the American Association of State Highway and Transportation officials, Washington, D.C. Used by permission.

Appendices

APPENDIX 5.D
Superpave Mix Design Procedural Outline

I. Selection of Materials
  A. Select asphalt binder.
    1. Determine project weather conditions using weather database.
    2. Select reliability.
    3. Determine design temperatures.
    4. Verify asphalt binder grade.
    5. Determine temperature-viscosity relationship for lab mixing and compaction.
  B. Select aggregates.
    1. consensus properties
      a. combined gradation
      b. coarse aggregate angularity
      c. fine aggregate angularity
      d. flat and elongated particles
      e. clay content
    2. agency and other properties
      a. specific gravity
      b. toughness
      c. soundness
      d. deleterious materials
      e. other
  C. Select modifiers.
II. Selection of Design Aggregate Structure
  A. Establish trial blends.
    1. Develop three blends.
    2. Evaluate combined aggregate properties.
  B. Compact trial blend specimens.
    1. Establish trial asphalt binder content.
      a. Superpave method
      b. engineering judgment method
    2. Establish trial blend specimen size.
    3. Determine N_{initial} and N_{design}.
    4. Batch trial blend specimens.
    5. Compact specimens and generate densification tables.
    6. Determine mixture properties (G_{mm} and G_{mb}).
  C. Evaluate trial blends.
    1. Determine percentage of G_{mm} at N_{initial} and N_{design}.
    2. Determine percentage of air voids and voids in mineral aggregate.
    3. Estimate asphalt binder content to achieve 4% air voids.
    4. Estimate mix properties at estimated asphalt binder content.
    5. Determine dust-asphalt ratio.
    6. Compare mixture properties to criteria.
  D. Select most promising design aggregate structure for further analysis.
III. Selection of Design Asphalt Binder Content
  A. Compact design aggregate structure specimens at multiple binder contents.
    1. Batch design aggregate structure specimens.
    2. Compact specimens and generate densification tables.
  B. Determine mixture properties versus asphalt binder content.
    1. Determine percentage of G_{mm} at N_{initial}, N_{design}, and N_{max}.
    2. Determine volumetric properties.
    3. Determine dust-asphalt ratio.
    4. Graph mixture properties versus asphalt binder content.
  C. Select design asphalt binder content.
    1. Determine asphalt binder content at 4% air voids.
    2. Determine mixture properties at selected asphalt binder content.
    3. Compare mixture properties to criteria.
IV. Evaluation of Moisture Sensitivity of Design Asphalt Mixture Using AASHTO T238

Reprinted with permission from the Asphalt Institute, *Superpave Mix Design (SP-2)*, 2001, App. C.

Appendices

Index

INDEX - A

D

E

F

G

INDEX - G

INDEX - L

M

N

INDEX - P

Q

R

INDEX - S

T

INDEX - T

U

V

W

INDEX - W

Y

Z